W9-DIF-850

Trace Elements in Human and Animal Nutrition

Fourth Edition

Trace Elements in Human and Animal Nutrition

Fourth Edition

ERIC J. UNDERWOOD

Department of Animal Science and Production
Institute of Agriculture
University of Western Australia
Nedlands, Western Australia

ACADEMIC PRESS New York San Francisco London 1977

A Subsidiary of Harcourt Brace Jovanovich, Publishers

ACADEMIC PRESS, INC.
111 Fifth Avenue, New York, New York 10003

United Kingdom Edition published by
ACADEMIC PRESS, INC. (LONDON) LTD.
24/28 Oval Road, London NW1

Library of Congress Cataloging in Publication Data

Underwood, Eric John, Date
Trace elements in human and animal nutrition.

Includes bibliographies.
1. Trace elements in nutrition. 2. Trace elements in animal nutrition. I. Title.
QP534.U5 1976 591.1'33 76-13950
ISBN 0-12-709065-7

PRINTED IN THE UNITED STATES OF AMERICA

To my wife, Erica, for her help and, remarkably,
her continued encouragement

Contents

6 NICKEL

7 MANGANESE

8 ZINC

9 CADMIUM

10 CHROMIUM

11 IODINE

Preface

The Preface to the third edition began with the words: "the nine-year period since the second edition was written has witnessed a surge of interest and activity in almost every aspect of trace element research." This Preface to the fourth edition might well begin with the same words, except for the substitution of six-year period for nine-year period. During this time developments have been so rapid that again considerable expansion of the coverage has been necessary, and some modifications in the method of presentation have been made in an effort to prevent the book from becoming unmanageably big. The short historical introductions previously presented at the beginning of each of the chapters on individual elements have been eliminated and historical aspects of trace element research transferred to an enlarged first chapter appropriately titled, Introduction. This has provided the opportunity for some economy of words. However, the three elements, mercury, lead, and arsenic, previously presented as sections in an omnibus chapter entitled "Other Elements," have achieved the status of individual chapters as a tribute to their increased importance as environmental contaminants and their metabolic interactions with other trace elements. In addition, short sections on antimony, lithium, and silver have been added to the "Other Elements" chapter.

The above changes have not significantly altered the general approach to the subject employed in earlier editions, and the overall aim of the book has remained, as before, to enable those interested in human and animal nutrition to obtain a balanced and detailed appreciation of the physiological roles of the trace elements, of their needs and tolerances and interactions with each other and with other nutrients and compounds, and of the biochemical and pathological changes that result from deficient, toxic, or imbalanced intakes by animals and man. Means of diagnosing and of preventing or overcoming such aberrant intakes of trace elements, and their principal sources, are given particular consideration. Complete citation of all the vast and growing literature on trace elements in human and animal nutrition would have been unrealistic and indeed impossible. It is hoped, however, that the hundreds of references which are cited constitute an adequate basis for the needs of those with specialized interests in particular facets of an element and that they provide proper support for the statements that appear in the text.

This work could not have been accomplished without the help and encouragement of many people in many countries to whom I owe a great debt. In particular I should mention Drs. E. H. Morgan, J. McC Howell, C. F. Mills, W. Mertz, and K. O. Godwin who read and constructively criticized my draft chapters on iron, copper, zinc, chromium, and selenium, respectively. Their help is greatly appreciated, although the sins of omission and commission that may still exist in those chapters are, of course, my responsibility, not theirs. It is a pleasure to record the debt I owe to Mrs. Sandra Bowker for her skill and care in typing the manuscript and checking references and to the Executive of The Commonwealth Scientific and Industrial Research Organization (C.S.I.R.O.) for generously providing me with this vital secretarial assistance for the final six months of manuscript preparation.

Eric J. Underwood

Trace Elements in Human and Animal Nutrition

Fourth Edition

1

Introduction

I. THE NATURE OF TRACE ELEMENTS

Many mineral elements occur in living tissues in such small amounts that the early workers were unable to measure their precise concentrations with the analytical methods then available. They were therefore frequently described as occurring in "traces" and the term trace elements arose to describe them . This designation has remained in popular usage despite the fact that virtually all the trace elements can now be estimated in biological materials with great accuracy and precision. It is retained here because it is brief and has become hallowed by time.

It is difficult to find a meaningful classification for the trace elements or even to draw a completely satisfactory line of demarcation between those so designated and the so-called major elements. At the present time 26 of the 90 naturally occurring elements are known to be essential for animal life. These consist of 11 major elements, namely, carbon, hydrogen, oxygen, nitrogen, sulfur, calcium, phosphorus, potassium, sodium, chlorine, and magnesium, and 15 elements generally accepted as trace elements. These are iron, zinc, copper, manganese, nickel, cobalt, molybdenum, selenium, chromium, iodine, fluorine, tin, silicon, vanadium, and arsenic. In addition, boron is essential for the higher plants but has not yet been shown to be necessary for animals.

It is clear that evolution has selected certain elements for the essential functioning of living organisms and has, on present evidence, rejected or ignored others. The molecular basis for this selection or rejection is far from clear. In this connection it should be noted that only three of the 27 elements now known to be essential for life, i.e., iodine, tin, and molybdenum, have an atomic number above 34, and a considerable proportion of the essential trace elements occupy positions in the Periodic Table between atomic numbers 23 and 34. This atomic number interval includes two elements—gallium and germanium—for which no vital roles are known. These two elements, therefore, clearly deserve further critical investigation as possible additional essential trace elements. Similarly, bromine warrants further investigation, since it has an atomic number of 35 and there is already suggestive evidence of its essentiality as discussed in Chapter 19.

An element is considered by Mertz (28) to be essential if its deficiency consistently results in impairment of a function from optimal to suboptimal. Cotzias (9) states the position more completely. He maintains that a trace element can be considered essential if it meets the following criteria: (*i*) it is present in all healthy tissues of all living things; (*ii*) its concentration from one animal to the next is fairly constant; (*iii*) its withdrawal from the body induces reproducibly the same physiological and structural abnormalities regardless of the species studied; (*iv*) its addition either reverses or prevents these abnormalities; (*v*) the abnormalities induced by deficiency are always accompanied by pertinent, specific biochemical changes; and (*vi*) these biochemical changes can be prevented or cured when the deficiency is prevented or cured.

Some 20 to 30 trace elements which do not meet the above criteria occur more or less constantly in variable concentrations in living tissues. They include aluminum, antimony, cadmium, mercury, germanium, rubidium, silver, lead, gold, bismuth, titanium, zirconium, and others. They are believed to be acquired by the animal body as environmental contaminants and to reflect the contact of the organism with its environment. Skewed (log-normal) distribution patterns have been reported for the concentrations of these elements in human organs, whereas the essential elements have a normal distribution (22,45). Liebscher and Smith (22) have proposed that the shape of the distribution curve for a trace element in tissue could be used as a method of determining whether the element is essential. They have supported this proposal with evidence obtained from a study of the levels of several essential and nonessential trace elements in tissues from healthy adults who died as a result of violence and who had no known industrial exposure to the elements in question. For the essential elements an internal control mechanism was postulated, leading to a normal or symmetrical distribution. For the nonessential elements, external control of tissue concentration arising from contamination would occur, leading to a distribution pattern similar to the environmental level.

Classification of the trace elements into a further group, known as toxic elements, is perhaps justified for a few elements such as lead, cadmium, and mercury, because their biological significance is so far confined to their toxic or potentially toxic properties at relatively low concentrations. This classification has limited value because all the trace elements are toxic if ingested or inhaled at sufficiently high levels and for long enough periods. This was recognized many years ago by Bertrand, who formulated such dose dependence into Bertrand's Law (4). Venchikov (53) has expanded this concept and presented the dose response in the form of a curve with two maxima. The first part of the curve, showing an increasing effect with increasing concentrations until a plateau is reached, expresses the *biological* action of the element, and the plateau expresses optimal supplementation and normal function. The width of the plateau is determined by the homeostatic capacity of the animal or system. With further increasing doses the element enters a phase of irritation and stimulation of some function, expressing its *pharmacological* action. In this phase the element acts as a drug independent of a deficiency state. At still higher doses this is followed by the appearance of signs of toxicity, expressing the *toxicological* action of the element. The intakes or dose levels at which these different phases of action become apparent, and the width of the optimal plateau, vary widely among the trace elements and can be markedly affected by the extent to which various other elements and compounds are present in the animal body and in the diet being consumed. More recently Venchikov (54), on the basis of numerous experiments with several animal species and tissues, has refined the concept to establish three zones of action of trace elements, named (*i*) biological action zone, (*ii*) inactive zone, and (*iii*) pharmacotoxicological action zone.

II. DISCOVERY OF TRACE ELEMENTS

Interest in trace elements in animal physiology began over a century ago with the discovery in living organisms of a number of special compounds which contained various metals not previously suspected to be of biological significance. These included turacin, a red porphyrin pigment occurring in the feathers of certain birds and containing no less than 7% copper (7); hemocyanin, another copper-containing compound found in the blood of snails (16); sycotypin, a zinc-containing blood pigment in Mollusca (27); and a vanadium-containing respiratory compound present in the blood of sea squirts (18). Such discoveries did little to stimulate studies of the possible wider significance of the component elements. These were to come from the early investigations of Bernard (3) and MacMunn (24) on cell respiration and iron and oxidative processes, which pointed the way to later studies of metal–enzyme catalysis and of metalloenzymes and which were to greatly illuminate our understanding of trace element functions within the tissues. This early period also saw (*i*) the discovery by Raulin (32) of the essentiality of zinc in the nutrition of *Aspergillus niger*; (*ii*) the remarkable observations of the French botanist Chatin (6) on the iodine content of soils, waters, and foods and the relationship of the occurrence of goiter in man to a deficiency of environmental iodine; and (*iii*) the demonstration by Frodisch in 1832 that the iron content of the blood of chlorotics is lower than that of healthy individuals, following an earlier discovery that iron is a characteristic constituent of blood (see Fowler, 14).

During the first quarter of the present century further studies were initiated on iron and iodine in human health and nutrition. These included the isolation by Kendall (21) of a crystalline compound from the thyroid gland containing 65% iodine which he claimed was the active principle and which he named thyroxine, and the successful control of goiter in man and animals in several goitrous areas by the use of supplemental iodine.

During this period, the advent of emission spectography permitted the simultaneous estimation of some 20 elements in low concentrations and the "distributional" phase in trace element research began. Extensive investigations of trace element levels in soils, plants, animal, and human tissues were undertaken at that time (12, 41, 57) and later, as analytical techniques improved (15, 31, 45). These and other distribution studies referred to in later chapters were responsible for (*i*) defining the wide limits of concentration of many of the trace elements in foods and tissues; (*ii*) illuminating the significance of such factors as age, location, disease, and industrial contamination in influencing those concentrations; (*iii*) stimulating studies of the possible physiological significance of several elements previously unsuspected of biological potentiality; and (*iv*) discriminating between those elements most likely to have such potential and those which were more likely to be environmental contaminants.

The second quarter of this century was notable for spectacular advances in our knowledge of the nutritional importance of trace elements. These advances came from basic studies with laboratory species aimed at enlarging our understanding of total nutrient needs and from investigations of a number of widely separated naturally occurring nutritional maladies of man and his domestic livestock. In the 1920's Bertrand of France and McHargue of the United States pioneered a purified diet approach to animal studies with these elements. The diets employed were so deficient in other nutrients, particularly vitamins, that the animals achieved little growth or survived only for short periods, even with the addition of the element under study (5,26). Vitamin research had not then progressed to the point where these vital nutrients could be supplied in pure or semipure form. The Wisconsin school, led by E. B. Hart, initiated a new era when they showed in 1928 that supplementary copper, as well as iron, is necessary for growth and hemoglobin formation in rats fed a milk diet (17). Within a few years the same group, using the special diet and purified diet techniques with great success, first showed that manganese (20, 55) and then zinc (46) were dietary essentials for mice and rats. These important findings were soon confirmed and extended to other species. Nearly 20 years elapsed before the purified diet technique was again successful in identifying further essential trace elements. These were first molybdenum (10, 33) in 1953, then selenium (30, 35) in 1957, and chromium (36) in 1959. Conclusive evidence that selenium is a nutritionally essential element, with a role beyond that of a substitute for a normal intake of vitamin E, did not emerge until 1969 from the critical studies of Thompson and Scott (44) with chicks.

During the 1930's a wide range of nutritional disorders of man and farm stock were found to be caused by deficient or excessive intakes of various trace elements from the natural environment. In 1931, mottled enamel in man was shown by three independent groups to result from the ingestion of excessive amounts of fluoride from the drinking water (8, 39, 52). In the same year, copper deficiency in grazing cattle was demonstrated in parts of Florida (29) and Holland (37). Two years later, in 1933 and 1935, "alkali disease" and "blind staggers" of stock occurring in parts of the Great Plains region of the United States were established as manifestations of chronic and acute selenium toxicity, respectively (1, 34). In 1935, cobalt deficiency was shown to be the cause of two wasting diseases of grazing ruminants occurring in localized areas in South Australia (25) and Western Australia (48). In 1936 and 1937, a dietary deficiency of manganese was found to be responsible for perosis or "slipped tendon" (56) and nutritional chondrodystrophy (23) in poultry. In 1937, enzootic neonatal ataxia in lambs in Australia was shown to be a manifestation of copper deficiency in the ewe during pregnancy (2). Finally, in the following year excessive intakes of molybdenum from the pastures were found to cause the debilitating diarrhea affecting grazing cattle confined to certain areas in England (13).

In the investigations just outlined attention was initially focused on discovering the cause and devising practical means of prevention and control of acute disorders with well-marked clinical and pathological manifestations of the nutritional abnormalities. It soon became evident that a series of milder maladies involving the trace elements also existed, which had less specific manifestations and affected more animals and greater areas than the acute conditions which prompted the original investigations. The deficiency or toxicity states were often found to be ameliorated or exacerbated, i.e., conditioned by the extent to which other elements, nutrients, or compounds were present or absent from the environment. The importance of dietary interrelationships of this type was highlighted by the discovery of Dick (11), while investigating chronic copper poisoning in sheep in southeastern Australia in the early 1950's, that a three-way interaction exists between copper, molybdenum, and inorganic sulfate and that the ability of molybdenum to limit copper retention in the animal can only be expressed in the presence of adequate sulfate. Metabolic interactions among the trace elements were subsequently shown to be of profound nutritional importance and to involve a wide variety of elements. These are considered in the appropriate chapters that follow and have been the subject of two recent reviews (43, 47).

The third quarter of this century has seen many other notable advances in our understanding of the nutritional physiology of the trace elements, besides the great significance of metabolic interactions mentioned previously. These advances have mostly been highly dependent on concurrent developments in analytical techniques, among which atomic absorption, neutron activation, and microelectron probe procedures have been particularly prominent. With such procedures the concentrations and distribution of most elements in the tissues, cells, and even the organelles of the cells can be determined with ever-increasing sensitivity and precision and the metabolic movements and the kinetics of those movements followed with the aid of radioactive isotopes of suitable half-life. At the same time many metalloproteins with enzymatic activity have been discovered, allowing the identification of basic biochemical lesions related causally to the diverse manifestations of deficiency or toxicity in the animal. The elucidation of the role of copper in elastin biosyntheses and its relation to the cardiovascular disorders that arise in copper-deficient animals provide a classical example of progress in this area (see Hill *et al.,* 19).

During the last decade there has been a rapid increase in the number of trace elements shown to be essential and a remarkable surge of interest and activity in their significance or potential significance in human health and nutrition. During this period six trace elements, namely, tin, silicon, fluorine, nickel, vanadium, and arsenic, have been added to the list of dietary essentials, thus giving a whole new dimension to trace element research.* An appraisal of these discoveries and their nutritional significance is given in later chapters. Progress in this area was made possible by the use of the plastic isolator technique developed initially by Smith and Schwarz (38) and employed with outstanding success by Schwarz and his collaborators. In this technique the animals are isolated in a system in which plastics are used for all component parts and there is no metal, glass, or rubber. The unit has an airlock to facilitate passage in and out of the so-called trace element sterile environment and two air filters which remove almost all dust down to a particle size of 0.35 μm. The virtual exclusion of dust is the critical final step in the whole technique, because rats fed the same highly purified crystalline vitamin and amino acid-containing diets and drinking water *outside* the isolator system may grow adequately and show none of the clinical signs of deficiency that appear in the animals maintained in the isolator.

The possibility that contamination of the air, water supply, and foods with trace elements, arising from modern agricultural and industrial practices and

*Author's note: Evidence that a further element, cadmium, may be essential for growth in the rat has now appeared. [Schwarz, K., and Spallholz, J., *Fed. Proc.* **35**, Abstr. 255 (1976).]

from the increasing motorization and urbanization of sections of the community, may have deleterious effects on the long-term health and welfare of human populations has stimulated increasing interest in the concentrations and movements of these elements in the environment and in the maximum permissible intakes by man. Such investigations involving lead, cadmium, and mercury have been particularly prominent. At the same time it has become evident that changes in dietary patterns involving increased consumption of refined and processed foods can lead to deficient or marginally deficient intakes or other trace elements, notably zinc and chromium, in substantial sections of western communities. The nature and extent of actual and potential deficiencies and toxicities of this sort are considered in later chapters.

III. MODE OF ACTION OF TRACE ELEMENTS

The only property that the essential trace elements have in common is that they normally occur and function in living tissues in low concentrations. These normal tissue concentrations vary greatly in magnitude and are characteristic for each element. They are usually expressed as parts per million (ppm), μg/g, or 10^{-6}, or with some, such as iodine, chromium, nickel, and vanadium, as parts per billion (ppb), ng/g, or 10^{-9}. It should be noted that certain of the nonessential elements, such as bromine and rubidium, occur in animal tissues in concentrations well above those of most of the essential trace elements.

The characteristic concentrations and functional forms of the trace elements must be maintained within narrow limits if the functional and structural integrity of the tissues is to be safeguarded and the growth, health, and fertility of the animals are to remain unimpaired. Continued ingestion of diets, or continued exposure to total environments that are deficient, imbalanced, or excessively high in a particular trace element, induces changes in the functioning forms, activities, or concentrations of that element in the body tissues and fluids so that they fall below or rise above the permissible limits. In these circumstances biochemical defects develop, physiological functions are affected, and structural disorders may arise in ways which differ with different elements, with the degree and duration of the dietary deficiency or toxicity, and with the age, sex, and species of the animal involved. A protective mechanism may be brought into play which can delay or minimize the onset of such diet-induced changes. With some elements, such as fluorine, the protective mechanisms can be extremely effective over prolonged periods, but to prevent ultimately deleterious changes the animal must be supplied with a diet that is palatable and nontoxic, as well as containing the required elements in adequate amounts, in proper proportions, and in available forms.

The trace elements act primarily as catalysts in enzyme systems in the cells, where they serve a wide range of functions. In this respect their roles range from weak ionic effects to highly specific associations known as metalloenzymes. In the metalloenzymes the metal is firmly associated with the protein and there is a fixed number of metal atoms per molecule of protein. These atoms cannot be replaced by any other metal. However, Vallee (50) has shown that cobalt and cadmium can be substituted for the native zinc atoms in several zinc enzymes while the enzyme remains active. This worker has stated further that "the key to the specificity and catalytical potential of metalloenzymes seems to lie in the diversity and topological arrangement both of their active site residues and of their metal atoms, all of which interact with their substrates." He has emphasized the importance of spectra as probes of active sites of metalloenzymes, which can reveal geometric and electronic detail pertinent to the functions of such systems (51). His basic studies in this area provide important beginnings to an understanding of the molecular mechanisms involved in the metalloenzymes and the nature of the metal ion specificity in their reactions.

Stiefel (40) has investigated the five known Mo enzymes (nitrogenase, nitrate reductase, aldehyde oxidase, xanthine oxidase, and sulfite oxidase) with their diverse functions. The reactions catalyzed by these enzymes each find the product differing from the substrate by two electrons and two protons, or some multiple thereof. A simple molecular mechanism, embodying coupled electron–proton transfer to and from the substrate compatible with the coordination chemistry of molybdenum, is presented for each of the enzymes, and reasons for the use of molybdenum, as distinct from most other metals, are suggested.

Evidence is accumulating that the protein–metal interactions not only enhance the catalytic activity of enzymes but also may increase the stability of the protein moiety to metabolic turnover. In this connection Harris (16a) has recently shown that copper is a key regulator of lysyl oxidase activity in the aorta of chicks and may be a major determinant of the steady-state levels of the enzyme in that tissue. The tissue levels and activities of many of metalloenzymes have now been related to the manifestations of deficiency or toxicity states in the animal, as discussed later. However, many clinical and pathological disorders in the animal as a consequence of trace element deficiencies or excesses cannot yet be explained in biochemical or enzymatic terms. This suggests either that there are many metabolically significant trace element-dependent enzymes which have still to be discovered and their precise loci identified, or that those elements participate in the activity and structure of other vital compounds in the tissues.

IV. TRACE ELEMENT NEEDS AND TOLERANCES

The minimum requirements of animals and man for the essential trace elements are commonly expressed in proportions or concentrations of the total dry diet consumed daily. The maximum intake of these and other elements that can be safely tolerated are usually expressed similarly. These requirements, or tolerances, are arrived at by relating the growth, health, fertility, or other relevant criteria in the animal to varying dietary mineral concentrations. The latter are found by the application of analytical techniques that measure the total amounts of the elements in the diets or their component foods. Such analyses are not normally affected by variations in the chemical forms of the element as they occur in foods of different types or from different environments. Since the availability of mineral elements to the animal is affected by the chemical form in which it is ingested, it is obvious that gross dietary intakes do not necessarily reflect minimum requirements or maximum tolerances of wide or universal applicability.

Trace element requirements and tolerances expressed as concentrations such as parts per million of the dry diet also carry the assumption that the whole diet is otherwise adequate and well balanced for the purpose of which it is fed, and that it is effectively free from other toxic factors capable of adversely affecting the animal's health, appetite, or utilization of the element concerned. The question of appetite is especially important since the capacity of a particular dietary concentration of an element to supply the needs of an animal will clearly depend on the amount of the diet consumed daily or over a given period. Equally important is the level of other minerals or other nutrients which influence the availability or utilization of the element in question. A "true" or basic minimum requirement can thus be conceived as one in which all the dietary conditions affecting the element in this way are at an optimum. A series of minimum requirements therefore exist depending on the extent to which such interacting factors are present or absent from the whole diet.

Similarly, a series of "safe" dietary levels of potentially toxic trace elements exist, depending on the extent to which other elements which affect their absorption and retention are present. These considerations apply to all the trace elements to varying degrees, but with some elements such as copper they are so important that a particular level of intake of this element can lead to signs either of copper deficiency or of copper toxicity in the animal, depending on the relative intakes of molybdenum and sulfur, or of zinc and iron. The many mineral interactions of this type which exist are discussed in later chapters. However, it is appropriate to mention at this point the experiments of Suttle and Mills (42) on copper toxicity in pigs, which strikingly illustrate the importance of trace element dietary balance in determining the "safe" intake of a particular element. These workers showed that dietary copper levels of 425 and 450 ppm caused severe toxicosis, all signs of which were prevented by simultaneously providing an additional 150 ppm Zn plus 150 ppm Fe to the diets.

Estimates of adequacy or safety also vary with the criteria employed. As the amount of an essential trace element available to the animal becomes insufficient for all the metabolic processes in which it participates, as a result of inadequate intake and depletion of body reserves, certain of these processes fail in the competition for the inadequate supply. The sensitivity of particular metabolic processes to lack of an essential element and the priority of demand exerted by them vary in different species and, within species, with the age and sex of the animal and the rapidity with which the deficiency develops. In the sheep, for example, the processes of pigmentation and keratinization of wool appear to be the first to be affected by a low-copper status, so that at certain levels of copper intake no other function involving copper is impaired. Thus, if wool quality is taken as the criterion of adequacy, the copper requirements of sheep are higher than if growth rate and hemoglobin levels are taken as criteria. In this species it has also been shown that the zinc requirements for testicular growth and development and for normal spermatogenesis are significantly higher than those needed for the support of normal live-weight growth and appetite (49). If body growth is taken as the criterion of adequacy, which would be the case with rams destined for slaughter for meat at an early age, the zinc requirements would be lower than for similar animals kept for reproductive purposes. The position is similar with manganese in the nutrition of pigs. Ample evidence is available that the manganese requirements for growth in this species are substantially lower than they are for satisfactory reproductive performance. Recent evidence relating zinc intakes to rate of wound healing also raises important questions on the criteria for adequacy to be employed in assessing the zinc requirements of man.

Criteria of adequacy and tolerance present particular problems with fluorine. Indisputable evidence is available that fluoride, additional to that consumed by man in most areas, confers improved resistance to dental caries, and there is some evidence that additional fluoride is also necessary to assist in the maintenance of a normal skeleton and in the reduction of osteoporosis in the mature adult population. If a reduced incidence of dental caries and osteoporosis is taken as the criterion of adequacy, then clearly the human fluoride requirements are higher than if it is not. With farm animals, certain levels of fluoride intake are tolerated for prolonged periods without any measurable decline in growth, appetite, well-being, fertility, or productivity, despite elevated bone fluoride levels and mild dental and skeletal adnormalities. Based on the usual criterion applied to farm stock, namely, performance, such levels should therefore not be ruled as excessive. On the other hand, similar fluoride intakes could be considered intolerable if applied to younger animals or if consumed for still longer periods.

REFERENCES

1. Beath, O.A., Eppson, H.F., and Gilbert, C.S., *Wyo, Agr. Exp. Stn., Bull.* **206** (1935).
2. Bennetts, H.W., and Chapman, F.E., *Aust. Vet. J.* **13**, 138 (1937).

3. Bernard, C., "Leçons sur les effets des substances toxiques et médicamenteuses." Baillière, Paris, 1857.
4. Bertrand, G., *Proc. Int. Congr. Appl. Chem., 8th, 1912* Vol. 28, p. 30 (1912).
5. Bertrand, G., and Benson, R., *C.R. Acad. Sci.* **175**, 289 (1922); Bertrand, G., and Nakamura, H., *ibid.* **186**, 480 (1928).
6. Chatin, A., *C.R. Acad. Sci.* **30–39** (1850–1854).
7. Church, A.W., *Phil. Trans. R. Soc. London* **159**,627 (1869).
8. Churchill, H.N., *Ind. Eng. Chem.* **23**, 996 (1931).
9. Cotzias, G.C., *Trace Subst. Environ. Health–Proc. Univ. Mo. Annu. Conf., 1st, 1967,* p. 5. (1967).
10. DeRenzo, E.C., Kaleita, E., Heyther, P., Oleson, J.J., Hutchings, B.L., and Williams, J.H., *J. Am. Chem. Soc.* **75**, 753 (1953); *Arch. Biochem. Biophys.* **45**, 247 (1953).
11. Dick, A.T., *Aust. J. Agric. Res.* **5**, 511 (1954).
12. Dutoit, P., and Zbinden, C., *C.R. Hebd. Seances Acad. Sci.* **188**, 1628 (1929).
13. Ferguson, W.S., Lewis, A.H., and Watson, S.J., *Nature (London)* **141**, 553 (1938); *Jealott's Hill Bull.* No. 1 (1940).
14. Fowler, W.M., *Ann. Med. Hist.* 8, 168 (1936).
15. Hamilton, E.J., Minski, M.J., and Cleary, J.J., *Sci. Total Environ.* **1**, 341 (1972–1973).
16. Harless, *E., Arch. Anat. Physiol. (Leipzig)* p. 148 (1847).
16a. Harris, E.D., *Proc. Natl. Acad. Sci. U.S.A.* **73**, 371 (1976).
17. Hart, E.B., Steenbock, H., Waddell, J., and Elvehjem, C.A., *J. Biol. Chem.* **77**, 797 (1928).
18. Henze, M., *Hoppe-Seyler's Z. Physiol. Chem.* **72**, 494 (1911).
19. Hill, C.H., Starcher, B., and Kim, C., *Fed. Proc., Fed. Am. Soc. Exp. Biol.* **26**, 129 (1968).
20. Kemmerer, A.R., Elvehjem, C.A., and Hart, E.B., *J. Biol. Chem.* **94**, 317 (1931).
21. Kendall, E.C., *J. Biol. Chem.* **39**, 125 (1919).
22. Liebscher, K., and Smith, H., *Arch. Environ. Health* **17**, 881 (1968).
23. Lyons, M., and Insko, W.M., *Ky. Agric. Exp. Stn., Bull.* **371** (1937).
24. MacMunn, C.A., *Phil. Trans. R. Soc. London* **177**, 267 (1885).
25. Marston, H.R., *J. Counc. Sci. Ind. Res. (Aust.)* **8**, 111 (1935); Lines, E.W., *ibid.* p. 117.
26. McHargue, J.S., *Am. J. Physiol.* 77, 245 (1926).
27. Mendel, L.B., and Bradley, H.C., *Am. J. Physiol.* **14**, 313 (1905).
28. Mertz, W., *Fed. Proc., Fed. Am. Soc. Exp. Biol.* **29**, 1482 (1970).
29. Neal, W.M., Becker, R.B., and Shealy, A.L., *Science* **74**, 418 (1931).
30. Patterson, E.L., Milstrey, R., and Stokstad, E.L.R., *Proc. Soc. Exp. Biol. Med.* **95**, 621 (1957).
31. Perry, H.M., Tipton, I.H., Schroeder, H.A., and Cook, M.J., *J. Lab. Clin. Med.* **60**, 245 (1962).
32. Raulin, J., *Ann. Sci. Nat., Bot. Biol. Veg.* [2] **11**, 93 (1869).
33. Richert, D.A., and Westerfeld, W.W., *J. Biol. Chem.* **203**, 915 (1953).
34. Robinson, W.O., *J. Assoc. Off. Agric. Chem.* **16**, 423 (1933).
35. Schwarz, K., and Foltz, C.M., *J. Am. Chem. Soc.* **79**, 3293 (1957).
36. Schwarz, K., and Mertz, W., *Arch. Biochem. Biophys.* **85**, 292 (1959).
37. Sjollema, B., *Biochem. Z.* **267**, 151 (1933).
38. Smith, J.C., and Schwarz, K., *J. Nutr.* **93**, 182 (1967).
39. Smith, M.C., Lantz, E.M., and Smith, H.V., *Ariz. Agric. Exp. Stn., Tech. Bull.* **32** (1931).
40. Stiefel, E.J., *Proc. Natl. Acad. Sci. U.S.A.* **70**, 988 (1973).

41. Stitch, S.R., *Biochem. J.* **67**, 97 (1957).
42. Suttle, N.F., and Mills, C.F., *Br. J. Nutr.* **20**, 135 and 149 (1966).
43. Suttle, N.F., *in* "Trace Elements in Soil–Plant–Animal Systems" (D.J.D. Nicholas and A.R. Egan, eds.), pp. 271–289. Academic Press, New York, 1975.
44. Thompson, J.N., and Scott, M.L., *J. Nutr.* **97**, 335 (1969); **100**, 797 (1970).
45. Tipton, I.H., and Cook, M.J., *Health Phys.* **9**, 103 (1963).
46. Todd, W.R., Elvehjem, C.A., and Hart, E.B., *Am. J. Physiol* **107**, 146 (1934).
47. Underwood, E.J., *in* "Heavy Metals in the Environment" (F.W. Oehme, ed.). Dekker, New York, 1975.
48. Underwood, E.J., and Filmer, J. F., *Aust. Vet. J.* **11**, 84 (1935).
49. Underwood, E.J., and Somers, M., *Aust. J. Agric. Res.* **20**, 889 (1969).
50. Vallee, B.L., *in* "Newer Trace Elements in Nutrition" (W. Mertz and W.E. Cornatzer, eds.), p. 33. Dekker, New York, 1971.
51. Vallee, B.L., *in* "Trace Element Metabolism in Animals" (W.G. Hoekstra *et al.*, eds.), Vol. 2, p. 5. Univ. Park Press, Baltimore, Maryland, 1974.
52. Velu, H., *C.R. Seances Soc. Biol. Ses. Fil.* **108**, 750 (1931); **127**, 854 (1938).
53. Venchikov, A.I., *Vopr. Pitan.* **6**, 3 (1960).
54. Venchikov, A.I., *in* "Trace Element Metabolism in Animals" (W.G. Hoekstra *et al.*, eds.), Vol. 2, p. 295. Univ. Park Press, Baltimore, Maryland, 1974.
55. Waddell, J., Steenbock, H., and Hart, E.B., *J. Nutr.* **4**, 53 (1931).
56. Wilgus, H.R., Norris, L.C., and Heuser, G.F., *Science* **84**, 252 (1936); *J. Nutr.* **14**, 155 (1937).
57. Wright, N.C., and Papish, J., *Science* **69**, 78(1929).

2

Iron

I. IRON IN ANIMAL TISSUES AND FLUIDS

1. Total Content and Distribution

The total iron content of the animal body varies with age, sex, nutrition, state of health, and species. Normal adult man is estimated to contain 4–5 g of iron (109) or 60–70 ppm of the whole body of a 70-kg individual. The adult rat contains approximately 50 ppm Fe in the whole body (272), while levels of 40 ppm or less are normal for suckling rats (172). Most of the body iron exists in complex forms bound to protein, either as porphyrin or heme compounds, particularly hemoglobin and myoglobin, or as nonheme protein-bound compounds such as ferritin and transferrin. In certain disease states large amounts of iron may also be present as hemosiderin, as discussed later. The hemoprotein and flavoprotein enzymes together constitute less than 1% of the total body iron. Free ionic iron is present in negligible quantities.

Hemoglobin iron occupies a dominant position in all healthy animals, although in myoglobin-rich species, such as the horse and dog, the proportion is lower than in man. Thus Hahn (117) estimates blood hemoglobin iron to be 57% and myoglobin iron to be 7% of total body iron in the adult dog, compared with 60–70% and 3%, respectively, in adult man. Approximately 50% of total body iron consists of heme iron in normal suckling rats (172).

Species differences in total body iron concentrations occur in the newborn but become much less pronounced in the adult, as shown in Table 1. The

TABLE 1
Iron Content of Bodies of Different Species[a]

Age	Fe (ppm) of fat-free tissue						
	Human	Pig	Cat	Rabbit	Guinea pig	Rat	Mouse
Adult	74	90	60	60	–	60	–
Newborn	94	29	55	135	67	59	66

[a]From Widdowson (306).

differences at birth reflect differences in liver iron stores and blood hemoglobin levels. For example, the pig has relatively little iron in its body at birth because it is normally born with low liver iron stores and has no polycythemia of the newborn as does the human infant. The newborn rabbit, by contrast, has an exceptionally high total body iron concentration because of its large liver iron stores (306). Manipulation of maternal iron metabolism, creating either iron-loaded or iron-depleted mothers during pregnancy and lactation, does not appear to greatly affect the iron status of the suckling young (210, 241).

Female rats have a higher total body iron than males and accumulate more iron in their livers on the same diet (229). Female mice and birds also carry greater concentrations of liver iron than males but no such sex difference is apparent in rabbits or guinea pigs (307). Lower values for total body iron would be expected in women than in men because of their normally lower blood hemoglobin, muscle myoglobin levels, and body iron stores.

Among the body organs the liver and spleen usually carry the highest Fe concentrations, followed by the kidney, heart, skeletal muscles, and brain, which contain only one-half to one-tenth the levels in the liver and spleen. Individual variation in the Fe levels in liver, kidney, and spleen can be very high. In most species the liver has a remarkably high storage capacity for iron. Large increases in the total iron content of the liver, up to a total of 10 g, occur in cases of human malignancy and chronic infection (140). In the final stages of hemochromatosis as much as 50 g of iron may accumulate in the human body (75). Iron overload in storage organs such as the liver and spleen is also characteristic of copper deficiency (170). The iron content of these organs, and of the bone marrow, is reduced below normal in dietary iron deficiency and in hemorrhagic anemia.

2. Iron in Blood

Iron occurs in blood as hemoglobin in the erythrocytes and as transferrin in the plasma in a ratio of nearly 1000:1. Small quantities of iron as ferritin are present in the erythrocytes of human blood (26, 86, 279a), in serum, and in the leukocytes, especially the normocytes (279a). The levels of ferritin in serum vary with the iron status of the individual and with certain disease states, as will be seen (2, 148, 312). The iron present in this form represents only 0.2–0.4% of the serum iron normally present in the adult.

a. Hemoglobin. Hemoglobin is a complex of globin and four ferroprotoporphyrin or "heme" moieties. A three-dimensional picture of the molecule and the nature of the bond between iron and globin were established in 1956 (143). This union stabilizes the iron in the ferrous state and allows it to be reversibly bonded to oxygen, thus permitting hemoglobin to function as an oxygen carrier. The molecule is similar in size (molecular weight close to 65,000) in all animal species and has an iron content close to 0.35%.

The synthesis of heme and its attachment to globin take place in the later stages of red cell development in the bone marrow. The biosynthesis of heme includes the condensation of two molecules of δ-aminolevulinic acid (ALA) to form porphobilinogen (248), a process that involves the enzyme ALA-dehydrogenase (144). Iron is carried to the bone marrow in the ferric form as transferrin, reduced to the ferrous form, and detached from transferrin with the aid of reducing substances, thus facilitating its transfer to protoporphorin (113, 248). The process of transport between iron uptake at the membrane surface of the reticulocyte and its incorporation into heme is considered in Section II, 4.

The normal ranges of blood hemoglobin (Hb) levels in the adults of different species are as follows: man and rat, 13–17; dog, 13–14; cow and rabbit, 11–12; and pig, sheep, goat, and horse, 10–11 g Hb/100 ml. The total hemoglobin in mammals is proportional to body weight. In five species examined by Drabkin (75) the mean proportion was found to be 12.7 g Hb/kg body weight.

The levels of hemoglobin in the blood vary with age, sex, nutrition, pregnancy, lactation, altitude, and disease. In man, the level falls rapidly from about 18 to 19 g/100 ml at birth to about 12 g at 3–4 months, at which level it usually remains until the child is about a year old, when a slow rise to adult values normally begins. A striking rise occurs at puberty in males and the higher hemoglobin levels of the male continue throughout the life-span. This is evidently a real sex difference, since a significant rise in the level of females does not take place after the menopause or a hysterectomy, when menstrual blood losses no longer occur (293). Sex differences similar to those in man do not occur in rats (157), or in cattle (49).

Blood hemoglobin levels decline in late pregnancy in normal rats and in healthy women but not in ewes (196). In both species the decline arises primarily from a rise in plasma volume greater than the rise in red cell volume, but iron deficiency can be a contributing factor. This point is discussed later when considering iron deficiency in human adults. The whole problem of the "anemia" of pregnancy has been assessed by de Leeuw *et al.* (171) and by Hall (119).

b. Serum Transferrin. The iron of the serum is bound to a specific protein, designated transferrin or siderophilin (136, 257). Transferrin occurs in the blood of all vertebrate species (151). Transferrin is a glycoprotein with two identical or almost identical Fe-binding sites each capable of binding one atom of ferric iron (79). The exact molecular weight is uncertain but 76,000 daltons is considered the most acceptable figure for human transferrin at the present time (88). The total carbohydrate content is 5.3% and is arranged in the form of two identical branched side chains, each branch ending in a molecule of *N*-acetylneuraminic acid (sialic acid), giving four sialic acid residues per molecule (150). Transferrins from other species have similar but not identical physical and chemical properties to those of human transferrin (89). Many genetically controlled transferrins occur in human blood (23).

Transferrin serves as the principal carrier of iron in the blood and therefore plays a central role in iron metabolism, although serum ferritin is also believed to serve a transport function (267). Transferrin also has a second important function, that of participating in the defense mechanisms of the body against infection (45). In normal individuals only 30–40% of the transferrin carries iron, the remainder being known as the latent iron-binding capacity. Serum iron and total and latent iron-binding capacity vary greatly among species, among individuals of the same species, and in various disease states. In man a well-marked diurnal rhythm occurs. Vahlquist (292) found 15 normal men to have a mean serum iron of 135 ± 10.6 mg/100 ml at 8:00 A.M. and 99.9 ± 9.2 mg/100 ml at 6:00 P.M. High morning and low evening values have since been observed by several investigators, except in night workers in which the diurnal rhythm is reversed and diminished (126, 232). Pigs display a unique diurnal variation in serum iron, with the highest values at midnight and 3:00 A.M. (101). Diurnal fluctuations in plasma iron in dogs appear to be the result of variable partitioning of iron between the rapid early and slower late phases in the release of iron from the reticuloendothelial cells (93).

Individual variability in serum iron and total iron-binding capacity (TIBC) is high in all species studied (236). Species differences are small, with some evidence of higher average values in pigs, sheep, and cattle (101, 102, 236, 282, 291) than in man. In zebu cattle both serum iron and TIBC values are at a maximum during early adulthood and decline to lower levels between 8 and 17 years of age (282). In male birds and chicks both serum iron and TIBC concentrations are similar to those of man. In hens and ducks during the laying season serum iron is markedly elevated by a factor of almost five. TIBC levels are also raised, although not to the same extent, and the iron-binding capacity is fully saturated (235, 237). The transport of iron in the laying fowl presents special characteristics involving the existence of mechanisms auxiliary to that of transferrin (235, 237). The elevation of plasma iron in hens during egg-laying is due to the appearance of a specific phosphoprotein, phosvitin, which binds about two-thirds of the iron present in plasma and transports iron to the ovocytes and egg yolk (5, 231).

Representative values for serum iron, TIBC, and percentage saturation in several species are presented in Table 2. The differences in human serum iron that appear in the table can be understood if the iron of the serum is conceived as a pool into which iron enters, leaves, and is returned at varying rates for the synthesis and resynthesis of hemoglobin, ferritin, and other iron compounds. For instance, in iron deficiency low serum Fe levels result from low intake, depletion of body stores, and reduced hemoglobin destruction accompanying the anemia. In fact, there is evidence that the serum Fe level begins to fall before the iron stores are completely mobilized (191). The high serum and percentage saturation values characteristic of pernicious anemia and aplastic anemia (56) can be associated with increased iron absorption and a bone marrow block to hemoglobin synthesis in the presence of adequate iron stores. The high serum iron and percentage saturation of the iron-binding protein in hemochromatosis

TABLE 2
Serum Iron, Total Iron-Binding Capacity (TIBC), and Percentage Saturation in Various Species

Species and condition	No. of cases	Serum iron (μg/100 ml)	TIBC (μg/100 ml)	Mean saturation (%)	Ref.
Human adults					
Normal male	35	127 (67–191)	333 (253–416)	33	55
Normal female	35	113 (63–202)	329 (250–416)	37	55
Iron deficiency anemia	35	32 (0–78)	482 (204–705)	7	55
Late pregnancy	106	94 (22–185)	532 (373–712)	18	55
Hemochromatosis	14	250 (191–290)	263 (205–330)	96	55
Infections	11	47 (30–72)	260 (182–270)	20	55
Bovine adults					
Normal cows	10	146 (89–253)	553 (388–724)	26	291
Normal bulls	10	145 (92–270)	432 (332–521)	33	291
Ovine adults					
Normal ewes	12	182 (102–304)	353 (278–456)	51	291
Normal rams	12	152 (114–191)	353 (248–455)	43	291
Castrate males	65	180 (108–268)	331 (264–406)	56	103
Adult pigs	91	123 (49–197)	540 (374–635)	22	101
Birds					
Chicks	90	102 ± 6	239 ± 4	43	235
Nonlaying hens	40	158	258	61	235
Laying hens	38	516	333	100	235

can be related to excessive absorption and deposition of iron in the body. The low serum iron and TIBC levels in the nephrotic syndrome (55) may be explained by losses of bound iron in the urine in the presence of considerable proteinuria (57).

Not all the changes in serum iron and TIBC levels can be explained so easily. For instance, the low serum iron and TIBC levels of malignancy and infections, other than hepatitis, where serum iron levels are elevated (32), present a problem, although it is known that the decrease in serum iron is not due to the reduction in iron-binding capacity (24, 25, 56). Beisel *et al.* (24) have produced evidence indicating that infections induce a redistribution of iron (and zinc) within the body, initiated by a hormonelike protein factor which is released from phagocytizing cells. This factor, leukocytic endogenous mediator (LEM), stimulates the liver to take up iron (and zinc) from serum.

The reasons for the variations in TIBC levels in various disease states and for the rise in this component in the third trimester of human pregnancy are far from clear. Such changes are not necessarily correlated with the absorption of iron, the magnitude of the iron stores, or the need of the body for greater or lesser transport of iron. However, Morgan (213) has demonstrated a good inverse relationship between TIBC and hemoglobin in the rat. He suggests that "the main factor regulating plasma transferrin concentration is the balance between tissue supply and requirements for oxygen, i.e. relative oxygen supply."

c. Serum Ferritin. The development of an immunoradiometric method for quantitating serum ferritin (2) has led to several studies of serum ferritin concentrations in normal men and women and in patients with anemia and disorders of iron metabolism. The mean concentration in men is 2–3 times that in women, suggesting that the level reflects total body stores (148, 303). A high degree of correlation has since been demonstrated between serum ferritin concentration and storage iron in normal subjects (64, 176, 303) (see Fig. 1), and serum ferritin assay has been shown to be a useful tool in the evaluation of iron status in man, particularly in children. Thus Siimes and co-workers (267) found the serum ferritin concentration in 573 infants and children to parallel known changes in iron stores during development. The median ferritin concentration was 101 ng/ml at birth, rose to 356 ng/ml at 1 month, and then fell rapidly to a median value near 30 ng/ml (range 7–142) between 6 months and 15 years of age. Median values for normal adults reported in this investigation were 39 ng/ml in the female and 140 ng/ml in the male. In 13 children with iron deficiency anemia the serum ferritin was 9 ng/ml or less. Lipschitz *et al.* (176) obtained a geometric mean value of 4 ng/ml for 32 patients with uncomplicated iron deficiency anemia and 2930 ng/ml for 23 patients with iron overload. Among subjects with anemia from causes other than iron deficiency the mean serum ferritin level increased to 180 ng/ml.

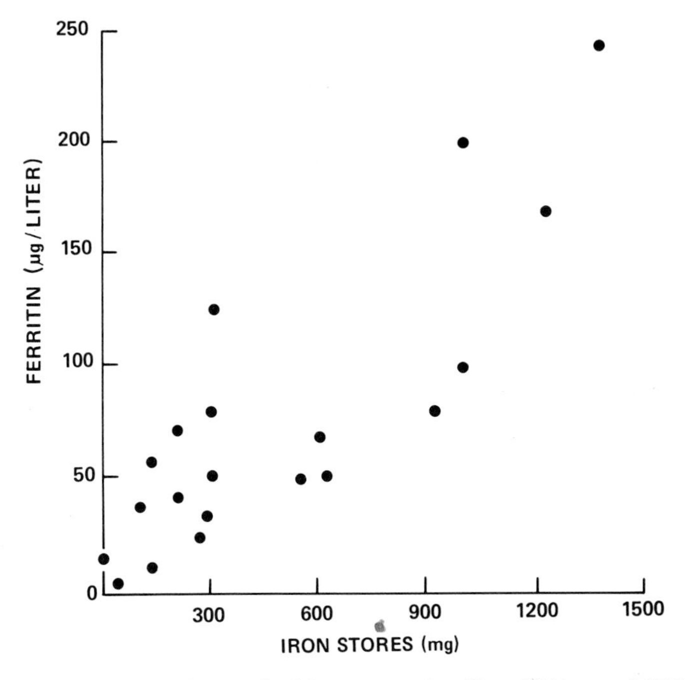

Fig. 1 Iron stores and serum ferritin concentration. From Walters *et al.*(303).

Liver disease and increased red cell turnover elevates serum ferritin concentration to a degree disproportionate to iron stores, as shown in Table 3. The data of Walters *et al.* (303) obtained with normal subjects in which iron was measured by quantitative phlebotomy suggest that 1 ng/ml of serum represents about 8 mg of storage iron. In a series of 75 healthy males and 44 healthy females in which the mean serum ferritin concentrations were 69 and 35 ng/ml, respectively (148), the mean iron stores would thus be 552 mg in the men and 280 in the women. Somewhat higher mean normal serum ferritin values were obtained by Cook *et al.* (64) in a group of subjects in whom iron-deficient erythropoiesis was excluded on the basis of transferrin saturation and red cell protoporphyrin. The serum ferritin was log-normally distributed with a geometric mean of 94 ng/ml in 174 males and 34 ng/ml in 152 females. Serum ferritin levels are markedly elevated in patients with leukemia (312). This increase in ferritin probably derives from the leukemic cells themselves since these cells, especially the monocytes, are high in ferritin compared with the red cells (279a). However, ferritin is modified during its entry into the plasma, and even in cases of iron overload the Fe content of serum ferritin may be low (313).

TABLE 3
Relation between Serum Ferritin and Bone Marrow Hemosiderin

	Control		Patients with inflammation		Patients with liver disease	
Marrow iron	Serum no.	Ferritin[a] (ng/ml)	Serum no.	Ferritin[a] (ng/ml)	Serum no.	Ferritin[a] (ng/ml)
Absent	12	6 (1–37)	2	21 (16–28)	6	61 (25–91)
Diminished	8	51 (21–163)	6	146 (47–296)	2	182 (91–368)
Moderate	5	159 (60–253)	9	581 (129–1338)	8	622 (304–1257)
Increased	2	589 (442–669)	8	922 (290–1358)	7	1631 (1201–2077)

[a]Geometric mean, Lipschitz *et al.* (176).

3. Storage Iron Compounds

The reserve or storage iron of the body occurs predominantly as the two nonheme compounds ferritin and hemosiderin. These occur widely in the tissues, with the highest concentrations normally present in the liver, spleen, and bone marrow. The two compounds are chemically dissimilar although intimately related in function. Chemical methods for their estimation are available based on the fact that ferritin is soluble in water while hemosiderin is insoluble (102, 157). Results obtained by such means are supported by immunochemical and radiotracer techniques (102).

In the crystalline state ferritin is a brown compound containing up to 20% Fe. It consists of a central nucleus of iron stored in six micelles arranged at the corners of a regular octahedron and surrounded by a shell of protein approximately spherical in shape (218). The colorless, iron-free protein, apoferritin, is a physicochemically homogeneous globulin with a molecular weight of 460,000 (110, 254). Hemosiderin is a relatively amorphous compound which may contain up to 35% Fe, consisting mainly of ferric hydroxide condensed into an essentially protein-free aggregate (263). It exists in the tissues as a brown, granular, readily stainable pigment.

Histochemical examination of aspirated samples of bone marrow for hemosiderin provides a useful index of body iron stores (276) and is a valuable aid in diagnosing iron deficiency anemia (29). Shoden and Sturgeon (264) have emphasized the value of using both staining and chemical methods which estimate the unstainable soluble (ferritin) as well as the insoluble (hemosiderin) forms of storage iron. In a study of 130 human necropsies involving histological and chemical estimates of hepatic and splenic storage iron, Morgan and Walters (216) found a general agreement between the two estimates, but there was considerable variation. Histological examination of these tissues gave only a very approximate idea of storage iron levels. The situation is similar in respect to histochemical examinations of bone marrow (159).

The factors affecting the amounts and proportions of ferritin and hemosiderin in the liver and spleen of rats, rabbits, and man have been extensively studied (212, 216, 262, 264, 265). Up to certain levels and rates of iron storage, iron is deposited in the liver and spleen readily and in roughly the same amounts as ferritin and hemosiderin. Iron is utilized equally readily from these two compounds for the demands of erythropoiesis and placental iron transfer to the fetus. A threefold increase in liver and spleen total iron storage was achieved in rats without a change in the relative proportions of ferritin and hemosiderin, compared with those of normal rats not fed additional iron (212). Moreover, the ferritin Fe:hemosiderin Fe ratio changed little in these organs in induced chronic or acute hemolytic anemia. Davis *et al.* (71) similarly observed little change after the first week in the ferritin:hemosiderin ratio in the livers of chicks during a depletion period on an iron-deficient diet.

The main factor affecting the relative distribution of iron between ferritin and hemosiderin in mammals is the total storage iron concentration. When the level in the liver and spleen of rats and rabbits increases beyond about 2000 μg Fe/g, hemosiderin begins to predominate. At levels beyond 3000–4000 μg Fe/g additional storage iron is deposited quantitatively as hemosiderin (212, 265). From their study of iron storage in human necropsies, Morgan and Walters (216) concluded that with total storage iron in liver and spleen below 500 μg/g of tissue, more iron was stored as ferritin than as hemosiderin. With levels above 1000 μg/g more was stored as hemosiderin. The situation may not be comparable in avian species. Larger amounts of hemosiderin than ferritin exist in the livers of chicks at much lower levels of total iron than those quoted for the rat and man (71). In human diseases such as hemochromatosis and transfusional siderosis, which are characterized by extremely high levels of iron in the tissues, most of the iron is present as hemosiderin (75, 216). A similar situation also exists with respect to the heavy iron deposits in the liver and spleen in nutritional siderosis in cattle (127) and Cu and Co deficiencies in sheep (209, 289).

The ratio of ferritin Fe to hemosiderin Fe is affected further by the rate of storage. When iron is injected at very high rates or administered in a form such as saccharated iron which is rapidly cleared from the serum, hemosiderin is deposited rather than ferritin. With equivalent injections of iron dextran, which remains in the serum for a relatively long time, ferritin production is greater and hemosiderin smaller (263). This suggests that there is a limit to the amount of apoferritin which is present in, or can be produced by, the liver of the rabbit in a given interval of time. These findings apply only to very high rates of iron storage. Over a wide range of iron depletion and storage rates and levels, the distribution of iron between ferritin and hemosiderin remains relatively constant, and the iron moves readily from one storage form to the other.

The bone marrow and the muscles contain considerable amounts of nonheme Fe. The storage iron concentration of the bone marrow in normal man is given by Hallgren (124) as about 100 μg/g. If the total active bone marrow weight is taken as 3000 g, it can be calculated that the bone marrow of the human body would contain approximately 300 mg of storage iron, or about one-third to one-fifth of the total estimated storage iron. The concentration of nonheme Fe in the muscles is low (124, 285), but because of their large mass, the total amount is high. Torrance *et al.* (285) have shown that in human subjects with normal body stores, the total amount of nonheme iron in the muscles is at least equal to that in the liver. In subjects with iron overload the Fe concentrations in muscles are raised but to a much smaller extent than in the liver. In rats this muscle iron represents a relatively nonmiscible pool which responds little to acute changes in the iron environment.

4. Iron in Milk

The iron content of milk varies with the species and stage of lactation and is highly resistant to changes in the level of dietary iron. Individual variation within the species is high but some of the reported variation probably reflects analytical inadequacies and insufficient care to avoid contamination. Contamination from metal receptacles can more than double the iron content of cow's milk (90, 145).

The average Fe concentration is very similar in human, cow's, and goat's milk. A high proportion of the most acceptable values falls between 0.3 and 0.6 μg/ml, with a mean close to 0.5 μg/ml (58, 90). The level of iron in colostrum is three–five times higher than that of true milk. After the initial fall there is no evidence of a significant decline throughout lactation in the milk of women (87), cows (85), or sows (301). The milk of sows and of several other species is appreciably richer in iron than that of the species just cited. In an early study, values ranging from 1.4 to 2.4 μg/ml were reported for the milk of sows in mid-lactation (302). Similar mean levels of 1.4 and 1.2 μg Fe/ml were later obtained for the milk of sows iron-supplemented and unsupplemented, respectively, during the first 3 weeks of lactation (242). Rabbit's milk is still higher in iron, with levels mostly lying between 2 and 4 μg/ml (283). Rat's milk is exceptionally rich in this element and so is the milk of the Australian marsupial *Setonyx brachyurus* (179). In all three species the iron content of milk is highest at the beginning of lactation and is considerably reduced toward the end (179).

A marked effect of stage of lactation on milk iron is apparent in rats. In one study the Fe concentration declined from the remarkably high mean levels of 13.5 to 8.1 μg/ml in the first 4 days and then more slowly to a mean of 3.0 μg/ml at 24 days, with no change thereafter (85, 158). Similar findings were reported by Loh (178).

Administration of supplementary iron to lactating cows, sows, and women is ineffective in raising the Fe content of their milk to levels above normal (162, 242, 301). This is of interest because of the low iron content of the milk of those species relative to the needs of the suckling. In the rat the level of iron in the milk can be raised by iron-loading of the lactating animal (85). Whether the iron content of milk is reduced below normal in the Fe-deficient animal is not clear, although Ezekiel and Morgan (85) observed a decrease in the Fe level of the milk of rats whose iron stores and hemoglobin levels were depleted by repeated bleeding.

Iron occurs in milk in combination with several proteins. An iron protein compound named ferrilactin or "red protein" occurs in milk (114) and in many other body fluids including saliva and sweat (184). The milk from several species contains transferrin as well as lactoferrin (185). The nature of the iron-binding sites and their reactions with iron are very similar in transferrin and lactoferrin (3) but the two proteins differ in immunological properties (73), amino acid and peptide composition (107, 275), and electrophoretic mobility (185).

Rabbit milk whey is remarkable for its extremely high Fe-binding capacity, 5–10 times that of serum, due to the presence of transferrin (18, 154). The physiological role of the high concentration of transferrin in rabbit milk is obscure. It does not appear to aid the transfer of iron from serum to milk in the lactating rabbit or to aid iron absorption by the suckling young (154). Perhaps, as Baker and co-workers (17) have suggested, it possesses some antibacterial function. A variable portion of the total iron in milk is not carried by the specific iron-binding proteins lactoferrin and transferrin. In cow's milk a fat globule membrane carries a significant fraction of the "non-whey" iron (247), and a high proportion of the iron of whole rat milk occurs in association with the casein (180). Loh and Kaldor (179) have shown that in the rat and marsupial (*Setonix brachyurus*), two species with particularly high milk iron content, the amount of iron in the cream and whey fractions is small. The remainder of the iron, which gives the milk its iron-rich character, is associated with casein.

5. Iron in the Avian Egg

An average hen's egg contains close to 1 mg of iron (74, 284), or approximately 20 ppm of the edible portion. A high proportion of this iron is present in the yolk. The presence in egg white of the Fe-containing glycoprotein conalbumin was demonstrated some years ago (4). Conalbumin has since been shown to differ from chicken transferrin only in its carbohydrate content (308) and to have similar reactions with iron (89). It probably acts as a bacteriostatic agent but its physiological functions, if any, remain to be determined.

6. Iron in Hair and Wool

The level of iron in human hair in healthy adults is given as 40–60 ppm in females and 50–70 ppm in males (277). A lower mean level of 29 ± 10 ppm Fe is given for normal women by Eatough *et al.* (77). Levels generally less than 20 ppm were obtained for women in late pregnancy with hematocrits below 38. Baumslag *et al.* (21a) report mean levels of 22.1 and 30.9 μg Fe/g for maternal and neonatal scalp hair, respectively.

The average Fe concentration of cleaned wool from 50 fleeces was found to be 50 ppm (48). Apparently a great deal of this iron is in the grease and the suint, because separate analyses of eight of these fleeces gave the following mean concentrations: wool fiber, 10; grease, 700; and suint, 600 ppm Fe.

II. IRON METABOLISM

1. Factors Affecting Iron Absorption

Because of the limited capacity of the body to excrete iron (192), iron hemeostasis is maintained primarily by adjusting iron absorption to bodily needs. The absorption of iron is affected by (*i*) the age, iron status, and state of health of the animal; (*ii*) conditions within the gastrointestinal tract; (*iii*) the amount and chemical form of the iron ingested; and (*iv*) the amounts and proportions of various other components of the diet, both organic and inorganic.

In monogastric species Fe absorption takes place mainly in the duodenum (42) and in the ferrous form (225). Absorption does not depend on valency since some ferric compounds are more available than some ferrous compounds (99). Inorganic forms of iron and iron–protein compounds need to be reduced to the ferrous state and released from conjugation for effective absorption, whereas the iron in heme compounds is absorbed as the heme moiety into the mucosal cells of the intestine, without the necessity of release from its bound form (120). The absorption of food iron thus involves two independent systems. The reduction and release from conjugation are accomplished by the gastric juice and other digestive secretions. Normal gastric secretion is necessary for optimal absorption of iron by rats (221), and the administration of HCl can increase Fe absorption in achlorhydrics (149). Inorganic iron is able to form complexes with normal gastric juice at a low pH (147). These complexes remain soluble when the pH is raised to neutrality and enable the iron to be available in a suitable state for absorption.

The absorption of inorganic iron is much more sensitive to changes in the intestinal milieu than the absorption of heme iron. Ascorbic acid, which can both reduce and chelate iron, increases iron absorption under a wide range of conditions (40, 43, 74a, 164, 296). Ascorbic acid has no such effect on the absorption of hemoglobin iron (120), or the iron in wheat by the rat (216a). Ascorbic acid is particularly effective in elevating Fe absorption under conditions of Fe deficiency (43, 298) (see Table 4). Other organic acids, including citric, lactic, pyruvic, and succinic, also enhance iron absorption (40, 239). Desferrioxamine, by contrast, is a chelate with a potent depressing effect on nonheme Fe absorption (164).

Numerous other dietary components and compounds influence the absorption of iron. Thus Amine and Hegsted (8) found the retention of iron to be affected by various carbohydrates in the order lactose > sucrose > glucose > starch. However, Garretson and Conrad (104) found no effect of starch or sucrose on Fe absorption. Such simple sugars as fructose (146) and sorbitol (132) can by contrast increase Fe absorption. Several amino acids are similarly effective in enhancing ferrous iron uptake from isolated intestinal segments

TABLE 4
Effect of Dietary (0.5%) Ascorbic Acid on Iron Retention by Iron-Deficient and Iron-Adequate Rats[a]

	Percentage retention (9 days)	
Ascorbic acid	Fe-adequate	Fe-deficient
−	41.4	52.5
+	44.5	81.7

[a]From Van Campen (298).

(163), and histidine, lysine, and cysteine increase the absorption of ferric iron (146, 239, 296, 297, 299). The ability of these amino acids to form tridentate chelates is essential to their effectiveness in enhancing iron uptake (296). In healthy and Fe-deficient children absorption of iron was enhanced by valine, histidine, and ascorbic acid but not by cysteine or glutamic acid (79a).

Hemoglobin iron is well absorbed in all animal species studied (20, 120, 305). Hallberg and Sölvell (120) found that the iron of hemoglobin and ferrous sulfate were equally well absorbed by normal human subjects when equivalent doses were given. The absorption of iron from the ferrous sulfate was increased by ascorbic acid and reduced by phytate, whereas the absorption of hemoglobin iron was unaffected by either. The addition of sodium phytate can reduce iron absorption in man (120, 139). Hussain and Patwardhan (139), in 8-day balance studies on healthy men, found that the addition of sodium phytate to the diet reduced the average retention of iron from 2.5 to 0.17 mg. Added sodium phytate does not necessarily reduce Fe absorption in rats (65), and it is doubtful if the normal phytates of foods are of great practical significance with respect to Fe absorption, either in man (302) or in chicks (71).

Morris and Ellis (216a) have recently shown that the major fraction of the iron in whole wheat consists of monoferric phytate, and that the iron in this compound has a very high biological availability to the rat. No significant difference was observed between the availability of the iron in wheat and its milling fractions–bran, germ, and shorts–and the iron in the reference substance employed, ferrous ammonium sulfate. The extent to which these important findings with the rat can be extrapolated to man is unknown.

Several varieties of clay reduce iron absorption from $^{59}FeSO_4$ and ^{59}Fe-labeled hemoglobin in normal and Fe-deficient subjects (202); the greater the cation exchange capacity of the clays the greater their effect on Fe absorption. This effect could be a contributing factor in the production of anemia in geophagia.

High levels of phosphate reduce Fe absorption (41, 130) and may block it almost completely in man (233). High intakes of Co, Zn, Cd, Cu, and Mn also interfere with Fe absorption through competition for absorption binding sites. These interactions with iron are considered in the appropriate chapters that follow. The mutual antagonism between cobalt and iron at the absorptive level reported by Forth and Rummel (97) is of particular interest. These workers found that a 10-fold Co excess reduced ^{59}Fe absorption from jejunal loops of the rat by nearly two-thirds, and a 100-fold excess depressed Fe absorption almost completely. It appears that cobalt shares with iron at least part of the same intestinal mucosal transport pathway—a pathway in which acceleration of transport of both elements is apparently governed by the same mechanism. For example, Co absorption as well as Fe absorption is significantly enhanced in iron-deficient rats (238), in iron deficiency in man (295), and in patients with portal cirrhosis with iron overload, and in those with idiopathic hemochromatosis (294, 295). The findings of Settlemire and Matrone (260) with rats fed diets very high in zinc are also pertinent. These workers obtained evidence that the zinc reduced Fe absorption by interfering with the incorporation of iron into or release from ferritin. The increase in iron requirement brought about by the high-Zn intakes was enhanced by a shortening of the life of the red blood cells, resulting in a faster turnover of iron.

The efficiency of iron absorption changes in various disease and deficiency states. Increased absorption occurs in iron deficiency (258), hemolytic anemia, aplastic anemia, pernicious anemia, pyridoxine deficiency, and hemochromatosis (56) and is related to increased erythropoiesis (304), depletion of body iron stores (201), and hypoxia (200). Decreased iron absorption occurs in transfusional polycythemia (36) and has been related to tissue iron overload (62). Markedly reduced iron absorption occurs in severely Cu-deficient rats, presumably as a response to the raised iron stores in the animals (258), and also occurs in nickel deficiency in rats, as discussed in Chapter 6.

Iron is poorly absorbed from most diets, with better absorption from foods of animal than plant origin. Only some 5–15% of food iron is absorbed by adult man from ordinary mixed diets (208, 234). Absorption of food iron may be increased to twice this level or more in children and in iron deficiency. In the rat the high absorption of iron during suckling drops abruptly at weaning to the low level characteristic of the adult (96). Josephs (155) reported that 2–20% of an oral dose of radioiron was absorbed in normal human subjects, compared with 20–60% in patients with iron deficiency anemia. Iron absorption values obtained in another study were 27 ± 9.9% for subjects with no hematological abnormality and 62 ± 11.8% for those with iron deficiency anemia (271). More recent studies employing the double radioiron isotope method of Hallberg and Bjorn-Rasmussen (121) have disclosed lower Fe absorption levels than those just described. In a study of young men consuming Swedish diets, 2–4% of the iron was reported to be absorbed (31). Iron absorption from a diet of rice, vegetables, and spices consumed by Thai subjects and containing almost 10 mg Fe approximated only 0.4% in normal individuals. When fruit or meat or both was added, the absorption increased but to values still below 2% in normal subjects. In a few Fe-deficient individuals Fe absorption up to nearly 10% was observed (122). A satisfactory explanation of such low values has not appeared. Since the diets were not abnormally high in phytate and the subjects had no malabsorption of an iron salt, the authors suggest that the diets contained some unknown compound or compounds markedly inhibiting the absorption of nonheme iron from the diets.

2. Mechanism and Regulation of Iron Absorption

The mechanisms by which the body increases the efficiency of iron absorption during periods of iron need and decreases this absorption during times of iron overload, or in other words regulates iron absorption in accordance with body iron needs, are not completely understood. According to the mucosal block theory advanced in 1943 by Hahn and co-workers (118) and elaborated by Granick (108), the intestinal mucosa absorbs iron during periods of need and rejects it when stores are adequate. This was explained as follows: iron taken into mucosal cells is converted into ferritin, and when the cells become physiologically saturated with ferritin, further absorption is impeded until the iron is released from ferritin and transferred to plasma. Crosby and Conrad (61, 62) later provided evidence that in the rat, the ultimate regulator of iron absorption is the iron concentration in the epithelial cells of the upper intestine. In normal rats with moderate intestinal Fe concentrations only a small part of the ingested iron taken up by mucosal cells is transferred to the bloodstream and retained by the animal; the remainder stays in the mucosal cells and is lost into the gut lumen when the cells are sloughed from the tips of the intestinal villi. In Fe-deficient rats with decreased intestinal iron concentration, most of the ingested iron is absorbed directly into the bloodstream with very little remaining in the mucosal cells. In rats given excessive iron stores parenterally the epithelial cells are "loaded from the rear" and are therefore unable to accept the ingested iron. This mechanism apparently does not apply to man (7). The nonheme iron concentrations of duodenal mucosa obtained from normal subjects did not differ from those of Fe-deficient or Fe-loaded subjects. Furthermore, these concentrations showed no significant correlations with simultaneous measurements of serum iron levels or with radioiron absorption. Recent evidence points to the existence of an iron transport system involving binding of iron to iron receptor sites in the plasma membrane of the intestinal epithelial cells and interaction of plasma transferrin with these sites with the release of the iron (83).

Differences in the mechanism of iron absorption also occur within as well as among species depending on the chemical form in which the iron is ingested. As mentioned earlier, the absorption of the iron of hemoglobin, unlike inorganic iron, is not affected by ascorbic acid, phytate, or nonabsorbable chelating agents. This suggests that iron is not released from heme in the gut lumen. Instead the complex is taken up directly by the intestinal epithelial cell, subsequently appearing in the plasma in nonheme form (63, 120). Weintraub *et al.* (305) demonstrated in duodenal mucosal homogenates from the dog the presence of an enzymelike substance which is capable of releasing iron from hemoglobin *in vitro*. The rate at which the heme-splitting substance works *in vito* appeared to be increased by the removal of the nonheme iron end product from the epithelial cell to the plasma. It seems, therefore, that the labile nonheme content of the intestinal cell determines its ability to accept heme from the lumen in dogs, as well as ionized iron from the lumen in rats.

Two other processes, both involving digestive secretions, have been invoked in attempts to throw further light on the mechanisms available to the body for regulating Fe absorption. Davis and associates (72) found an iron-binding protein (gastroferrin) in normal gastric juice and observed that it was absent in hemochromatosis. In iron deficiency anemia caused by blood loss, the concentration of gastroferrin in gastric juice was reduced and returned to normal levels when hemoglobin values were restored (181). The hypothesis was advanced that gastroferrin production is involved in the regulation of Fe absorption—normal levels acting to inhibit the absorption of excessive Fe intakes and reduced levels permitting enhanced absorption in Fe deficiency. It was further proposed that failure to produce gastroferrin through an inborn error of metabolism is a causal factor in hemochromatosis. Other workers have proposed that increased Fe absorption in iron deficiency or hemochromatosis is not due to gastroferrin deficiency but to a factor in the gastric juice that increases absorption (220, 287). The excess absorption and deposition of iron characteristic of hemochromatosis has also been related to a primary pancreatic defect. Davis and Biggs (70) showed that iron absorption can be significantly reduced in hemochromatosis by the addition of a pancreatic extract with the oral dose of iron. A similar inhibitory effect of a pancreatic extract on iron absorption was demonstrated in the intact rat and with the isolated loop of the rat jejunum (30, 70). The actual importance of the gastric and pancreatic secretions to the regulation of iron absorption remains to be determined.

3. Excretion of Iron

The limited ability of the body to excrete iron has been abundantly confirmed since the original observations of McCance and Widdowson (192). Even in hemolytic anemia and in the treatment of polycythemia with phenylhydrazine, when large amounts of iron are liberated in the body from the destruction of red cells, less than 0.5% of this iron appears in the urine and feces (193). Although absorbed iron is retained with great tenacity and, in the absence of bleeding, excretion is very small, the amounts lost cannot be neglected and are of nutritional importance, as Moore (208) has stressed.

The total iron in the feces of normal human adults usually lies between 6 and 16 mg/day, depending on the amounts ingested. Most of this consists of unabsorbed food iron. True excretory iron is estimated at about 0.2 mg/day by chemical balance studies (142) and 0.3–0.5 mg/day by a radioiron technique (76). This iron is derived from desquamated cells and bile. Iron occurs in the bile, mostly from hemoglobin breakdown, to the extent of about 1 mg/day. Most of this is reabsorbed and does not reach the feces. The mean urinary excretion of iron by normal adult men and women was reported many years ago to be 0.2–0.3 mg/day (21). Only about half this amount has recently been observed in young New Zealand women (250). These quantities can be greatly increased, to as much as 10 mg/day, by the injection of various chelating agents, and this technique can be used to increase iron loss in excess iron storage diseases (92).

In addition to the iron excreted in the urine and feces, there is a continual dermal loss in the sweat, hair, and nails. Most of this occurs in desquamated cells but cell-free sweat contains some iron. In one investigation an average iron content of 0.3 μg/ml was found in sweat low in cells and 7.1 μg/ml for sweat high in cells (1). Hussain and Patwardhan (140) collected the sweat from the forearm of healthy Indian men and women and from Fe-deficient anemic women. The "cell-rich" sweat of the healthy individuals averaged 1.15 and 1.61 μg Fe/ml compared with 0.34 and 0.44 μg/ml for the "cell-free" sweat of men and women, respectively. In the anemic women the average Fe content of the cell-rich sweat was 0.44 μg/ml and the cell-free fraction had no detectable iron. The total amount of iron lost daily in the sweat will depend on the individual, the ambient temperature, and the cell content. Losses as high as 6.5 mg/day have been estimated in some circumstances (1, 98). The average loss of iron through the skin of a healthy adult has been assessed as about 0.5 mg/day (203). Iron lost by this route can be much greater in the tropics where the volume of sweat can be as much as 5 liters/day. Such losses have been proposed as a factor contributing to the high incidence of iron deficiency anemia in tropical areas (98).

The total quantity of iron lost in the urine, feces, and sweat amounts to 0.6–1.0 mg/day in most individuals. Bothwell and Finch (33) estimate these basal losses of iron as 14 μg/kg/day, which is close to 0.8–1.0 mg/day for women and men of average size. A loss of this magnitude is appreciable when it is realized that the average amount of iron absorbed from ordinary mixed diets is only 1.0–1.5 mg/day and may be much lower for some Asian diets (122). In women the problem of iron balance is more precarious because they are subject to regular additional losses in the menstrual blood, from the menarche to the menopause, apart from that lost from time to time with the newborn infant and its adnexa and in the milk during lactation. Despite the low-Fe content of milk, lactation is of some significance with respect to iron loss. If the average concentration in human milk is taken as 0.5 μg Fe/ml and 800 ml is secreted daily, the loss of iron would amount to 0.4 mg/day. In some women the loss of iron in the milk would, of course, be much greater.

Menstrual blood losses are extremely variable among women, with a much smaller variation from period to period in the same women (60, 123, 255). Average losses of 35–70 ml containing 16–32 mg of iron have been reported (153). In a study of 12 young women over 12 months the mean individual losses ranged from 4 to 26 mg Fe per period (123). Higher losses have been reported in other investigations (106), with 60–80 ml of blood (27–37 mg Fe) regarded as the upper limit of normal loss (255). It is evident that most women between puberty and the menopause lose 0.5–0.8 mg Fe/day as a consequence of the menstrual flow and some lose considerably more.

4. Intermediary Metabolism of Iron

A high proportion of absorbed iron is continuously redistributed throughout the body in several metabolic circuits of which the cycle, plasma→erythroid marrow→red cell→senescent red cell→plasma, is quantitatively the most important. Subsidiary metabolic circuits exist, including the cycles, plasma→ferritin and hemosiderin→plasma and plasma→myoglobin and iron-containing enzymes→ plasma. The iron of the plasma (transferrin) provides the link between the cycles and regulates the distribution of iron in the body.

The hemoglobin cycle dominates the intermediary metabolism of iron. Normally more than 70% of plasma iron turnover goes to the bone marrow (34). In the pregnant animal near term a relatively large proportion of plasma iron turnover may also go to the placenta (35). Some 21–24 mg of endogenous iron is liberated daily from the destruction of hemoglobin, if the survival period of hemoglobin is taken as the same as that of the erythrocyte in man, namely, 120–125 days. This far transcends the amount of iron absorbed daily from ordinary diets and indicates the magnitude of the hemoglobin cycle. The removal of hemoglobin or nonviable erythrocytes, the breakdown of the heme moiety and release of iron, and the return of this iron to the plasma are the responsibility of the reticuloendothelial (RE) cells of the liver, spleen, and bone marrow. The recircuiting of iron from senescent red cells is an important link in internal iron exchange. After an initial processing period within the RE cell the iron is either rapidly returned to the circulation ($t_{1/2}$, 34 min) or transferred to a slowly exchanging pool of storage iron within the RE cell ($t_{1/2}$ release to plasma, 7 days) (93, 177).

In adult man 25–40 mg of total iron is transported in the plasma every 24 hr, even though only 3–4 mg is present in the whole plasma volume at any one time. This iron has a rapid turnover rate with a normal half-time of only 90–100 min (240). The large and rapid flow of plasma iron to the bone marrow is reflected in the completeness and promptness with which tracer doses of iron are used for hemoglobin synthesis. Tagged hemoglobin can be identified in the peripheral blood within 4–8 hr of administration of tracer iron. Within 7–14 days 70–100% of the isotope is found in circulating hemoglobin (207).

During the process of incorporation of Fe into hemoglobin in the reticulocyte, the first step is the binding of the transferrin in the cells (152, 215). Reticulocytes, but not mature erythrocytes, bind iron transferrin to the surface and return apotransferrin to the medium (152, 214). Most of this iron is incorporated into heme in the mitochondria (256). Transferrin molecules move into reticulocytes by a two-stage process involving first binding at membrane receptor sites and then movement into the reticulocyte by a slower temperature-dependent process, the probable mechanism of uptake being pinocytosis (13). Fielding and Speyer (91, 274) have obtained several Fe-binding components from human reticulocytes by chromatographic fractionation and investigated their role in intracellular iron transport. Their results suggest that iron moves initially from iron transferrin receptor complex B2 (mol. wt. 230,000) to membrane component B1 (mol. wt. of the order of 10^6), from which it may diverge into membrane component A of high particle size, or follow the main pathway through cytosol component C to hemoglobin.

The rapid process of incorporation of plasma iron into ferritin in the cells of the liver, spleen, and bone marrow is dependent on energy-yielding reactions for the continued synthesis of ATP which, together with ascorbic acid, reduces the ferric iron of transferrin to the ferrous state, thus releasing it from its bond to protein and rendering it available for incorporation into ferritin (188–190). It has been claimed that the reverse process, the release of iron from hepatic ferritin to the plasma, is mediated by the Mo- and Fe-containing enzyme xanthine oxidase acting as a dehydrogenase (189, 190, 227). However, Osaki and Sirivech (228) found that in liver homogenates xanthine oxidase substrates did not produce a significant release of iron from ferritin, nor did milk xanthine oxidase or chicken liver xanthine dehydrogenase. Recent investigations with horse spleen ferritin have shown that only the reduced riboflavin and riboflavin derivatives can reduce ferritin Fe(III) at a rate and to an extent that is likely to be significant physiologically (268). In the reduced state thus produced the iron of ferritin dissociates from its bond to protein and is accepted by transferrin.

Mobilization of iron from iron stores also requires the presence of the Cu-containing enzyme of the plasma ceruloplasmin (ferroxidase I), as discussed in Chapter 3. The demonstration of the iron oxidase activity of ceruloplasmin (227) led to the suggestion that this enzyme may be important in normal iron metabolism. This hypothesis is supported by *in vivo* experimental evidence (253) and by studies carried out by Osaki and co-workers (226) with perfused porcine and canine livers. These workers concluded that "ferroxidase activity results in the substantial elimination of Fe (II), generating a maximum concentration gradient from the iron stores to the capillary system, thus promoting a rapid iron efflux from the iron storage cells of the perfused liver." The redistribution of body iron involving increased movement of iron to the liver from the serum that occurs in acute and chronic infections was discussed earlier in Section I, 2,b.

The iron of plasma is deposited in the liver and spleen as freely as ferritin as hemosiderin and is released just as readily from both compounds for utilization by the tissues. This occurs over a wide range of rates of iron deposition and depletion. Shoden and Sturgeon (266) showed that the iron of transferrin is not incorporated directly into hemosiderin as it is into ferritin, at least in rabbits. In this species ferritin has first to be formed, so that liver ferritin may be regarded as the immediate percursor of liver parenchymal cell hemosiderin.

Placental iron transport is unidirectional (35) and increases rapidly as pregnancy progresses, with the plasma iron turnover of the fetus in late pregnancy being much greater than that of the newborn (197, 198). In animals with the hemochorial type of placenta, including the rat, rabbit, guinea pig, and human, the rate of iron transfer across the placenta from maternal plasma transferrin is sufficient to account for all iron accumulated by the fetus (34). Wong and Morgan (310) have shown that this is an active process dependent on cellular metabolism in which maternal plasma transferrin is taken up by the placenta, followed by removal of iron from the transferrin and transfer to the fetal blood. In animals with endotheliochorial types of placentas, such as the cat and the dog, the rate of iron transfer from maternal plasma to fetus is much less than that required for fetal needs (16, 259). In the cat, for example, transfer from maternal plasma is insignificant, and the major source of fetal iron is the maternal erythrocytes (305). This process probably occurs, as explained by Wong and Morgan (311), "by extravasation of maternal erythrocytes into the uterine lumen, followed by their phagocytosis and digestion by paraplacental chorionic epithelial cells, and transfer of the iron released from hemoglobin to fetal capillaries in the chorionic membrane and thence to the fetuses." Since each milliliter of blood contains 300–500 times as much hemoglobin iron as transferrin iron this is clearly an effective and economical transfer process.

III. IRON DEFICIENCY

1. Manifestations of Iron Deficiency

Iron deficiency in human adults is manifested clinically by listlessness and fatigue, palpitation on exertion, and sometimes by a sore tongue, angular stomatitis, dysphagia, and koilonychia. In children anorexia, depressed growth, and decreased resistance to infection are commonly observed as in other young, growing, iron-deficient animals, but the oral lesions and nail changes are rare. A significant reduction in physical activity, performance (78), and resistance to infection with *Salmonella typhimurium* (14) in Fe-deficient anemic rats and prolonged cardiorespiratory recovery periods after exercise in anemic women (11) have been observed.

Iron deficiency results in the development of an anemia of the hypochromic microcytic type accompanied by a normoblastic, hyperplastic bone marrow containing little or no hemosiderin. Serum iron and serum ferritin levels are subnormal, total iron-binding capacity is above normal, and there is a decreased saturation of transferrin. Studies of patients with iron deficiency anemia have shown that 16% saturation of plasma transferrin or less implies an inadequate supply of iron to the erythroid tissue and is associated in time with hypochromic, microcytic anemia (15). The percent saturation of transferrin is thus a good criterion of iron-deficient erythropoiesis. The value of serum ferritin levels as an indicator of body iron stores has been discussed earlier (Section I,2,c).

Lipid abnormalities with elevated serum triglyceride levels associated with iron deficiency of nutritional origin have been observed in rats (116, 173) and chicks (9). Whether the accumulated lipid reflects an increased rate of synthesis or a decrease in the rate of clearing is not entirely understood. The latter seems more probable in light of the observation by Lewis and Iammarino (173) that serum and tissue lipoprotein lipase activity is decreased in Fe-deficient rats. Blood lipid alterations may occur in several types of human anemias not limited to iron deficiency, but no consistent trend is apparent as with rats.

Iron deficiency anemia is further characterized by low serum folic acid levels, which can be restored by iron therapy (249, 286). The decrease in serum folic acid is not due to reduced absorption and is apparently secondary to increased requirements for this vitamin, consequent on a significant decrease in red cell survival and a nearly threefold increase in heme catabolism in the Fe-deficient animals (286). Decreased red cell survival (138, 166) and ineffective erythropoiesis (44, 251) have both been reported in iron deficiency.

Abnormalities of the gastrointestinal tract, including gastric achlorhydria and varying degrees of gastritis and atrophy, can occur in iron deficiency anemia (128, 169). A high incidence of blood loss and loss of plasma proteins into the intestinal lumen (309) and a diffuse and reversible enteropathy (222) have been reported in infants and children with iron deficiency. Gastric achlorhydria and impaired absorption of xylose and vitamin A, together with duodenitis and mucosal atrophy, were observed. Most of the abnormalities returned to normal following treatment with iron and were generally absent from children with anemias not due to iron deficiency.

Muscle myoglobin, long believed to remain inviolate in iron deficiency, can be reduced. A profound depression in myoglobin concentration has been observed in young iron-deficient piglets and puppies (115), chicks (71), and rats (67, 69). Young animals are more susceptible to loss of myoglobin than mature ones (115), and skeletal muscles appear to be more susceptible to this loss than cardiac or diaphragmatic muscle (67).

A reduction in the levels and activities of the heme enzymes in the blood and tissues of iron deficient animals is well documented. Beutler (27, 28) was the first to show clearly that the body does not necessarily accord priority to the heme enzymes over hemoglobin under conditions of restricted iron supply. In Fe-deficient rats a marked decrease in cytochrome c was found in the liver and kidneys, in succinic dehydrogenase activity in the cardiac muscle and kidneys but not in the liver, and in cytochrome oxidase activity in the kidneys but not in the heart. Dallman (69) showed that cytochrome c can be reduced to as low as half the normal concentrations in the skeletal muscles, heart, liver, kidney, and intestinal mucosa of young Fe-deficient rats, with a smaller reduction in the brain. More recently Grassman and Kirchgessner (111,112) reported a marked reduction in the catalase activity of the blood of Fe-deficient rats, piglets, and calves. The reduction in enzyme activity was greater than the fall in hemoglobin, indicating that, during the development of the anemia, priority for iron use was given to hemoglobin synthesis over catalase synthesis.

2. Iron Deficiency in Human Infants

Iron deficiency anemia can arise in the rapidly growing suckling animal as a consequence of inadequate amounts of storage iron at birth and low-iron concentrations in maternal milk. Even in the rat, whose milk is richer in iron than that of women and most other species, such an anemia develops (84). The anemia of human infants, which occurs most frequently between 4 and 24 months of age, is an expression of Fe deficiency since it usually responds to iron and is characterized by the typical abnormalities of erythrocyte morphology (278). A depletion of iron reserves occurs during the period of rapid growth of the infant, despite the considerable "store" of iron contained in the plethora of hemoglobin in the blood at birth. The blood of the newborn child contains 18–19 g Hb/100 ml which, if the blood volume is taken as 300 ml, represents 180–190 mg Fe or about six times the amount usually stored in the liver (194). At 6 months of age the average baby weighs 7 kg and has a blood hemoglobin level of about 12 g/100 ml. It therefore contains about 280 mg Fe in the form of hemoglobin, of which 180–190 mg was present at birth. If this iron plus the amount stored in the liver and absorbed from the milk were all retained in the body, it is difficult to visualize iron deficiency arising at all, except in premature, low birth weight infants with low body stores. However, Cavell and Widdowson (58), in a study of babies at one week of age, found significant negative Fe balances due to the excretion of approximately 10 times as much iron in the feces as was ingested daily in the milk. If such excess iron excretion were to continue beyond the early neonatal period, it could be an important cause of iron depletion.

In several studies of infants in the United States an incidence of iron deficiency anemia has been reported from less than 5% (100) and 8% (22) to as high as 64% (278). Some of this variation undoubtedly stems from differences in the socioeconomic groups under investigation and the criteria of anemia employed. Several workers have demonstrated significant increases in hemoglobin levels in infants receiving oral iron or intramuscular injections of iron dextran (269). The regular need for supplementation with iron or with iron-fortified foods is stressed by Sturgeon (278), even with infants born of nonanemic mothers who have received ample iron during pregnancy. However, in the study of Fuerth (100) none of the objective criteria revealed any difference between the iron-supplemented group and the children receiving the placebo. In a further study by Burman (47) an iron supplement of 10 mg Fe/day in the form of colloidal ferric hydroxide raised the hemoglobin levels of infants of lower birth weight, presumably due to their lower body iron stores, but made no difference to the incidence of infection. This worker contends that efforts to raise hemoglobin levels should not be made unless it can be shown that maximal hemoglobin is beneficial to infants. He maintains further that "levels above 9.5 or 10 g Hb/100 ml should be considered nonpathological not only at 6–8 weeks of age but also for the remainder of the first 2 years." There is, nevertheless, a definite need for iron supplementation, especially during early infancy, in premature babies, and in babies born of anemic mothers (194).

3. Iron Deficiency in Human Adults

Iron deficiency is probably the most prevalent deficiency state affecting human populations. In adult men and postmenopausal women the principal cause is chronic blood loss due to infections, malignancy, bleeding ulcers, and hookworm infestation. However, anemia of dietary origin is not uncommon among the elderly, especially the institutionalized elderly (174, 199). Because of their iron losses in menstruation, pregnancy, parturition, and lactation, as discussed earlier, iron deficiency is much more common in women during their fertile years than in men (19, 50, 160).

Surveys in many countries have revealed a widely varying incidence of low body iron stores (19) and anemia in women, the latter attributed to iron deficiency (50, 160, 199, 255). In developing countries where the population relies heavily on vegetable foods and where infections and excessive sweating are common, the incidence of iron deficiency anemia is generally higher (219, 300) than the 20–25% reported in Swedish studies (255), or the 10–20% in English studies (50, 160). In the former circumstances an extreme paucity of iron stores has been revealed in some population groups (19). In U.S. surveys of pregnant women iron deficiency anemia has ranged in incidence from 15 (6) to 58% (243). Blood hemoglobin levels of 11.5 or 12.0 g/100 ml or less are usually regarded as evidence of anemia, although seriously depleted iron stores can exist without manifest anemia (95, 273). In view of the value of serum ferritin as an indicator of body iron stores (Section I,2,c), further surveys of serum ferritin levels are necessary.

While there is no doubt that iron deficiency anemia and low body iron stores are common in women in many populations and that an increase in hemoglobin is usually obtained in response to appropriate iron therapy, it is less certain that such increases, in cases of mild anemia, confer significant benefits in terms of improved well-being or activity, or lower morbidity or mortality (50, 81, 281).

Normal adult females escape menstruation only by pregnancy. From the standpoint of evading iron loss this can be an unprofitable exercise because some 350–450 mg Fe is lost in the fetus and its adnexa, compared with 200–300 mg yearly in the menstrual flow. Some compensation for the iron demands of pregnancy is achieved by increased absorptive efficiency, but this is insufficient to prevent signs of iron deficiency in many women in late pregnancy. In the third trimester of pregnancy hemoglobin levels normally fall due mainly to a rise in plasma volume greater than the rise in red cell volume (see Section I,2,a). Inadequate intakes of iron or intakes of low availability can be a contributory cause, and supplementary iron often increases blood hemoglobin levels (171, 211, 279). The intramuscular injection of a single dose of 1000 mg of iron dextran or the daily oral administration of 78 mg of ferrous iron from 24 weeks prepartum is equally effective for this purpose (171). The therapeutic value of iron preparations containing ascorbic acid, which potentiates the absorption of ferrous sulfate, has been stressed (195), while ferric fructose is particularly effective in the treatment of anemia in pregnant women (77).

4. Iron Deficiency in Pigs

Iron deficiency anemia occurs in baby pigs and in older animals fed rations very high in copper to promote growth. Piglets denied access to sources of iron other than sow's milk develop anemia within 2–4 weeks of birth. Blood hemoglobin levels fall from a normal of about 10 g/100 ml to as low as 4 g/100 ml. Breathing becomes labored and spasmodic, hence the name "thumps," appetite declines and growth is poor, or the piglets lose weight and some die. Surviving piglets begin a slow spontaneous recovery at 6–7 weeks, when they begin to eat the sow's food and undertake such foraging as is permitted by their conditions of housing. The condition is an uncomplicated iron deficiency responsive to iron (301) and with no significant additional response from copper (131, 161) or vitamin B_{12} (161). Piglet anemia is often associated with *Escherichia coli* infection, inducing piglet edema disease. Decreased resistance to the endotoxin of *E. coli* has been demonstrated in anemic piglets (137).

The baby pig is particularly liable to iron deficiency because of its high growth rate and poor endowment with iron at birth. Piglets normally reach four–five times their birth weight at the end of 3 weeks and eight times their birth weight by 8 weeks. A growth rate of this magnitude imposes iron demands much greater than can be supplied by the sow's milk. At birth the pig has relatively low concentrations of total body iron and liver iron stores (Table 1) and has no polycythemia of the newborn. These sources of endogenous iron are therefore denied to the baby pig, and anemia is inevitable unless the piglet has access to additional iron.

Feeding supplementary iron to the sow before or after farrowing is ineffective (161), because such treatment does not significantly increase the iron stores of the piglet at birth (131, 301), or the iron content of the sow's milk (242). Successful treatment involves direct increase in the iron intake of the piglets. Oral or parenteral administration of iron to the piglets within the first few days of life is now routine practice, where pigs are maintained free from soil and grass. Injection of such compounds as iron dextran, iron fumarate, or dextrin ferric oxide (Table 5) is effective in promoting maximum hemoglobin levels (175, 230, 241).

When an iron complex containing 100 mg Fe is injected at 2–4 days, a second injection within 2 weeks is necessary to increase hemoglobin levels at 3–4 weeks of age (175, 230). Similarly, oral administration of iron tablets or solutions containing 300–400 mg Fe within 4 days of birth promotes growth and prevents mortality, but a second such dose is required for maximum hemoglobin levels. A single oral dose of 150 mg Fe given as Fe tartrate on the third day (12), or 315

TABLE 5
Treatment of Piglet Anemia by Injection of Iron Complexes[a]

Treatment	No. of pigs	Hemoglobin (g %)		Body weight (lb)	
		0	3 wk	0	3 wk
Untreated	24	7.5	4.8	3.8	9.6
Iron dextran at 2–4 days (100 mg Fe)	21	7.5	9.0	3.6	11.3
Dextrin ferric oxide at 2–4 days (100 mg Fe)	22	8.0	7.8	3.6	11.9
Dextrin ferric oxide at 2 and 14 days (100 mg Fe each dose)	16	8.3	10.5	3.8	11.5

[a]From Linkenheimer *et al.* (175).

mg Fe as iron dextran tablets on the fourth day (161), maintained the hemoglobin level of the piglets at greater than 10 g/100 ml for a period of 15–17 days following administration.

Anemia due to iron deficiency can occur in older pigs fed rations very high in copper to promote growth and increase the efficiency of feed use, unless these rations are supplemented with iron at levels well above those that are otherwise adequate (46, 105). Gipp and co-workers (105) showed that the hypochromic, microcytic anemia induced by high dietary copper is due to an impairment of iron absorption and that this impairment is ameliorated by ascorbic acid.

5. Iron Deficiency in Lambs and Calves

There is no convincing evidence that iron deficiency ever occurs in sheep or cattle grazing under natural conditions, except in circumstances involving blood loss or disturbance in iron metabolism as a consequence of parasitic infestation or disease. Heavy infestation with helminth intestinal parasites results in an Fe deficiency type of anemia in lambs and calves (53, 246). Iron deficiency anemia also occurs in young calves reared for veal on milk-based rations and in lambs similarly raised. Intramuscular injections of 150 mg Fe as iron dextran into newborn lambs, or 12 mg Fe/lb body weight into newborn calves, can produce significant improvements in hemoglobin levels and some improvement in body weight over several weeks (54, 245). Oral iron supplements can be equally effective. Thus Bremner and Dalgarno (39) found that adding iron at the rate of 30 μg Fe/kg to a fat-supplemented skim milk ration improved the hematological status of calves provided that the iron was supplied as ferrous sulfate, ferric citrate, or ferric ethylenediaminetetraacetate. The iron of phytate was less available.

IV. IRON REQUIREMENTS

1. Man

The demands of the body for iron are greatest during three periods—the first two years of life, the period of rapid growth and hemoglobin increase of adolescence, and throughout the child-bearing period in women. The physiological requirement of the infant for iron averages approximately 0.6 mg/day, but may be as high as 1 mg/day during the first year of life (156). At age 15–16 years in males the requirement increases temporarily due to the rapid growth and hemoglobin accretion that occurs during this time. To provide the iron needs of 95% of normal menstruating women enough iron must be consumed to

TABLE 6
Recommended Daily Dietary Allowance for Iron[a]

	Age (yr)	Weight (kg)	Amount (mg)
Infants	0.0–0.5	6	10
	0.5–1.0	9	15
Children	1–3	13	15
	4–6	20	10
	7–10	30	10
Males	11–14	44	18
	15–18	61	18
	19–51+	67–70	10
Females	11–50	44–58	18
	51+	58	10
Pregnant			18+[b]
Lactating			18

[a]From National Academy of Sciences (223).
[b]This increased requirement cannot be met by ordinary diets. The use of supplemental iron is therefore recommended.

permit absorption of 2 mg/day (206). Recommended daily dietary allowances for iron have recently been given by the U.S. National Academy of Sciences (see Table 6) (223).

Conversion of the physiological requirements for iron into dietary requirements is made difficult by variations among individuals in absorptive capacity and among foods and various combinations of foods in the availability of their iron for absorption. The position is further complicated by the ability of the body to increase Fe absorption in Fe deficiency. Normal subjects commonly absorb 5–10% of the iron of mixed diets, and iron-deficient individuals 15–20% or more of this iron, but considerable divergence from these values can occur. Numerous workers have shown that iron absorption is significantly lower, and dietary iron requirement therefore higher, in vegetal diets than in foods from animal sources, mixed diets, and iron salts (122, 141, 167). This is particularly apparent from a study of Venezuelan diets (167) in which absorption of 3–4 mg of vegetal iron was found to be increased about twice by the addition of 50 g of meat, about three times by 100 g fish, and about five times by 150 g papaya containing 66 mg ascorbic acid. The ingredients of meals and diets clearly affect dietary iron requirements profoundly through their effects on absorption. This aspect is considered further in Section V.

2. Pigs

The minimum iron requirements of pigs for satisfactory growth and hemoglobin formation cannot be given with any precision. Braude and co-workers (38) estimated that the piglet must retain 21 mg Fe/kg body weight increase to maintain a satisfactory level of iron in the body. This is equivalent to less than the 7–11 mg Fe/day for this purpose obtained in an earlier study (301). The dietary concentrations of iron required to supply these quantities vary with the percentage absorption of the iron in the diet and with the levels of other elements in the diet, such as copper, zinc, and manganese, which affect this absorption. Ullrey and co-workers (288) found 125 ppm Fe to be necessary for full growth and hemoglobin production in baby pigs, while the results of Matrone *et al.* (187) indicate that 60 ppm is adequate. More recently Hitchcock *et al.* (134) have shown that 50 ppm of supplemental iron will support maximum gain, while 100 ppm results in significantly higher hemoglobin and serum iron levels in both germfree and conventionally raised baby pigs fed condensed cow's milk containing 2 ppm Fe. The marked increase in the iron requirements of older pigs brought about by feeding high-Cu rations is discussed in Chapter 3.

3. Poultry

The iron requirements of chicks during the first 4 weeks of life have been estimated as 75–80 ppm of the diet (71). These estimates were based on growth, blood data, myoglobin levels, liver iron stores, and succinic dehydrogenase levels, and are appreciably higher than the 50 ppm Fe suggested earlier as adequate for chicks by Hill and Matrone (133).

The demands of egg production for iron are very large compared with those of the nonlaying mature bird. Since an average hen's egg contains about 1 mg Fe, a heavy layer will lose close to 6 mg/week in the eggs alone, so that more than this needs to be absorbed from the diet. With the onset of laying the efficiency of Fe absorption is increased and serum Fe levels are raised. This is usually accompanied by a small fall in blood hemoglobin levels but no significant reduction in storage iron (244). The increased requirements for iron imposed by egg-laying are apparently met by these means by all normal laying rations without the need for iron supplements.

4. Cattle and Sheep

Little is known of the iron requirements of adult sheep and cattle based on definitive experiments. In male calves raised on a milk diet containing 1 ppm Fe dry basis (d.b.), normal growth and hemoglobin levels were maintained for 40 weeks from birth with supplemental iron at the rate of either 30 mg or 60 mg Fe/day (186). This suggests that the minimum iron requirement for calves is not more than 30 mg/day. A later study with male calves reared on fat-supplemented skim milk indicates that a dietary intake of 40 μg Fe/g of dry diet is sufficient to prevent all but a very mild anemia, provided the supplemental iron is presented in a soluble form (39). Experiments with growing–finishing lambs indicate that 10 ppm Fe is inadequate and that their minimum dietary Fe requirements lie between 25 and 40 ppm (165).

V. SOURCES OF IRON

1. Iron in Human Foods and Dietaries

The iron content of most foods varies greatly from sample to sample, as a reflection of varietal differences and differences in the soil and climatic conditions under which the foods are grown or produced. The richest sources of total iron are the organ meats (liver and kidney), egg yolk, dried legumes, cocoa, cane molasses, and parsley. Poor sources include milk and milk products, unless contaminated in processing, white sugar, white flour and bread (unenriched), polished rice, sago, potatoes, and most fresh fruit. Foods of intermediate iron content are the muscle meats, fish and poultry, nuts, green vegetables, and wholemeal flour and bread. Boiling in water can reduce the levels of iron in vegetables by as much as 20% (270), while milling of wheat greatly lowers the iron content in the resulting white flour. For example, in a study of North American wheats and the flours milled from those wheats, the mean iron content of the whole grain was 43 ppm and that of the flour 10.5 ppm (68). The effect of refining processes on the iron content of foods is further evident from a study of the concentration of various elements in different types of cane sugars (125). The Fe concentrations reported were Barbados brown sugar, 49; Demerara sugar, 8; refined sugar, 11; and granulated (white) sugar, 0.1 ppm dry basis.

Overall dietary iron intakes vary greatly with the total amounts of foods and beverages consumed and with the proportion of iron-rich and iron-poor foods that they contain. The degree of contamination with iron to which the foods and beverages are exposed in processing, storing, and cooking may also be important. Average U.S. dietaries were reported 40 years ago to supply 14–20 mg Fe per man value daily (261) and Australian diets 20–22 mg Fe per adult male daily (224). Daily intakes by Japanese adults as a whole have been estimated to be close to 19 mg (314). In a much more recent study of total daily intakes from English diets a mean level of 23.2 ± 1.1 mg Fe/day was obtained (125). Lower total as well as available iron intakes are common from the diets, high in milled rice or corn and low in meat, of underprivileged groups in many of the developing countries (122, 129, 167).

The importance of total calorie intake in determining iron intake is evident from two North American studies. Monsen and co-workers (206) found that the iron intakes of 13 young women averaged only 9.2 mg/day when consuming their normal diets. These diets were adequate in protein, calcium, vitamins A and C, and the B vitamins, but provided only 1600 calories/day. Diets designed for 16- to 19-year-old boys, providing no less than 4200 calories daily by contrast, supplied an average of 35.6 mg Fe/day (315). Typical Western diets provide about 6 mg total iron per 1000 calories (199).

Great caution needs to be exercised in relating total dietary intakes of iron to deficiency or sufficiency of this element because of the marked differences in iron availability from different foods. The amounts of iron ingested, therefore, do not necessarily reflect the net amounts absorbed by individuals. The lower absorption of iron from most vegetable foods than from most foods of animal origin was mentioned earlier when considering iron absorption. The two types of food interact in the ingesta so that the inclusion of foods of animal origin raises the absorption of iron from foods of plant origin and vice versa. The better absorption of the iron of cereals and comparable foods when combined with foods of animal origin, other than egg (82), than when given alone is of great practical importance (122, 167, 168, 183). The iron of egg yolk is very poorly utilized except when combined with sufficient ascorbic acid (217). Liver is a particularly valuable dietary component because of its high-iron content, high absorbability, and enhancing effect on the absorption of vegetal iron (183). The inclusion of fruit juices rich in ascorbic acid to improve iron absorption from such diets, or from bread, can also be important (51, 82, 122). In fact the regular provision of some meat, fish, or fruit in the diet is of paramount importance to ensure a reasonable utilization of nonheme iron.

With normal men and postmenopausal women there is little reason for concern about iron intakes. Infants and women in their fertile period are in a more precarious position because of their larger iron needs and smaller calorie consumption. For some women an otherwise adequate diet may be, and often is, inadequate in iron. Furthermore there is some evidence that iron intakes are falling as modern food handling processes reduce the opportunities for contamination. Some form of iron supplementation or food fortification is therefore necessary. In the infant this can readily be accomplished through fortification of special food items, the use of which is confined to infancy. An intake of 10 mg Fe/day from proprietary milk and infant cereals fortified with iron is very effective in preventing anemia in infants (204).

For adults the foods chosen for fortification need to be widely used and in sufficient quantity to make a worthwhile contribution to iron intakes. The most

TABLE 7
Estimated Absorption of Iron in United Kingdom if Supplemented to U.S. Standard[a]

			mg Fe absorbed	
	oz/person/week	mg Fe/day	With meat	Without meat
Bread (all kinds)	34.68	7.7	0.55	0.16
Buns, cakes, biscuits	10.72	2.2	0.16	0.05
Flour	5.39	1.9	0.13	0.04
Total		11.8	0.84	0.25

[a]From Callender (50).

popular vehicle chosen for iron fortification has therefore been white flour, although the potential for fortification of cane sugar with iron and ascorbic acid has recently been encouragingly explored (74a).* The original intention was to restore the level of iron in white bread to that of wholemeal bread. The iron enrichment of flour with 15 mg Fe/lb (about 10 mg/lb in bread) was started in the United States in 1945. This has had little effect in reducing the incidence of anemia in women in that country, or in England where a similar enrichment program was initiated. This is due at least in part to the low availability of some of the iron compounds, such as reduced iron, employed (see Elwood, 80). Since April 1974 an iron fortification level of 40 mg Fe/lb in flour and 25 mg Fe/lb in bread must be met for the products to bear the label "enriched." This move has not been without its opponents on the grounds of efficacy and safety. In respect to efficacy, Callender (50) has calculated that at the American level of supplementation and the English level of consumption of bread, flour, biscuits, and cake, only 0.84 mg of absorbed iron/day would be contributed from these sources if meat was taken at each meal, and only 0.25 mg/day if little or no meat was consumed or eggs were included in the diet (Table 7). In her words such a fortification program "cannot be expected to do more than prevent some women who are marginally deficient from developing iron deficiency." With respect to the question of safety many physicians are of the opinion that fortification at the U.S. level could be hazardous for individuals with idiopathic hemochromatosis, Laennec's cirrhosis, and hereditary iron-loading anemias, or with unusually high bread intakes (50, 66). Swiss and Beaton (280) reached conclusions essentially similar to the above workers in a prediction study of the effects of iron fortification based on Canadian data. They concluded that an appreciable reduction in the risk of iron deficiency in the target population

*See also Layrisse *et al.*, *Am. J. Clin. Nutr.* **29**, 8 (1976).

(menstruating women) could be achieved by iron fortification of bread, without an exorbitant increase in iron intakes by other groups; but reduction of risk of deficiency to zero in the target population by iron fortification alone would require much higher levels of fortification "than would seem either prudent or technologically feasible."

2. Iron in Animal Feeds

The iron content of animal feeds is highly variable. The level of iron in herbage plants is basically determined by the species and type of soil on which the plants grow and can be greatly affected by contamination with soil and dust. In a recent study of New Zealand pastures iron levels ranged from 111 to 3850 ppm d.b. (mean 581 ± 163), with the highest levels probably due to soil contamination (52). Leguminous pasture plants usually contain 200–400 ppm Fe (d.b.). Values as high as 700–800 ppm have been recorded for uncontaminated lucerne (alfalfa) and as low as 40 ppm or less for some grasses grown on sandy soils (see Underwood, 290). Most cereal grains contain 30–60 ppm and species differences appear to be small. The leguminous and oil seeds are richer in iron than the cereal grains and may contain 100–200 ppm Fe.

Feeds of animal origin other than milk and milk products, such as skim milk, whey, and buttermilk powders, are rich sources of iron. Meat meals and fish meals commonly contain 400–600 ppm and blood meals more than 3000 ppm Fe. Ground limestone, oyster shell, and many forms of calcium phosphate used as mineral supplements frequently contain 2000–5000 ppm Fe.

All the above figures refer to total iron. Little is known of the availability of this iron to herbivorous animals, particularly ruminants with a digestive and absorptive apparatus very different from that of man or other monogastric species. However, orally administered ferrous sulfate, ferrous carbonate, and ferric chloride are equally available to calves and sheep on the basis of tissue ^{59}Fe deposition and the iron in ferric oxide significantly less available on this basis (10). It seems, therefore, that the solubility of the chemical form of iron affects its availability to ruminant species, as it does in other species.

VI. IRON TOXICITY

Data which permit the delineation of toxic levels of dietary iron in a manner analogous to iron requirement levels are unavailable. Intakes of 50 mg Fe/day (205) or 25–75 mg Fe/day (94) by man have been cited as safe, but many individuals have taken iron medication at these or higher levels for extended periods without reported harm. More information on maximum safe human tolerances to iron in different forms is obviously needed.

Long-term iron overload has been observed in malnourished Bantus in South Africa (Bantu siderosis) (182). The native diet may supply 200 mg Fe daily due to contamination from iron cooking vessels and a high consumption of Kaffir beer. The beer is prepared in iron pots and has been reported to contain from 15 to 120 mg Fe/liter (37, 182). This iron is in a soluble form and may supply as much as 2–3 mg of absorbed iron daily (59). The highest dietary intakes of iron appear to be in Ethiopia, where the staple cereal teff has a high-iron content and becomes heavily contaminated with iron in grinding, storing, and cooking. An average iron intake of some 470 mg/day has been estimated, of which about three-quarters comes from contamination (252). This iron must be largely unavailable because tissue siderosis is not common, and examination of the storage iron of individuals dying from accidents has shown no difference from that found in a comparable Swedish population (135).

The importance of the availability of the iron in a particular diet in determining maximum safe levels, or the minimum levels at which toxic effects can be expected, is further apparent from numerous studies with animals involving varying intakes of other elements with which iron interacts at the absorptive level. High intakes of cobalt, copper, zinc, manganese, and deficient intakes of nickel, depress iron absorption, as discussed elsewhere. Levels of iron intake which would produce signs of toxicity in the animal would clearly need to be much higher under conditions of abnormally high intakes of one or more of these interacting elements than when such intakes are low or normal. This is merely an expression for iron of the general principle for trace elements enunciated in Chapter 1, namely, that "a series of safe dietary levels of potentially toxic trace elements exist, depending on the extent to which other elements which affect their absorption and retention are present."

REFERENCES

1. Adams, W.S., Leslie, A., and Levin, M.H., *Proc. Soc. Exp. Biol. Med.* **74,** 46 (1950).
2. Addison, G.M., Beamish, M.R., Hales, C.N., Hodkins, M., Jacobs, A., and Llewelin, P., *J. Clin. Pathol.* **25,** 326 (1972).
3. Aisen, P., and Leibman, A., *Biochim. Biophys. Acta* **257,** 314 (1972).
4. Alderton, G., Wald, W.H., and Fevold, H.L., *Arch. Biochem.* **11,** 9 (1946).
5. Ali, K.E., and Ramsay, W.N.M., *Biochem. J.* **110,** 36P (1968).
6. Allaire, B.I., and Campagna, A., *Obstet. Gynecol.* **17,** 605 (1961).
7. Allgood, J.W., and Brown, E.B., *Scand. J. Haematol.* **4,** 217 (1967).
8. Amine, E.K., and Hegsted, D.M., *J. Nutr.* **101,** 927 (1971).
9. Amine, E.K., and Hegsted, D.M., *J. Nutr.* **101,** 1575 (1971).
10. Ammerman, C.B., Wing, J.M., Dunavant, B.G., Robertson, W.K., Feaster, J.P., and Arrington, L.R., *J. Anim. Sci.* **26,** 404 (1947).
11. Anderson, H.T., and Barkue, H., *Scand. J. Clin. Lab. Invest.* **25,** Suppl., 114 (1970).

12. Anke, M., Hennig, A., Hoffman, G., Dittrich, G., Grun, M., Ludke, H., Groppel, B., Gartner, P., Schuler, D., and Schwarz, S., *Arch. Tierernaehr.* **22**, 357 (1972).
13. Appleton, T.C., Morgan, E.H., and Baker, E., *Proc. Int. Symp. Erythropoieticum*, 1970 (T. Travnicek and J. Neuwert, eds.), p. 310. Univ. Karlova, Prague, 1971.
14. Baggs, R.B., and Miller, S.A., *J. Nutr.* **103**, 1554 (1973).
15. Bainton, D.F., and Finch, C.A., *Am. J. Med.* **37**, 62 (1964).
16. Baker, E., and Morgan, E.H., *J. Physiol. (London)* **232**, 485 (1973).
17. Baker, E., Jordan, S.M., Tuffery, A.A., and Morgan, E.H., *Life Sci.* **8**, 89 (1969).
18. Baker, E., Shaw, D.C., and Morgan, E.H., *Biochemistry* **7**, 1371 (1968).
19. Banerji, L., Hood, S.K., and Ramalingaswami, V., *Am. J. Clin. Nutr.* **21**, 1139 (1968).
20. Bannerman, R.M., *J. Lab. Clin. Med.* **65**, 944 (1965).
21. Barer, A.P., and Fowler, W.M., *J. Lab. Clin. Med.* **23**, 148 (1937).
21a. Baumslag, N., Yeager, D., Levin, L., and Petering, H.G., *Arch. Environ. Health* **29**, 186 (1974).
22. Beal, V.A., Meyers, A.J., and McCammon, R.W., *Pediatrics* **30**, 518 (1962).
23. Bearn, A.G., and Parker, W.C., *in* "Glycoproteins, Their Structure and Function" (A.Gottschalk, ed.), p. 413. Elsevier, Amsterdam, 1966.
24. Beisel, W.R., Pekarek, R.S., and Wannemacher, R.W., *in* "Trace Element Metabolism in Animals" (W.G. Hoekstra *et al.*, eds.), Vol. 2, p. 217. Univ. Park Press, Baltimore, Maryland, 1974.
25. Benstrup, P., *Acta Med. Scand.* **145**, 315 (1953).
26. Bernard, J., Boiron, M., and Paoletti, C., *Rev. Fr. Etud. Clin. Biol.* **3**, 367 (1958).
27. Beutler, E., *Am. J. Med. Sci.* **234**, 517 (1957); *Acta Haematol.* **21**, 317 (1959); *J. Clin. Invest.* **38**, 1605 (1959).
28. Beutler, E., and Blaisdell, R.K., *J. Lab. Clin. Med.* **52**, 694 (1958); *Blood* **15**, 30 (1960).
29. Beutler, E., Robson, M.J., and Buttenweiser, E., *Ann. Intern. Med.* **48**, 60 (1958).
30. Biggs, J.C., and Davis, A.E., *Lancet* **1**, 814 (1963); *Australas Ann. Med.* **15**, 36 (1966).
31. Bjorn-Rasmussen, E., Hallberg, L., Isaksson, B., and Arvidsson, B., *J. Clin. Invest.* **53**, 247 (1974).
32. Bolin, T., and Davis, A.E., *Am. J. Dig. Dis.* **13**, 16 (1968).
33. Bothwell, T.H., and Finch, C.A., *Symp. Swed. Nutr. Found.* **6**, (1968), cited by Hallberg *et al.* (122).
34. Bothwell, T.H., and Finch, C.A., "Iron Metabolism." Churchill, London, 1962.
35. Bothwell, T.H., Pribilla, W.F., Mebust, W., and Finch, C.A., *Am. J. Physiol.* **193**, 615 (1958).
36. Bothwell, T.H., Pirzio-Biroli, G., and Finch, C.A., *J. Lab. Clin. Med.* **51**, 24 (1958).
37. Bothwell, T.H., Seftel, H., Jacobs, P., Torrance, J.D., and Brauneslog, N., *Am J. Clin. Nutr.* **14**, 47 (1964).
38. Braude, R., Chamberlain, A., Kotarbinski, M., and Mitchell, K.G., *Br. J. Nutr.* **16**, 427 (1962).
39. Bremner, I., and Dalgarno, A.C., *Br. J. Nutr.* **29**, 229 (1973).
40. Brise, H., and Hallberg, L., *Acta Med. Scand.* **171**, Suppl. 376, 51, and 59 (1962).
41. Brock, A.B., and Diamond, L.M., *J. Pediatr.* **4**, 445 (1934).
42. Brown, E.B., and Justus, B.W., *Am. J. Physiol.* **194**, 319 (1958).
43. Brown, E.B., and Rother, M.L., *J. Lab. Clin. Med.* **62**, 804 (1963).
44. Brunstrom, G.M., Karabus, C., and Fielding, J., *Br. J. Haematol.* **14**, 525 (1968).
45. Bullen, J.J., Rogers, H.J., and Griffiths, E., *Br. J. Haematol.* **23**, 389 (1972).

46. Bunch, R.J., Speers, V.C., Hays, V.M., and McCall, J.T., *J. Anim. Sci.* **22,** 56 (1963).
47. Burman, D., *Arch. Dis. Child.* **47,** 261 (1972).
48. Burns, R.H., Johnston, A., Hamilton, J.W., McColloch, R.J., Duncan, W.E., and Fisk H.G., *J. Anim. Sci.* **23,** 5 (1964).
49. Byers, J.H., Jones, I.R., and Haag, J.R., *J. Dairy Sci.* **35,** 661 (1953).
50. Callender, S.T., *in* "Nutritional Problems in a Changing World" (D. Hollingsworth and M. Russell, eds.), p. 205. Appl. Sci. Publ., London, 1973.
51. Callender, S.T., and Warner, G.T., *Am. J. Clin. Nutr.* **21,** 1170 (1968).
52. Campbell, A.G., Coup, M.R., Bishop, W.H., and Wright, D.E., *N. Z. J. Agric. Res.* **17,** 393 (1974).
53. Campbell, J.A., and Gardiner, A.C., *Vet. Rec.* **72,** 1006 (1960).
54. Carlson, R.H., Swenson, M.J., Ward, G.M., and Booth, N.H., *J. Am. Vet. Med. Assoc.* **139,** 457 (1961).
55. Cartwright, G.E., and Wintrobe, M.M., *J. Clin. Invest.* **28,** 86 (1949).
56. Cartwright, G.E., and Wintrobe, M.M., *in* "Modern Trends in Blood Diseases" (J.F. Wilkinson, ed.), p. 183. Butterworth, London, 1954.
57. Cartwright, G.E., Gubler, C.J., and Wintrobe, M.M. *J. Clin. Invest.* **33,** 685 (1954).
58. Cavell, P.A., and Widdowson, E.M., *Arch. Dis. Child.* **39,** 496 (1964).
59. Carlton, R.W., Bothwell, T.H., and Seftel, H.C., *Clin. Haematol.* **2,** 383 (1973).
60. Cole, S.K., Thomson, A.M., Billewicz, W.Z., and Black, A.E., *J. Obstet. Gynaecol. Brit. Commonw.* **78,** 933 (1971); Cole, S.K., W.Z., Billewicz, and Thomson, A.M., *ibid* **79,** 994 (1972).
61. Conrad, M.E., and Crosby, W.H., *Blood* **22,** 406 (1963).
62. Conrad, M.E., Weintraub, L.R., and Crosby, W.H., *J. Clin. Invest.* **43,** 963 (1964).
63. Conrad, M.E., Weintraub, L.R., Sears, D.A., and Crosby, W.H., *Am. J. Physiol.* **211,** 1123 (1966).
64. Cook, J.D., Lipschitz, D.A., Miles, L.E., and Finch, C.A., *Am. J. Clin. Nutr.* **27,** 681 (1974).
65. Cowan, J.W., Esfahani, M., Salji, J.P., and Azzam, S.A., *J. Nutr.* **90,** 423 (1966).
66. Crosby, W.H., *Arch. Intern, Med.* **126,** 911 (1970).
67. Cusack, R.P., and Brown, W.D., *J. Nutr.* **86,** 383 (1965).
68. Czerniejewski, C.P., Shank, C.W., Bechtel, W.G., and Bradley, W.B., *Cereal Chem.* **41,** 65 (1964).
69. Dallman, R.P., *J. Nutr.* **97,** 475 (1969).
70. Davis, A.E., and Biggs, J.C., *Australas. Ann. Med.* **13,** 201 (1964); *Gut* **6,** 140 (1965).
71. Davis, P.N., Norris, L.C., and Kratzer, J.H., *J. Nutr.* **94,** 407 (1968).
72. Davis, P.S., Luke, C.G., and Deller, D.J., *Lancet* **2,** 1143 (1966); *Nature (London)* **214,** 1126 (1967).
73. Derechin, S.S., and Johnson, P., *Nature (London)* **194,** 473 (1962).
74. Dewar, W.A., Teague, P.W., and Downie, J.N., *Br. Poultry Sci.* **15,** 119 (1974).
74a. Disler, P.B., Lynch, S.R., Charlton, R.W., Bothwell, T.H., Walker, R.B., and Mayet, F., *Br. J. Nutr.* **34,** 141 (1975).
75. Drabkin, D.L., *Physiol. Rev.* **31,** 345 (1951).
76. Dubach, R., Moore, C.V., and Callender, S.T., *J. Lab. Clin. Med.* **45,** 599 (1955).
77. Eatough, D.J., Mineer, W.A., Christensen, J.J., Izatt, R.M., and Mangelson, N.F., *in* "Trace Element Metabolism in Animals" (W.G. Hoekstra *et al.*, eds.), Vol. 2, p. 659. Univ. Park Press, Baltimore, Maryland, 1974.
78. Edgerton, V.R., Bryant, S.L., Gillespie, C.A., and Gardner, G.W., *J. Nutr.* **102,** 381 (1972).

79. Ehrenberg, A., and Laurell, C.B., *Acta Chem. Scan.* **9,** 68 (1955).
79a. El-Howary, M.F.S., El-Shobaki, F.A., Kholeif, T., Sakr, R., and El-Bassoussy, M., *Br. J. Nutr.* **33,** 351 (1975).
80. Elwood, P.C., *Br. Med. J.* **1,** 224 (1963).
81. Elwood, P.C., *Am. J. Clin. Nutr.* **26,** 958 (1973).
82. Elwood, P.C., Newton, D., Eakins, J.D., and Brown, D.A., *Am. J. Clin. Nutr.* **21,** 1162 (1968).
83. Evans, G.W., and Grace, C.I., *Proc. Soc. Exp. Biol. Med.* **147,** 687 (1974).
84. Ezekiel, E., *J. Lab. Clin. Med.* **70,** 138 (1967).
85. Ezekiel, E., and Morgan, E.H., *J. Physiol. (London),* **165,** 336 (1963).
86. Faber, M., and Falbe-Hansen, I., *Nature (London),* **184,** 1043 (1959).
87. Fabiano, A., *Ann. Obstet. Gynecol.* **78,** 1043 (1959).
88. Feeney, R.E., and Allison, R.G., "Evolutionary Biochemistry of Proteins." Wiley (Interscience), New York, London, 1969.
89. Feeney, R.E., and Komatsu, S.K., *Struct. Bonding (Berlin)* **1,** 149 (1966).
90. Fenillon, Y.M., and Plumier, M., *Acta Pediatr. (Stockholm)* **41,** 138 (1952).
91. Fielding, J., and Speyer, B.E., *Biochim. Biophys. Acta* **363,** 387 (1974).
92. Figueroa, W.G., *in* "Metal-Binding in Medicine" (M.J. Seven and L.A. Johnson, eds), p. 146. Lippincott, Philadelphia, Pennsylvania, 1960.
93. Fillet, G., Cook, J.D., and Finch, C.A., *J. Clin. Invest.* **53,** 1527 (1974).
94. Finch, C.A., and Monsen, E.R., *J. Am. Med. Assoc.* **219,** 1462 (1972).
95. Finch, S., Haskins, D., and Finch, C.A., *J. Clin. Invest.* **29,** 1078 (1950).
96. Forbes, G.B., and Reina, J.C., *J. Nutr.* **102,** 647 (1972).
97. Forth, W., and Rummel, R., *in* "Intestinal Absorption of Metal Ions, Trace Elements and Radionuclides" (S.C., Skoryna and D. Waldron–Edward, eds.), p. 173. Pergamon, Oxford, 1971.
98. Foy, H., and Kondi, A., *J. Trop. Med. Hyg.* **60,** 105 (1957).
99. Fritz, J.C., Pla, G.W., Roberts, T., Boehne, J.W., and Hove, E.L., *J. Agric. Food Chem.* **18,** 647 (1970).
100. Fuerth, J., *J. Pediatr.* **80,** 974 (1972).
101. Furugouri, K., *J. Anim. Sci.* **37,** 667 (1971).
102. Gabrio, B.W., Shoden, A.W., and Finch. C.A., *J. Biol. Chem.* **204,** 815 (1953).
103. Gardiner, M.R., *J. Comp. Pathol.* **75,** 397 (1965).
104. Garretson, F.D., and Conrad. M.E., *Proc. Soc. Exp. Biol. Med.* **126,** 304 (1967).
105. Gipp, W.F., Pond, W.G., Kallfelz, F.A., Tasker, J.B., VanCampen, D.R., Krook, L., and Visek, W.J., *J. Nutr.* **104,** 532 (1974).
106. Goltner, E., and Gailer, H.J., *Zentralbl. Gynaekol.* **34,** 1177 (1964).
107. Gordon, W.G., Groves, M.L., and Basch, J.J., *Biochemistry* **2,** 817 (1963).
108. Granick, S., *Physiol. Rev.* **31,** 489 (1951).
109. Granick, S., *in* "Trace Elements" (C.A. Lamb, O.G., Bentley, and J.M. Beattie, eds.), p. 365. Academic Press, New York, 1958.
110. Granick, S., and Michaelis, L., *Science* **95,** 439 (1942).
111. Grassman, E., and Kirchgessner, M., *Zentralbl, Veterinaermed., Reihe, A* **20,** 481 (1973).
112. Grassman, E., and Kirchgessner, M., *Z. Tierphysiol., Tierernaehr. Futtermittel Kd.* **31,** 38 (1973).
113. Green, S., Saha, A.K., Carleton, A.W., and Mazur, A., *Fed. Proc., Fed. Am. Soc. Exp. Biol.* **17,** 233 (1958).
114. Groves, M.L., *J. Am. Chem. Soc.* **82,** 3345 (1960).

115. Gubler, C.J., Cartwright, G.E., and Wintrobe, M.M., *J. Biol. Chem.* **224,** 533 (1958).
116. Guthrie, H.A., Froozani, M., Sherman, A.R., and Barron, G.P., *J. Nutr.* **104,** 1273 (1974).
117. Hahn, P.F., *Medicine (Baltimore)* **16,** 249 (1937).
118. Hahn, P.F., Bale, W.F., Ross, J.F., Balfour, W.M., and Whipple, G.H., *J. Exp. Med.* **78,** 169 (1943).
119. Hall, M.H., *Br. Med. J.* **2,** 661 (1974).
120. Hallberg, L., and Sölvell, L., *Acta Med. Scand.* **181,** 335 (1967).
121. Hallberg, L., and Bjorn-Rasmussen, E., *Scand. J. Haematol.* **9,** 193 (1972).
122. Hallberg, L., Garby, L., Suwanik, R., and Bjorn-Rasmussen, E., *Am. J. Clin. Nutr.* **27,** 826 (1974).
123. Hallberg, L., and Nilsson, L., *Acta Obstet. Gynecol. Scand.* **43,** 352 (1964).
124. Hallgren, B., *Acta Soc. Med. Up. [N.S.]* **59,** 79 (1954).
125. Hamilton, E.I., and Minski, M.J., *Sci. Total Environ.* **1,** 375 (1972–1973).
126. Hamilton, L.D., Gubler, C.J., Cartwright, G.E., and Wintrobe, M.M., *Proc. Soc. Exp. Biol. Med.* **73,** 65 (1950).
127. Hartley, W.J., Mullins, J., and Lawson, E.M., *N. Z. Vet. J.* **7,** 99 (1959).
128. Hawksley, J.C., Lightwood, R., and Bailey, U.M., *Arch. Dis. Child.* **9,** 359 (1934).
129. Health Bull. No. 23. Govt. India Press, Delhi, 1951.
130. Hegsted, D.M., Finch, C.A., and Kinney, T.D., *J. Exp. Med.* **90,** 147 (1949).
131. Hemingway, R.G., Brown, N.A., and Luscombe, J., *in* "Trace Element Metabolism in Animals" (W.G. Hoekstra *et al.,* eds.), Vol. 2, p. 601. Univ. Park Press, Baltimore, Maryland, 1974.
132. Herndon, J.G., Rice, T.G., Tucker, R.G., Van Loon, E.J., and Greenberg, S.M., *J. Nutr.* **64,** 615 (1958).
133. Hill, C.H., and Matrone, G., *J. Nutr.* **73,** 425 (1961).
134. Hitchcock, J.P., Ku, P.K., and Miller, E.R., *in* "Trace Element Metabolism in Animals" (W.G. Hoekstra *et al.*, eds.), Vol. 2, p. 598. Univ. Park Press, Baltimore, Maryland, 1974.
135. Hofvander, G., *Acta Med. Scand., Suppl.* **494,** (1968).
136. Holmberg, C.G., and Laurell, C.B., *Acta Chem. Scand.* **1,** 944 (1947).
137. Horvath, Z., *in* "Trace Element Metabolism in Animals" (C.F. Mills, ed.), Vol. 1, p. 328. Livingstone, Edinburgh, 1970.
138. Huser, J.H., Rieber, E.E., and Berman, A.R., *J. Lab. Clin. Med.* **69,** 405 (1967).
139. Hussain, R., and Patwardhan, V.N., *Indian J. Med. Res.* **47,** 676 (1959).
140. Hussain, R., and Patwardhan, V.N., *Lancet* **1,** 1073 (1959); Hussain, R., Patwardhan, V.N., and Sriranachari, S., *Indian J. Med. Res.* **48,** 235 (1960).
141. Hussain, R., Walker, R.B., Layrisse, M., Clark, P., and Finch, C.A., *Am. J. Clin. Nutr.* **20,** 842 (1967).
142. Ingalls, R.L., and Johnston, F.A., *J. Nutr.* **53,** 351 (1954).
143. Ingram, D.J.E., Gibson, J.F., and Peratz, M.F., *Nature (London)* **178,** 906 (1956).
144. Iodice, A.A., Richert, D.A., and Schulman, M.P., *Fed. Proc. Fed. Am. Soc. Exp. Biol.* **17,** 248 (1958).
145. Itzerott, A.G., *J. Aust. Inst. Agric. Sci.* **8,** 119 (1942).
146. Jacobs, A., and Miles, P.M., *Br. Med. J.* **4,** 778 (1969).
147. Jacobs, A., and Miles, P.M., *Gut* **10,** 226 (1969).
148. Jacobs, A., Miller, F., Worwood, M., Beamish, M.R., and Wardrop, C.A., *Br. Med. J.* **4,** 206 (1972).
149. Jacobs, P., Bothwell, T., and Charlton. R.W., *J. Appl. Physiol.* **19,** 187 (1964).
150. Jamieson, G.A., Jett, M., and DeBernardos, S.L., *J. Biol. Chem.* **246,** 3686 (1971).

151. Jandl, J.H., Inman, J.K., Simmons, R.L., and Allen, D.W., *J. Clin. Invest.* **38,** 161 (1959).
152. Jandl, J.H., and Katz, J.H., *J. Clin. Invest.* **42,** 314 (1963).
153. Johnston, F.A., and McMillan, T.J., *J. Am. Diet. Assoc.* **28,** 633 (1952).
154. Jordan, S.M., Kaldor, I., and Morgan, E.H., *Nature (London)* **215,** 76 (1967).
155. Josephs, H.W., *Blood* **13,** 1 (1958).
156. Josephs, H.W., *Acta Paediatr. (Stockholm)* **48,** 403 (1959).
157. Kaldor, I., *Aust. J. Exp. Biol. Med. Sci.* **32,** 437 (1954).
158. Kaldor, I., and Ezekiel, E., *Nature (London)* **196,** 175 (1961).
159. Kerr, L.M.H., *Biochem. J.* **67,** 627 (1957).
160. Kilpatrick, G.S., *in* "Iron Deficiency: Pathogenesis, Clinical Aspects, Therapy" (L. Hallberg, H.G. Harwerth, and A. Vannotti, eds.), p. 441. Academic Press, New York, 1970.
161. Kirchgessner, M., and Pallauf, J., *Sonderdruck Zuechtungskd.* **45,** 245 and 249 (1973); Kirchgessner, M., and Weigand, E., *Z. Tierphysiol. Tierernaehr. Futtermittelkd.* **34,** 205 (1975).
162. Kraus, W.E., and Washburn, R.G., *J. Biol. Chem.* **114,** 247 (1936).
163. Kroe, D., Kinney, T.D., Kaufman, N., and Klavins, J.V., *Blood* **21,** 546 (1963).
164. Kuhn, I.N., Layrisse, M., Roche, M., Martinez-Torres, C., and Walker, R.B., *Am. J. Clin. Nutr.* **21,** 1184 (1968).
165. Lawlor, M.J., Smith, W.H., and Beeson, W.M., *J. Anim. Sci.* **24,** 742 (1965).
166. Layrisse, M., Linares, J., and Roche, M., *Blood* **25,** 73 (1965).
167. Layrisse, M., Martinez-Torres, C., and Gonzalez, M., *Am. J. Clin. Nutr.* **27,** 152 (1974).
168. Layrisse, M., Martinez-Torres, C., and Roche, M., *Am. J. Clin. Nutr.* **21,** 1175 (1968).
169. Lee, F., and Rosenthal, F.D., *Q. J. Med.* **27,** 19 (1958).
170. Lee, G.R., Nacht, S., Lukens, J.N., and Cartwright, G.E., *J. Clin. Invest.* **47,** 2058 (1968).
171. Leeuw, N.K. de, Lowenstein, L., and Hsieh, Y.S., *Medicine (Baltimore)* **45,** 219 (1966).
172. Leslie, A.J., and Kaldor, I., *Br. J. Nutr.* **26,** 469 (1971).
173. Lewis, M., and Iammarino, R.M., *J. Lab. Clin. Med.* 78, 547 (1971).
174. Lind, D.E., *J. Geriatr.* **4,** 19 (1973).
175. Linkenheimer, W.H., Patterson, E.L., Milstrey, B.A., Brochman, J.A., and Johnston, D.D., *J. Anim. Sci.* **19,** 763 (1960).
176. Lipschitz, D.A., Cook, J.D., and Finch, C.A., *N. Engl. J. Med.* **290,** 1213 (1974).
177. Lipschitz, D.A., Simon, M.O., Lynch, S.R., Dugard, J., Bothwell, T.H., and Charlton, R.W., *Br. J. Haematol.* **21,** 289 (1971).
178. Loh, T.T., *Proc. Soc. Exp. Biol. Med.* **134,** 1070 (1970).
179. Loh, T.T., and Kaldor, I., *Comp. Biochem. Physiol. B* **44,** 337 (1973).
180. Loh, T.T., and Kaldor, I., *J. Dairy Sci.* **50,** 339 (1974).
181. Luke, C.G., Davis, P.S., and Deller, D.J., *Lancet* **1,** 926 (1967).
182. MacDonald, R.A., Becker, B.J.P., and Picket, G.S., *Arch. Intern,. Med.* **3,** 315 (1963).
183. Martinez-Torres, C., Leets, I., Renzi, M., and Layrisse, M., *J. Nutr.* **104,** 983 (1974).
184. Masson, P.L., and Heremans, J.F., *Protides Biol. Fluids, Proc. Collog.* **14,** 115 (1966).
185. Masson, P.L., and Heremans, J.F., *Comp. Biochem. Physiol. B* **39,** 119 (1971).
186. Matrone, G., Conley, C., Wise, G.H., and Waugh, R.K., *J. Dairy Sci.* **40,** 1437 (1957).
187. Matrone, G., Thomason, E.L., and Bunn, C.R., *J. Nutr.* **72,** 459 (1960).
188. Mazur, A., and Carleton, A., *Blood* **26,** 317 (1965).
189. Mazur, A., Green, S., and Carleton, A., *J. Biol. Chem.* **235,** 595 (1960).

190. Mazur, A., Green, S., Saha, A., and Carleton, A., *J. Clin. Invest.* **37,** 1809 (1958).
191. McCall, M.G., Newman, G.E., O'Brien, J.R., and Witts, L.J., *Br. J. Nutr.* **16,** 305 (1962).
192. McCance, R.A., and Widdowson, E.M., *Lancet* **2,** 680 (1937); *J. Physiol. (London)* **94,** 138 (1938).
193. McCance, R.A., and Widdowson, E.M., *Nature (London)* **152,** 326 (1943).
194. McCance, R.A., and Widdowson, E. M., *Br. Med. Bull.* **7,** 297 (1951).
195. McCurdy, P.R., and Dern, R.J., *Am. J. Clin. Nutr.* **21,** 284 (1968).
196. McDougal, E.I., *J. Agric. Sci.* **37,** 337 (1947).
197. McLaurin, L.P., and Cotter, J.R., *Am. J. Obstet. Gynecol.* **98,** 931 (1967).
198. McLean, F.W., Cotter, J. R., Blechner, J.N., and Noyes, W.D., *Am. J. Obstet. Gynecol.* **106,** 699 (1970).
199. McLennan, W.J., Andrews, G.R., Macleod, C., and Caird, F.I., *Q. J. Med.* **42,** 1 (1973).
200. Mendel, G.A., *Blood* **18,** 727 (1961).
201. Mendel, G.A., Wiler, R.J., and Mangalik, A., *Blood* **22,** 450 (1963).
202. Minnich, V., Okcuoglu, A., Tarcon, Y., Arcasoy, A., Cin, S., Yorukoglu, O., Renda, F., and Demirag, B., *Am. J. Clin. Nutr.* **21,** 78 (1968).
203. Mitchell, H.H., and Edman, M., *Am. J. Clin. Nutr.* **10,** 163 (1962).
204. Moe, P.J., *Acta Paediatr. (Stockholm), Suppl.* **150,** (1963).
205. Monsen, E.R., *J. Nutr. Educ.* **2,** 152 (1971).
206. Monsen, E.R., Kuhn, I.N., and Finch, C.A., *Am. J. Clin. Nutr.* **20,** 842 (1967).
207. Moore, C.V., *Harvey Lect.* **55,** 67 (1951).
208. Moore, C.V., *Am. J. Clin. Nutr.* **3,** 3 (1955).
209. Moore, H.O., *Aust., C. S. I. R. O. Bull.* **133,** (1938).
210. Morgan, E.H., *J. Physiol. (London)* **158,** 573 (1961).
211. Morgan, E.H., *Lancet* **1,** 9 (1961).
212. Morgan, E.H., *Aust. J. Exp. Biol. Med.* **39,** 361 and 371 (1961); *J. Pathol. Bacteriol.* **84,** 65 (1962).
213. Morgan, E.H., *Q. J. Exp. Physiol. Cogn. Med. Sci.* **47,** 57 (1962); **48,** 170 (1963).
214. Morgan, E.H., *Br. J. Haematol.* **10,** 442 (1964).
215. Morgan, E.H., and Laurell, C.B., *Br. J. Haematol.* **9,** 471 (1963).
216. Morgan, E.H., and Walters, M.N.I., *J. Clin. Pathol.* **16,** 101 (1963).
216a. Morris, E.R., and Ellis, R., *J. Nutr.* **106,** 753 (1976).
217. Morris, E.R., and Greene, F.E., *J. Nutr.* **102,** 901 (1972).
218. Muir, A.R., *Q. J. Exp. Physiol. Cogn. Med. Sci.* **45,** 192 (1960).
219. Mukherjee, C., and Mukherjee, S.K., *J. Indian Med. Assoc.* **22,** 345 (1953).
220. Murray, J., and Stein, N., *Lancet* **1,** 614 (1968).
221. Murray, J., and Stein, N., *in* "Trace Element Metabolism in Animals" (C.F. Mills, ed.), Vol. 1, p. 321. Livingstone, Edinburgh, 1970.
222. Naiman, J.L., Oski, F.A., Diamond, L., Vawter, G.F., and Schwachman, H., *Pediatrics* **33,** 83 (1964).
223. National Academy of Sciences (U.S.), "Recommended Dietary Allowances," 8th ed. Natl. Acad. Sci., Washington, D.C., 1974.
224. National Health, *Natl. Health Med. Res. Counc. (Aust.),* Spec. Rep. No. 1 (1945).
225. Niccum, W.L., Jackson, R.L., and Stearns, G., *Am. J. Dis. Child.* **86,** 553 (1953).
226. Osaki, S., Johnson, D.A., and Frieden, E., *J. Biol. Chem.* **246,** 3018 (1971).
227. Osaki, S., Johnson, D.A., and Frieden, E., *J. Biol. Chem.* **241,** 2746 (1966).
228. Osaki, S., and Sirivech, S., *Fed. Proc. Fed. Am. Soc. Exp. Biol.* **30,** Abstr. 1292 (1971).

229. Otis, L., and Smith, M.C., *Science* **91,** 146 (1940).
230. Pallauf, J., and Kirchgessner, M., *Sonderdruck Zuechtungskd.* **45,** 119 (1973).
231. Panic, B., *in* "Trace Element Metabolism in Animals" (C.F. Mills, ed.), Vol. 1, p. 324, Livingstone, Edinburgh, 1970.
232. Patterson, J.C.S., Marrack, D., and Wiggins, H.S., *J. Clin. Pathol.* **6,** 105 (1953).
233. Peters, T., Apt, L., and Ross, J.F., *Gastroenterology* **61,** 315 (1971).
234. Pirzio-Biroli, G., Bothwell, T.H., and Finch, C.A., *J. Lab. Clin. Med.* **51,** 37 (1958).
235. Planas, J., *Nature (London)* **215,** 289 (1967).
236. Planas, J., and de Castro, S., *Nature (London)* **187,** 1126 (1960).
237. Planas, J., de Castro, S., and Recio, J.M., *Nature (London)* **189,** 668 (1961).
238. Pollack, S., George, J.N., Raba, R.C., Kaufman, R.M., and Crosby, W.H., *J. Clin. Invest.* **44,** 1470 (1965).
239. Pollack, S., Kaufman, R.M., and Crosby, W.H., *Blood* **24,** 557 (1964).
240. Polycove, M., *in* "Iron in Clinical Medicine" (R.O. Wallerstein and S.R. Mettier, eds.), p. 43. Univ. of California Press, Berkley, 1958.
241. Pond, W.G., Lowrey, R.L., Maner, J.H., and Loosli, J.K., *J. Anim. Sci.* **19,** 1286 (1960).
242. Pond, W.G., Veum T.L., and Lazar, V.A., *J. Anim. Sci.* **24,** 668 (1965).
243. Pritchard, J.A., and Hunt, C.F., *Surg., Gynecol. Obstet.* **106,** 516 (1958).
244. Ramsay, W.N.M., and Campbell, E.A., *Biochem. J.* **58,** 313 (1958).
245. Rice, R.W., Nelms, G.E., and Schoonover, C.O., *J. Anim. Sci.* **26,** 613 (1967).
246. Richard, R.M., Shumard, R.F., Pope, A.L., Phillips, P.H., Herrick, C.A., and Bohstedt, G., *J. Anim. Sci.* **13,** 274 and 674 (1954).
247. Richardson, T., and Guss, P.L., *J. Dairy Sci.* **48,** 523 (1965).
248. Rimington, C., *Br. Med. Bull.* **15,** 19 (1959).
249. Roberts, P.D., St. John, D.J.B., Sinha, R., Stewart, J.S., Baird, I.M., Coghill, N.F., and Morgan, J.O., *Br. J. Haematol.* **20,** 165 (1971).
250. Robinson, M.F., McKenzie, J.F., Thomson, C.D., and van Rij, A.L., *Br. J. Nutr.* **30,** 195 (1975).
251. Robinson, S.H., *Blood* **33,** 909 (1969).
252. Roe, D.A., *N.Y. State J. Med.* **66,** 1233 (1966).
253. Roeser, H.P., Lee, G.R., Nacht, S., and Cartwright, G.E., *J. Clin. Invest.* **49,** 2408 (1970).
254. Rothen, A., *J. Biol. Chem.* **152,** 679 (1944).
255. Rybo, G., *Acta Obstet. Gynecol. Scand.* **45,** Suppl. 7 (1966).
256. Sano, S., Inoue, S., Tanabe, Y., Sumiya, C., and Koike, S., *Science* **129,** 275 (1959).
257. Schade, A.L., Reinhart R.W., and Levy, H., *Arch. Biochem.* **20,** 170 (1949).
258. Schwarz, F.J., and Kirchgessner, M., *Int. J. Vitam. Nutr. Res.* **44,** 116 (1974).
259. Seal, U.S., Sinha, A.A., and Doe, R.P., *Am. J. Anat.* **134,** 263 (1972).
260. Settlemire, C.T., and Matrone, G., *J. Nutr.* **92,** 153 and 159 (1967).
261. Sherman, H.C., "Chemistry of Food and Nutrition." Macmillan, New York, 1935.
262. Shoden, A., and Sturgeon, P., *Am. J. Pathol.* **34,** 113 (1958).
263. Shoden, A., and Sturgeon, P., *Acta Haematol.* **23,** 376 (1960); **27,** 33 (1962); *Nature (London)* **189,** 846 (1961).
264. Shoden, A., and Sturgeon, P., *Proc. Int. Congr. Histochem. Cytochem 1st, 1960.*
265. Shoden, A., Gabrio, B.W., and Finch, C.A., *J. Biol. Chem.* **204,** 823 (1953).
266. Shoden, A., and Sturgeon, P., *Br. J. Haematol.* **9,** 471 (1963).
267. Siimes, M.A., Addiego, J.E., Jr., and Dallman, P.R., *Blood* **43,** 581 (1974); Siimes, M.A., and Dallman, R.P., *Br. J. Haematol.* **28,** 7 (1974).
268. Sirivech, S., Frieden, E., and Osaki, S., *Biochem. J.* **143,** 311 (1974).

269. Sisson, T.R.C., *Fed. Proc., Fed. Am. Soc. Exp. Biol.* **23,** 879 (1964).
270. Skeets, O., Frazier, E., and Dickins, D., *Miss. Agric. Exp. Stn. Bull.* **291** (1931).
271. Smith, M.D., and Mallet, B.J., *Clin. Sci.* **16,** 23 (1957).
272. Smythe, C.V., and Miller, R.C., *J. Nutr..* **1,** 209 (1929).
273. Sood, S.K., Banerji, L., and Ramalingaswami, V., *Am. J. Clin. Nutr.* **21,** 1149 (1968).
274. Speyer, B.E., and Fielding J., *Biochim. Biophys. Acta* **332,** 192 (1974).
275. Spik, G., and Montreuil, J., *C.R. Seances Soc. Biol. (Ses Fil.)* **160,** 94 (1966).
276. Stevens, A.R., Coleman, D.H., and Finch, C.A., *Ann. Intern. Med.* **38,** 199 (1953).
277. Strain, W.H., private communication (1974).
278. Sturgeon, P., *Pediatrics* **17,** 341 (1956); **18,** 267 (1956).
279. Sturgeon, P., *Br. J. Haematol.* **5,** 31 (1959).
279a. Summers, M., Worwood, M., and Jacobs, A., *Br. J. Haematol.* **28,** 19 (1974).
280. Swiss, L.D., and Beaton, G.H., *Am. J. Clin. Nutr.* **27,** 373 (1974).
281. Takkunen, H., and Aromaa, A., *Am. J. Clin. Nutr.* **27,** 323 (1974).
282. Tartour, G., *Res. Vet. Sci.* **15,** 389 (1973).
283. Tarvydas, H., Jordan, S.M., and Morgan, E.H., *Br. J. Nutr.* **22,** 565 (1968).
284. Tolan, A., Robertson, J., Orton, C.R., Head, M.J., Christie, A.A., and Millburn, B.A., *Br. J. Nutr.* **31,** 185 (1974).
285. Torrance, J.D., Charlton, R.W., Schmaman, A., Lynch, S.R., and Bothwell, T.R., *J. Clin. Pathol.* **21,** 498 (1968).
286. Toskes, P.P., Smith, G.W., Bensinger, T.A., Giannella, R.A., and Conrad, M.E., *Am. J. Clin. Nutr.* **27,** 335 (1974).
287. Turnberg, L.A., *Lancet* **1,** 921 (1968).
288. Ullrey, D.E., Miller, E.R., Thompson, O.A., Ackerman, I.M., Schmidt, D.A., Hoefer, J.A., and Luecke, R.W., *J. Nutr.* **70,** 187 (1960).
289. Underwood, E.J., *Aust. Vet. J.* **10,** 87 (1934).
290. Underwood, E.J., "The Mineral Nutrition of Livestock," FAO/CAB. Publ. Central Press, Aberdeen, 1966.
291. Underwood, E.J., and Morgan, E.H., *Aust. J. Exp. Biol. Med. Sci.* **41,** 247 (1963).
292. Vahlquist, B., *Acta Paediatr. (Stockholm)* **28,** Suppl. 5 (1941).
293. Vahlquist, B., *Blood* **5,** 874 (1950).
294. Valberg, L.S., *in* "Intestinal Absorption of Metal Ions, Trace Elements and Radionuclides" (S.C. Skoryna and D. Waldron-Edward, eds.), p. 257. Pergamon, Oxford, 1971.
295. Valberg, L.S., Ludwig, J., and Olatunbosun, D., *Gastroenterology* **56,** 241 (1969).
296. Van Campen, D., *J. Nutr.* **102,** 165 (1972).
297. Van Campen, D., *J. Nutr.* **103,** 139 (1973).
298. Van Campen, D., *Fed. Proc., Fed. Am. Soc. Exp. Biol.* **33,** 100 (1974).
299. Van Campen, D., and Gross, E., *J. Nutr.* **99,** 68 (1969).
300. Venkatachalam, P.S., *Bull. W. H. O.* **26,** 193 (1962); *Am. J. Clin. Nutr.* **21,** 1156 (1968).
301. Venn, J.A.J., McCance, R.A., and Widdowson, E.M., *J. Comp. Pathol. Ther.* **57,** 314 (1947).
302. Walker, A.R.P., Fox, F.W., and Irving, J.T., *Biochem. J.* **42,** 252 (1948).
303. Walters, G.O., Miller, F.M., and Worwood, M., *J. Clin. Pathol.* **26,** 770 (1973).
304. Weintraub, L.R., Conrad, M.E., and Crosby, W.H., *Br. J. Haematol.* **11,** 432 (1965).
305. Weintraub, L.R., Conrad, M.E., and Crosby, W.H., *J. Clin. Invest.* **47,** 531 (1968).
306. Widdowson, E.M., *Nature (London)* **166,** 626 (1950); Spray, C.M., and Widdowson, E.M., *Br. J. Nutr.* **4,** 332 (1951); Widdowson, E.M., and Spray, C.M., *Arch. Dis. Child.* **26,** 205 (1951).

307. Widdowson, E.M., and McCance, R.A., *Biochem. J.* **42,** 488 (1948).
308. Williams, J., *Biochem. J.* **83,** 355 (1962).
309. Wilson, J.F., Heiner, D.C., and Lahey, M.E., *J. Pediatr.* **60,** 787 (1962).
310. Wong, C.T., and Morgan, E.H., *Q. J. Exp. Physiol. Cogn. Med. Sci.* **58,** 47 (1973).
311. Wong, C.T., and Morgan, E.H., *Aust. J. Exp. Biol. Med. Sci.* **52,** 413 (1974).
312. Worwood, M., Summers, M., Miller, F., Jacobs, A., and Whittaker, J.A., *Br. J. Haematol.* **28,** 27 (1974).
313. Worwood, M., Aherne, W., Dawkins S., and Jacobs, A., *Clin. Sci. Mol. Med.* **48,** 441 (1975).
314. Yamagata, N., and Yamagata, T., *Bull. Inst. Public Health, Tokyo* **13,** 11 (1964).
315. Zook, E.G., and Lehmann, J., *J. Assoc. Off. Agric. Chem.* **48,** 850 (1965).

3

Copper

I. COPPER IN ANIMAL TISSUES AND FLUIDS

1. Copper in the Body

The healthy adult body has been estimated to contain 80 mg of total copper (75). Newborn and very young animals are normally richer in copper per unit of body weight than adults of the same species (302, 342)(Table 8). The newborn levels are largely maintained throughout the suckling period, followed by a steady fall during growth to the time when adult values are reached. However, in a study of the copper content of 17 tissues from each of 4 newborn calves and

TABLE 8
Concentrations[a] of Copper in the Whole Bodies of Various Species

	Human	Pig	Cat	Rabbit	Rat	Guinea pig
Newborn	4.7	3.2	2.9	4.0	4.3	6.9[b]
Adult	1.7	2.5	1.9	1.5	2.0	–[c]

[a] ppm of fat-free tissue.
[b] From Widdowson (342).
[c] From Spray and Widdowson (302).

two adult cows, the Cu concentrations of all tissues other than the eye, skin, tongue, bone, and liver were found to be comparable (34).

The distribution of total body copper among the tissues varies with the species, age, and copper status of the animal. Cartwright and Wintrobe (75) studied the distribution of copper in the tissues of five normal humans. They found a total of 23 mg Cu in the liver, heart, spleen, kidneys, brain, and blood, of which 8 mg was present in the liver and 8 mg in the brain. Smith (295) reported data pointing to a smaller proportion of total body Cu in the brain, although the liver and the brain contained similar Cu concentrations. The data obtained by Cunningham (85) for a range of species also indicate that a higher proportion of total body Cu exists in the liver than in the brain. In ruminants, which have a high capacity for hepatic copper storage, the proportion in the liver can be very high. Thus Dick (100) found the total body copper of two adult sheep with very high liver Cu concentrations to be distributed as follows: liver, 72–79%; muscles, 8–12%; skin and wool, 9%; and skeleton, 2%. It is further apparent from a study of neonate and adult bovines that by far the largest proportion of total body copper occurs in the liver in that species (34).

The pituitary, thyroid, thymus, and prostate glands and the ovary and testis are examples of organs low in copper; the pancreas, skin, muscles, spleen, and bones represent tissues of intermediate Cu concentration; and the liver, brain, kidneys, heart, and hair those of relatively high-Cu concentration (65, 295). The values obtained in two recent studies may be cited to illustrate typical tissue Cu concentrations. Hamilton *et al.* (149) found the following mean concentrations in adult humans: liver, 14.7 ± 3.9; brain, 5.6 ± 0.2; lung, 2.2 ± 0.2; kidney, 2.1 ± 0.4; ovary, 1.2 ± 0.3; testis, 0.8 ± 0.2; lymph nodes, 0.8 ± 0.06; and muscle, 0.7 ± 0.02 μg Cu/g wet weight. The following mean values for four newborn calves calculated from the data of Bingley and Dufty (34) are given in μg Cu/g dry weight: liver, 490; heart, 14.9; kidney, 13.8; brain, 9.4; lung, 5.8; muscle, 4.6; and bone, 1.2.

Individual tissues differ greatly in their susceptibility to variations in dietary Cu intakes. The liver, kidney, blood, spleen, lungs, brain, and bones are particularly responsive to such changes, while the endocrine glands, the muscles, and heart are much less so (157, 212, 276). Hennig *et al.* (157) found that a Cu concentration of 9–18 ppm in the brain was normal for cows. Signs of Cu deficiency were apparent in cows with brain Cu levels below 9 ppm. Cows and sheep with brain Cu levels below 9 ppm displayed decreased Cu concentrations in the liver, kidneys, ribs, blood serum, and hair, compared with animals with higher brain Cu levels. The influence of dietary Cu status and other factors on levels in the liver, blood, and hair are considered more fully in Sections 1, 2, 3, and 5 to follow.

Exceptionally high-Cu concentrations occur in the pigmented parts of the eye (38, 39, 315). Differences among species are large, but in all species studied the eye tissues can be placed in a similar descending order of Cu concentration as follows: iris, choroid, vitreous humor, aqueous humor, retina (minus pigmented epithelium), optic nerve, cornea, sclera, and lens. Levels as high as 105 ppm for the iris and 88 ppm (dry basis) for the choroid of the eyes of freshwater trout and 50 and 13.5 ppm for these tissues in sheep's eyes were reported. The copper is associated particularly with the melanins and is largely bound to protein. The role of copper in these sites is not clear.

The Cu concentrations in the inner and outer layers of human dental enamel have been reported as 11.3 and 9.5 μg/g, respectively (249). In a later study (214) of human dental enamel the exceedingly wide range of 0.07–208 μg Cu/g was obtained, with a median value of 0.7 and a mean of 6.8 ± 4.0 μg/g.

2. Copper in the Liver

Liver Cu concentrations vary with the species and age of the animal, the chemical composition of the diet, and in various disease conditions (Table 9). There is no effect of sex on liver Cu concentration, except with the Australian salmon (*Arripis trutta*), in which the female carries higher levels than the male (23). Individual variation is high in all species.

a. Effect of Species. The livers of normal adults of most species contain 10–50 ppm Cu (dry basis), with a high proportion containing 15–30 ppm (23, 213). These levels apply to species as diverse and unrelated in their environments as man, rats, rabbits, cats, dogs, foxes, pigs, kangaroos, whales, snakes, crocodiles, domestic fowls, turkeys, sharks, and sea herring. Sheep, cattle, ducks, frogs, and certain fish by contrast exhibit consistently higher liver Cu levels, with a normal range of 100–400 ppm. These differences cannot be explained on the basis of differing Cu intakes since domestic fowls, turkeys, and ducks consume similar diets, but the first two species carry much lower liver Cu concentrations than ducks. The characteristic species difference between fowls and ducks is maintained even when dietary Cu intakes are raised 2- to 5-fold (25). It seems probable that sheep and cattle have a superior capacity to bind copper in the liver, because blood Cu levels do not rise in these species with increased Cu intakes as they do in rats, except at very high intakes (240).

b. Effect of Age. With most species, including man, liver Cu concentrations are higher in the newborn than in adults. Sheep and cows provide exceptions. In the former they rise continuously from birth and in the latter they change little from birth to old age. The extent of intrauterine Cu storage and the time of maximum Cu concentration in the fetal liver vary with the species. In the rat, rabbit, guinea pig, dog, and man the peak occurs at, or very shortly after, birth (48, 230). In the pig the peak occurs slightly earlier in embryonic life (347),

TABLE 9
The Influence of Species, Age, and Copper Intake on the Concentration of Copper in the Liver

Species	Age and treatment	No. of animals	Copper concentration[a]	Ref.
Man	Newborn (0–7 wk)–normal	–	230	48
Man	Adult–normal diet	–	35	48
Rat	Newborn–normal diet	30	58 ± 4.0	213
Rat	Mature–normal diet	10	9 ± 0.4	213
Rabbit	Newborn–normal diet	30	37 ± 6.7	213
Rabbit	Mature–normal diet	10	23 ± 3.6	213
Pig	Newborn–normal diet	–	233	85
Pig	Mature–normal diet	12	19 (12–48)	85
Sheep	Newborn–normal diet	27	168 (74–430)	85
Sheep	Newborn–Cu-deficient	29	13 (4–34)	85
Sheep	Mature (aged)–normal diet	44	599 (186–1374)	85
Sheep	Mature (aged)–Cu-deficient	35	27 (7–106)	85
Cattle	Newborn–normal	41	381 (143–655)	85
Cattle	Newborn–Cu-deficient	20	55 (8–109)[b]	85
Cattle	Mature–normal	23	200 (23–409)	85
Cattle	Mature–Cu-deficient	41	11.5 (3–32)	85
Domestic fowl	Mature–normal	51	14.8 (10–31)	23
Domestic duck	Mature–normal	34	153 (37–555)	23

[a] ppm on the dry matter.
[b] Much lower levels have been obtained for Cu-deficient calves in Western Australia.

while in the bovine liver Cu concentrations do not rise greatly during gestation (268).

The decline in whole liver Cu concentration that occurs as the rat matures is accompanied by changes in Cu distribution among the subcellular fractions (139). At birth over 80% of the total is present in the nuclear and mitochondrial fractions, while in the adult rat the supernatant contains about one-half the total Cu content of the liver. Evans (110) has similarly shown that in adult mammals, the major portion of the total hepatic copper is located in the cytosol, where the metal is bound to the enzyme superoxide dismutase and a low molecular weight protein similar to metallothionein (109).

A protein subfraction containing more than 4% Cu (neonatal hepatic mitochondrocuprein) has been isolated from the mitochondrial fraction of newborn bovine liver (264) and a similar substance, although of slightly lower Cu content, from normal newborn human liver (265). Significant amounts of this compound are not found in adult human or bovine liver, except in patients with Wilson's disease (289). A copper thionein with a Cu-binding constant four times greater than that of normal subjects has been demonstrated in the livers of such subjects

(111). Neonatal hepatic mitochondrocuprein has a very high cystine content (263) and an amino acid composition similar to that of metallothionein. This led Porter (262) to suggest a relationship between the two compounds. Rupp and Weser (277) have produced evidence that copper thionein is possibly the low molecular weight form of neonatal hepatic mitochondrocuprein.

The forms and distribution of copper in the liver in both rats and ruminants have been further illuminated by the studies of Bremner with animals of varying Cu and Zn status, considered in more detail in Chapter 8. The presence of several low molecular weight Cu- and Zn-binding fractions was demonstrated in rat and ruminant liver cytosol, including a "metallothionein-like" protein. These vary in amounts and metal contents with varying Cu and Zn status of the animal (see Bremner and Davies, 45) and are probably involved in cellular metal detoxification mechanisms. This topic has been admirably reviewed by Bremner (44).

c. Effect of Diet. Liver Cu concentrations are sensitive to low-Cu intakes and provide useful aids in the diagnosis of Cu deficiency. Subnormal liver Cu levels occur in rats and pigs suffering from milk anemia (286, 350), in Cu-deficient rats (104, 240), chicks (276), and dogs (18, 334), and in sheep and cattle grazing Cu-deficient pastures (16, 30, 232). The minimum liver Cu level necessary to maintain a normal plasma Cu level has been estimated to be approximately 40 ppm dry weight in cattle (78a). Bennetts and Beck (30) found the liver copper of five ataxic lambs from Cu-deficient ewes to range from 4 to 8 ppm (d.b.), compared with 120–350 ppm for normal lambs from ewes receiving adequate copper. Supplementary copper administered during pregnancy raises fetal Cu levels to normal, where the mother is Cu deficient, but has little effect on liver Cu storage in the newborn, where the mother is already receiving adequate copper (7).

Ruminants and nonruminant animals differ in their response to high dietary Cu intakes. Dick (100) studied liver copper storage in sheep ingesting graded increments of copper from 3.6 to 33.6 mg/day for 177 days. The liver Cu levels increased steadily from 562 ppm (dry wt) at the lowest to 2340 ppm at the highest Cu intake. The proportion of the ingested copper stored in the liver was uniform at intakes up to 18.6 mg/day and the increase in liver storage was linear. Copper supplementation of the normal diets of rats has no comparable effects on liver copper storage until high intakes are reached. At this threshold, which has been reported as 1 mg/day or 200 ppm Cu of the ration (240), liver Cu levels increase rapidly, apparently due to overloading of the excretory mechanism, but do not reach those given above for sheep.

In studies of copper-loading in the rat the nuclei and mitochondria were found to hold most of the excess copper, with the microsomes and cytoplasm accumulating much less (139). Under different conditions, a linear increase in the amount of copper in each intracellular fraction with the total amount in the liver was observed (240). The relative amount in the mitochondria and the soluble fraction remained essentially constant, while the relative amount in the microsomes increased and in the debris decreased from the Cu-depleted to the Cu-supplemented groups.

Liver copper levels are affected by other dietary factors that influence Cu retention in the body through their effects on Cu absorption, excretion, or both. The storage of copper in the livers of sheep and cattle can be reduced significantly by an increase in dietary molybdenum (102), provided that dietary sulfate intakes are adequate (100, 101). Copper retention in the liver and other tissues is also influenced by the levels of zinc, cadmium, iron, and calcium carbonate in the diet, as discussed later in this chapter and in the appropriate chapters that follow. A marked Zn–Cu antagonism is evident both when copper is limiting (159) and in copper toxicosis (313). A significant inverse correlation between hepatic iron and copper concentrations has been demonstrated in rats (301). Rats consuming an Fe-deficient diet accumulated high liver Cu concentrations in 7–8 weeks, and rats fed a Cu-deficient diet accumulated excessive amounts of iron.

d. Effect of Disease. Abnormally high liver Cu levels are characteristic of a number of diseases in man. These include Mediterranean anemia, hemochromatosis, cirrhosis and yellow atrophy of the liver, tuberculosis, carcinoma, severe chronic diseases accompanied by anemia, and Wilson's disease (hepatolenticular degeneration) (72). In rats, depletion of liver copper has been demonstrated in acute and chronic infections, due to increased hepatic synthesis and secretion of ceruloplasmin by that organ (260), but similar evidence has not been obtained in man (294). Extremely high liver copper levels, as high as 4000 ppm Cu (dry, fat-free basis), can occur in chronic copper poisoning in sheep (132).

3. Copper in Blood

a. Forms and Distribution. Some 60% or more of total red cell copper occurs as a nearly colorless copper protein, generally called erythrocuprein, although the name hemocuprein is often used (291). Erythrocuprein was first isolated from bovine erythrocytes and liver and shown to contain approximately 0.34% Cu and to have a molecular weight of about 35,000 (217). The isolation and identification of erythrocuprein from normal human erythrocytes was later achieved (291), and a series of similar soluble Cu-containing proteins having a molecular weight of 32,000 ± 2000 was isolated from various tissues. These were variously named cerebrocuprein, hepatocuprein, and cytocuprein (see Weser, 341). The identity of these copper proteins from human tissues was established by Carrico and Deutsch (71), and a second metallic component consisting of two atoms of zinc was found to be present, in addition to the two Cu atoms.

The erythrocuprein content of erythrocytes remains constant under a wide range of conditions in man (75). Erythrocuprein functions as a superoxide dismutase (226), i.e., it has the ability to catalyze the dismutation of monovalent superoxide anion radicals into hydrogen peroxide and oxygen as follows:

$$O_2^- + O_2^- + 2H^+ \rightarrow O_2 + H_2O_2$$

An additional function of erythrocuprein, namely, the scavenging of singlet oxygen in metabolism, was later proposed (10). The nonerythrocuprein erythrocyte copper is more loosely bound to protein and is in a more labile form (291).

The copper in plasma also occurs in two main forms, one firmly and one loosely bound. The former consists of the blue copper protein ceruloplasmin, first shown by Holmberg and Laurell (164) to be an α_2-globulin with a molecular weight of 151,000 containing eight atoms of Cu per molecule. More recent studies indicate a molecular weight of 160,000 and seven atoms of Cu per molecule (see Frieden 123). In normal mammals about 90% of the plasma copper exists as ceruloplasmin (72), so that highly significant correlations have been demonstrated between ceruloplasmin levels and plasma, serum, and whole blood copper (227, 321). In chicks (304) and turkeys (343) a much smaller proportion of the normally low level of plasma copper is present as ceruloplasmin.

Ceruloplasmin is a true oxidase (ferroxidase) involved in iron utilization and in promoting the rate of iron saturation of transferrin in the plasma (253). It does not play a significant role in Cu transport because the amount of ceruloplasmin Cu exchanged daily is small compared with the amounts of copper absorbed from the intestinal tract. The albumin-bound, "direct-reacting" copper of the plasma constitutes true transport copper (141). In addition to ceruloplasmin and albumin-bound copper, a small proportion of the plasma Cu exists in combination with amino acids (247) and as the copper enzymes in concentrations that vary with the Cu status of the animal. At high-Mo intakes the plasma Cu also exists, at least in the sheep and guinea pig, in the form of stable Cu–Mo protein compounds (294a), considered further in Chapter 4.

In mammalian blood the Cu concentration is normally higher in the plasma than it is in the erythrocytes (35, 59, 72, 202). The amount of copper in the individual human erythrocyte has been calculated to be 65 ± 10.8 pg and the amount in the individual leukocyte and platelet about one-quarter that figure (201). The numbers of leukocytes and platelets are so much smaller than the number of erythrocytes that they contain an insignificant proportion of the total blood copper. Since plasma Cu is more labile than corpuscular Cu it is a better indicator of the Cu status of an animal than whole blood copper. The marked effect of high doses of molybdenum and sulfate on the level of copper in the blood and on the distribution of this copper in the blood and the plasma is considered later.

b. Normal Blood Copper Levels. The normal range of concentration of copper in the blood of healthy animals can be given as 0.5–1.5 μg/ml, with a high proportion of values lying between 0.8 and 1.2 μg/ml (25). A narrower normal range, with lower mean values approximating one-half those given above for mammals, has been found by Beck (25) for poultry, fish, frogs, and marsupials (Table 10).

Within-species variation can be accounted for more by individual differences than by diurnal or day-to-day variations in man (72) and in sheep (59). Significant breed differences have been demonstrated in sheep (152, 344, 345). Thus Hayter *et al.* (152) found Finnish Landrace sheep to have markedly lower plasm Cu concentrations than merinos–the mean difference after adjustment for other factors varied from 16 to 54 μg/100 ml, with first crosses displaying levels halfway between those of the parental breeds. An association between hemoglobin type and blood Cu concentration in sheep has also been demonstrated (346). Corpuscular Cu levels, as distinct from plasma levels, were less affected by Hb type differences. Plasma copper does not increase after meals or decrease during fasting, and there appears to be no cyclic pattern of variation in man (72). Violent exercise induces a significant increase in whole blood copper in sheep (99). In most species whole blood and plasma Cu levels are similar in males and females, but plasma Cu is slightly higher in human females than in males.

TABLE 10
Copper Concentration in the Blood of Different Species

Species	Age and condition	Mean copper concentration[a]	Ref.
Human	Healthy–adult male	1.10 ± 0.12[c]	248
Human	Healthy–adult female	1.23 ± 0.16[c]	248
Human	Pregnant female at delivery	2.69 ± 0.49[c]	248
Human	Female–late pregnancy	1.92 ± 0.05	147
Ovine	Healthy–mature	1.01 ± 0.96[b]	25
Ovine	Healthy–mature	0.91	86
Bovine	Healthy–mature	0.93	86
Guinea pig	Healthy–mature	0.50 ± 0.006[b]	25
Domestic fowl	Healthy–mature	0.23 ± 0.008[b]	25
Domestic duck	Healthy-mature	0.35 ± 0.007[b]	25

[a]Measured in μg/ml.
[b]Whole blood.
[c]Serum.

Thus Cartwright (72) reported normal values of 105.5 ± 5.03 μg Cu/100 ml for men and 114 ± 4.67 for women. Hambidge and Droegemueller (147) reported 91.6 μg/100 ml for men and 107.4 for women. Serum copper levels are greatly elevated in women taking oral contraceptives. In three separate studies the following values were obtained: 3.0 ± 0.7 μg Cu/ml in 10 such women, compared with 1.18 ± 0.2 in 20 controls (145a); 2.16 μg/Cu/ml, compared with 1.24 μg/ml in control women and 1.07 μg/ml in normal men (79); and 221.4 ± 13.8 μg Cu/100 ml, compared with 107.4 ± 5.1 for female controls (147). This is apparently an estrogen effect since the administration of estradiol in man (181) and stilbestrol in rats (324) significantly increases plasma Cu levels. The use of the copper intrauterine device (IUD) is not associated with a significant increase in plasma Cu concentrations in women, although the Cu levels in the endometrium and cervical mucus increase (314), and decreases in endometrial protein and RNA content have been reported (158).

An estrogen effect is further apparent from plasma Cu changes at sexual maturity and the onset of lay in pullets. Hill (163a) obtained plasma Cu values of 11–15 μg/100 ml in 19-week-old immature pullets. The levels rose to 25–31 μg/100 ml during the following 6 weeks, with no further increase when egg-laying began just after 25 weeks of age.

c. Effect of Pregnancy and Neonatal Growth. Elevated plasma Cu levels coupled with normal erythrocyte Cu levels during human pregnancy are well documented (116, 145a, 147, 248). For example Nielsen (248), in a study of 31 pregnant women, observed serum Cu to increase from the third month to an average of 2.7 μg/ml, compared with nonpregnant levels of 1.2 μg/ml. Hambidge and Droegemueller (147) found the plasma Cu of 20 normal women with uncomplicated pregnancies to average 162.4 ± 6.1 μg Cu/100 ml at 16 weeks and 192.1 ± 5.4 at 38 weeks gestation. These levels returned to normal in the first few weeks postpartum.

In the normal full-term newborn human infant serum Cu and ceruloplasmin levels are about one-third those of the normal adult range (116, 156, 282). Henkin *et al.* (156) found these levels to rise gradually during the first week of life, fall to below adult levels at 2 months of age, rise to within the adult range at 3 months of age, and to rise still higher above the adult range at 8 months of age, at which levels the values persisted throughout the remainder of infancy. It was shown further that these changes related mainly to nondiffusible, i.e., ceruloplasmin copper.

A different pattern of blood Cu distribution from that just described for women and babies is apparent in ewes and lambs. Butler (59) found that ewes maintained on a constant diet exhibited a decline in whole blood, plasma, ceruloplasmin, and erythrocyte Cu during pregnancy. Studies with grazing sheep have revealed a similar decline in pregnancy in the levels of whole blood and plasma Cu (4, 60), and in ceruloplasmin levels (170). Unfortunately, erythrocyte copper was not determined in these latter investigations, so that Butler's finding remains unconfirmed. Howell and co-workers (170) observed blood Cu and ceruloplasmin levels to rise in the ewe at parturition, reaching the highest levels recorded 1 week after lambing. In the lambs, blood Cu and ceruloplasmin levels were low at birth and 24 hr later but were within the normal adult range by 1 week of age. In the bovine, whole blood and plasma Cu levels are lower and erythrocyte Cu levels higher in newborn calves than in their mothers (35). Ceruloplasmin is absent from the serum of baby pigs at birth. It is synthesized during the first 3 days of the life of the piglet and shows no significant difference from that synthesized by the adult (77).

d. Effect of Diet. Subnormal levels of dietary copper are reflected in subnormal blood Cu concentrations in all species studied. In sheep and cattle, values consistently below 0.6 μg Cu/ml in whole blood or plasma are indicative of Cu deficiency. Levels as low as 0.1 μg Cu/ml whole blood have been reported in these species grazing Cu-deficient pastures (30), and 0.2 μg/ml in Cu-deficient pigs (202). Erythrocyte copper was also reduced in the Cu-deficient pigs, although to a lesser extent than in the plasma. In Cu-deficient rats a decline in erythrocyte copper and the ceruloplasmin and direct-reacting copper of the plasma has been observed (240), but no such decline in erythrocyte copper was apparent in this species in the studies of Dreosti and Quicke (104), even in severe copper deficiency. Ceruloplasmin estimations possess advantages over blood or plasma Cu determinations for detecting Cu deficiency because of the relative stability of the enzyme, the technical conveniences of the assay, the smaller serum sample size required, and the avoidance of Cu contamination problems (321).

Elements such as zinc, cadmium, and iron that depress Cu absorption can reduce plasma Cu concentrations when ingested at high dietary levels. The effect of molybdenum and sulfate depends on the status of the animal with respect to these nutrients and copper (138). Under some circumstances Mo can reduce blood Cu levels, or it can maintain them in the face of Cu deficiency (87, 220). At high and prolonged intakes of molybdenum and sulfate striking changes in the levels of copper in the blood and its distribution among the blood components can occur in sheep (33). Sheep on a daily intake of 120 mg Mo and 7.4 g sulfate for 29 months maintained plasma total Cu concentrations at twice the preexperimental level and plasma direct-reacting Cu at 10 times this level. Plasma ceruloplasmin and ultrafilterable Cu levels were not significantly increased, but the Cu concentration in the red cells was reduced to one-tenth the normal level. All the copper in the plasma from the sheep on the high-Mo + SO_4 intakes could be accounted for in terms of direct-reacting, ceruloplasmin, and ultrafilterable Cu. It is apparent that there was a marked redistribution of blood Cu from erythrocytes to direct-reacting Cu in the plasma.

Moderate additions of copper to normal diets have little effect on blood Cu levels. Milne and Weswig (240) observed no increase in plasma Cu in rats when the Cu content of the diet was raised from 10 to 50 ppm, whereas a Cu intake of 100 ppm doubled the plasma concentration from 1.13 to 2.34 μg Cu/ml. Copper intakes of 100 ppm or more are similarly necessary in pigs to produce a significant elevation of plasma copper. At highly toxic intakes, such as 750 ppm Cu, a severe hypercupremia develops in pigs which can largely be prevented by the concurrent administration of 500 ppm zinc (313). Hypercupremia occurs in other species as a consequence of extremely high dietary Cu intakes (86, 107), while during the terminal stages of copper poisoning, i.e., within 24–48 hr of the hemolytic crisis in sheep, blood levels as high as 10–14 μg Cu/ml have been reported (322).

e. Effect of Disease. Elevated plasma Cu levels are characteristic of most acute and chronic infections. Hypocupremia is associated with nephrosis and Wilson's disease (Table 11). Hypocupremia also occurs in kwashiorkor and cystic fibrosis, associated with low protein intakes. In one study the mean plasma Cu of nephrotics was found to be 0.6 ± 0.2 μg/ml, compared with 1.2 μg/ml in normal subjects (73). The mean total serum Cu of 36 patients with Wilson's disease was 0.61 ± 0.21 μg/ml, and that of 205 normal subjects 1.14 ± 0.17 μg/ml. In this disease serum Cu levels are highly positively correlated with ceruloplasmin concentrations. Almost all patients have less than 23 mg ceruloplasmin/100 ml serum, which can be taken as the lower limit of normality (74). However, the correlation between ceruloplasmin activity and the duration and severity of the clinical manifestations is poor, and some individuals without Wilson's disease exhibit abnormally low levels of ceruloplasmin (74). Such low levels of total plasma copper and ceruloplasmin have been demonstrated in patients with Menke's kinky hair syndrome (92). This effect, together with degenerative changes in the brain and arteries, hair changes, and bone lesions, have been shown by Danks and co-workers (90, 91) to be consistent with Cu deficiency arising from an X-linked defect in copper transport from the intestinal mucosal cells to the blood. A primary defect in Cu absorption from the mucosa leading to low serum, brain, and liver copper levels has recently been reported in X-linked mottled mutant mice (175).

Hypercupremia is evident in most acute and chronic infections of man from bacterial and viral organisms and in leukemia, Hodgkin's disease, various anemias, "collagen" disorders, hemochromatosis, and myocardial infarctions (28, 258, 259, 350) (Table 11). The increases in concentration of both serum Cu and its

TABLE 11
Blood Copper in Various Clinical Conditions in Man[a]

Condition	No. of subjects	Whole blood Cu (μg/%)	Plasma Cu (μg/%)	Cell Cu (μg/%)	Volume of packed RBC (ml/100 ml)	Plasma Fe (μg/%)
Normal	63	98 ±13	109 ± 17	115 ± 22	47	115 ±42
Pregnancy	30	169	222	130	37	91
Infection	37	141	167	116	41	57
Acute leukemia	19	195	236	98	27	171
Chronic leukemia	21	119	148	101	39	113
Hodgkin's disease	14	142	171	109	40	78
Pernicious anemia	10	111	121	98	27	173
Aplastic anemia	8	130	152	86	28	203
Iron deficiency anemia						
Adults	9	114	132	109	30	26
Infants	24	155	168	152	28	31
Hemochromatosis	14	103	134	–	–	234
Wilson's disease	3	79	55	110	42	64
Nephrosis	3	70	80	119	44	62

[a]From Wintrobe *et al.* (350).

binding protein develop concomitantly at the onset of symptomatic illness during an infection (28). They are associated with a redistribution of the metal from the liver to the blood, believed to be initiated by a leukocytic endogenous mediator (LEM) which stimulates the liver to synthesize additional quantities of ceruloplasmin (26). Serum copper is also significantly elevated in pellagrins (166.6 ± 7.95 μg/100 ml), compared with normal values (103.9 ± 5.95 μg/100 ml) following treatment of the subjects studied (199). The activity of ceruloplasmin was unaltered in the disease.

Elevated ceruloplasmin levels have been observed by McCosker (228) in several disease conditions in sheep, and infection of chicks with *Salmonella gallinarum* can produce a sixfold increase in ceruloplasmin activity (304). A similar increase was observed with other stressors, including ACTH and hydrocortisone. A 2- to 5-fold increase in ceruloplasmin levels has also been obtained by intravenous injections of endotoxin preparations from three strains of *E. coli* into pathogen-free fowls (61). Starcher and Hill (304) maintain that any stress or any condition resulting in elevated corticosteroid levels could increase ceruloplasmin concentrations. On the other hand, Henkin (155) demonstrated an inverse relationship between plasma cortisol and serum Cu concentration in human patients and cats and a direct relationship between plasma cortisol and urinary excretion of this metal.

4. Copper in Milk

The copper content of milk varies with the species, stage of lactation, and Cu nutriture of the animal. Rat's milk is exceptionally rich in copper, as it is in iron. Values ranging from 2.8 to 3.8 μg/ml from Cu-sufficient and 1.7 to 2.8 μg/ml from Cu-deficient rats have been reported (103). In all species colostrum is substantially richer in copper than milk and there is generally a decline throughout lactation. Beck (21) found the milk of normal ewes to decline from 0.2–0.6 μg Cu/ml in early lactation to 0.04–0.16 μg/ml several months later. Similar values and a comparable decline have been observed in the milk of some cows, but not in others (21, 187). A lactation decline also occurs in mares and women (22, 209). The level in mare's milk fell from a mean of 0.36 μg Cu/ml in the first week of lactation to 0.17 μg/ml several weeks later (22). In two women the level fell from 0.62–0.89 μg Cu/ml in the first few weeks to 0.15–0.17 several months later. Cavell and Widdowson (76) also reported a mean of 0.62 (range 0.51–0.77) μg Cu/ml for the milk of 10 women at the end of the first week of lactation. In two more recent studies of human milk in which the stages of lactation are not given, the following values are reported: for 22 women, 0.24 ± 0.08 μg Cu/ml (244); and for 25 women, 0.11 to 0.34 μg Cu/ml (261). It appears that human milk is appreciably richer in copper than cow's milk. For example, Murthy *et al.* (245) give a range of 0.044 to 0.190 μg Cu/ml and a national average of 0.086 μg Cu/ml for market cow's milk from U.S. cities.

Subnormal Cu levels in the milk of ewes and cows grazing Cu-deficient pastures, with values as low as 0.01–0.02 μg/ml, have been reported by Beck (21). Adding copper to diets already adequate in this element has little effect on the Cu content of the milk of cows, goats (108), and women (243). However, Dunkley *et al.* (105) obtained a substantial elevation of milk copper for at least 4 weeks following subcutaneous injections of the cows with 300 mg Cu as copper glycinate. This was achieved without any increase in the incidence of spontaneous oxidized flavor in the milk, probably because of the small amount of copper associated with the milk fat in early lactation. After 2–4 weeks of lactation only about 15% of the copper in cow's milk is associated with the fat, whereas after 15 weeks the proportion so associated rises to some 35% (188).

5. Copper in Hair and Wool

The level of copper in the hair of man and other species has been studied as a potential index of the Cu status of individuals or groups. High individual variation, differences in sample preparation, and exogenous contamination have limited the usefulness of hair Cu determinations for this purpose in the past. For example, Hambidge (146) found that the proximal sections, ranging in length from 1 to 5 cm, of the hair of 27 healthy human subjects averaged 11.8 ppm, while the most distal sections averaged 20.7 ppm ($p < 0.005$). The higher mean concentration of that part of the hair shaft that was exposed to the external environment for the longest duration suggests that exogenous copper contributes to the hair content of this element. Hambidge maintains that "interpretation of analytical data on hair copper requires great caution" and that "analyses should be limited to recently grown hair within 1–2 cm of the scalp."

From the results of a study of human hair as a biopsy material for the assessment of Cu status, Klevay (192) concluded that only age and sex-matched individuals or groups should be compared. He observed a fall in the Cu content of hair in the first decade of life, while in older children and adults female hair contained more copper than male hair. Reinhold and co-workers (271, 272) found that the Cu content of human hair does not rise significantly with age, as occurs in the hair of rats. Earlier claims that the Cu concentration of pigmented hair is higher than that of unpigmented hair are not supported by later studies (89, 134, 204). It seems also that kwashiorkor is not necessarily accompanied by subnormal hair Cu levels (204). The Cu content of the hair of pregnant women remains within normal limits (11–12 μg g dry wt), despite significantly elevated plasma Cu concentrations (147). High hair Cu levels have been reported in women taking oral contraceptives (46).

In three studies of cattle hair, average values of 7.8 (9), 10 (332), and 9 ppm Cu (89) were reported. Van Koetsveld (332) observed that levels below 8 ppm Cu were associated with signs of Cu deficiency in the cattle, but Cunningham and Hogan (89) found no relationship between the Cu content of the hair and the levels of copper in the diet or the liver. A reduced Cu content was apparent with increased dietary molybdenum. In a further study with rats (103) there was no significant decline of hair copper in Cu deficiency.

The copper concentration of wool appears to be extremely variable. Burns *et al.* (54) reported a mean level of 25 ppm Cu for wool from 50 washed fleeces, with a coefficient of variation of 100%. Cunningham and Hogan (89) obtained levels ranging from 8.3 to 13.3 ppm and Healy *et al.* (153) 42–147 ppm Cu for wool from small groups of New Zealand sheep. By contrast Bingley (33) found the washed wool of two housed experimental sheep on a good diet to contain only about 2–4 ppm Cu (d.b.), and these levels were reduced to less than one-third when the animals were liberally dosed with molybdate and sulfate. It seems that the level of copper in wool is highly dependent on the nature of the diet. Contamination is probably a further factor contributing to the reported variability.

6. Copper in the Avian Egg

The total copper content of the average hen's egg is about 30 mg. In a recent study of eggs from hens on good practical rations a mean of 31.4 ± 0.82 mg Cu was obtained (98). The extent to which this amount is affected by varying dietary Cu intakes has apparently not been studied, although Hill (163a) found that the ^{64}Cu level of egg yolk was positively correlated with plasma ^{64}Cu level in hens following ^{64}Cu intramuscular injections. This worker also reported considerable deposition of the copper in the egg white. The Cu concentration in the yolk was about two-thirds and in the white one-third the total Cu concentration in the whole egg.

II. COPPER METABOLISM

1. Absorption

Copper is absorbed from the stomach and all portions of the small intestine, particularly the upper small intestine (255, 323, 331). In the sheep considerable net absorption of copper takes place in the large intestine (135). In most species dietary copper is poorly absorbed (80). The extent of this absorption is influenced by the amount and chemical form of the copper ingested, by the dietary level of several other metal ions and organic substances, and by the age of the animal. For example, mature sheep normally utilize less than 10% of the copper they ingest, while young lambs prior to weaning utilize four–seven times this proportion of dietary copper (308a).

Limited evidence exists that Cu absorption from the intestine is regulated in accordance with bodily Cu needs, at least at low dietary Cu levels. In rats and mice Cu absorption from the intestine is higher in Cu-depleted than in Cu-adequate animals and, this absorption is higher at low than at higher Cu doses (84, 131, 287, 329). Using isolated rat jejunum segments Schwarz and Kirchgessner (287) demonstrated further that both Cu uptake by the intestinal wall and its transfer to the serosal solution were elevated in Cu deficiency. At higher dose levels, absorption seems to parallel the dose for a wide range of dose sizes until a point is reached, probably corresponding to tissue saturation, when further increase in Cu intake results in no further absorption (329).

Copper occurs in foods in many chemical forms and combinations which affect its availability to the animal. Many years ago, it was shown that anemic rats are unable to utilize the copper of copper sulfide or copper porphyrin for hemoglobin synthesis, whereas the oxide, hydroxide, iodide, glutamate, glycerophosphate, aspartate, citrate, and pyrophosphate were well utilized for this purpose (286). Pigs also absorb the copper of cupric sulfide much less efficiently than that of copper sulfate (37). Lassiter and Bell (203) studied the availability to sheep of various compounds labeled with ^{64}Cu. The copper in copper wire was largely unavailable and the copper in copper oxides less available than that in water-soluble salts or the carbonate. Later Chapman and Bell (78) tested the uptake of ^{64}Cu from several inorganic compounds by beef cattle. The relative appearance of ^{64}Cu in the blood was in the following order: $CuCO_3$ > $Cu(NO_3)_2$ > $CuSO_4$ > $CuCl_2$ > Cu_2O > CuO (powder) > CuO (needles) > Cu (wire).

Changes in the chemical forms of copper in plants affecting availability must occur because fresh green herbage is less effective in promoting body copper stores than hay or dried herbage of equivalent total Cu content (151). Apparently changes in the chemical forms of the copper occur during the curing or drying process which improve their absorption. Mills (235, 236) found that fresh herbage contains the greater part of its copper as neutral or anionic complexes. When fed to Cu-deficient rats these complexes induced a more rapid response and greater liver Cu storage than equivalent amounts of copper as copper sulfate. Mills therefore suggested that copper may be transported through the intestinal mucosa both as ionic Cu and in the form of complexes such as those encountered in herbage. This suggestion received support from the work of Kirchgessner and co-workers (189, 190), who showed that the affinity of Cu ions for inorganic and organic ligands in the diet can reduce the rate of absorption, depending on the size and stability of the resulting complexes. Small stable Cu complexes may be superior in absorbability to copper sulfate. With single amino acids as ligands the rate of Cu absorption depends on the type of amino acid, its configuration, and the degree of polymerization. Differences of this type may provide at least a partial explanation of the variations in copper availability arising from diets containing different amounts and types of protein (215, 313).

Phytates form very stable complexes with copper (335) and reduce the assimilation of this element (96). High levels of ascorbic acid significantly depress ^{64}Cu absorption when placed into a ligated segment of rat intestine along with the radio-copper. Whole body retention studies indicate that the depressing effect of ascorbic acid on Cu retention is achieved primarily by reduced intestinal absorption (330). High dietary ascorbic acid has been shown to increase the severity of Cu deficiency in chicks (66, 160) and rabbits (173).

Several inorganic dietary factors markedly affect Cu absorption, retention, and distribution within the body. These include particularly calcium, cadmium, zinc, iron, lead, silver, and molybdenum plus sulfur. Less copper was absorbed by mice from a high-Ca than from a low-Ca diet due to an increase in intestinal pH (323), and diets high in calcium enhance copper toxicity in pigs, presumably due to a lowering of zinc availability (313). Similarly, calcium can alleviate meat anemia in mice through increasing the availability of copper by reducing the Zn–Cu antagonism. The metabolic antagonism at the absorptive level between copper and iron was considered further in Chapter 2, between copper and molybdenum plus sulfur in Chapter 4, and between copper and zinc and cadmium in Chapters 8 and 9, respectively. In the present context it is important to appreciate that high dietary levels of cadmium depress Cu uptake, and even a

TABLE 12
Effects of Dietary Supplements of Cadmium and Zinc on Plasma Ceruloplasmin Activity (U/Liter) in Rats Maintained for 9 Weeks on Diets Providing 2.6 mg Cu/kg Dry Matter[a]

	Cadmium (mg/kg) diet[b]			
Zinc (mg/kg)	0.16	1.5	6.1	18
30	43.5	18.2	9.3[d]	6.5
300	27.4[c]	17.0[d]	5.2[d]	4.9
1000	9.5[d]	5.0[d]	5.1[d]	3.7

[a]From Campbell and Mills (63).
[b]Ceruloplasmin activity of all groups is significantly lower than that of control group (30 mg Zn and 0.16 mg Cd/kg).
[c]$p < 0.01$.
[d]$p < 0.001$.

relatively small increase in cadmium intake can adversely affect Cu metabolism when Cu intakes are marginal. In fact Campbell and Mills (63) showed that a dietary Cd intake at the rate of 6.1 mg/kg was as effective in reducing plasma ceruloplasmin activity over a 9-week period in rats as 1000 mg/kg of dietary zinc (Table 12).

The relation of silver and mercury to copper uptake and utilization is less clear than is the relation of zinc, cadmium, and iron. Hill and co-workers (161) reported that silver, but not mercury, accentuated the effects of Cu deficiency in chicks, while mercury had an adverse effect on Cu-adequate chicks. Van Campen (328), on the other hand, found that mercury produced a moderate, but not significant, lowering of ^{64}Cu uptake from the intestine of the rat, whereas silver had very little effect. Each of these elements induced significant and different changes in the distribution of the retained Cu among the tissues of the body. Further evidence of a metabolic interaction between silver and copper has been presented by Jensen and co-workers (180a). These workers found that the toxic effects produced by adding 900 ppm Ag to practical turkey poult diets was completely prevented by 50 ppm of supplementary copper.

The mechanisms which regulate Cu absorption are little understood, although it now seems clear that metal-binding components are involved and that the inhibition of Cu absorption brought about by various metals, as described previously, results from competition for protein metal-binding sites. In 1969 Starcher (303) identified a single metal-binding protein in chick duodenum and demonstrated that the protein would bind copper, as well as cadmium and zinc.

Subsequently Evans and Hahn (112) found that orally administered copper becomes associated with a variety of metal-binding ligands and macromolecules in the intestine of the rat. In the intestinal lumen copper was complexed with a protein similar to metallothionein, but whether this compound is actually involved in copper transport from the intestine to the blood is not yet clear. The results of investigations into Menke's kinky hair disease in children (91), and a mouse mutant with defective Cu metabolism (175), indicate the existence of two different mechanisms, one involving Cu transport from the intestinal lumen into the mucosal cells and one from the mucosal cells to the plasma. In these two conditions copper readily enters the mucosal cells where it builds up, but enters the bloodstream in inadequate quantities due to an undefined genetic defect or defects in this phase of Cu absorption.

2. Intermediary Metabolism

Copper entering the blood plasma from the intestine becomes loosely bound to serum albumin and amino acids, in which forms it is widely distributed to the tissues and can pass readily into the erythrocytes (58). This serum Cu pool also receives copper from the tissues. The copper in ceruloplasmin is not so readily available for exchange or transfer, and has a turnover time too slow to permit ceruloplasmin to be the primary transport form (306).

The copper reaching the liver, the main storage organ of the body for copper and a key organ in the metabolism of this element, is incorporated into the mitochondria, microsomes, nuclei, and soluble fraction of the parenchymal cells in proportions which vary with the age (139, 266), strain (317), and the Cu status of the animal (139, 240). The copper is either stored in these sites or released for incorporation into erythrocuprein, ceruloplasmin, and the numerous Cu-containing enzymes of the cells. Ceruloplasmin is synthesized in the liver (218) and secreted into the blood serum. Erythrocuprein is synthesized in normoblasts in the bone marrow (58). The liver also provides the major pathway of copper excretion via the bile.

The profound changes in serum and liver Cu levels that can occur in infections and various disease states and that result from hormonal influences were discussed earlier. Changes in copper excretion through such influences are considered in the following section.

3. Excretion

In all species studied a high proportion of ingested copper appears in the feces. Most of this is normally unabsorbed copper but active excretion occurs via the bile. Cartwright and Wintrobe (75) have estimated that of the 2–5 mg Cu ingested daily by man, 0.6–1.6 mg (32%) is absorbed, 0.5–1.3 mg is excreted in the bile, 0.1–0.3 mg passes directly through the bowel, and 0.01–0.06 mg appears in the urine. The biliary system is also the major pathway of Cu excretion in pigs and dogs (37, 216), mice (131), and poultry (25). Intravenous injection of copper, resulting in elevated blood and tissue Cu levels, increases Cu excretion in the bile and hence the feces but does not normally raise urinary Cu output (216). In patients with liver cirrhosis accompanied by biliary obstruction urinary Cu excretion is increased, but this does not occur in Laennec's cirrhosis without significant biliary obstruction (20). In Wilson's disease bile Cu concentrations are not greatly raised, despite the presence of a markedly elevated hepatic Cu pool (57).

Reported values for human urinary Cu concentrations, and for the small amounts excreted daily in the urine, are extremely variable. Thus Giorgio *et al.* (129) estimated the mean urinary excretion of 20 normal adults to be 21 ± 5.2 μg Cu/day. This value compares favorably with the mean level of 30 μg Cu/day reported much earlier by Van Ravesteyn (333), but is much lower than the 60 μg Cu/day given for the average urinary output of adult man given by Schroeder and co-workers (283). In Wilson's disease urinary levels may reach 1500 μg Cu/day. In nephrosis the excess urinary copper occurs as a nondialyzable Cu–protein complex, partly as ceruloplasmin. The amount of urinary Cu in nephrotics can be correlated with the high urinary excretion of zinc, iron, and protein resulting from the firm binding of the metals to protein in the plasma.

Changes in Cu metabolism occur in patients with abnormalities of adrenocorticosteroid metabolism. Henkin (155) showed that patients with untreated adrenal cortical insufficiency (ACI) exhibit significantly higher serum Cu (and Zn) concentrations, accompanied by significantly lower urinary Cu (and Zn) excretion. Both of these returned to normal following hormonal replacement with ACTH. In normal volunteers given ACTH the reverse situation was demonstrated, i.e., reduced serum Cu concentrations and increased urinary Cu excretion. Increased serum and biliary copper with decreased urinary Cu excretion were observed in adrenalectomized and hypophysectomized cats, and again the situation was reversed with hormone therapy. A direct relationship between plasma cortisol and urinary Cu excretion is apparent from these studies.

Negligible amounts of copper are lost in the sweat (150) and comparatively small amounts in the normal menstrual flow. An average of 0.5 mg Cu per period has been estimated (210). This is slightly less than 0.02 (0.5 × 13/365) mg/day, or only one twenty-fifth or less of the average loss of iron per day through menstruation. Since the average intake of dietary copper is close to one-fifth that of iron, it is clear that the menstruating woman is in a much more favorable position with respect to copper than iron. The loss in the milk at the height of human lactation is approximately 0.4 mg Cu/day, but there is no evidence that this imposes any nutritional stress.

III. COPPER DEFICIENCY AND FUNCTIONS

The manifestations of copper deficiency vary with the age, sex, and species of the animal and with the severity and duration of the deficiency. As the copper available to the animal becomes insufficient for all the metabolic processes involving copper as a result of inadequate intake, depletion of body reserves, or interaction with metabolic antagonists, certain of the processes fail in the competition for the inadequate supply. In the sheep the processes of pigmentation and keratinization of wool are the first to be affected by a lowered Cu status, so that at certain levels of Cu intake these defects can develop without any other signs of Cu deficiency being apparent. Neonatal ataxia occurs readily in lambs from Cu-deficient ewes in some areas but occurs only rarely in calves in the same areas. Further, the cardiovascular lesions affecting the major blood vessels in Cu-deficient pigs and chicks have not been observed in Cu-deficient cattle. Acceptable biochemical explanations of these differences in the expressions of Cu deficiency have yet to emerge.

1. Anemia and Iron Metabolism

Anemia is a common expression of Cu deficiency in all species where the deficiency is severe or prolonged. A blood level of 0.10–0.12 μg Cu/ml limits hematopoiesis in the sheep (21, 222), and 0.2 μg Cu/ml is suggested as the minimum level at which hematopoiesis can take place in the pig (202). If such low levels are maintained for any length of time anemia is inevitable. In rats, rabbits, and pigs this anemia is hypochromic and and microcytic (202, 299, 300). In lambs the anemia is also hypochromic and microcytic (30), but in ewes and cattle it is hypochromic and macrocytic (30, 86). In Cu-deficient chicks (224) and dogs (18) the anemia has been reported as normocytic and normochromic.

The bone marrow does not undergo a normoblastic hyperplasia in the dog as it does in the Cu-deficient pig, and maturation of the erythrocytes is defective (18). Pigs and dogs with Cu deficiency anemia contain a lower percentage of reticulocytes in their blood than those with Fe deficiency anemia (18, 202), and the addition of copper to the diet of rats, rabbits, and pigs suffering from a combined Fe and Cu deficiency anemia elicits a marked and persistent reticulocyte response, whereas iron has little effect (202, 299, 300). In addition, the survival time of erythrocytes is shorter than normal in the Cu-deficient pig (56). Copper is therefore an essential component of adult red cells, and a certain minimum of copper must be available for their production and for the maintenance of their integrity in the circulation. Copper does not appear to be involved in the heme biosynthetic pathway. In fact as the anemia developed in Cu-deficient pigs, Lee *et al.* (205) observed an increase in the activity of the heme biosynthetic enzymes, indicating that Cu could not be a cofactor in the enzymes studied.

In 1952 Gubler and co-workers (142) produced evidence that Cu-deficient pigs have an impaired ability to absorb iron, mobilize it from the tissues, and utilize it in hemoglobin synthesis. These findings were confirmed and extended to include the demonstration of increased amounts of stainable iron in sections of the duodenal mucosa and impairment of iron transfer to the plasma (206). When the iron was given intramuscularly to the Cu-deficient pigs, increased amounts were found in the reticuloendothelial system, the hepatic parenchymal cells, and the normoblasts. The release of iron to the plasma from these sources is dependent on the presence of adequate amounts of ceruloplasmin. This Cu compound was shown by Osaki *et al.* (253) to greatly accelerate the oxidation of ferrous iron (hence the suggested name "ferroxidase") and to be rate-determining in the formation of Fe^{3+} transferrin. In the absence of ferroxidase the rate of iron oxidation in the plasma is not adequate to meet the Fe^{3+} transferrin demands of the erythropoietic tissues. The metabolism of iron as affected by Cu deficiency and ceruloplasmin (Cp) is summarized by Frieden (123) as follows: "Iron absorption by the mucosal cells appears to be unaffected in copper deficiency but release of iron into the plasma is impaired. The release of iron from the liver parenchymal cells is also reduced in copper deficiency. The most significant pathway, quantitatively, is the mobilization of iron from the R-E system into the plasma. This is lowered in copper deficiency and can be restored rapidly by injection of Cp, or more slowly, by copper administration. The formation of Fe(III)-transferrin is dependent on Cp. The iron in the Fe(III)-transferrin is contributed directly to the developing reticulocyte in the bone-marrow. When the red cell has run its course, it is scavenged by the R-E cells, principally in the spleen and the liver, and the iron cycle is renewed. Cp can affect the mobilization of iron, particularly from R-E and/or liver systems." The critical influence of copper, through ceruloplasmin, in mobilizing absorbed iron for hemoglobin synthesis has been clearly demonstrated by Evans and Abraham (113) in Cu-depleted and Cu-repleted rats.

2. Bone Disorders

Skeletal abnormalities occur in Cu-deficient rabbits (174), chicks (68, 125, 276), pigs (202, 316), dogs (18, 334), and foals (29). Poor mineralization of bones responsive to Cu supplementation also occurs in mice fed a meat diet (143, 144). In young dogs rendered severely Cu-deficient, a gross bone disorder develops with fractures and severe deformities (18, 334). The skeletal changes are a specific effect of the Cu deficiency unrelated to the concurrent anemia. The histological changes in affected bones and in those of Cu-deficient pigs (122) were thinned cortices, broadened epiphyseal cartilage, and a low level of osteoblastic activity. The bones were normal in ash, Ca, P, and CO_2 contents (18). The ash content of the bones of Cu-deficient chicks and the Ca, P, and Mg contents of this ash were similarly found to be normal (276). The fragile bones from Cu-deficient chicks fracture with less deformation and torque than bone from Cu-deficient controls, probably as a result of decreased bone collagen cross-linking (276).

Gross skeletal abnormalities have not been observed in growing rats (238) and rabbits (173) given Cu-deficient diets and showing other signs of Cu deficiency, and bone defects are not a conspicuous feature in ruminants (166). A low incidence of spontaneous bone fractures occurs in sheep and cattle grazing Cu-deficient pastures (87, 94). The fractured bones were reported to be almost normal in appearance and shaft thickness and to exhibit a mild degree of osteoporosis. More recently, definite evidence has been obtained of osteoporosis in bones from 10-week-old lambs which were born of and had suckled ewes given Cu-deficient diets for 2 years (310). The lesion was most severe in the central metaphyseal region and was consistent with a matrix osteoporosis arising from a reduction in or cessation of osteoblastic activity. It differed from that described in the Cu-deficient pig, chick, and dog in that the epiphyseal cartilage was not thickened and there were no gross bone deformities. The authors concluded that osteoblastic activity is one of the first functions to be impaired in lambs born of Cu-depleted ewes. The osteoblast is apparently more sensitive to Cu deficiency in the fetal and neonatal lamb than in older Cu-deficient lambs, since Suttle and co-workers (310) reported no evidence of osteoporosis or depressed osteoblastic activity in weaned lambs rendered hypocupremic at 5 months of age and Cu depleted for 8 months further. Other signs of Cu deficiency, including loss of wool crimp, were apparent in these lambs.

The primary biochemical lesion in the bones of Cu-deficient animals is probably a reduction in the activity of the copper enzyme amine or lysyl oxidase, leading to diminished stability and strength of bone collagen as a result of impaired cross-linkage of their polypeptide chains, in a manner analogous to the effect on aortic elastin described in Section III, 7. A marked reduction in amine oxidase activity occurs in the bones of Cu-deficient chicks, and collagen extracted from such bones is more easily solubilized than collagen from control bones (276). Since the solubility of collagen, like elastin, is inversely related to the degree of cross-linking (55), there seems little doubt that the structural integrity of the collagen is impaired through a failure of establishment of the required cross-linkages.

3. Neonatal Ataxia

A nervous disorder of lambs characterized by uncoordination of movement has long been recognized in various parts of the world and given local names such as "swayback" and "lamkruis." All these conditions are pathologically identical and the term enzootic neonatal ataxia can properly be applied to them. In 1937 Bennetts and Chapman (31) showed that the ataxia of lambs occurring in Western Australia was associated with subnormal levels of copper in the pastures and in the blood and tissues of both ewes and affected lambs, and could be prevented by copper supplementation of the ewe during pregnancy (30). Subsequent investigations in several countries confirmed the efficacy of Cu supplements to the ewe in preventing the ataxia (4). In some areas the incidence of ataxia could not always be explained in terms of a simple dietary Cu deficiency (177, 290). Thus in the swayback areas of England the Cu content of the pastures can be in the normal range, and in Scotland no correlation was found between the Cu status of the animal and the severity of the ataxia (16), although many of the reported values for copper in the pastures, blood, and tissues of affected lambs and their mothers are close to the Cu deficiency levels observed in Australia. The possibility that swayback is a Cu deficiency induced by molybdenum received little support from the study of Allcroft and Lewis (5), but more recently Alloway (8), in an investigation of Cu and Mo in swayback pastures, has produced evidence which suggests that in some cases a Mo-induced hypocuprosis may be a contributory factor in the incidence of the disease. Under experimental conditions neonatal ataxia has been produced by maintaining ewes of low-Cu status on high molybdate and sulfate diets (119, 311) and by feeding stall-fed ewes a semipurified Cu-deficient diet, adequate and "normal" in other respects (211).

Above-normal lead intakes may sometimes be involved in the incidence of swayback, as suspected earlier (177), especially since recent studies with rats have demonstrated a reciprocal antagonism between Pb and Cu at the absorptive level (191).

Neonatal ataxia in Cu-deficient goats (256, 285) and lesions in the CNS (central nervous system) in Cu-deficient newborn guinea pigs (115) have been observed. Ataxia was previously observed in young pigs, associated with low liver Cu levels and "demyelination of all areas of the spinal cord except the dorsal areas" (182, 348), and a subclinical myelopathy resembling that of swayback in lambs has been observed in Cu-deficient miniature swine (64). Demyelination of the spinal cord, extending into the medulla and cerebellum in young pigs, with ataxia and subnormal liver Cu concentrations have been described (64, 231), and Carlton and Kelly (67) reported neural lesions in the offspring of female rats fed a Cu-deficient diet, accompanied by behavioral disturbances in some animals.

Two types of ataxia occur in lambs under field conditions: the neonatal form in which the lambs are affected when born, and a delayed type in which signs may not appear for some weeks. Uncoordinated movements of the hind limbs, a staggering gait, and swaying of the hind quarters are evident as the disease develops. Some of the lambs affected at birth are unable to rise and soon die.

Neonatal ataxia (swayback) in lambs was initially characterized as a demyelinating encephalopathy, with cavitation of the white matter of the cerebral hemispheres leading to collapse of these structures (177). Barlow *et al.* (16) found no cerebral lesions in half the ataxic lambs examined, whereas cell necrosis and nerve fiber degeneration occurred in the brainstem and spinal cord of all the affected animals. The pathological criteria were given as "cavitation or gelatenous lesions of the cerebral white matter and/or a characteristic picture of chromatolysis, neurone necrosis and myelin degeneration in the brain stem and spinal cord." The concept that the significant changes of swayback are those of necrosis and degeneration of neurons accompanied by changes in spinal cord nerve fibers has been supported by Howell (165). Later investigations always revealed chromatolysis and degenerative changes in the large nerve cells of the brainstem and spinal cord of swayback sheep (13–15). Mills and co-workers (119, 239) found the ataxia to be associated with degeneration of the nuclei of the large motor neurons of the red nucleus in the brainstem. Howell *et al.* (169) also observed changes in the neurons of the brain and spinal cord of ataxic lambs but products of degenerating myelin were not detected. It was suggested that the lesion in the white matter of the spinal cord may be one of myelin aplasia.

Myelin aplasia, rather than myelin degeneration, is compatible with the biochemical evidence from ataxic lambs. The primary lesion in this condition is now generally accepted as a low-Cu content in the brain leading to a deficiency of cytochrome oxidase, the Cu-containing terminal respiratory enzyme, in the motor neurons (118). Howell and Davison (168) first demonstrated a significant lowering of Cu content and cytochrome oxidase in the brain of swayback lambs. Mills and Williams (239) also found a lower cytochrome oxidase activity in the brainstem of swayback than normal lambs and provided evidence that brain Cu levels are of greater significance to the integrity and function of the CNS than are liver Cu levels. None of the 18 ataxic lambs they examined had a brain content as high as 3 μg Cu/g dry weight, whereas of the 37 animals in the normal group only two had less than 3 μg Cu/g. The greatest reduction in cytochrome oxidase activity occurs in those groups of nerve cells, i.e., the large motor neurons of the red nucleus and ventral horns of the gray matter in the spinal cord, which show the morphological lesions of the disease (13, 118).

In studies of Cu deficiency in rats, Gallagher *et al.* (126) demonstrated depressed cytochrome oxidase activity in the brain, as well as in the heart and liver, together with a depression in phospholipid synthesis. It was concluded that the loss of cytochrome oxidase activity results from a failure of synthesis of its prosthetic group, heme, and that this must be regarded as a basic function of copper. Positive linear and/or quadratic responses in brain, liver, and tibia ossification center cytochrome oxidase activities were later demonstrated by Evans and Brar (114) when graded increments of dietary copper were given to rats receiving a Cu-low diet.

Gallagher and Reeve (127) have reported a direct causal relationship between the loss of cytochrome oxidase and impaired phospholipid synthesis. Liver mitochondria were found to be unable to synthesize phospholipids significantly under anaerobic conditions with or without added electron transfer systems unless ATP was added. It appears that the loss of cytochrome oxidase in Cu deficiency leads to depressed phospholipid synthesis by liver mitochondria, by interfering with the provision of endogenous ATP to maintain an optimal rate of phospholipid synthesis. On this basis the particular sensitivity of the lamb to neonatal ataxia could be explained in the following terms: at the critical period in late gestation when myelin is being laid down most rapidly in the fetal lamb, Cu deficiency causes a depression in cytochrome oxidase activity which leads to inhibition of aerobic metabolism and phospholipid synthesis. This leads to inhibition of myelin synthesis, since myelin is composed largely of phospholipids.

4. Pigmentation of Hair and Wool

Achromotrichia is one of the manifestations of Cu deficiency in rats, rabbits, guinea pigs, cats, dogs, cattle, and sheep, but has not been observed in the pig. Lack of pigmentation in the fur of the rabbit (298) and in the wool of sheep is a more sensitive index of Cu deficiency than anemia. The pigmentation process is so susceptible to changes in the Cu status of the sheep that once a Cu-deficient condition has been established, alternating bands of pigmented and unpigmented wool fibers can be produced accordingly as copper is added to or withheld from the diet (Fig. 2). Even on fairly high-Cu intakes it is possible to block the functioning of copper in the pigmentation process within two days by raising the Mo and sulfate intakes sufficiently. Dick (99) suggests that this effect may take place within hours of giving the first dose of these compounds. A breakdown in the conversion of tyrosine to melanin is the probable explanation of this failure of pigmentation, since this conversion is catalyzed by Cu-containing polyphenyloxidases (269).

5. Impaired Keratinization

Changes in the growth and appearance of hair, fur, and wool have been noted in Cu-deficient rats, rabbits, guinea pigs, dogs, cattle, and sheep, in mutant mice, and in Menke's kinky hair syndrome in children arising from a genetic defect in Cu metabolism.

A reduction in the quantity and quality of wool produced by sheep in Cu-deficient areas was noted early in the Australian investigations (29, 223). The lowered wool weights are probably an expression of an inadequate supply of substrate to the wool follicles, consequent on a reduced feed intake by the

Fig. 2 Wool from black-wooled sheep showing unpigmented bands when the animals were suffering from copper deficiency. (Photo courtesy of H.J. Lee.)

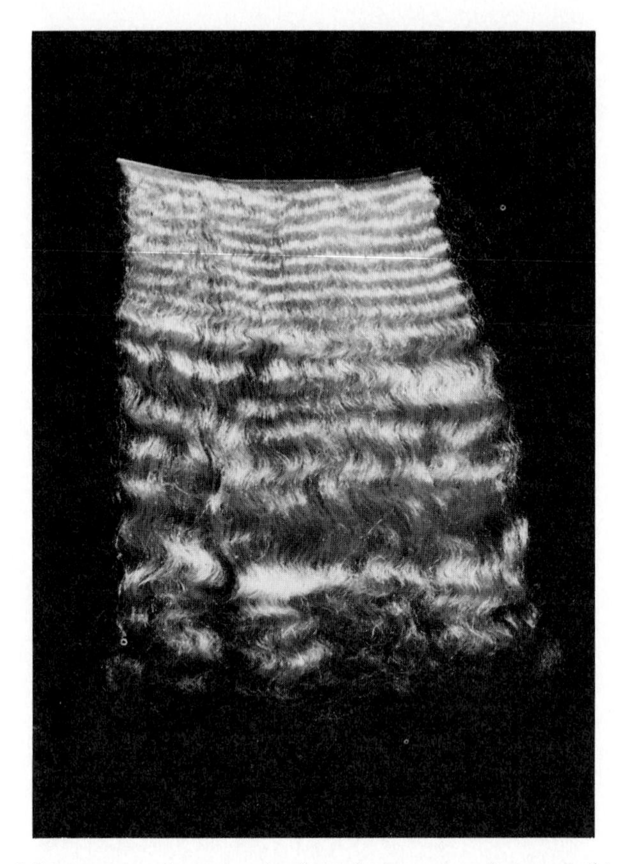

Fig. 3 Wool from merino sheep showing loss of crimp when diet was lacking in copper, and restoration of crimp with the provision of supplemental copper. (Photo courtesy of H.J. Lee.)

Cu-deficient animals. The deterioration in the process of keratinization, signified by the failure to impart crimp to the fibers, is a specific effect of Cu deficiency.

As the sheep's reserves of copper are depleted and its blood Cu level falls, the crimp in the wool becomes less distinct in the newly grown staple, until the fibers emerge as almost straight hairlike growths, to which the descriptive terms "stringy" or "steely" wool have been given (Fig. 3). The tensile strength of steely wool is reduced, the elastic properties are abnormal, and it tends to set permanently when stretched (219). A spectacular restoration of the crimp and physical properties of the wool occurs when Cu supplements are given. These abnormalities are more obvious in merino wool, which is normally heavily crimped, than in British breed wool. However, Lee (207) has observed fleece abnormalities in four British breeds comparable with those encountered in

Cu-deficient merinos. Furthermore, loss of wool crimp has been reported in crossbred ewes as an early indication of Cu deficiency (312).

The characteristic physical properties of wool, including crimp, are dependent on the presence of disulfide groups that provide the cross-linkages or bonding of keratin and on the alignment or orientation of the long-chain keratin fibrillae in the fiber. Both of these are adversely affected in Cu deficiency (219). Straight steely wool has more sulfhydryl groups and fewer disulfide groups than normal (219). An essentially similar pattern has been demonstrated in the hair of patients with Menke's kinky hair syndrome (90). It seems therefore that copper is required for the formation or incorporation of disulfide groups in keratin synthesis. Furthermore, wool from Cu-deficient sheep contains more *N*-terminal glycine and alanine, and sometimes more *N*-terminal serine and glutamic acid, than normal wool, indicating that a lack of copper can interfere with the arrangement of the polypeptide chain in keratin synthesis (53).

The phenotypic similarities, especially of hair growth, between mice homozygous for the autosomal recessive mutant gene *crinkled* (*cr*) and Cu-deficient animals led Hurley and Bell (176) to examine the effect of Cu supplementation during pregnancy, and lactation on the expression of the gene in homozygous mutants. The feeding of a high-Cu diet (500 ppm Cu) during this time was found to double the survival of the mutant mice to 30 days of age, prevent their characteristic lag in pigmentation, and produce near normal skin and hair development. The interaction of a gene and a trace metal in development is strikingly illustrated by this finding.

6. Infertility

In female rats and guinea pigs, Cu deficiency results in reproductive failure due to fetal death and resorption (106, 145, 171, 172). The estrous cycles remain unaffected and conception appears to be uninhibited. Normal fetal development in Cu-deficient rats has been shown to cease on the thirteenth day of pregnancy when the fetal tissues were disintegrating (171). Necrosis of the placenta became apparent on the fifteenth day of gestation.

Hens fed a severely Cu-deficient diet (0.7–0.9 ppm Cu) for 20 weeks displayed reduced egg production and subnormal levels of copper in the plasma, liver, and eggs, while hatchability dropped rapidly and approached zero in 14 weeks (279a). The embryos from these hens exhibited anemia, retarded development, a high incidence of hemorrhage after 72–96 hr of incubation, and a reduction in monamine oxidase activity. The anemia, hemorrhages, and mortality are probably due to a defect in red cell and connective tissue formation during early embryonic development.

Low fertility in cattle grazing Cu-deficient pastures, associated with delayed or depressed estrus, has been observed in several areas (6, 32, 332), and infertility, associated in some cases with aborted small dead fetuses, has been demonstrated in experimental Cu deficiency in ewes (172).

7. Cardiovascular Disorders

The first evidence of cardiac lesions in Cu deficiency emerged from studies of a disease of cattle occurring in Western Australia known locally as "falling disease" (32). The essential lesion of this disease is a degeneration of the myocardium with replacement fibrosis. The morbid process is slow and progressive, commencing with the presence of occasional areas of small-celled infiltration and proceeding to the replacement of large areas of degenerate myocardium by dense collagenous tissue. The sudden deaths are believed to be due to acute heart failure, usually after mild exercise or excitement. The disease is completely preventable by copper therapy of the animals or by treatment of the pastures with copper compounds to raise their inherently very low-Cu levels (1–3 ppm d.b.) to normal. Falling disease has never been observed in sheep or horses grazing the same untreated areas and occurs only rarely in cattle in Cu-deficient areas elsewhere (94).

Sudden cardiac failure associated with cardiac hypertrophy has been reported in Cu-deficient pigs (140). In Cu-deficient rats cardiac hypertrophy and enlargement of the spleen have similarly been observed (1, 132), together with a marked increase in the mitochrondrial area of the heart muscles (132). The cytochrome oxidase activity of the heart muscle was more reduced, or more rapidly susceptible to Cu deficiency, and recovered more slowly in Cu repletion than the cytochrome oxidase of the brain (1). The cardiac hypertrophy and splenomegaly also recovered more slowly in Cu repletion than the anemia or cytochrome oxidase. Cytochrome oxidase and ceruloplasmin displayed a striking resemblance in their pattern of recovery. This supports the proposal of Broman (47) that copper is incorporated into cytochrome oxidase only if it is presented to the cell as ceruloplasmin.

In 1961 O'Dell and co-workers (251) demonstrated a derangement in the elastic tissue of the aortas of Cu-deficient chicks, and found the mortality in these animals to be caused by a rupture of the major blood vessels. At about the same time the Utah group independently described cardiovascular changes in Cu-deficient pigs which resulted in ruptures of the major blood vessels and death (70, 83). Subsequently, aortic rupture with degeneration of the elastic membrane was demonstrated in Cu-deficient chicks by others (65, 292). Extensive internal hemorrhage with a high incidence of aortic aneurysms were also observed in young Cu deficiency guinea pigs (115).

Studies in several laboratories have combined to elucidate the role of copper in elastin (and collagen) biosynthesis (see Carnes, 69). The most significant findings may be given as follows. The elastin content of the aortas of Cu-deficient pigs (340) and chicks (305) is decreased. The elastin from such animals contains an elevated content of lysine and less desmosine and isodesmosine than that of normal animals (234, 250). Desmosine, a tetracarboxylic tetraamino acid, and its isomer isodesmosine are the key cross-linkage groups in elastin (257), and at least two and possibly four lysine residues condense to form these substances (318). For this reaction to take place the ϵ-amino group of the lysine residues needs to be removed and the carbon oxidized to an aldehyde—a reaction catalyzed by amine or lysyl oxidases, which are Cu-containing enzymes (49, 163). The amine oxidase of the plasma of Cu-deficient pigs is lower than that of pigs given a Cu supplement (36), and that of ewes and lambs with low plasma Cu levels is also subnormal (237). The chick is hatched without detectable amine oxidase activity in the aorta or liver. When the diet contains copper the activity appears in both tissues by the third day at levels which are essentially maintained thereafter. When the diet is Cu deficient the lysyl oxidase activity remains undetectable in the aorta for 2–3 weeks, and that in the liver remains below that of Cu-supplemented chicks (162) (see Table 13). The aortas of Cu-supplemented chicks are more capable of incorporating lysine into desmosine than Cu-deficient chicks, and addition of amine oxidase to the culture medium of Cu-deficient aortas decreases the lysine to desmosine ratio to approximately that of the aortas of Cu-supplemented chicks (162). Harris (150a) has recently shown that Cu deficiency in chicks severely depresses lysyl oxidase activity in aortic tissue, and that copper is a key regulator of lysyl oxidase activity in the aorta and may be a major determinant of the steady-state levels of the enzyme in that tissue.

The role of copper in the formation of aortic elastin can best be stated in the words of Hill *et al.* (162) as follows: "The primary biochemical lesion is a reduction in amine-oxidase activity of the aorta. This reduction in enzymatic activity results, in turn, in a reduced capacity for oxidatively deaminating the epsilon-amino group of the lysine residues in elastin. The reduction in oxidative

TABLE 13
Effect of Copper Deficiency on Amine Oxidase Activity of Chick Aorta and Liver Mitochonidria[a]

		Specific activity at[b]					
Tissue	Treatment	1 day	3 days	5 days	10 days	17 days	26 days
Aorta	+Cu	0	15.3	20.7	19.4	14.8	22.2
	–Cu	0	0	0	0	0	4.0
Liver	+Cu	0	9.6	8.8	7.3	8.4	6.8
	–Cu	0	0	1.5	5.6	0	1.9

[a]From Hill *et al.* (161).
[b]Expressed as ΔOD/min/g protein.

deamination results, in turn, in less lysine being converted to desmosine. The reduction in desmosine, which is the cross-linkage group of elastin, results in fewer cross-linkages, in this protein, which in turn results in less elasticity of the aorta."

Copper deficiency in the sheep is associated with an increase in proteoglycans in the elastin-rich neck ligament and aorta (246). The whole of this increase was accounted for in terms of the larger proteoglycan complexes, with no change in those of small molecular weight. Changes in the pattern of glycosaminoglycans in each of the proteoglycan fractions were also observed. Hyaluronic acid was reduced and chondroitin sulfates A and C and dermatan sulfate increased in the Cu-deficient tissues.

8. Diarrhea (Scouring) of Cattle

Diarrhea is not a common manifestation of Cu deficiency in most species and does not always occur in cattle in Cu-deficient areas (94). Intermittent scouring in cattle occurs in the severely Cu-deficient areas in Australia, where Mo pasture levels are not high (32), and has been reported in England (6). Scouring was such a prominent feature of a Cu-responsive disease occurring in parts of Holland that it was designated "scouring disease" (293). Molybdenum has been incriminated as the primary causal factor in "peat scours" in New Zealand (87), but it seems unlikely that a high dietary Mo:Cu ratio can account for the occurrence of scouring in cattle in other Cu-low, Mo-normal areas. Of interest in this connection are the findings of Fell and co-workers (117) with young Friesian steers showing clinical signs of Cu deficiency. A marked depletion of cytochrome oxidase in the epithelium of the duodenum, jejunum, and ileum with partial villus atrophy in the duodenum and jejunum was found. Enterocytes from these parts of the small intestine showed mitochondrial abnormalities ranging from slight swelling to marked localized dilation. While no clear relationship between the histochemical and ultrastructural changes and diarrhea was apparent, these observations represent the first indications of pathological changes which could be of significance to the incidence of diarrhea in Cu-deficient cattle.

9. Fatty Acid Metabolism

In the young Cu-deficient male rat decreased monounsaturated:saturated ratios for C_{16} and C_{18} fatty acids from subcutaneous adipose tissue and decreased desaturase activity in liver microsomes have been demonstrated (336). When the animals were repleted with copper, the indices of Cu status and desaturase activity for liver microsomes returned to normal values, indicating an involvement of Cu in the desaturase reaction. The authors suggest that the site of this involvement could be the terminal component of the microsomal electron transport chain.

IV. COPPER REQUIREMENTS

Copper absorption and retention is so strongly influenced by a number of other mineral elements and dietary components that a series of minimum Cu requirements exist, depending on the extent to which these influencing factors are present or absent from the diet, and on the criteria of sufficiency employed.

1. Rats and Guinea Pigs

Copper deficiency develops rapidly in young rats and more slowly in adult rats on diets containing 1 ppm Cu or less. Mills and Murray (238), employing a diet containing only 0.3–0.4 ppm Cu, established the following tentative minimum requirements for rats of about 70 g live weight fed 10 g diet daily: 1 ppm for hemoglobin production; 3 ppm for growth; and 10 ppm for melanin production in hair. The minimum requirements for reproduction and lactation were not investigated, but at a level of 50 ppm Cu these requirements were stated to be fully met through at least one generation. Spoerl and Kirchgessner (301a) have recently shown that the Cu requirements for pregnancy and lactation in rats are fully met on a starch–casein diet containing 8 ppm Cu.

Everson and co-workers (115) found that young female guinea pigs fed a diet providing 0.5–0.7 ppm Cu grew well at first, and their reproduction was equal to that of controls receiving a 6-ppm Cu diet. Eventually they displayed a mild anemia, their hair coats became wiry and depigmented, and the growth of their offspring began to slow down at about the twelfth day. Surviving animals maintained on the Cu-deficient diet were markedly stunted by 50 or 60 days of age.

2. Pigs

No differences due to treatment were observed in baby pigs fed diets containing 6, 16, or 106 ppm Cu (325). It was therefore assumed that 6 ppm Cu is adequate for growth in such animals. It was later concluded that 4 ppm of dietary Cu is adequate for growing pigs up to 90 kg live weight (2). All ordinary swine rations contain appreciably higher levels of copper than the 4 or 6 ppm just quoted. Cereal grains contain 4–8 ppm Cu and the leguminous seeds and oilseed meals provided as protein supplements generally contain 15–30 ppm.

In 1955 Barber and associates (11) found that the addition of copper sulfate to a normal ration at the rate of 250 ppm Cu resulted in increased rate of weight gains in growing pigs. In 1967 Braude (41) assessed the results of trials carried out in Great Britain and concluded that daily live-weight gain can be accelerated by about 8% and the efficiency of feed use by about 5.5% by the inclusion of 250 ppm Cu in pig rations. In a large trial carried out at 21 centers such supplementation was found to increase the mean daily live weight by 9.7% and the efficiency of feed use by 7.9% (43). The results of a further trial in which copper sulfate supplements at 150, 200, and 250 mg Cu/kg diet were compared revealed significant improvements from all three levels, but the best performance and the highest increase in profitability came from the 250-ppm treatment (42). The response in weight gain is related to the amount of copper in the gut and copper sulfate is more effective than copper sulfide or oxide. However, it is the copper and not the sulfate radicle that is effective (12, 37). Copper apparently does not stimulate growth by suppressing or increasing bacterial multiplication in the gastrointestinal tract, because differences between Cu-supplemented and unsupplemented pigs in the bacterial flora of the feces (124), or throughout the alimentary canal (296), have not been observed. Kirchgessner and co-workers (188a) have demonstrated an improvement in the digestibility of proteins in young pigs after dosing with copper and have also shown that appropriate concentrations of cupric ions activate pepsin and raise peptic hydrolysis. These important findings could provide at least a partial explanation for the growth stimulating effect of the added dietary copper.

The extremely high-Cu requirements of pigs for maximum growth rate and efficiency of feed use apparent from the trials just described are not valid under all conditions. In several trials carried out elsewhere Cu supplementation of normal rations at the rate of 250 ppm Cu has been shown to (a) reduce live-weight gains and efficiency of feed use (17), (b) lead to anemia and Cu toxicosis in some animals (273, 338), (c) cause mortality and result in skin lesions similar to those of parakeratosis and rectifiable by zinc supplements (252), and (d) induce anemia rectifiable by increasing the iron content of the ration (52, 130).

It is apparent that the discrepancies in the reported responses to high-Cu levels relate primarily to differences in the zinc and iron levels of the basal rations, and that these high-Cu levels can only be satisfactorily and safely exploited if the zinc and iron contents of the pigs' diets are higher than those found adequate at lower Cu intakes. This results from a competitive antagonism between copper and iron and copper and zinc at the absorptive level as discussed in Sections II,1 and VI,2. Differences in the calcium content of the rations constitute a further factor of potential nutritional importance. High-Ca intakes reduce zinc availability, thus enhancing the possibility of Cu toxicity from Cu-supplemented pig rations and introducing a three-way interaction between

Ca, Zn, and Cu (313). Heavy Cu supplementation of swine rations is clearly a matter of considerable complexity. It may also pose environmental problems because dietary Cu is concentrated in animal wastes, resulting in substantial increases in soil and water Cu concentrations when the wastes are discharged into bodies of water or onto land (95). Chronic copper poisoning in sheep grazing herbage dressed with the liquid manure of pigs fed Cu-supplemented rations has been reported from Holland (326). The copper in slurry from a piggery where such high-Cu rations were fed is highly available and potentially toxic to sheep (267).

3. Poultry

Definitive data on the minimum Cu requirements of chicks for growth or hens for egg production have not been reported, nor is there any evidence that poultry have high-Cu requirements, relative to mammals, as they have for manganese. Diets containing 4–5 ppm Cu can be considered adequate, so long as these diets do not contain excessive amounts of elements that are metabolic antagonists of copper, such as iron, zinc, cadmium, and molybdenum. All poultry rations composed of normal feeds are likely to contain 5 ppm Cu or more.

The results of experiments in which poultry are fed Cu supplements at rates equivalent to those used to stimulate the growth of pigs (100–250 ppm Cu) are equivocal. Some studies have shown no significant improvement in growth rate while others have demonstrated a positive response between 75 and 225 ppm Cu, with an inhibition above 300 ppm (see Smith, 297). More recently King (185) observed a slight but nonsignificant increase in growth rate from 100 ppm Cu as copper sulfate in broiler chicks over a 9-week period and a significant increase in the growth rate of ducklings 8–63 days of age from the same level of Cu supplementation (186). A feature of this work is the finding that the ceca of the Cu-treated birds are significantly smaller and thinner, both as a proportion of body weight and as the weight of unit length (186), and that the intestinal wall is thinned (185). It was suggested that this contributes to the growth-promoting effect by facilitating the uptake of nutrients or by exposing more glandular tissue, thus allowing a more rapid contact of the ingesta by digestive ferments. The extent to which variations in dietary intakes of the interacting elements zinc and iron contribute to the varying results reported with chicks is not clear, but the source of the protein can be important. Jenkins and co-workers (180) obtained significant increases in chick growth when 250 ppm Cu as copper sulfate was added to a diet composed mainly of wheat and fish meal with added tallow, but observed either no growth response or a depression when this amount of copper was added to a maize–soybean diet.

4. Sheep and Cattle

Under appropriate conditions crossbred sheep can be maintained in Cu balance at intakes as low as 1 mg Cu day, which is equivalent to about 1 ppm Cu in the dry diet (100). In most environments conditions exist that impose a substantially higher Cu requirement on sheep (and cattle) than 1 ppm. Thus Marston and associates (221, 222) found that a supplement of 5 mg Cu/day, bringing the total dietary Cu intake to about 8 ppm, was insufficient to ensure normal blood Cu levels and wool keratinization in all merino sheep grazing pastures grown on the calcareous soils of South Australia. Under these conditions, where there is a high consumption of calcium carbonate from the environment and moderate intakes of Mo and sulfate from the herbage, the minimum Cu requirements of wool sheep are close to 10 mg/day, or about 10 ppm of the dry diet. In Western Australia pastures containing 4–6 ppm Cu (d.b.) and with Mo concentrations generally below 1.5 ppm provide sufficient copper for the full requirements of cattle and British breed and crossbred sheep (24). To avoid defective keratinization of the wool, merino sheep require a minimum of 6 ppm Cu in the dry pastures. However, even the relatively low levels of Mo in these pastures may be having some effect on the Cu requirements, as Dick (100) has shown that Mo intakes as low as 0.5 mg/day can adversely affect Cu retention in the sheep, provided the sulfate intake is higher. More recently Suttle (309) also demonstrated effects on Cu metabolism from Mo intakes much smaller than those commonly conceived as significant. These findings and the importance of the Cu:Mo dietary ratio are considered in Chapter 4.

The critical significance of the *relative* intakes of Cu and Mo is further apparent from recent studies by Thornton and co-workers (320) in England. They found evidence of Cu deficiency, responsive to copper therapy, in cattle grazing pastures "normal" in Cu content (7–14 ppm) and with Mo contents within the 3–20 ppm range. Many of the animals showed no marked clinical signs of hypocuprosis, although unthriftiness and poor fertility were stated to be common in the affected area. All these investigations emphasize the difficulty of assigning precise minimum Cu requirements and the impossibility of basing adequacy on Cu intakes alone.

The most widely practiced means of providing adequate copper to grazing stock in many areas is to apply Cu-containing fertilizers to the pastures. Australian experience indicates that 5–7 kg/hectare of copper sulfate, or its Cu equivalent in copper ores, is sufficient to raise the Cu in the herbage to adequate levels for several years. In some areas significant increases in total herbage production are coincidentally achieved. Under range conditions where fertilizers are not normally applied, or on calcareous soils where Cu absorption by plants is poor, salt licks containing 0.5–1.0% copper sulfate may be supplied. Such licks are usually consumed by sheep and cattle in Cu-deficient areas in sufficient quantities and with sufficient frequency to maintain adequate Cu intakes.

Intramuscular or subcutaneous injections of copper glycinate or CuCa edetate (copper calcium ethylenediaminetetraacetate) at intervals of 3–6 months, in doses of 30–40 mg Cu for sheep or 120–240 mg for cattle, have been found satisfactory (62, 88, 242, 308). A single injection of 100 mg of CuCa edetate in Cu-deficient cattle was found adequate to promote and sustain satisfactory blood Cu levels for 6 months (320), and a single injection in mid-pregnancy of 40–50 mg Cu, either as copper glycinate, methionate, or CuCa edetate, was similarly found effective in preventing swayback in the lambs and in maintaining satisfactory blood and liver Cu concentrations in the ewes and the lambs (154). Some workers have reported deaths of ewes given subcutaneous CuCa edetate even when the injections are at the recommended levels (3, 179).

V. COPPER IN HUMAN HEALTH AND NUTRITION

1. Copper Deficiency in Infants

Copper deficiency has been implicated in the etiology of three distinct clinical syndromes in the human infant. In the first of these, anemia, hypoproteinema and low serum iron and serum Cu levels are present and combined iron and copper therapy is necessary to promote complete recovery (284, 307). The hypocupremia results from an inability of the infants to obtain sufficient copper from their Cu-low milk diets to prevent Cu depletion in the face of the increased loss of the plasma Cu protein into the bowel (75). Even normal breast-fed infants are often unable to obtain sufficient copper from the milk to prevent some Cu depletion (76). However, hypocupremia does not necessarily arise in infants fed exclusively on milk diets. For example, Wilson and Lahey (349) detected no differences in weight gains, hemoglobin values, serum protein, and plasma Cu levels in two groups of premature infants, one of which received a milk diet supplying only 14 μg Cu/kg body weight/day for 7–10 weeks and one supplying five–six times this amount of copper.

The second syndrome is one affecting malnourished infants being rehabilitated on high-calorie, low-Cu diets and exhibiting anemia, marked neutropenia, chronic or recurrent diarrhea, "scurvylike" bone changes, and hypocupremia (81, 82, 137), all of which are responsive to copper therapy. The Cu requirements of rapidly growing infants with poor body stores was estimated at between 42 and 135 μg/kg body weight/day (82). Cartwright and Wintrobe (75) estimate the dietary Cu requirements of infants and children to be 50–100 μg/kg body weight/day, a requirement that is met by most diets.

The third clinical condition in infants involving copper is Menke's kinky hair syndrome. This disease, discussed earlier in Section I,3,e, is not due to a dietary Cu deficiency. It is caused by a genetically determined defect in Cu absorption from the intestinal mucosa to the blood (91) and is characterized by subnormal Cu levels in blood, liver, and hair, progressive mental deterioration, hypothermia, defective keratinization of hair, metaphyseal lesions, and degenerative changes in aortic elastin (92).

Subnormal levels of copper in the blood and liver also occur in most (208, 229, 270) but not in all (279) cases of kwashiorkor and marasmus. It is doubtful if these changes have any primary clinical significance. Signs of Cu deficiency, which responded promptly to Cu supplementation, have also been observed in an infant with ileal atresia and a postoperative short bowel syndrome maintained by total parenteral nutrition (183).

2. Copper in Human Foods and Dietaries

The richest sources of copper in human dietaries are crustaceans and shellfish, especially oysters, and the organ meats, especially lamb or beef liver, followed by nuts, dried legumes, dried vine, and dried stone fruits and cocoa. These items can range in Cu content from 20–30 to 300–400 ppm (225, 283). The poorest sources are the dairy products, white sugar, and honey, which rarely contain more than 0.5 ppm Cu. The nonleafy vegetables, most fresh fruits, and refined cereals generally contain up to 2 ppm. On an "as cooked and served basis," sheep's liver, calf's liver, oysters, many species of fish, and green vegetables have recently been classed as "usually good sources of copper (> 100 μg Cu/100 kcal), and cheese (except Emmental), milk, beef, mutton, white or brown bread, and many breakfast cereals as relatively poor sources (< 50 μg Cu/100 kcal)" (351).

The refining of cereals for human consumption results in significant losses of copper, although this loss is not as severe as it is with Fe, Mn, and Zn. In a recent study of North American wheat and wheat products (352) the following mean values for copper were reported: common hard wheat, 5.1 ± 0.5; common soft wheat, 4.5 ± 0.5; baker's patent flour, 1.9 ± 0.2; soft patent flour, 1.6± 0.3; and white bread, 2.1 ± 0.2 ppm, all on the dry basis. The refining of sugar similarly results in a significant lowering of the Cu concentration of the product. For example, Hamilton and Minski (148) reported 3 μg Cu/g in Barbados brown sugar, compared with 0.08 μg Cu/g in granulated cane sugar.

Most Western-style mixed diets supply adults with 2–4 mg Cu/day. This is evident from studies in England (148, 323), New Zealand (274), and the United States (210, 283). Lower estimates have been made for certain Dutch (27) and poor Scottish diets (93), while Indian adults consuming rice and wheat diets have been shown to ingest 4.5–5.8 mg Cu/day (97). Copper intakes at the above levels are apparently fully adequate for all normal individuals, since Cu deficiency has never been reported in human adults, even in areas where the soils and herbage are low in this element and Cu deficiency occurs in livestock not receiving Cu supplements. In fact, Scheinberg (280) contends that most adult diets supply a substantial excess of copper. The reverse situation can apply to adults receiving total parenteral nutrition (TPN). It has recently been shown that patients receiving TPN for 3–17 weeks exhibited subnormal serum Cµ and ceruloplasmin levels and a decline in hemoglobin levels which responded rapidly, in most cases, to oral feeding (121).

The importance of the drinking water, particularly soft water, as a source of copper to man was stressed by Schroeder *et al.* (283). These workers observed a progressive increase in the copper in water from brook to reservoir to hospital tap, as well as a considerable increment in soft water, compared with hard water, from private homes. They maintained that some soft waters, with their capacity to corrode metallic copper, could raise intakes as much as 1.4 mg Cu/day, whereas hard waters would reduce this increment to 0.05 mg Cu or less. Robinson *et al.* (274), working in New Zealand, have also shown that soft water can contribute appreciable quantities of copper. They calculated that if the beverages consumed by one individual had been made up with soft water from the cold tap, 0.4 mg Cu/day would have been obtained from this source alone and if from the hot tap no less than 0.8 mg Cu/day. It is obvious that the drinking water cannot be neglected as a source of dietary copper. Indeed, the possible occurrence of chronic Cu poisoning in infants as a result of contamination of water supplied through the use of copper pipes has been the subject of two recent reports (278, 337).

3. The Zinc:Copper Dietary Ratio as a Risk Factor in Coronary Heart Disease

Kelvay (193) has hypothesized that an imbalance in zinc and copper metabolism contributes to the risk of coronary heart disease (CHD). This hypothesis was initially based on the results of experiments with rats consuming a cholesterol-free diet in which the intakes of Zn and Cu were varied by varying the ratio of salts of these elements in the drinking water. Water with a Zn:Cu ratio of 40 consistently and significantly produced higher concentrations of cholesterol in the plasma than did water with a ratio of 5. Subsequent investigations by this worker has given added substance to this hypothesis and has related variations in the Zn:Cu ratio to other risk factors in CHD. For example, a relationship between the amount of fat and the ratio of zinc to copper of foods ($x^2 = 13.5, p < 0.001$) was demonstrated (194). Some U.S. meals and diets of considerable variability were shown to have Zn:Cu ratios in excess of those which produce hypercholesteremia in rats (196), and the mortality rate for CHD and the ratio of zinc to copper in milk of 47 cities in the United States were correlated ($r = 0.354$, $p < 0.02$) (195). Attention was also drawn to the higher Zn:Cu ratio in cow's milk than in breast milk. It was suggested that one of the benefits that might be gained from a return to breast-feeding is a reduction in CHD. In this connection it is pertinent to mention the work of Osborne (254), who studied the coronary arteries of 109 people up to the age of 20 and concluded that the arteries of those who had not been breast fed had the greatest reductions in luminal size. Such an association by chance was calculated to be unlikely ($p < 0.005$) (195).

The above findings suggest that dietary Zn:Cu ratios must be given further critical consideration in future epidemiological and experimental studies of the etiology of CHD. However, Walravens and Hambidge (339) found no effect on the plasma cholesterol levels of infants fed on a milk formula in which the Zn:Cu ratio was raised from 5:1 to 17:1.

4. Hepatolenticular Degeneration (Wilson's Disease)

Wilson's disease is an expression of an inborn error of metabolism that affects copper homeostasis. The abnormality is inherited as an autosomal recessive characteristic occurring in either sex but is rare because both parents must carry the abnormal gene. The pattern of inheritance, clinical features, and pathological changes associated with the disease have been well described (19, 281). The genetic abnormality leads to excessive Cu accumulation in the liver, brain, kidneys, and cornea, together with low serum ceruloplasmin levels, high levels of serum Cu not bound to ceruloplasmin, increased urinary Cu, and decreased fecal Cu. The progress of the disease can be arrested by giving copper-chelating agents which mobilize Cu from the tissues and promote its excretion in the urine. The most valuable of these is penicillamine, which can be given orally.

It has been 20 years since Uzman *et al.* (327) first suggested that the primary defect in Wilson's disease is the synthesis of an abnormal protein with a high affinity for copper. Such a compound, copper thionein, obtained from the livers of Wilson's disease patients, has recently been shown to have a Cu-binding constant four times as great as that of the protein from control subjects (111). The authors explain the probable role of an abnormal protein with a high affinity for copper in the defects in Cu homeostasis of Wilson's disease in the following terms; "the increased binding affinity of the temporary storage protein in the hepatocyte of Wilson's disease patients probably shifts the normal equilibrium of the hepatic Cu pool, which results in both depressed biliary Cu excretion and decreased incorporation of Cu into ceruloplasmin. As the disease progresses, the binding sites on the storage protein become saturated and the excess metal is ingested by the hepatic lysosomes. Saturation of the hepatic Cu-binding sites results in a decreased uptake of the metal with a concomitant elevation in plasma Cu not bound to ceruloplasmin. Whether the increased deposition of copper in extra-hepatic tissues results from the elevated noncerulo-plasmin Cu or from the presence of the protein in these organs is not known."

VI. COPPER TOXICITY

1. General

Chronic copper poisoning may occur in animals (a) under natural grazing conditions, (b) as a consequence of excessive consumption of Cu-containing salt licks or mixtures, (c) from the unwise use of Cu-containing drenches, (d) from contamination of feeds with Cu compounds from agricultural or industrial sources, and (e) with pigs given Cu supplements as growth stimulants if the basal diet is not suitably balanced with other minerals with which copper interacts.

In all animals the continued ingestion of copper in excess of requirements leads to some accumulation in the tissues, especially in the liver. The capacity for hepatic copper storage varies greatly among species, and differences among species in tolerance to high-Cu intakes are also great. Sheep are particularly susceptible to high-Cu intakes, whereas rats appear to be extremely tolerant. For example, Boyden *et al.* (40) reported normal growth and health in rats maintained on diets containing 500 ppm Cu, or about 100 times normal, despite a 14-fold increase in liver copper. However, the actual liver Cu levels attained were much lower than those characteristic of chronic copper poisoning in sheep.

Copper interacts metabolically with so many other elements, such as Zn, Fe, Cd, and Mo, as considered elsewhere, that it is impossible to give maximum safe or minimum tolerable dietary Cu levels based on copper alone. A series of such levels exist depending on the extent to which these interacting substances are present or absent from the diet. This is strikingly apparent from studies with pigs described next.

2. Pigs

Signs of Cu poisoning can occur in pigs fed certain diets supplemented with copper at the rate of 250 ppm Cu, unless the zinc and iron levels in these diets are suitably increased (52, 273, 313). Dietary Cu levels of 425, 450, 600, and 750 ppm were shown by Suttle and Mills (313) to cause severe toxicosis in pigs, manifested by a marked depression in feed intake and growth rate, hypochromic microcytic anemia, jaundice, and marked increases in liver and serum Cu levels and serum aspartate aminotransferase (AAT) activities. All these signs of Cu toxicosis were eliminated by providing simultaneously an additional 150 ppm Zn + 150 ppm Fe to the diets containing 425 or 450 ppm Cu. With the diets containing 750 ppm Cu the addition of 500 ppm Zn or 750 Fe eliminated the jaundice and produced normal serum Cu and AAT values, but only iron afforded protection against the anemia. These effects can best be explained by competition between the metal ions for protein-binding sites in the absorptive process and by a Zn-induced alteration in Cu-binding to hepatic proteins (44), considered in Chapter 8.

3. Sheep and Cattle

Sheep are more susceptible than cattle to copper toxicosis on the basis of field evidence, although the experimental evidence for cattle is conflicting, probably because of differences in the basal diets. Adult cows have been fed 1.2–2.0 g copper sulfate daily for 5–18 weeks without harm (120) and 0.8–5.0 g copper sulfate fed daily for 9 months produced no observable ill effects (86). On the other hand, Kidder (184) observed generalized icterus, hemoglobinuria, and death in a steer fed 5 g copper sulfate/day for 122 days, and Shand and Lewis (288) induced typical signs of chronic Cu poisoning in calves by feeding a milk substitute powder averaging 115 ppm Cu. Hemoglobinemia and hemoglobinuria associated with marked jaundice were apparent and very high liver Cu levels were found in fatal cases.

Losses of sheep from chronic Cu poisoning, with hemolytic icterus and hemoglobinuria as characteristic signs, have been reported from the ingestion of herbage in orchards and vineyards previously sprayed with copper compounds (200) and from pastures sprayed with copper sulfate as a molluscicide (136). Chronic Cu poisoning may also occur on dry feed (198) and at pasture (197) from continued free-choice consumption of a mineral salt mixture containing a recommended amount of copper. In parts of Australia chronic Cu poisoning occurs in sheep on natural grazing under the following sets of conditions: (a) when the Cu content of the soils and pastures are abnormally high, (b) when these are normal but the Mo levels are very low, and (c) in association with liver damage due to poisoning by the plant *Heliotropium europaeum*.

The copper poisoning that arises under the first of the above conditions is due to a high gross Cu intake from the abnormally high-Cu content of the pastures growing on the cupriferous soils of the area, and to the copper taken in by the animals from soil and dust. Some plant species growing on these soils contain as much as 50–60 ppm Cu (d.b.). The disease that occurs under the second set of conditions is usually seasonal in occurrence, appears less in merinos than in British breeds or crosses, occurs only on the more acid soils of the region, and is favored by dominance of the pastures with the clover *Trifolium subterraneum*. This plant generally contains 10–15 ppm Cu and extremely low-Mo levels that rarely exceed 0.1–0.2 ppm. These circumstances favor the development of a high-Cu status in the sheep and lead to copper poisoning. Providing Mo-containing salt licks (102) or dosing of the animals with ammonium molybdate and sodium sulfate (275) is highly effective in reducing liver Cu levels and mortality from the disease.

The disease that occurs in association with the plant *Heliotropium europaeum* has been named "yellows," enzootic jaundice, and hemolytic jaundice. This plant contains hepatotoxic alkaloids, including heliotrine and lasiocarpine (51), which initiate an irreversible change in liver parenchyma cells reflected in an increase in size of the cells and their nuclei. These enlarged cells (megalocytes) have a shortened life and are not replaced by new liver cells. As the megalocytes disappear the liver becomes atrophic and the attempted regeneration results in extensive new bile duct formation (51). Copper retention in the heliotrope-damaged livers is extremely high, usually over 1000 ppm on the dry, fat-free basis, due to an increased avidity of the cells for this element. This increases the susceptibility of the sheep to death from chronic Cu poisoning.

Chronic Cu poisoning in sheep has also been reported from the consumption of toxic, alkaloid-containing lupins (128). The copper intakes of the sheep from the lupins are low to normal, but marked increases in the blood and liver Cu levels and in 24-hr urinary Cu excretions were observed in sheep consuming the lupin diets. It was suggested that the disease, lupinosis, is in part a conditioned form of Cu poisoning and that hemolytic factors in toxic lupins and hemolysis resulting from release of liver copper may both operate to cause the jaundice of lupinosis.

Early in the investigations of chronic Cu poisoning in sheep, Bull (50) postulated that the critical factor in the precipitation of the hemolytic crisis is a high concentration of mobile or active copper. Copper becomes toxic to the cell, it was stated, when it is liberated in sufficient concentration from organic combination within the cell that toxic concentrations may be reached from such liberation, either through calls on the food reserves of the liver, or directly by the cell taking up sufficient copper by assimilation from a favorable diet. Since that time, studies of experimental chronic Cu poisoning in sheep, particularly by Howell and by Todd and their colleagues, have thrown considerable light on the pathological changes and sequence of biochemical events that occur prior to, during, and following the hemolytic crisis. The underlying mechanism of the Cu-induced hemolysis is still not clearly established. However, Metz and Sagone (233) have recently proposed that the injury to the red blood cells stems primarily from accelerated oxidation of glutathione, with resulting oxidative injury to hemoglobin and the cell membrane.

During the first phase of the two-phase process of the disease, when copper is accumulating in the tissues, a significant rise in serum transaminases and lactic dehydrogenase occurs, preceding the hemolytic crisis, the second phase of the disease, by several weeks (275, 322). Determination of serum glutamic oxalo-acetic transaminase (SGOT) levels has therefore been proposed as an aid in the

TABLE 14
Blood Changes in Sheep during the Hemolytic Crisis of Chronic Copper Toxicity[a]

Sheep	Blood copper (μg/100 ml)	Serum GOT[b] (IU/liter)	Serum CPK[c] (IU/liter)
33	798	5975	14,160
37	720	23725	41,000
Dorset	370	3580	23,350
Normals	70–120	0–140	0–125

[a]From Thompson and Todd (319).
[b]Glutamic oxaloacetic transaminase.
[c]Creatine phosphokinase.

early detection of chronic Cu poisoning and in the assessment of the rate of recovery. The hemolytic crisis is accompanied by a marked rise in blood Cu levels (Table 14) and falls in hemoglobin and glutathione concentrations, while methemoglobin levels rise (321). There is also a sudden and dramatic rise in creatine phosphokinase (CPK) levels in the serum (319) (Table 14), suggesting that changes occur in the muscle cell membrane as well as in the liver, kidney, and brain.

Some 6 weeks before the hemolytic crisis, swelling and necrosis of isolated hepatic parenchymal cells are apparent, together with swollen Kupffer cells rich in acid phosphatase and containing PAS-positive diastase-resistant material and copper (178). In addition the liver parenchyma exhibits a reduced activity of adenosine triphosphatase, nonspecific esterase, succinic tetrazolium reductase, and glutamic dehydrogenase, all pointing to a functional disturbance of the liver occurring some time before the crisis. Hemolytic crises have, in fact, been shown to be associated with extensive focal necrosis of liver tissue, neutrophilia, and high blood urea levels. Postmortem examination disclosed markedly elevated liver and kidney Cu levels and slightly elevated Cu levels in the spinal cord (178).

Prior to hemolysis, Cu levels in the kidneys and liver rise significantly in sheep suffering from experimental Cu toxicosis brought about by daily drenching with about 20 mg $CuSO_4 \cdot 5H_2O$/kg body weight (133). Eosinophilic intracytoplasmic granules become numerous in the epithelium of the proximal convoluted tubules (PCT), but histochemical changes are not seen and kidney function is not impaired. In animals that developed hemolysis, degeneration, necrosis, and loss of enzyme activity from the cells of the PCT become apparent, together with a marked functional impairment at this time.

Studies of the brain by both light and electron microscopy have revealed significant changes in chronic Cu poisoning in sheep (167, 241). Spongy transformation of the white matter was observed by Morgan (241), arising from the formation of intramyelinic vacuoles with splitting of lamellae at the intraperiod line. There was no significant involvement of neurons, glia, or blood vessels and no apparent increase in the extracellular space. Howell *et al.* (167) observed astrocytic changes in the hemolytic and posthemolytic animals, with the volume of astrocytic nuclei in the brain significantly greater than that of the controls. Vacuolation of white matter was seen as a terminal phenomenon in some, but not all, of the sheep. The authors suggest that the changes that occur in the brain of animals in chronic Cu poisoning may be due to the effects of altered metabolic processes on glial transport mechanisms.

REFERENCES

1. Abraham, P.A., and Evans, J.L., *Trace Subs. Environ. Health–5, Proc. Univ. Mo. Annu. Conf., 5th, 1971* p. 335.
2. Agricultural Research Council (Great Britain), *Nutr. Requir. Farm. Livestock* No. 3 (1967).
3. Allcroft, R., Buntain, D., and Rowlands, W.T., *Vet. Rec.* **77**, 634 (1965).
4. Allcroft, R., Clegg, F.G., and Uvarov, O., *Vet. Rec.* **71**, 884 (1959).
5. Allcroft, R., and Lewis, G., *J. Sci. Food Agric.* **8**, Suppl., S96 (1957).
6. Allcroft, R., and Parker, W.H., *Br. J. Nutr.* **3**, 205 (1949).
7. Allcroft, R., and Uvarov, O., *Vet. Rec.* **71**, 797 (1959).
8. Alloway, B.J., *J. Agric. Sci.* **80**, 521 (1973).
9. Anke, M., *Arch. Tierernaehr* **17**, 1 (1967); *Chem. Abstr.* **66**, 8673 (1967).
10. Arneson, R.M., *Arch. Biochem. Biophys.* **136**, 352 (1970).
11. Barber, R.S., Braude, R., Mitchell, K.G., and Cassidy, J., *Chem. Ind. (London)* **21**, 601 (1955); Barber, R.S., Braude, R., and Mitchell, K.G., *Br. J. Nutr.* **9**, 378 (1955).
12. Barber, R.S., Braude, R., Mitchell, K.G., and Rook, J.F., *Br. J. Nutr.* **11**, 70 (1957).
13. Barlow, R.M., *J. Comp. Pathol.* **73**, 51 and 61 (1963).
14. Barlow, R.M., and Cancilla, P., *Acta Neuropathol.* **6**, 175 (1960).
15. Barlow, R.M., Field, A.C., and Ganson, N.C., *J. Comp. Pathol. Ther.* **74**, 530 (1964).
16. Barlow, R.M., Purves, D., Butler, E.J., and McIntyre, I.J., *J. Comp. Pathol. Ther.* **70**, 396 and 411 (1960).
17. Bass, B., McCall, J.T., Wallace, H.D., Combs. G., Palmer, A.Z., and Carpenter, J.E., *J. Anim. Sci.* **15**, 1230 (1956).
18. Baxter, J.H., and Van Wyk, J.J., *Bull. Johns Hopkins Hosp.* **93**, 1 (1953); Baxter, J.H., Van Wyk, J.J., and Follis, R.H., *ibid.* p.25.
19. Bearn, A.G., *in* "The Metabolic Basis of Inherited Disease" (J.B. Stanbury, J.B. Wyngaarden, and D.S. Fredrickson, eds.), 3rd ed., p. 1033. McGraw-Hill, New York, 1972.
20. Bearn, A.G., and Kunkel, H., *J. Clin. Invest.* **33**, 400 (1954).
21. Beck, A.B., *Aust. J. Exp. Biol. Med. Sci.* **19**, 145 and 249 (1941).
22. Beck, A.B., personal communication (1961).
23. Beck, A.B., *Aust. J. Zool.* **4**, 1 (1956).
24. Beck, A.B., and Harley, R., *West. Aust. Dep. Agric., Leafl.* No. 678 (1951).
25. Beck, A.B., *Aust. J. Agric. Res.* **12**, 743 (1961).

26. Beisel, W., Pekarek, R.S., and Wannemacher, R.W., *in* "Trace Element Metabolism in Animals" (W.G. Hoekstra *et al.* eds.), Vol. 2, p. 217. Univ. Park Press, Baltimore, Maryland, 1974.
27. Belz, R., *Voeding* **6**, 236 (1960).
28. Bendstrip, P., *Acta Med. Scand.* **145**, 315 (1953).
29. Bennetts, H.W., *Aust. Vet. J.* **8**, 137 and 183 (1932).
30. Bennetts, H.W., and Beck, A.B., *Aust. C. S. I. R. O. Bull.* **147**, (1942).
31. Bennetts, H.W., and Chapman, F.E., *Aust. Vet. J.* **13**, 138 (1937).
32. Bennetts, H.W., and Hall, H.T.B., *Aust. Vet. J.*, **15**, 52 (1939); Bennetts, H.W., Harley, R., and Evans, S.T., *ibid.* **18**, 50 (1942); Bennetts, H.W., Beck, A.B., and Harley, R., *ibid.* **24**, 237 (1948).
33. Bingley, J.B., *Aust. J. Agric. Res.* **25**, 467 (1974).
34. Bingley, J.B., and Dufty, J.H., *Res. Vet. Sci.* **13**, 8 (1972).
35. Bingley, J.B., and Dufty, J.H., *Clin. Chim. Acta* **24**, 316 (1969).
36. Blaschko, H., Buffoni, F., Weismann, N., Carnes, W.H., and Coulson, W.F., *Biochem. J.* **96**, 4c (1965).
37. Bowland, J.P., Braude, R., Chamberlain, A.C., Glascock, R.F., and Mitchell, K.G., *Br. J. Nutr.* **15**, 59 (1961).
38. Bowness, J.M., and Morton, R.A., *Biochem. J.* **51**, 530 (1952).
39. Bowness, J.M., Morton, R.A., Shakir, M.H., and Stubbs, A.L., *Biochem. J.* **51**, 521 (1952).
40. Boyden, R., Potter, V.R., and Elvehjem, C.A., *J. Nutr.* **15**, 397 (1938).
41. Braude, R., *World Rev. Anim. Prod.* **3**, 69 (1967).
42. Braude, R., and Ryder, K., *J. Agric. Sci.* **80**, 489 (1973).
43. Braude, R., Townsend, M.J., Harrington, G., and Rowell, J.G., *J. Agric. Sci.* **58**, 251 (1962).
44. Bremner, I., *Q. Rev. Biophys.* **7**, 1 (1974).
45. Bremner, I., and Davies, N.T., *Rep. Rowett Inst.* **29**, 126 (1973).
46. Briggs, M.H., Briggs, M., and Wakatama, A., *Experientia* **28**, 406 (1972).
47. Broman, L., *Acta Soc. Med. Ups.* **69**, Suppl. 7 (1964), cited by Abraham and Evans (1).
48. Bruckmann, S., and Zondek, S.G., *Nature (London)* **146**, 3 (1940).
49. Buffoni, F., and Blaschko, H., *Proc. R. Soc. London, Ser. B.* **161**, 153 (1964).
50. Bull, L.B., *Br. Commonw. Spec. Conf. Agric., 1949,* p. 300 (1951).
51. Bull, L.B., Culvenor, C.C.J., and Dick, A.T., "The Pyrrolizidine Alkaloids." North-Holland Publ., Amsterdam, 1968.
52. Bunch, R.J., Speers, V.C., Hays, V.M., and McCall, J.T., *J. Anim. Sci.* **22**, 56 (1963).
53. Burley, R.W., and de Koch, W.T., *Arch. Biochem. Biophys.* **68**, 21 (1957).
54. Burns, R.H., Johnston, A., Hamilton, J.W., McColloch, R.J., Duncan, W.E., and Fisk, H.G., *J. Anim. Sci.* **23**, 5 (1964).
55. Burnstein, P., Kang, A.H., and Piez, K.A., *Proc. Natl. Acad. Sci. U.S.A.* **55**, 417 (1966).
56. Bush, J.A., Jensen, W.N., Atkens, J.W., Ashenbrucker, H., Cartwright, G.E., and Wintrobe, M.M., *J. Exp. Med.* **103**, 701 (1956).
57. Bush, J.A., Mahoney, J.P., Markowitz, M., Gubler, C.J., Cartwright, G.E., and Wintrobe, M.M., *J. Clin. Invest.* **34**, 1766 (1955).
58. Bush, J.A., Mahoney, J.P., Gubler, C.J., Cartwright, G.E., and Wintrobe, M.M., *J. Lab. Clin. Med.* **47**, 898 (1956).
59. Butler, E.J., *Comp. Biochem. Physiol.* **9**, 1 (1963).
60. Butler, E.J., and Barlow, R.M., *J. Comp. Pathol. Ther.* **73**, 107 (1963).

61. Butler, E.J., Curtis, M.J., Harry, E.G., and Deb, J.R., *J. Comp. Pathol.* **82,** 299 (1972).
62. Camargo, W.V., Lee, H.J., and Dewey, D.W., *Proc. Aust. Soc. Anim. Prod.* **4,** 12 (1962).
63. Campbell, J.K., and Mills, C.F., *Proc. Nutr. Soc.* **33,** 15A (1974).
64. Cancilla, P.A., and Barlow, R.M., *J. Comp. Pathol.* **80,** 315 (1970).
65. Carlton, W.W., and Henderson, W., *J. Nutr.* **81,** 200 (1963).
66. Carlton, W.W., and Henderson, W., *J. Nutr.* **85,** 67 (1965).
67. Carlton, W.W., and Kelly, W.A., *J. Nutr.* **97,** 42 (1969).
68. Carlton, W.W., and Henderson, W., *Avian Dis.* **8,** 48 (1964).
69. Carnes, W.H., *Fed. Proc. Fed. Am. Soc. Exp. Biol.* **30,** 995 (1971).
70. Carnes, W.H., Shields, G.S., Cartwright, G.E., and Wintrobe, M.M., *Fed. Proc. Fed. Am. Soc. Exp. Biol.* **20,** 118 (1961).
71. Carrico, R.J., and Deutsch, H.J., *J. Biol. Chem.* **244,** 6087 (1969); **245,** 723 (1970).
72. Cartwright, G.E., *in* "Symposium on Copper Metabolism" (W.D. McElroy and B. Glass, eds.), p. 274. Johns Hopkins Press, Baltimore, Maryland, 1950.
73. Cartwright, G.E., Gubler, C.J., and Wintrobe, M.M., *J. Clin. Invest.* **33,** 685 (1954).
74. Cartwright, G.E., Markovitz, H., Shields, G.S., and Wintrobe, M.M., *Am. J. Med.* **28,** 555 (1960).
75. Cartwright, G.E., and Wintrobe, M.M., *Am. J. Clin. Nutr.* **14,** 224 (1964); **15,** 94 (1964).
76. Cavell, P.A., and Widdowson, E.M., *Arch. Dis. Child.* **39,** 496 (1964).
77. Chang, I.C., Lee, T.-P., and Matrone, G., *J. Nutr.* **105,** 624 (1975).
78. Chapman, H.L., Jr., and Bell, M.C., *J. Anim. Sci.* **22,** 82 (1963).
78a. Claypool, D.W., Adams, F.W., Pendell, H.W., Hartman, N.A., Jr., and Bone, J.F., *J. Anim. Sci.* **41,** 911 (1975).
79. Clemetson, A., *Aust. Vet. J.* **42,** 34 (1966).
80. Comar, C.L., *in* "Symposium on Copper Metabolism" (W.D. McElroy and B. Glass, eds.), p. 191. Johns Hopkins Press, Baltimore, Maryland, 1950.
81. Cordano, A., Baertl, J.M., and Graham, G.G., *Pediatrics* **34,** 324 (1964).
82. Cordano, A., Placko, R.P., and Graham, G.G., *Blood* **28,** 280 (1966).
83. Coulson, W.F., and Carnes, W.H., *Am. J. Pathol.* **43,** 945 (1963).
84. Crampton, R.F., Matthews, D.M., and Poisner, R., *J. Physiol. (London)* **178,** 111 (1965).
85. Cunningham, I.J., *Biochem. J.* **25,** 1267 (1931).
86. Cunningham, I.J., *N. Z. J. Sci. Technol., Sect. A* **27,** 372 and 381 (1946).
87. Cunningham, I.J., *in* "Symposium on Copper Metabolism" (W.D. McElroy and B. Glass, eds.), p. 246. Johns Hopkins Press, Baltimore, Maryland, 1950.
88. Cunningham, I.J., *N. Z. Vet. J.* **7,** 15 (1959).
89. Cunningham, I.J., and Hogan, K.G., *N. Z. J. Agric. Res.* **1,** 841 (1958).
90. Danks, D.M., Campbell, P.E., Mayne, V., and Cartwright, E., *Pediatrics* **50,** 188 (1972).
91. Danks, D.M., Cartwright, E., Stevens, B.J., and Townley, R.R.W., *Science* **179,** 1140 (1973).
92. Danks, D.M., Stevens, B.J., Campbell, P.E., Gillespie, J.M., Walker-Smith, J., Blomfield, J., and Turner, B., *Lancet* **1,** 1100 (1972).
93. Davidson, L.S.P., Fullerton, H.W., Howie, J.W., Croll, J.M., Orr, J.B., and Godden, W.M., *Br. Med. J.* **2,** 685 (1933).
94. Davis, G.K., *in* "Symposium on Copper Metabolism" (W.D. McElroy and B. Glass, eds.), p. 216. Johns Hopkins Press, Baltimore, Maryland, 1950.

95. Davis, G.K., *Fed. Proc., Fed. Am. Soc. Exp. Biol.* **33**, 1194 (1974).
96. Davis, P.N., Norris, L.C., and Kratzer, F.H., *J. Nutr.* **77**, 217 (1962).
97. De, H.N., *Indian J. Med. Res.* **37**, 301 (1949).
98. Dewar, W.A., Teague, P.W., and Downie, J.N., *Br. Poult. Sci.* **15**, 119 (1974).
99. Dick, A.T., Doctoral Thesis, University of Melbourne, Australia (1954).
100. Dick, A.T., *Aust. J. Agric. Res.* **5**, 511 (1954).
101. Dick, A.T., *Soil Sci.* **81**, 229 (1956).
102. Dick, A.T., and Bull, L.B., *Aust. Vet. J.* **21**, 70 (1945).
103. Dreosti, I.E., and Quicke, G.V., *S. Afr. J. Agric. Sci.* **9**, 365 (1966).
104. Dreosti, I.E., and Quicke, G.V., *Br. J. Nutr.* **22**, 1 (1968).
105. Dunkley, W.J., Ronning, M., and Voth, J., *J. Dairy Sci.* **46**, 1059 (1963).
106. Dutt, B., and Mills, C.F., *J. Comp. Pathol. Ther.* **70**, 120 (1960).
107. Eden, A., *J. Comp. Pathol. Ther.* **52**, 429 (1939); **53**, 90 (1940).
108. Elvehjem, C.A., Steenbock, H., and Hart, E.B., *J. Biol. Chem.* **83**, 27 (1929).
109. Evans, G.W., *Nutr. Rev.* **29**, 195 (1971).
110. Evans, G.W., *Physiol. Rev.* **53**, 535 (1973).
111. Evans, G.W., Dubois, R.S., and Hambidge, K.M., *Science* **181**, 1175 (1973).
112. Evans, G.W., and Hahn, C.J., *in* "Protein-Metal Interactions" (M. Friedman, ed.). Plenum, New York, 1974.
113. Evans, J.L., and Abraham, P.A., *J. Nutr.* **103**, 196 (1973).
114. Evans, J.L., and Brar, B.S., *J. Nutr.* **104**, 1285 (1974).
115. Everson, G.J., Tsai, M.C., and Wang, T., *J. Nutr.* **93**, 533 (1967).
116. Fay, J., Cartwright, G.E., and Wintrobe, M.M., *J. Clin. Invest.* **28**, 487 (1949).
117. Fell, B.F., Dinsdale, D., and Mills, C.F., *Res. Vet. Sci.* **18**, 274 (1975).
118. Fell, B.F., Mills, C.F., and Boyne, R., *Res. Vet. Sci.* **6**, 10 (1965).
119. Fell, B.F., Williams, R.B., and Mills, C.F., *Proc. Nutr. Soc.* **20**, xxvii (1961).
120. Ferguson, W.S., *J. Agric. Sci.* **33**, 116 (1943).
121. Fleming, C.R., Hodges, R.E., and Hurley, L.S., *Am. J. Clin. Nutr.* **29**, 70 (1976).
122. Follis, R.H., Jr., Bush, J.A., Cartwright, G.E., and Wintrobe, M.M., *Bull. Johns Hopkins Hosp.* **97**, 405 (1955).
123. Frieden, E., *Adv. Chem. Ser.* **100**, (1971).
124. Fuller, R., Newland, L.G.M., Briggs, C.A.E., Braude, R., and Mitchell, K.G., *J. Appl. Bacteriol.* **23**, 195 (1960).
125. Gallagher, C.H., *Aust. Vet. J.* **33**, 311 (1957).
126. Gallagher, C.H., Judah, J.D., and Rees, K.R., *Proc. R. Soc. London, Ser. B* **145**, 134 and 195 (1956).
127. Gallagher, C.H., and Reeve, V.E., *Aust. J. Exp. Biol. Med. Sci.* **49**, 21 (1971).
128. Gardiner, M.R., *J. Comp. Pathol.* **76**, 107 (1966).
129. Giorgio, A.P., Cartwright, G.E., and Wintrobe, M.M., *Am. J. Clin. Pathol.* **41**, 22 (1964).
130. Gipp, W.F., Pond, W.G., Kallfelz, F.A., Tasker, J.B., Van Campen, D.R., Krook, L., and Visek, W.J., *J. Nutr.* **104**, 532 (1974).
131. Gitlin, D., Hughes, W.L., and Janeway, C.A., *Nature (London)* **188**, 150 (1960).
132. Goodman, J.R., Warshow, J.B., and Dallman, R.P., *Pediatr. Res.* **4**, 244 (1970).
133. Gopinath, C., Hall, G.A., and Howell, J. McC., *Res. Vet. Sci.* **16**, 57 (1974).
134. Goss, H., and Green, M.M., *Science* **122**, 330 (1955).
135. Grace, N.D., *Br. J. Nutr.* **34**, 73 (1975).
136. Gracey, J., and Todd, J.R., *Br. Vet. J.* **116**, 405 (1960).
137. Graham, G.G., and Cordano, A., *Johns Hopkins Med. J.* **124**, 139 (1969).
138. Gray, L.F., and Daniel, L.J., *J. Nutr.* **84**, 31 (1964).

139. Gregoriadis, G., and Sourkes, T.L., *Can. J. Biochem.* **45,** 1841 (1967); Lal, S., and Sourkes, T.L., *Toxicol. Appl. Pharmacol.* **18,** 562 (1971).
140. Gubler, C.J., Cartwright, G.E., and Wintrobe, M.M., *J. Biol. Chem.* **244,** 533 (1957).
141. Gubler, C.J., Lahey, M.E., Cartwright, G.E., and Wintrobe, M.M., *J. Clin. Invest.* **32,** 405 (1953).
142. Gubler, C.J., Lahey, M.E., Chase, M.S., Cartwright, G.E., and Wintrobe, M.M., *Blood* **7,** 1075 (1952).
143. Guggenheim, K., *Blood* **23,** 786 (1964).
144. Guggenheim, K., Tal, E., and Zor, V., *Br. J. Nutr.* **18,** 529 (1964).
145. Hall, G.A., and Howell, J. McC., *Br. J. Nutr.* **23,** 41 (1969).
145a. Halsted, J.A., Hackley, B.M., and Smith, J.C., *Lancet* **2,** 278 (1968).
146. Hambidge, K.M., *Am. J. Clin. Nutr.* **26,** 1212 (1973).
147. Hambidge, K.M., and Droegemueller, W., *Obstet. Gynecol.* **44,** 666 (1974).
148. Hamilton, E.I., and Minski, M.J., *Sci. Total Environ.* **1,** 375 (1972-1973).
149. Hamilton, E.I., Minski, M.J., and Cleary, J.J., *Sci. Total Environ.* **1,** 341 (1972–1973).
150. Hamilton, T.S., and Mitchell, H.H., *J. Biol. Chem.* **178,** 345 (1949).
150a. Harris, E.D., *Proc. Natl. Acad. Sci. (U.S.A.)* **73,** 371 (1976).
151. Hartmans, J., and Bosman, M.S.M., *in* "Trace Element Metabolism in Animals, (C.F. Mills, ed.), Vol. 1, p. 362. Livingstone, Edinburgh, 1970.
152. Hayter, S., Wiener, G., and Field, A.C., *Anim. Prod.* **16,** 261 (1973).
153. Healy, W.B., Bate, L.C., and Ludwig, T.G., *N. Z. J. Agric. Res.* **7,** 603 (1964).
154. Hemingway, R.G., MacPherson, A., and Ritchie, N.S., *in* "Trace Element Metabolism in Animals" (C.F. Mills, ed.), Vol. 1, p. 264. Livingstone, Edinburgh, 1970.
155. Henkin, R.I., *in* "Trace Element Metabolism in Animals" (W.G. Hoekstra *et al.* eds.), Vol. 2, p. 647. Univ. Park Press, Baltimore, Maryland, 1974.
156. Henkin, R.I., Schulman, J.D., Schulman, C.B., and Bronzert, D.A., *J. Pediatr.* **82,** 831 (1973).
157. Hennig, A., Anke, M., Groppel, B., and Ludke, H., *in* "Trace Element Metabolism in Animals" (W.G. Hoekstra *et al.*, eds.), Vol. 2, p. 726. Univ. Park Press, Baltimore, Maryland, 1974.
158. Hicks, J.J., Hernandez-Perez, O., Aznar, R., Mendez, J.D., and Rosado, A., *Am. J. Obstet. Gynecol.* **121,** 981 (1975).
159. Hill, C.H., and Matrone, G., *World's Poult. Congr., Proc. 12th, 1962* p. 219 (1962).
160. Hill, C.H., and Starcher, B., *J. Nutr.* **85,** 271 (1965).
161. Hill, C.H., Starcher, B., and Matrone, G., *J. Nutr.* **83,** 107 (1964).
162. Hill, C.H., Starcher, B., and Kim, C., *Fed. Proc., Fed. Am. Soc. Exp. Biol.* **26,** 129 (1968).
163. Hill, J.M., and Mann, P.G., *Biochem. J.* **85,** 198 (1962).
163a. Hill, R., *in* "Trace Element Metabolism in Animals" (W.G. Hoekstra *et al.*, eds.), Vol. 2, p. 632. Univ. Park Press, Baltimore, Maryland, 1974.
164. Holmberg, C.G., and Laurell, C.B., *Acta Chem. Scand.* **1,** 944 (1947); **2,** 250 (1948); **5,** 476 (1951).
165. Howell, J. McC., *in* "Trace Element Metabolism in Animals" (C.F. Mills, ed.), Vol. 1, p. 103. Livingstone, Edinburgh, 1970.
166. Howell, J. McC., *Vet. Rec.* **83,** 226 (1968).
167. Howell, J. McC., Blakemore, W.F., Gopinath, C., Hall, G.A., and Parker, J.H., *Acta Neuropathol.* **29,** 9 (1974).
168. Howell, J. McC., and Davidson, A.N., *Biochem. J.* **72,** 365 (1959).
169. Howell, J. McC., Davidson, A.N., and Oxberry, J., *Res. Vet. Sci.* **5,** 376 (1964).

170. Howell, J. McC., Edington, N., and Ewbank, R., *Res. Vet. Sci.* **9,** 160 (1968).
171. Howell, J. McC., and Hall, G.A., *Br. J. Nutr.* **23,** 47 (1969).
172. Howell, J. McC., and Hall, G.A., *in* "Trace Element Metabolism in Animals" (C.F. Mills, ed.), Vol. 1, p. 106. Livingstone, Edinburgh, 1970.
173. Hunt, C.E., and Carlton, W.W., *J. Nutr.* **87,** 385 (1965).
174. Hunt, C.E., Carlton, W.W., and Newberne, P.M., *Fed Proc., Fed. Am. Soc. Exp. Biol.* **25,** 432 (1966).
175. Hunt, D.M., *Nature (London)* **249,** 852 (1974).
176. Hurley, L.S., and Bell, L.T., *Proc. Soc. Exp. Biol. Med.* **149,** 830 (1975).
177. Innes, J.R.M., and Shearer, G.D., *J. Comp. Pathol. Ther.* **53,** 1 (1940).
178. Ishmael, J., Gopinath, C., and Howell, J. McC., *Res. Vet. Sci.* **12,** 358 (1971); **13,** 22 (1972).
179. Ishmael J., Howell, J. McC., and Treeby, P.J., *Vet. Rec.* **85,** 205 (1969); *in* "Trace Element Metabolism in Animals" (C.F. Mills, ed.), Vol. 1, p. 268. Livingstone, Edinburgh, 1970.
180. Jenkins, N.K., Morris, T.R., and Valamotis, D., *Br. Poult. Sci.* **11,** 241 (1970).
180a. Jensen, L.S., Peterson, R.P., and Falen, L., *Poult. Sci.* **53,** 57 (1964).
181. Johnson, N.C., Kheim, T., and Kountz, W.B., *Proc. Soc. Exp. Biol. Med.* **102,** 98 (1959).
182. Joyce, J.M., *N. Z. Vet. J.* **3,** 157 (1955).
183. Karpel, J.T., and Peden, V.H., *J. Pediatr.* **80,** 32 (1972).
184. Kidder, R.W., *J. Anim. Sci.* **8,** 623 (1949).
185. King, J.O.L., *Br. Poult. Sci.* **13,** 61 (1972).
186. King, J.O.L., *Br. Poult. Sci.* **16,** 409 (1975).
187. King, R.L., and Dunkley, W.J., *J. Dairy Sci.* **42,** 420 (1959).
188. King, R.L., and Williams, W.F., *J. Dairy Sci.* **46,** 11 (1963).
188a. Kirchgessner, M., and Giessler, H., *Z. Tierphysiol., Tierernaehr. Futtermittelkd.* **16,** 297 (1961); Kirchgessner, M., Beyer, M.G., and Steinhart, H., *Br. J. Nutr.* **36,** p. 15 (1976).
189. Kirchgessner, M., and Grassman, E., *in* "Trace Element Metabolism in Animals" (C.F. Mills, ed.), Vol. 1, p. 277. Livingstone, Edinburgh, 1970.
190. Kirchgessner, M., and Weser, U., *Z. Tierphysiol., Tierernaehr. Futtermittelkd.* **20,** 44 (1965).
191. Klauder, D.S., Murthy, L., and Petering, H.G., *Trace Subst. Environ. Health–6*, p. 131, Columbia, Missouri, 1972.
192. Klevay, L.M., *Am. J. Clin. Nutr.* **23,** 1194 (1970).
193. Klevay, L.M., *Am. J. Clin. Nutr.* **26,** 1060 (1973).
194. Klevay, L.M., *Nutr. Rep. Int.* **9,** 393 (1974).
195. Klevay, L.M., *Trace Subst. Environ. Health–8,* p. 9, 1974.
196. Klevay, L.M., *Nutr. Rep. Int.* **11,** 237 (1975).
197. Kowalczyk, T., Pope, A.L., Berger, K.C., and Muggenberg, B.A., *J. Am. Vet. Assoc.* **145,** 352 (1964).
198. Kowalczyk, T., Pope, A.L., and Sorensen, D.K., *J. Am. Vet. Assoc.* **141,** 362 (1962).
199. Krishnamachari, K.A.V.R., *Am. J. Clin. Nutr.* **27,** 108 (1974).
200. Lafenêtre, H., Monteil, L., and Galter, F., *Vet. Bull.* **5,** 516 (1957).
201. Lahey, M.E., Gubler, C.J., Cartwright, G.E., and Wintrobe, M.M., *J. Clin. Invest.* **32,** 322 and 329 (1953).
202. Lahey, M.E., Gubler, C.J., Chase, M.S., Cartwright, G.E., and Wintrobe, M.M., *Blood* **7,** 053 (1952).
203. Lassiter, J.W., and Bell, M.C., *J. Anim. Sci.* **19,** 754 (1960).

204. Lea, C.M., and Luttrell, V.A.S., *Nature (London)* **206**, 413 (1965).
205. Lee, G.R., Cartwright, G.E., and Wintrobe, M.M., *Proc. Soc. Exp. Biol. Med.* **127**, 977 (1968).
206. Lee, G.R., Nacht, S., Lukens, J.N., and Cartwright, G.E., *J. Clin. Invest.* **47**, 2058 (1968).
207. Lee, H.J., *J. Agric. Sci.* **47**, 218 (1956).
208. Lehmann, B.H., Hansen, J.D.L., and Warren, P.J., *Br. J. Nutr.* **26**, 197 (1971).
209. Lesne, E., Zizine, P., and Briskas, S.B., *Rev. Pathol. Comp. Hyg. Gen.* **36**, 1369 (1936).
210. Leverton, R.M., and Binkley, E.S., *J. Nutr.* **27**, 43 (1944).
211. Lewis, G., Terlecki, S., and Allcroft, R., *Vet. Rec.* **81**, 415 (1967).
212. Lindow, C.W., Peterson, W.H., and Steenbock, H., *J. Biol. Chem.* **84**, 419 (1929).
213. Lorenzen, E.J., and Smith, S.E., *J. Nutr.* **33**, 143 (1947).
214. Losee, F., Cutress, T.W., and Brown, R., *Trace Subst. Environ. Health–7*, 1973.
215. MacPherson, A., and Hemingway, R.G., *J. Sci. Food Agric.* **16**, 220 (1965).
216. Mahoney, J.P., Bush, J.A., Gubler, C.J., Moretz, W.H., Cartwright, G.E., and Wintrobe, M.M., *J. Lab. Clin. Med.* **46**, 702 (1955).
217. Mann, T., and Keilin, D., *Proc. R. Soc. London, Ser. B* **126**, 303 (1939).
218. Markowitz, H., Gubler, C.J., Mahoney, J.P., Cartwright, G.E., and Wintrobe, M.M., *J. Clin. Invest.* **34**, 1498 (1955).
219. Marston, H.R., *Proc. Symp. Fibrous Proteins, Soc. Dyers & Color. Leeds 1946.*
220. Marston, H.R., *Physiol. Rev.* **32**, 66 (1952).
221. Marston, H.R., and Lee, H.J., *Aust. J. Sci. Res., Ser. B* **1**, 376 (1948); *J. Agric. Sci.* **38**, 229 (1948).
222. Marston, H.R., Lee, H.J., and McDonald, I.W., *J. Agric. Sci.* **38**, 216 & 222 (1948).
223. Marston, H.R., Thomas, R.G., Murnane, D., Lines, E.W., McDonald, I.W., Moore, H.O., and Bull, L.B., *Aust. Commonw. Counc. Sci. Ind. Res., Bull.* **113** (1938).
224. Matrone, G., *Fed. Proc., Fed. Am. Soc. Exp. Biol.* **19**, 659 (1960).
225. McCance, R.A., and Widdowson, E.M., "The Chemical Composition of Foods." Chem. Publ. Co., New York, 1947.
226. McCord, J.M., and Fridovich, I., *J. Biol. Chem.* **244**, 6049 (1969).
227. McCosker, P.J., *Nature (London)* **190**, 887 (1961).
228. McCosker, P.J., *Res. Vet. Sci.* **9**, 91 and 103 (1968).
229. McDonald, I., and Warren, P.J., *Br. J. Nutr.* **15**, 593 (1961).
230. McFarlane, W.D., and Milne, H.I., *J. Biol. Chem.* **107**, 309 (1934).
231. McGavin, M.D., Ranby, P.D., and Tammemagi, L., *Aust. Vet. J.* **38**, 8 (1962).
232. McNaught, K.J., *N. Z. J. Sci. Technol., Sect. A* **30**, 26 (1948).
233. Metz, E.N., and Sagone, A.L., *J. Lab. Clin. Med.* **80**, 405 (1972).
234. Miller, E.J., Martin, E.R., Mecca, C.E., and Piez, K.A., *J. Biol. Chem.* **240**, 3623 (1965).
235. Mills, C.F., *Biochem. J.* **57**, 603 (1954); **63**, 187 and 190 (1956).
236. Mills, C.F., *Br. J. Nutr.* **9**, 398 (1955).
237. Mills, C.F., Dalgarno, A.C., and Williams, R.B., *Biochem. Biophys. Res. Commun.* **24**, 537 (1966).
238. Mills, C.F., and Murray, G., *J. Sci. Food Agric.* **9**, 547 (1960).
239. Mills, C.F., and Williams. R.B., *Biochem. J.* **85**, 629 (1962).
240. Milne, D.B., and Weswig, P.H., *J. Nutr.* **95**, 429 (1968).
241. Morgan, K.T., *Res. Vet. Sci.* **15**, 88 (1973).
242. Moule, G.R., Sutherland, A.K., and Harvey, J.M., *Queensl. J. Agric. Sci.* **18**, 93 (1959).

243. Munch-Petersen, S., *Acta Paediatr. (Stockholm)* **39,** 378 (1951).
244. Murthy, G.K., and Rhea, U.S., *J. Dairy Sci.* **54,** 1001 (1971).
245. Murthy, G.K., Rhea, U.S., and Peeler, J.T., *J. Dairy Sci.* **56,** 1666 (1972).
246. Muthiah, P., and Cleary, E.G., private communication (1975).
247. Neuman, P.Z., and Sass-Kortsak, A., *J. Clin. Invest.* **46,** 646 (1967).
248. Nielsen, A.L., *Acta Med. Scand.* **118,** 87 and 92 (1944).
249. Nixon, G.S., Smith, H., and Livingstone, H.D., *in* "Symposium on Nuclear Activation Techniques in the Life Sciences," IAEA, Vienna, 1967.
250. O'Dell, B.L., Bird, D.W., Ruggles, D.F., and Savage, J.E., *J. Nutr.* **88,** 9 (1966).
251. O'Dell, B.L., Hardwick, B.C., Reynolds, G., and Savage, J.E., *Proc. Soc. Exp. Biol. Med.* **108,** 402 (1961).
252. O'Hara, P.J., Newman, A.P., and Jackson, R., *Aust. Vet. J.* **36,** 225 (1960).
253. Osaki, S., Johnson, D.A., and Frieden, E., *J. Biol. Chem.* **241,** 2746 (1966).
254. Osborn, G.R., *in* "Le Rôle de la paroi artérielle dans l'athirogenese" (M.J. Lenegre, L. Scebat and J. Renais, eds.). CNRS, Paris, 1968.
255. Owen, C.A., Jr., *Am. J. Physiol.* **207,** 1203 (1964).
256. Owen, E.C., Proudfoot, R., Robertson, J.M., Barlow, R.M., Butler, E.J., and Smith, B.S.W., *J. Comp. Pathol.* **75,** 241 (1965).
257. Partridge, S.M., Elsden, D.F., Thomas, J., Dorfman, A., Telser, A., and Ho, P.L., *Biochem. J.* **93,** 30c (1964).
258. Pekarek, R.S., Burghen, G.A., Bartelloni, P.J., Calia, F.M., Bostian, K.A., and Beisel, W.R., *J. Lab. Clin. Med.* **76,** 293 (1970).
259. Pekarek, R.S., Kluge, R.M., Dupont, H.L., Wannemacher, R.W., Hornick, R.B., Bostian, K.A., and Beisel, W.R., *Clin. Chem.* **21,** 528 (1975).
260. Pekarek, R.S., Wannemacher, R.W., Powanda, M.C., and Beisel, W.R., *Clin. Res.* **21,** 608 (1973).
261. Picciano, M.F., and Guthrie, H.A., *Fed. Proc., Fed. Am. Soc. Exp. Biol.* **32,** 929 (1973).
262. Porter, H., *Biochem. Biophys. Res. Commun.* **56,** 661 (1974).
263. Porter, H., *Biochim. Biophys. Acta* **229,** 143 (1971).
264. Porter, H., Johnston, J., and Porter, E.M., *Biochim. Biophys. Acta* **65,** 66 (1962).
265. Porter, H., Sweeney, M., and Porter, E.M., *Arch. Biochem. Biophys.* **104,** 97 (1964).
266. Porter, H., Wiener, W., and Barker, M., *Biochim. Biophys. Acta* **52,** 419 (1961).
267. Price, J., and Suttle, N.F., *Proc. Nutr. Soc.* **34,** 9A (1975).
268. Pryor, W.J., *Res. Vet. Sci.* **5,** 123 (1964).
269. Raper, H.S., *Physiol. Rev.* **8,** 245 (1928).
270. Reiff, B., and Schneiden, H., *Blood* **14,** 967 (1959).
271. Reinhold, J.G., Kfoury, G.A., Galamber, N.A., and Bennett, J.C., *Am. J. Clin. Nutr.* **18,** 294 (1966).
272. Reinhold, J.G., Kfoury, G.A., and Thomas, T.A., *J. Nutr.* **92,** 173 (1967).
273. Ritchie, H.D., Luecke, R.W., Baltzer, B.V., Miller, E.R., Ullrey, D.E., and Hoefer, J.A., *J. Nutr.* **79,** 117 (1963).
274. Robinson, M.F., McKenzie, J.M., Thompson, C.D., and van Rij, A.L., *Br. J. Nutr.* **30,** 195 (1973).
275. Ross, D.B., *Vet. Rec.* **76,** 875 (1964); *Br. Vet. J.* **122,** 279 (1966).
276. Rucker, R.B., Parker, H.E., and Rogler, J.C., *J. Nutr.* **98,** 57 (1969); Rucker, R.B., Riggins, R.S., Laughlin, R., Chan, M.M., Chien, M., and Tom, K., *ibid* **105,** 1062 (1975).
277. Rupp, H., and Weser, U., *FEBS Lett.* **44,** 293 (1974).
278. Salmon, M.A., and Wright, T., *Arch. Dis. Child.* **46,** 108 (1971).

279. Sandstead, H.H., Shukry, A.S., Prasad, A.S., Gabr, M.K., El Hifney, A., Mokhtar, N., and Darby, W.J., *Am. J. Clin. Nutr.* **17,** 15 (1965).
279a. Savage, J.E., *Fed. Proc., Fed. Am. Soc. Exp. Biol.* **27,** 927 (1968).
280. Scheinberg, I.H., *Fed. Proc., Fed. Am. Soc. Exp. Biol.* **20**(Suppl. 10), 179 (1961).
281. Scheinberg, I.H., and Sternlieb, I., *Annu. Rev. Med.* **16,** 119 (1965).
282. Scheinberg, I.H., Cook, C.D., and Murphy, J.A., *J. Clin. Invest.* **33,** 963 (1954).
283. Schroeder, H.A., Nason, A.P., Tipton, I.H., and Balassa, J.J., *J. Chronic. Dis.* **19,** 1007 (1966).
284. Schubert, W.K., and Lahey, M.E., *Pediatrics* **24,** 710 (1959).
285. Schultz, K.C.A., Van der Merwe, P.K., Van Rensburg, P.J., and Swart, J.S., *Onderstepoort J. Vet. Res.* **25,** 35 (1961).
286. Schultze, M.O., Elvehjem, C.A., and Hart, E.B., *J. Biol. Chem.* **115,** 453 (1936); **16,** 93 and 107 (1936).
287. Schwarz, F.J., and Kirchgessner, M., *Int. J. Vitam. Nutr. Res.* **44,** 116 (1974).
288. Shand, A., and Lewis, G., *Vet. Rec.* **69,** 618 (1957).
289. Shapiro, J., Morell, A.G., and Scheinberg, I.H., *J. Clin. Invest.* **40,** 1081 (1971).
290. Shearer, G.D., Innes, J.R.M., and McDougal, E.I., *Vet. J.* **96,** 309, (1940).
291. Shields, G.S., Markowitz, H., Klassen, W.H., Cartwright, G.E., and Wintrobe, M.M., *J. Clin. Invest.* **40,** 2007 (1961).
292. Simpson, C.F., and Harms, R.H., *Exp. Mol. Pathol.* **3,** 390 (1964).
293. Sjollema, B., *Biochem. Z.* **267,** 151 (1933); **295,** 372 (1938).
294. Smallwood, R.A., Williams, H.A., Rosenoer, V.M., and Sherlock, S., *Lancet* **2,** 1310 (1968).
294a. Smith, B.S.W. and Wright, H., *J. Comp. Pathol.* **85,** 299 (1975); *Clin. Chim. Acta* **62,** 55 (1975).
295. Smith, H., *J. Forensic Sci. Soc.* **7,** 97 (1967).
296. Smith, H.W., and Jones, J.E.T., *J. Appl. Bacteriol.* **26,** 262 (1963).
297. Smith, M.S., *Br. Poult. Sci.* **10,** 97 (1969).
298. Smith, S.E., and Ellis, G.H., *Arch. Biochem.* **15,** 81 (1947).
299. Smith, S.E., and Medlicott, M., *Am. J. Physiol.* **141,** 354 (1944).
300. Smith, S.E., Medlicott, M., and Ellis, G.H., *Am. J. Physiol.* **142,** 179 (1944).
301. Sourkes, T.L., Lloyd, K., and Birnbaum, H., *Can. J. Biochem.* **46,** 267 (1968).
301a. Spoerl, R., and Kirchgessner, M., *Arch. Tierernaehr.* **26,** 25 (1976).
302. Spray, C.M., and Widdowson, E.M., *Br. J. Nutr.* **4,** 361 (1951).
303. Starcher, B.C., *J. Nutr.* **97,** 321 (1969).
304. Starcher, B.C., and Hill, C.H., *Comp. Biochem. Physiol.* **15,** 429 (1965).
305. Starcher, B.C., Hill, C.H., and Matrone, G., *J. Nutr.* **82,** 318 (1964).
306. Sternlieb, I., Morell, A.G., Tucker, W.D., Green, M.W., and Scheinberg, I.H., *J. Clin. Invest.* **40,** 1834 (1961).
307. Sturgeon, P., and Brubaker, C., *Am. J. Dis. Child.* **92,** 254 (1956).
308. Sutherland, A.K., Moule, G.R., and Harvey, J.M., *Aust. Vet. J.* **31,** 141 (1955).
308a. Suttle, N.F., *Proc. Nutr. Soc.* **32,** 24A (1973).
309. Suttle, N.F., *Trace Subst. Environ. Health–7, Proc. Univ. Mo. Annu. Conf., 7th, 1973,* p. 245 (1974).
310. Suttle, N.F., Angus, K.W., Nisbet, D.I., and Field, A.C., *J. Comp. Pathol.* **82,** 93 (1972).
311. Suttle, N.F., and Field, A.C., *J. Comp. Pathol.* **78,** 351 and 363 (1968).
312. Suttle, N.F., Field, A.C., and Barlow, R.M., *J. Comp. Pathol.* **80,** 151 (1970).
313. Suttle, N.F., and Mills, C.F., *Br. J. Nutr.* **20,** 135 and 149 (1966).
314. Tatum, H.J., *Am. J. Obstet. Gynecol.* **117,** 602 (1973).

315. Tauber, F.W., and Krause, A.C., *Am. J. Ophthalmol.* **26**, 260 (1943).
316. Teague, H.S., and Carpenter, L.E., *J. Nutr.* **43**, 389 (1951).
317. Thiers, R.E., and Vallee, B.L., *J. Biol. Chem.* **226**, 911 (1957).
318. Thomas, J., Elsden, D.F., and Partridge, S., *Nature (London)* **200**, 661 (1963).
319. Thompson, R.H., and Todd, J.R., *Res. Vet. Sci.* **16**, 97 (1974).
320. Thornton, I., Kershaw, G.F., and Davies, M.K., *J. Agric. Sci.* **78**, 157 and 165, (1972).
321. Todd, J.R., *in* "Trace Element Metabolism in Animals" (C.F. Mills, ed.), Vol. 1, p. 448. Livingstone, Edinburgh, 1970.
322. Todd, J.R., and Thompson, R.H., *Br. Vet. J.* **119**, 151 (1963).
323. Thompsett, S.L., *Biochem. J.* **34**, 961 (1940).
324. Turpin, R., Jerome, H., and Schmidt-Jubeau, H., *C.R., Seances Soc. Biol. Ses Fil.* **146**, 1703 (1952).
325. Ullrey, D.E., Miller, E.R., Thompson, O.A., Zutaut, C.L., Schmidt, D.A., Ritchie, H.D., Hoefer, J.A., and Luecke, R.W., *J. Anim. Sci.* **19**, 1298 (1960).
326. Ulsen, F.W. Van, *Tijdchr. Diergeneesk.* **97**, 735 (1972).
327. Uzman, L.L., Iber, F.L., Chalmers, T.C., and Knowlton, M., *Am. J. Med. Sci.* **231**, 511 (1956).
328. Van Campen, D., *J. Nutr.* **88**, 125 (1966).
329. Van Campen, D., *in* "Intestinal Absorption of Metal Ions, Trace Elements & Radionuclides" (S. Skoryna and D. Waldron-Edward, eds.), p. 211. Pergamon Press, Oxford, 1971.
330. Van Campen, D., and Gross, E., *J. Nutr.* **95**, 617 (1968).
331. Van Campen, D., and Mitchell, E.A., *J. Nutr.* **86**, 120 (1965).
332. Van Koetsveld, E.E., *Tijdsehr. Diergeneesk.* **83**, 229 (1958).
333. Van Ravesteyn, A.H., *Acta Med. Scand.* **118**, 163 (1944).
334. Van Wyk, J.J., Baxter, J.H., Akeroyd, J.H., and Motulsky, A.G., *Bull. Johns Hopkins Hosp.* **93**, 51 (1953).
335. Vohra, P., Gray, G.A., and Kratzer, F.H., *Proc. Soc. Exp. Biol. Med.* **120**, 447 (1965).
336. Wahle, K.W.J., and Davies, N.T., *Br. J. Nutr.* **34**, 105 (1975).
337. Walker-Smith, J., and Blomfield, J., *Arch. Dis. Child.* **48**, 476 (1973).
338. Wallace, H.D., McCall, J.T., Bass, B., and Combs, G., *J. Anim. Sci.* **19**, 1155 (1960).
339. Walravens, P.A., and Hambidge; K.M., *Am. J. Clin. Nutr.* (in press).
340. Weismann, N., Shields, G.S., and Carnes, W.H., *J. Biol. Chem.* **238**, 3115 (1963).
341. Weser, U., *Struc. Bonding (Berlin)* **17**, 1 (1973).
342. Widdowson, E.M., *Nature (London)* **166**, 626 (1960).
343. Wiederanders, R.E., *Proc. Soc. Exp. Biol. Med.* **128**, 627 (1968).
344. Wiener, G., and Field, A.C., *J. Comp. Pathol.* **79**, 7 (1969).
345. Wiener, G., and Field, A.C., *J. Agric. Sci.* **76**, 513 (1971); **83**, 403 (1975).
346. Wiener, G., Hall, J.G., Hayter, S., Field, A.C., and Suttle, N.F., *Anim. Prod.* **19**, 291 (1974).
347. Wilkerson, V.A., *J. Biol. Chem.* **104**, 541 (1934).
348. Wilkie, W.J., *Aust. Vet. J.* **35**, 203 (1959).
349. Wilson, J.F., and Lahey, M.E., *Pediatrics* **25**, 40 (1960).
350. Wintrobe, M.M., Cartwright, G.E., and Gubler, G.J., *J. Nutr.* **50**, 395 (1953).
351. World Health Organization, *W.H.O., Tech. Rep. Ser.* **532**, (1973).
352. Zook, E.G., Greene, F.E., and Morris, E.R., *Cereal Chem.* **47**, 720 (1970).

4

Molybdenum

I. MOLYBDENUM IN ANIMAL TISSUES AND FLUIDS

1. General Distribution

Molybdenum occurs in low concentrations, comparable with those of manganese, in all the tissues and fluids of the body. Species differences appear to be small (70) and there is very little accumulation in any particular organ with age (60, 110). Typical normal concentrations for the principal body organs for adult men, rats, and chicks are given in Table 15. Similar Mo concentrations for human tissues have more recently been reported by Hamilton and co-workers (54).

Changes in dietary Mo intakes are reflected in changes in the level of this element in the tissues (21, 25, 26). For example, Davis (26) increased the levels in the bones of rats from 0.2 to 9–12 ppm (d.b.) and in the livers from 1–2 to 11–12 ppm (d.b.) by raising the Mo content of the diet from below 1 to 30 ppm. The level of Mo in the tibia of chicks was increased 100-fold when a diet low in Mo was supplemented with 2000 ppm Mo as molybdate (25). Even higher concentrations were observed in the bones and soft tissues of rabbits and guinea pigs fed highly toxic or lethal doses of calcium molybdate or molybdemum trioxide (40).

The response of the tissues to changes in dietary Mo is greatly influenced by dietary inorganic sulfate levels. A reduction in Mo retention, and hence lower tissue Mo concentrations, was first demonstrated as a consequence of high-

TABLE 15
Normal Molybdenum Concentrations in Animal Organs[a]

Species	Liver	Kidney	Spleen	Lung	Brain	Muscle	Ref.
Adult man	3.2	1.6	0.20	0.15	0.14	0.14	110
Adult rats	1.8	1.0	0.52	0.37	0.24	0.06	60
Chickens	3.6	4.4	–	–	–	0.14	60

[a]ppm Mo on dry basis.

sulfate intakes by Dick (28) in the sheep. A similar effect is apparent in cattle (22) and rats (85). High dietary intakes of tungstate also reduce the levels of Mo in the tissues of rats and chicks (25, 60). The magnitude of the sulfate effect on the retention of molybdenum in the tissues of the sheep is illustrated in Table 16. The amounts of Mo in all the tissues examined are smaller at high- than at low-sulfate intakes, both when dietary Mo levels are high and low. It is further apparent from the figures in Table 16 that more than half the total body Mo is

TABLE 16
Influence of Dietary Molybdenum and Inorganic Sulfate Intake on the Molybdenum Content of the Tissues of Sheep[a]

	Molybdenum content (mg)			
	0.3 mg/day Mo intake		20.9 mg/day Mo intake	
	0.9 g/day sulfate intake	6.3 g/day sulfate intake	0.9 g/day sulfate intake	6.3 g/day sulfate intake
Liver	1.58	0.48	5.79	1.93
Kidney	0.17	0.02	1.17	1.32
Spleen	0.52	0.02	0.57	0.14
Heart	0.18	0.01	0.94	0.04
Lung	0.65	0.09	3.96	0.42
Muscle	5.84	0.08	28.6	1.92
Brain	0.01	0.01	0.09	0.01
Skin	6.62	1.50	58.9	3.44
Wool	15.2	0.99	26.9	1.14
Small intestine	0.26	0.03	0.88	0.79
Cecum	0.24	0.01	1.64	0.58
Colon	0.63	0.04	4.21	0.62
Skeleton	61.0	13.0	164.0	16.0
Total body Mo (mg)	92.9	16.8	297.7	28.4

[a]From Dick (32).

present in the skeleton, with the next largest proportions in the skin, wool, and muscles, and only about 2% of the total in the liver.

Increasing dietary Mo intakes raise the Mo content of cattle hair and sheep's wool, just as they raise these levels in other tissues, except where sulfate intakes are also high (21). Healy and co-workers (59) obtained a mean concentration of 0.16 ppm Mo (range 0.03–0.58) for 20 samples of wool, compared with a lower mean level of 0.06 ppm (range 0.02–0.13) for human hair obtained from the same area in New Zealand (10).

Human dental enamel can be relatively rich in molybdenum. A mean concentration of 5.5 ± 0.71 (range 0.7–39) μg Mo/g dry weight has been reported (79).

2. Molybdenum in the Liver

On normal diets the Mo level in the liver mainly lies within the range 2–4 ppm Mo (d.b.) in a very wide range of animal species (13, 20, 49, 70, 110). Similar concentrations occur in the livers of newborn lambs, indicating that this metal is not normally stored in the fetal liver during pregnancy. However, molybdenum can readily pass the placental barrier because Mo concentrations 3–10 times normal were observed in the livers of newborn lambs from ewes receiving a high-Mo diet (20).

Adult sheep and cows retain Mo concentrations in their livers of 25–30 ppm, so long as they are ingesting large or moderately large amounts of the element. These high levels rapidly return to normal when administration of the extra Mo ceases. The extent of retention in the liver and human tissues and the rate at which the levels fall depend on the amount and proportion, relative to Mo, of the inorganic sulfate. The level of Mo in the liver therefore gives little indication of the animal's dietary Mo status and is of limited diagnostic value for this purpose, unless the dietary sulfur and protein status is also known.

3. Molybdenum in Blood

Sheep grazing pastures normal in Cu and low in Mo average 1 μg Mo/100 ml whole blood, compared with 6 μg/100 ml for the same sheep consuming a proportion of their diet as low-sulfate hay (11). A level of 6 μg Mo/100 ml is also normal for cattle grazing herbage normal in Cu and low in Mo (20). When the animals were dosed with molybdate equivalent to a dietary intake of 30 ppm Mo, the levels rose to 60–80 μg/100 ml in young cattle and 240–340 μg/100 ml in breeding ewes. When the Mo intake is increased in this way the blood Mo concentration immediately rises until a steady value is reached, the level of which depends on the amount of Mo ingested daily. Dick (29) found the steady values of sheep to increase from 2 to 495 μgMo/100 ml whole blood when the intake was raised from 0.4 to 96 mg Mo/day. This large increase was accomplished without a change in the ratio blood Mo:intake Mo.

Within certain limits blood Mo levels in sheep are also markedly and inversely dependent on the inorganic sulfate intakes (29). At low-Mo and sulfate intakes over 70% of blood Mo was present in the red cells. At high intakes of both a very small proportion of the much larger amount of blood Mo was present in the red cells, with most of the increase occurring in the plasma. In Mo-supplemented sheep and guinea pigs most of the plasma has recently been shown to exist in the form of protein–Cu–Mo compounds insoluble in trichloroacetic acid (101a). The possible significance of the formation of these compounds to the Cu–Mo interaction is discussed in Section V, 3.

In a study of the blood of male adults resident in 19 cities in different parts of the United States, Allaway *et al.* (1) found that more than 80% of the samples contained less than 0.5 μg Mo/100 ml and only about 3% contained more than 10 μg. Considerable variation from site to site was apparent, with little evidence of a broad geographical pattern. Bala and Liftshits (8) found the mean Mo contents of the whole blood of healthy human subjects to be 1.47 ± 0.12 μg/100 ml, evenly distributed between the plasma and the red cells. They state that the molybdenum is "firmly bound only to the erythrocyte and plasma proteins." In patients with leukemia, whole blood and red cell Mo was significantly increased, with no change in the plasma. In patients with various types of anemia, whole blood, red cell, and plasma Mo levels were all reduced. In posthemorrhagic anemia the blood Mo decreased solely as a result of a decrease in the number of red cells.

4. Molybdenum in Milk

The molybdenum content of milk is extremely susceptible to changes in dietary Mo intakes in cows (5, 58, 71), goats, (58), and ewes (62). Archibald (5) found "normal" cow's milk to average 73 μg Mo/liter (range 18–120). This concentration was raised to a mean of 371 μg/liter by feeding the same cows 500 mg Mo as ammonium molybdate daily. Hart *et al.* (58) obtained smaller increases, from 20–30 to 40–60 μg Mo/liter, by increasing the cow's Mo intakes by 50 and 100 mg Mo/cow/day. In a more recent study (117) little individual difference between cows was observed, or between forage Mo levels and the Mo content of milk, where the difference in forage levels was only twofold. Milk values all fell between 32 and 48 ppb Mo, or close to 32 and 48 μg Mo/liter.

The powerful effect of Mo intake on milk Mo concentration is equally apparent from studies with ewes. Hogan and Hutchinson (61) found the milk of ewes grazing a low-Mo pasture (less than 1 ppm) to average less than 10 μg Mo/liter, while that of ewes on a high-Mo pasture (13 ppm) averaged no less than 980 μg Mo/liter. These workers also demonstrated the remarkable effect of sulfate in reducing the Mo content of milk when this level is high. At the end of 3 days of dosing ewes grazing on a high-Mo pasture (25 ppm Mo) with 23 g sulfate ion per day, the concentration of molybdenum in the milk had fallen to 137 μg Mo/liter, compared with 1043 μg/liter in comparable ewes receiving no additional sulfate.

In ewe's milk a high proportion of the Mo is associated with the aqueous phase, i.e., it remains after separation of the fat by centrifugation and the casein and albumin by clotting (61). All the Mo of the milk of cows on normal low-Mo rations is bound to xanthine oxidase, so that the xanthine oxidase activity of such milk is proportional to its Mo content (54, 65). The rapid rise in the Mo content of milk that follows administration of molybdate to the cow is not accompanied by an increase in its xanthine oxidase activity (58, 71). The xanthine oxidase content of goat's milk is lower than that of cow's milk, and there is no correlation between the enzyme activity and the Mo content as there is in normal cow's milk (58). The ingestion of sodium tungstate by cows and goats reduces the xanthine oxidase activity of their milk (92).

The composition of milk xanthine oxidase approximates closely 2 molecules of FAD, 2 g-atoms Mo, and 8 g-atoms Fe/molecule of protein, with a molecular weight of about 275,000 (57). This and other purified xanthine oxidase samples are contaminated by significant amounts of an inactivated form containing less Mo (demolybdoxanthine oxidase). The authors contend that nutritional rather than genetic factors determine the relative amounts of the two forms. The chemical nature of the xanthine oxidase in cow's milk was shown to change with the composition of the diet, particularly with the amounts and proportions of Mo and Cu ingested. When cows were fed on pasture, copper as well as Mo and Fe were found by Kovalskii and co-workers (74) in the xanthine oxidase of the milk. At normal Cu:Mo ratio in the diet, these workers succeeded in dividing the xanthine oxidase of milk into electrophoretically different isoenzymes designated KO_2, KO_{4a}, and KO_{4b}; KO_2 contained only Mo (specific activity 5.0 U/mg protein) and KO_{4a} + KO_{4b} were isoenzymes which contained copper (total specific activity about 1.0). Feeding additional copper to the cows led to a decrease of Mo, or to its loss from the KO_2 isoenzyme, and to an increase in copper in the sum of the isoenzymes KO_{4a} + KO_{4b}. Feeding Mo salts increased the Mo content of KO_2 and decreased the Cu content of isoenzymes KO_{4a} + KO_{4b}. These findings provide an important example of the capacity of the organism to adapt to alterations in dietary conditions.

II. MOLYBDENUM METABOLISM

Molybdenum is readily and rapidly absorbed from most diets and inorganic forms of the element. The hexavalent water-soluble forms, sodium and ammonium molybdate, and the molybdenum of high-Mo herbage, most of which is water soluble, are particularly well absorbed by cattle (43). Even such insoluble compounds as MoO_3 and CaMoO4, but not MoS_2, are well absorbed by rabbits and guinea pigs when fed in large doses (40). Yearling cattle have been shown to absorb oral doses of ^{99}Mo much less rapidly than growing pigs (12). The urine is a major route of excretion of Mo in pigs (12, 100), rats (91), and man (96, 97, 112), but apparently not in cattle (12) or sheep on low-sulfate intakes (28, 98).

Studies of Mo metabolism are of limited value unless the sulfur status of the diet is controlled. Protein intakes can also be important since this can be a source of endogenous sulfate. The potent influence of sulfate on Mo excretion, retention, and the route of excretion of absorbed Mo is illustrated by the following results obtained by Dick (28). Sheep fed a diet of oaten chaff ($<0.1\%$ sulfate) plus 10 mg Mo/day excreted 63% of this molybdenum in the total excreta during a period of 4 weeks, of which 3–4.6% appeared in the urine. When fed a diet of lucerne chaff (0.3% sulfate) plus 10 mg Mo/day the recovery in the total excreta was 96%, of which 50–54% appeared in the urine. Fractionation of the lucerne revealed that inorganic sulfate was the factor responsible for the marked effect on molybdenum excretion. The administration of a single oral dose of potassium sulfate to the sheep on the cereal hay (low-sulfate) diet induced a rapid rise in urinary Mo excretion (Fig. 4). Sodium sulfate produced the same effect without the diuresis that accompanies the use of potassium sulfate, whereas potassium chloride induced diuresis had no effect on Mo excretion. Similar results were later obtained by Scaife (98), who fed sheep a low- and a high-sulfate diet plus 50 mg Mo in each case. On the low-sulfate diet only 5% of the Mo appeared in the urine compared with 30–40% on the high-sulfate diet.

Inorganic sulfate increases urinary Mo excretion in the marsupial *Setonix brachyurus* without significantly increasing urine volume (9) and alleviates Mo toxicity in all species studied. Whether the effect of sulfate on toxicity is mediated entirely through an influence on the pattern of Mo excretion is doubtful. In rats (85) and cattle (22), as in sheep, sulfate reduces Mo retention in the tissues, presumably through increased urinary excretion. No such reduction in tissue Mo retention by sulfate occurs in the chick, despite a marked beneficial effect on the manifestations of Mo toxicity (25).

In the sheep sulfate limits Mo retention both by reducing intestinal absorption and by increasing urinary excretion, the extent of each depending on the previous history of the animal with respect to Mo and sulfate intakes (29, 98). The increased urinary excretion is not a passive result of the greater urinary volume that occurs on high-sulfate diets (28). Several diuretics increase urine volume without increasing Mo excretion (20). The sulfate effect is highly specific and in sheep is not shared by such anions as tungstate, selenate, silicate, permanganate, phosphate, malonate, and citrate (29, 98). The capacity of sulfate

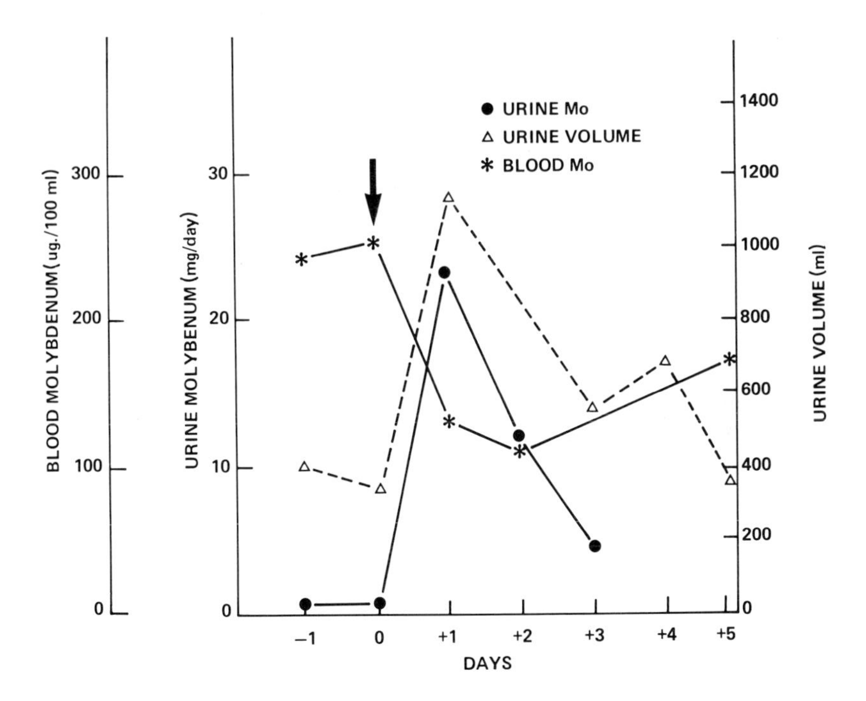

Fig. 4 Effect on blood and urine Mo of 11 g potassium sulfate given by mouth, at the time indicated by the arrow, to a sheep on a diet of chaffed hay and a Mo intake of 10 mg/day. From Dick (28).

to alleviate Mo toxicity in the rat is also not shared by citrate, tartrate, acetate, bromide, chloride, or nitrate (115). However, sulfate of endogenous origin can be just as effective as dietary inorganic sulfate. This is indicated by the effects of high protein diets, by the catabolic breakdown of body tissue, and by the administration of thiosulfate and methionine to sheep (29, 98). The protective action of methionine, cystine, and thiosulfate administration to rats fed high-Mo diets (48, 116) may also result from the production of endogenous sulfate.

The influence of sulfate on Mo absorption and excretion is explained by Dick (29) on the hypothesis that inorganic sulfate interferes with, and if its concentration is high enough, prevents the transport of Mo across membranes. Such an effect would increase Mo excretion through the rise in the sulfate concentration in the ultrafiltrate of the kidney glomerulus which follows high-sulfate intakes, impeding or blocking Mo reabsorption through the kidney tubule. The mechanism of this postulated interference with membrane transport is still unknown. The metabolic interaction between Mo and Cu and the mechanism of the profound effect of Mo plus sulfur on Cu retention are discussed in Section V, 3.

III. REQUIREMENTS AND FUNCTIONS OF MOLYBDENUM

1. Functions

The first indication of an essential role for molybdenum came in 1953 when two groups of workers independently discovered that the flavoprotein enzyme xanthine oxidase is a Mo-containing metalloenzyme which is dependent for its activity on the presence of this metal (93, 94). In the following year molybdenum was shown to be a component of aldehyde oxidase and to be required for its catalytic activity (81). Subsequently, sulfite oxidase was also found to be a Mo-containing metalloenzyme (19). A molecular mechanism for the action of Mo in these enzymes and in nitrogenase and nitrate reductase has recently been put forward, based on the coordination chemistry of the metal (104). The distributions, substrate specificities, and electron acceptor specificities of xanthine oxidase and aldehyde oxidase in mammalian tissues have also been assessed and compared (76).

The biochemical functions of Mo in animals, apart from its reactions with copper, are related to the formation and activities of xanthine oxidase, aldehyde oxidase, and sulfite oxidase. Xanthine oxidase may participate in the uptake and release of iron from ferritin, although its quantitative significance is now doubtful, as discussed in Chapter 2. Molybdenum participates in the reaction of xanthine oxidase with cytochrome c (80) and facilitates the reduction of cytochrome c by aldehyde oxidase (46). Tissue concentrations of this enzyme in experimental animals have been related to the dietary Mo intake (56).

In the early experiments with rats fed low-Mo diets intestinal xanthine oxidase levels were reduced below normal, but neither the growth nor the purine metabolism of the depleted animals was affected (93, 94). Similar low-Mo diets were equally well tolerated by chicks but when tungstate, a Mo antagonist, was added to give W:Mo ratios of 1000:1 and 2000:1, the birds exhibited reduced growth and tissue molybdenum and xanthine oxidase levels, and their capacity to oxidize xanthine to uric acid was lowered (60). All these effects were prevented by additional dietary molybdenum. These findings were confirmed by Leach and Norris (77) both with chicks fed a purified casein diet containing 0.5–0.8 ppm Mo plus added tungsten and with highly depleted chicks from hens fed a special low-mineral diet. Subsequently Richert and co-workers (95) showed that Mo depletion influences uric acid formation in the rat. The uric acid and allantoin normally excreted by this species were not formed in the absence of xanthine oxidase in the liver.

Administration of tungsten to rats at levels of 1–100 ppm in the drinking water was later shown to result in proportionate decreases in liver sulfite oxidase activity as well as liver xanthine oxidase activity, and in total liver Mo (69). The tungsten effect was also observed when the metal was injected intraperitoneally. Injection of Mo into rats depleted of hepatic sulfite oxidase and xanthine oxidase activities by tungsten treatment resulted in significant restoration of both activities within a few hours. Tungsten-treated rats, totally deficient in sulfite oxidase activity, appeared healthy but were markedly more susceptible than normal animals to toxicity from SO_2 and bisulfite ions. The single reported human patient with a deficiency in sulfite oxidase activity of genetic origin displayed severe physical and neurological abnormalities and died in childhood (66a).

2. Molybdenum Requirements

The minimum dietary requirements for molybdenum compatible with satisfactory growth and health cannot be given in even approximate terms for any animal species, including man, nor has Mo deficiency been observed under natural conditions with any species.

The rat and the chick have an extremely low-Mo requirement, so much so that they have been shown to grow normally, reproduce, and oxidize xanthine normally on diets containing only 0.2 ppm Mo (60). A significant growth response to molybdenum was obtained in lambs fed a semipurified diet containing 0.36 ppm Mo when this diet was supplemented with molybdate at the rate of 2 ppm Mo (37). Cellulose digestibility was significantly improved on the Mo-supplemented diets. This led the authors to suggest that Mo stimulates growth in lambs by increasing cellulose degradation by rumen organisms. A growth response from supplementary Mo in lambs fed a low-Mo purified diet was confirmed by Ellis and Pfander (38), but these interesting findings do not appear to have been followed up by other workers. Many pastures grazed regularly by sheep and cattle contain lower Mo concentrations than the 0.36 ppm present in the deficient diets just cited. Such low-Mo pastures favor the accumulation of copper in the tissues of sheep and under certain conditions can lead to chronic Cu poisoning, but there is no evidence that their consumption adversely affects growth or feed utilization or leads to any other disabilities in the animal. The minimum Mo requirements of ruminants and other animals and the dietary factors that affect these requirements clearly require further critical study.

Human Mo requirements are unknown and the results of short-term balance studies with this element are not very informative because of the widely varying amounts of Mo involved. Thus Robinson *et al.* (96) found young adult women to be in Mo equilibrium on intakes of 48–96 μg Mo/day, while Tipton *et al.* (111) found one of three subjects ingesting 100 μg Mo/day was in negative Mo balance. Molybdenum intakes from different foods and dietaries are considered in Section IV, 1.

3. Molybdenum and Dental Caries

Several experimental studies with rats and epidemiological studies in man have led to claims that Mo exerts a beneficial effect on the incidence and severity of dental caries, and that it can enhance the well-established effect of fluoride. Thus Buttner (17) reported that the combined administration of 25 or 50 ppm Mo and 50 ppm F in the drinking water was more effective in reducing caries incidence in rats than the ingestion of water containing 50 ppm F alone. Subsequently Malthus and co-workers (82) found no difference between the caries scores of animals fed a combination of Mo and F and these given F alone. Moreover, the uptake of fluoride by intact human enamel is not increased by the addition of Mo salts to solutions of NaF (47), and neither F retention in the bones of rats nor the F content of the saliva is increased when ^{99}Mo and ^{18}F are given simultaneously by stomach tube, compared with animals receiving only ^{18}F (39).

The earlier epidemiological studies were critically assessed by Hadjimarkos (51), who pointed out that the evidence for the claims for Mo were contradictory and inconclusive. In a later study in California the teeth of children residing in high-Mo areas where molybdenosis in cattle occurred were compared with those of children in "normal" or low-Mo areas (23). No difference between the two groups in the prevalence of caries was observed, despite the fact that the level of Mo in the enamel of teeth collected from children in the high-Mo area was about 10 times greater than that of children from the control area. In another study in England of two groups of 12-year-old children living continuously in high- or low-Mo areas, a 20% lower prevalence of caries was observed in the high-Mo area (4). However, as Hadjimarkos (52) pointed out, the lower prevalence of caries seen in children from the high-Mo area could have resulted from the presence of fluoride, since teeth from the children of this group were reported to contain 34% more fluoride than the teeth of the children from the low-Mo area, although the drinking water used by both groups was low in fluoride. The actual average F levels for the teeth examined were high-Mo area, 194; low-Mo area, 128 ppm.

IV. SOURCES OF MOLYBDENUM

1. Molybdenum in Human Foods and Dietaries

The Mo content of foods varies greatly within and among the different classes of foodstuffs comprising human dietaries, particularly in relation to the soil type from which they are derived. In fact Warren and co-workers (119), in a study of the trace mineral content of vegetables from different locations in British Columbia, reported that some vegetables had 500 times more molybdenum than others. In another report by Warren (118) cabbage was found to range from 0.1 to 25.0 (mean 2.8), carrots from 0.1 to 20.0 (mean 2.0), and potato from 0.1 to 6.0 (mean 0.8) ppm Mo, all on the dry basis. A similar wide variation is apparent from a recent French study (67) in which the Mo concentrations in broad beans was found to range remarkably from 0.1 to 31.0 ppm (d.b.). Narrower ranges of 0.2–4.7 ppm Mo for leguminous seeds and 0.12–1.14 ppm Mo for cereal grains were reported by Westerfeld and Richert (120). In a study of the mineral content of North American wheats and flours, the wheats averaged 0.48 ppm and the white flours 0.25 ppm Mo (18). Much higher Mo levels have been reported for the millet grain *sorghum vulgare* than for rice, or for wheat as previously cited (27).

Schroeder and co-workers (99) suggest that the probable Mo intake from standard good diets might average 350 μg/day for adults. This estimate appears high when compared with the 128 ± 34 μg Mo/day reported for adult diets in England (55) and the 48–96 μg Mo/day obtained for four young adult women in New Zealand (96). The Mo intakes of individuals in some areas of Armenia (Soviet Union) have been reported to be as high as 10–15 mg/day compared with 1–2 mg/day in nearby areas, where the population had a low incidence of gout (73). At Mo intakes up to 1.5 mg/day, which can occur in some areas in individuals with sorghum grain as a staple dietary item, uric acid metabolism is not affected but Cu excretion is greatly increased (27), as discussed in Section V, 1.

2. Molybdenum in Animal Feeds and Forages

The level of Mo in pasture herbage varies greatly, depending mainly on the soil conditions. Values as low as 0.10 ppm and as high as 100 ppm (d.b.) have been recorded. Whitehead (121) gives 0.1–4.0 mg Mo/kg dry matter as the "normal" range of English grassland herbage, and a range of 0.6–12.0 has been reported for British Columbian range forages (44). Miltmore and Mason (88) found the Mo concentrations of all the Canadian feeds examined to be low, with an overall mean of 1.7 ppm. Thirty-five percent of the feeds were below 1 ppm and 85% below 3 ppm. Corn silage and grains were in the lowest class. Todd (113) gives a mean of 0.25 ± 0.22 for oats and 0.30 ± 0.35 mg Mo/kg d.m. for barley. These values compare well with the mean of 0.48 ppm given in Section IV,1 for wheat, but are well below those obtained for sorghum as mentioned in the same section. The importance of the ratio of the levels of Mo in feeds to the levels of copper is discussed in Section V,3.

V. MOLYBDENUM TOXICITY

1. General Considerations

The effect on the animal of high dietary Mo intakes depends on the species and age of the animal; the amount and chemical form of the ingested Mo; the copper status and copper intake of the animal; the inorganic sulfate and total sulfur content of the diet and its content of substances such as protein, cystine, and methionine, capable of oxidation to sulfate in the body; and the level of intake of some other metals, including zinc and lead (50).

Species differences in tolerance to molybdenum are substantial. Cattle are by far the least tolerant, followed by sheep, while horses and pigs are the most tolerant of farm livestock. The high tolerance of horses is apparent from their failure to show signs of toxicity on "teart" pastures that severely affect cattle, and the high tolerance of pigs is evident from the report of Davis (26) that intakes of 1000 ppm Mo for 3 months induced no ill effects. This is 10–20 times the levels that result in drastic scouring in cattle. These species differences are not understood. The high tolerance of pigs cannot be due to poor absorption, judging by the results of experiments with ^{99}Mo which indicate rapid absorption and rapid excretion of the metal in this species (12, 100). Rats, rabbits, guinea pigs, and poultry are not so tolerant of Mo as pigs, but are much more tolerant than cattle. Intakes of 2000 ppm Mo induce severe growth depression in chicks, accompanied by anemia when the dietary level is raised to 4000 ppm Mo (7, 25, 75). At 200 ppm Mo some inhibition of chick growth occurs (48), and at 300 ppm of dietary Mo turkey poult growth is depressed (75).

The clinical manifestations of Mo toxicity also vary among different species. Growth retardation or loss of body weight is an invariable result of high-Mo intakes, but diarrhea is a conspicuous feature only in cattle. Diarrhea has not been reported in rabbits and guinea pigs ingesting sufficient Mo to produce marked loss of weight and death (40). In young rabbits the molybdenosis syndrome is characterized further by alopecia, dermatosis, and severe anemia, with a deformity in the front legs of some animals but no achromotrichia or diarrhea (6). In molybdenotic rats deficient lactation and male sterility associated with some testicular degeneration (68) and connective tissue changes resulting in mandibular or maxillary exostoses (115) have been observed. Connective tissue changes, associated mainly with the humerus, also occur in sheep grazing high-Mo pastures (5–8 ppm rising to 20 ppm d.b.) (62). The lesions were characterized by lifting and hemorrhage of the periosteum and the muscle insertions and sometimes spontaneous fractures. The condition was largely prevented by supplementary copper. Body weight gains, wool weights, and blood hemoglobin levels were also significantly improved when the sheep grazing the high-Mo pastures received supplementary copper. Supplementation with copper similarly prevents growth depression brought about by high-Mo intakes in rabbits (6) and chicks (7).

Supplementary methionine and cystine can be as effective as copper in alleviating Mo toxicity in rats (48) and sheep (98), and thiosulfate administration is equally effective in sheep (98). As these substances are all capable of oxidation to sulfate in the tissues, they probably act in the same manner as inorganic sulfate, by reducing Mo retention in the tissues by promoting its excretion in the urine, as discussed in Section II. An ameliorating effect of high-protein diets, which would also be expected to yield endogenous sulfate, has similarly been demonstrated in Mo toxicity in rats (84). The effect was associated with a depression in blood and liver Mo levels. In the chick neither methionine nor sulfate appears to reduce Mo retention in the tissues, although the latter is highly effective in preventing the growth depression induced by high-Mo diets (25).

The activities of several enzymes and thyroxine are affected in the molybdenotic animal. Molybdenosis, severe enough to kill some adult rabbits by 3–4 weeks, was accompanied by falls in plasma thyroxine (T_4) to 2.31 ± 0.34 μg/100 ml, compared with control values of 4.4 ± 0.34 μg/100 ml. The thyroid showed increased colloidal storage and epithelial inactivity and a decrease of epithelial cells from 25.8% in the controls to 12.8% in the Mo-treated rabbits. No such glandular changes were observed in the rabbits given a diet restricted to the intakes of the Mo-fed rabbits but without the molybdenum (122). Dinu and co-workers (34) demonstrated a significant fall in liver glucose 6-phosphate activity in rats given 1 and 10 mg Mo/kg, but the activities of acid and alkaline phosphatase, xanthine oxidase, aldolase, and succinic oxidase and dehydrogenase were unaffected by these relatively low-Mo supplements. In rats suffering from Mo toxicity alkaline phosphatase activity is decreased in the liver and increased in the kidney and intestine (114, 123). Alkaline phosphatase in the blood serum of young sheep grazing moderately high- to high-Mo pastures is also increased above those receiving additional copper (62).

Liver sulfide oxidase activity is depressed in molybdenotic rats but liver cysteine desulfhydrase and kidney aryl sulfatase activities remain unaffected. The oxidation of L-cysteine sulfinate by liver homogenates is also unaffected (114, 123). The depression in liver sulfide oxidase activity suggests that accumulation of sulfide in the tissues contributes to the syndrome of Mo toxicity. This is supported by the fact that feeding calcium sulfide (86) or large quantities of cystine, which could lead to excessive generation of sulfide in the liver through cysteine desulfhydrase activity, causes anemia, diarrhea, and death in rats consuming high-Mo diets (53). In both cases the disorders were prevented by supplying additional copper. The endogenous production of sulfide therefore appears to be a significant factor in molybdenosis in the rat and in the appearance of signs of copper deficiency.

The anorexia of molybdenosis results from a voluntary rejection of the diets, consequent on the development by the rats of an ability to recognize the presence of molybdenum in the diet (89). The ability to reject high-Mo diets through sensory, probably olfactory, recognition of the molybdate requires a learning or conditioning period. Since sulfate alleviates the diarrhea that develops on the high-Mo diet, Monty and Click (89) suggest that the rats learn to associate a gastrointestinal disturbance with a sensory attribute of diets containing toxic levels of molybdenum.

Responses to cold stress are increased by elevated dietary levels of molybdenum. Winston and co-workers (124) studied rats given Na_2MoO_4 in the drinking water at 10 and 0 ppm Mo from birth to 6–11 months of age and kept at 2°–3°C for 4 days, or at room temperature for this period. The elevated intakes of Mo had a significant effect on the response to cold, judging by the increased body weight loss, even for animals acclimated to the metal for some time. The basis for this effect has not yet been determined.

In man a high incidence of gout has been associated with abnormally high-Mo concentrations in the soils and plants in some parts of the Soviet Union (73). Humans and livestock exposed to these high-Mo intakes (10–15 mg Mo/day in man) displayed abnormally high serum uric acid levels and tissue xanthine oxidase activities. No increase in uric acid excretion was apparent in individuals consuming diets supplying lower Mo levels up to 1.5 mg/day, although these levels had a marked effect on urinary Cu excretion (27). This suggests that uric acid metabolism in man is altered only at very high-Mo intakes.

2. Molybdenosis in Cattle

Severe molybdenosis in cattle occurs under natural grazing conditions in many parts of the world (14, 20, 36, 42, 43, 90). In England, where the condition was first studied, it is known as "teart" and in New Zealand as "peat scours." All cattle are susceptible to molybdenosis, with milking cows and young stock suffering most. Sheep are much less affected and horses are not affected at all in teart areas. The characteristic scouring varies from a mild form to a debilitating condition so severe that cattle can suffer permanent injury or death. Within a few days of being placed on some teart pastures cattle begin to scour profusely and develop harsh, staring, discolored coats. They usually recover rapidly when transferred to normal pastures lower in molybdenum. Typical teart pastures contain 20 to as high as 100 ppm Mo d.b., compared with 3–5 ppm or less in nearby healthy herbage (78).

Following a report from Holland of a scouring condition in cattle responsive to copper therapy (15), teart was successfully treated with copper sulfate (42). Treatment at the very high rate of 2 g/day for cows and 1 g/day for young stock, or the intravenous injection of 200–300 mg/day, effectively controlled the

Mo-induced scouring and provided a practical means of field control. If the high-Mo intakes are prolonged without supplementary copper, a depletion of tissue Cu to deficiency levels occurs with associated hypocupremia, but the scouring and loss of condition of severe molybdenosis can occur without a concomitant hypocuprosis. In teart there is no inflammation or local damage to the intestines. Furthermore the scouring can be induced by intravenous injection of molybdate over 2–3 weeks (2). This molybdenum probably reaches the lumen of the intestine where it may disturb bacterial activity in this site, with resulting diarrhea. Dick and co-workers (33) have recently raised the possibility that the scouring that occurs in cattle, but not in horses, on teart pastures of high-Mo content may be the result of the formation of thiomolybdates in the rumen, and the curative effect of copper sulfate the result of the formation of insoluble copper thiomolybdates.

The possibility that the sulfate radicle may also be of some significance in the treatment of Mo toxicity with copper sulfate has been suggested by Dick (30). He showed that Mo-induced scouring in a calf can be corrected within 4 days by daily drenching with 2 g copper sulfate and within 6 days by drenching with an equivalent amount of sulfate as K_2SO_4.

Scouring and weight loss are such dominant manifestations of molybdenum toxicity in cattle at very high-Mo intakes that other disorders tend to be obscured. At moderately high levels of Mo in the herbage of some areas a disturbance of phosphorus metabolism occurs, giving rise to lameness, joint abnormalities, osteoporosis, and high serum phosphatase levels (26). Connective tissue changes, associated with the humerus, and some spontaneous bone fractures have been observed in sheep at comparable Mo intakes in the absence of supplementary Cu, as mentioned in the preceding section (62). Under such conditions, cows may conceive with difficulty and young male bovines exhibit a complete lack of libido. The testes reveal marked damage to interstitial cells and germinal epithelium with little spermatogenesis (109). These changes are comparable with those reported in molybdenotic rats (68, 115).

3. Molybdenum–Copper Interrelations

The quantitative nature of the physiological antagonism between Mo and Cu is apparent from a comparison of the conditions under which the two cattle diseases, teart and peat scours, occur and can be prevented or cured. Peat scours occurs in restricted areas in New Zealand where the pastures are (a) higher than normal in Mo but usually well below the levels typical of teart herbage and (b) subnormal in Cu. Under these circumstances the onset of scouring is delayed until the tissue Cu stores are depleted, and control of the scouring can be achieved merely by raising the Cu content of the pastures or the Cu intakes of the animals to normal levels (20). The Mo intakes from peat scours pastures are not high enough to require the very large Cu supplement needed to control teart.

The above field findings are supported by considerable experimental evidence which shows clearly that the toxicity of any particular level of dietary Mo is affected by the ratio of dietary Mo to dietary Cu. Miltmore and Mason (88) claim that the critical Cu:Mo ratio in animal feeds is 2.0 and that feeds or pastures with lower ratios would be expected to result in conditioned Cu deficiency. In a study of fodders and grains grown in British Columbia these workers found an extremely wide Cu:Mo ratio, ranging from 0.1 to 52.7 in individual samples. The results of a study of English pastures also indicate the importance of the Cu:Mo ratio to the incidence of hypocuprosis in sheep, but they point to a higher critical ratio than 2.0, perhaps closer to 4.0 (3).

Recent evidence obtained by Suttle (105) indicates that Mo can interfere with Cu metabolism at dietary levels below the 5 ppm or more that occur in teart and peat scours pastures, or the high-Mo levels that have usually been employed in experiments with laboratory species. Groups of hypocupremic ewes were repleted with a semipurified diet containing 8 mg Cu/kg and one of four dietary levels of Mo: 0.5, 2.5, 4.5, and 8.5 mg/kg. Using rate of recovery in plasma Cu as a measure of the efficiency of Cu utilization, the successive increments in dietary Mo were found to decrease that efficiency by 40, 80, and 40%, respectively (see Table 17). The results suggest that differences of 1 ppm in dietary Mo are of significance to ruminants with respect to Cu utilization. In a further experiment with Cu-depleted guinea pigs repleted with diets containing 8 mg Cu/kg and 0.6, 4.1, 26, or 104 mg Mo/kg, the lowest increments in Mo, 3.5 mg/kg, decreased the response in liver Cu as much as the highest, 78 mg/kg, a reduction of 23% being recorded. Some of the earlier data of Dick (28, 31) and Wynne and McClymont (125) with sheep had also shown an appreciable reduction in Cu retention from Mo intakes less than 0.5 mg/day or 8 ppm, respectively, when inorganic sulfate was added to the diets.

Balance studies in man provide further evidence that relatively low levels of dietary Mo can affect Cu metabolism. Deosthale and Gopalan (27) observed significant increases in urinary Cu excretion from 24 μg/day at a Mo intake of 160 μg/day to 42 and 77 μg/day at Mo intakes of 540 and 1540 μg/day, respectively. There was no effect on fecal Cu excretion but the high serum Cu levels suggest that the Mo induced increased tissue Cu mobilization. In this study the higher dietary Mo levels were obtained from the Mo present naturally in high-Mo samples of sorghum grain. This raises the possibility of a Mo-induced Cu deficiency arising in populations where such grains comprise the staple diet.

It is important to appreciate that Mo and sulfate can either increase or decrease the Cu status of an animal, depending on their intakes relative to that of copper. Thus as shown in Chapter 3, chronic Cu poisoning can occur in sheep with moderate Cu intakes and very low levels of Mo and sulfate. Conversely, depletion of the animal's copper reserves, to the extent of clinical signs of Cu

TABLE 17
Distribution of Copper in the Plasma of Initially Hypocupremic Sheep Given Copper-Supplemented Diets of Various Molybdenum Contents for 65 Days[a]

Dietary Mo (mg/kg)	Total plasma Cu	Ceruloplasmin Cu (mg Cu/liter plasma)	Direct-reacting Cu	Residual Cu[b]
0.5	424 ± 47[c]	238 ± 46	33 ± 4	153
2.5	160 ± 28	28 ± 15	24 ± 2	108
4.5	69 ± 34	3 ± 10	30 ± 4	36
8.5	139 ± 37	56 ± 27	104 ± 12	–21

[a]From Suttle (105).
[b]Determined by difference.
[c]Mean of 7 ewes ± SE.

deficiency, can arise on normal Cu and high Mo and sulfate intakes (125). Furthermore, sulfate can either aggravate or ameliorate the toxic effects of Mo, depending on the Cu status of the animal. Gray and Daniel (49) showed that when the Cu stores of rats were low and a Cu-deficient diet was fed, small amounts of Mo produced toxic symptoms that were intensified by the simultaneous addition of sulfate. By contrast, when the Cu stores and dietary Cu were adequate, larger amounts of Mo were required to produce molybdenosis, and sulfate completely prevented the harmful effects of Mo. Furthermore, several studies with pigs have failed to show significant reductions in tissue Cu levels from high dietary intakes of Mo and sulfate (24, 72), even at levels of 1500 ppm Mo, which substantially depressed growth and reduced plasma clearance of injected ^{64}Cu (103).

The Cu–Mo–S interaction can be modified by other dietary factors. Dick (29) showed that Mn intakes can block or antagonize the limiting effect of Mo on Cu retention in the sheep, even in the presence of adequate sulfate, and that when sheep are on a high-protein diet, Mo and Mn together exert a severely limiting effect on Cu retention. Copper deficiency has been produced in lambs by feeding high amounts of molybdate and sulfate to sheep fed grass cubes (41), whereas no such effect was obtained by Butler and Barlow (16) with a diet of hay and oats. Hogan *et al.* (64) were also unable to reduce the liver and brain Cu of lambs to deficiency levels by dosing their mothers with molybdate and sulfate for 9 months while grazing pastures normal to low in Cu content. Subsequently evidence was obtained of the presence of some dietary factor in the concentrate fed which facilitated excessive Cu storage in the liver (63). The effect of this unidentified factor on liver Cu storage was reduced but not eliminated by feeding additional molybdate and sulfate.

The mechanism of the complex effect of Mo plus sulfur on Cu retention has been extensively studied during the last 20 years. Suttle and Field (108) found that the addition of Mo to the diet of hypocupremic ewes on a low-Cu diet did not reduce the effectiveness of subcutaneous Cu injections in maintaining plasma Cu levels similar to those of Cu-supplemented ewes, and concluded that the primary site of the Cu–Mo interaction is located in the gut. Mills (87) contends that molybdate and sulfate restrict Cu utilization in sheep by depressing Cu solubility in the digestive tract through the precipitation of insoluble cupric sulfide. Higher sulfide levels were observed in the rumen of sheep fed Mo plus sulfate than in those fed sulfate alone, and the level of soluble copper in this site was inversely related to the sulfide level. Spais *et al.* (102) have also suggested that the sulfide formed in the rumen from high sulfate intakes acts by binding feed copper to insoluble copper sulfide. Gawthorne and Nader (45) have shown that sulfide concentrations can increase in the rumen when molybdate is infused, despite Mo inhibition of sulfide production, because of a second action of molybdate inhibiting the rate of apparent absorption of sulfide from the rumen. It is further apparent from the recent work of Huisingh *et al.* (66) that the effect of Mo on sulfide production by ruminal microorganisms in sheep depends on the source of the sulfur. Dietary Mo significantly inhibited the production of sulfide from sulfate but noticeably enhanced the production of sulfide from methionine.

Increased ruminal sulfide concentrations and a decreased flow of copper to the omasum have been demonstrated in sheep when the intake of sulfur from cystine or sulfate was increased under steady-state feeding conditions (13). Molybdenum has been shown by Suttle (107) to interact with organic S (cysteine and methionine) as well as inorganic S in limiting the utilization of dietary Cu by sheep. The interaction was so marked that variations within the normal ranges for Mo and S in herbage are believed by this worker to have significant effects on the Cu status of the grazing animal.* It was further stated that the interaction between Cu, Mo, and S takes place predominantly in the gut and may involve the formation of an unavailable compound containing Cu and Mo, similar to that described by Dowdy and Matrone (35).

A hypothesis put forward by Huisingh *et al.* (65) proposes that copper becomes unavailable by two routes: (*i*) through interaction with molybdate to form a biologically unavailable Cu–Mo complex called cupric molybdate [previously described (33)], which appears to be absorbed, transported, and excreted as a unit to make both Cu and Mo less available, and (*ii*) through the formation of an insoluble cupric sulfide in the rumen, intestine, or tissues. Finally, Dick *et al.* (33) have produced evidence which supports the hypothesis that there are three essential steps in the control of copper storage. These are (a) reduction in

*Authors note: Further evidence obtained by Suttle [*Br. J. Nutr.* **34**, 411 (1975)] indicates that total S rather than inorganic S is the more useful measurement in the context of the Cu–Mo–S interrelationship.

the rumen of sulfate to sulfide, (b) the reaction at relatively neutral pH of this sulfide with molybdate in the rumen to produce thiomolybdate, and (c) the reaction of the thiomolybdate with the copper to give the very insoluble copper thiomolybdate $CuMoS_4$. The concept of the formation of a thiomolybdate ion and the tight complexing of this ion with copper, which has also been independently proposed by Suttle (106), represents a significant step forward, involving for the first time a true three-way Cu–Mo–S interaction.

A metabolic interference of Cu by Mo is not confined to the gastrointestinal tract. Accumulation of sulfide in the tissues as a consequence of the depressed sulfide oxidase activity is known to occur in the molybdenotic rat (114, 123). This represents a possible further factor in the effect of Mo on Cu utilization. Such tissue sulfide accumulation could lead to a precipitation of insoluble unavailable cupric sulfide (53, 101, 102). Sulfide oxidase activity appears to be dependent on the *in vivo* supply of copper (101), so that high dietary Cu levels will help to maintain sulfide oxidase activity while the inhibiting effect of Mo on this enzyme is taking place. In this way an endogenous supply of sulfate will emerge which in turn will prevent Mo accumulation in the tissues. The protective action of copper against Mo toxicity could be explained to some degree by such a mechanism, the effectiveness of which will clearly depend on the intake of Cu relative to that of Mo. A metabolic interference with Cu uptake by the liver by Mo and SO_4 is also apparent from the experiments of Marcilese *et al.* (83), but the mechanism of this interference and the precise participation of the individual components of the Cu–Mo–S_4 nexus are not apparent from this study. However, Smith and Wright (101a) have recently shown that high- or relatively high-Mo intakes induce the formation of stable Cu–Mo protein compounds in the blood plasma of sheep and guinea pigs and suggest that this could explain the low tissue uptake of Cu that occurs in the presence of high plasma total and "direct-reacting" Cu concentrations. Such binding of Cu to plasma proteins therefore represents a further facet of the complex Cu–Mo interaction, although the reduction of Cu availability in the gut, as described in the preceding paragraph, probably represents the main means by which Mo and sulfate increase the Cu requirement. Whether the protein–Cu–Mo compounds are formed in plasma from normally existing plasma proteins or whether different plasma proteins which can enter into combination with Cu and Mo are formed in response to Mo is not yet known.

REFERENCES

1. Allaway, W.H., Kubota, J., Losee, F., and Roth, M., *Arch. Environ. Health* **16**, 342 (1967).
2. Allcroft, R., and Lewis, G., *Landbouwk. Tijdschr.* **68**, 711 (1956).

3. Alloway, B.J., *J. Agric. Sci.* **80,** 521 (1973).
4. Anderson, R.J., *Caries Res.* **3,** 75 (1969).
5. Archibald, J.G., *J. Dairy Sci.* **34,** 1026 (1951).
6. Arrington, L.R., and Davis, G.K., *J. Nutr.* **51,** 295 (1953).
7. Arthur, D., Motzok, I., and Branion, H.D., *Poult. Sci.* **37,** 1181 (1958).
8. Bala, Y.M., and Liftshits, V.M., *Fed. Proc., Fed. Am. Soc. Exp. Biol.* **25,** T370 (1966).
9. Barker, S., *Nature (London)* **185,** 41 (1960).
10. Bate, L.C., and Dyer, F.F., *Nucleonics* **23,** 74 (1965).
11. Beck, A.B., personal communication (1960).
12. Bell, M.C., Higgs, B.G., Lowrey, R.S., and Wright, P.L., *Fed. Proc., Fed. Am. Soc. Exp. Biol.* **23,** 873 (1964) (abstr.).
13. Bird, P.R., *Proc. Aust. Soc. Anim. Prod.* **8,** 212 (1970).
14. Britton, A.H., and Goss, H., *J. Am. Vet. Assoc.* **108,** 176 (1946).
15. Brouwer, F., Frens, A.M., Reitsma, P., and Kalesvaart, C., *Versl. Landbouwk. Onderz. C* **44,** 267 (1938).
16. Butler, E.J., and Barlow, R.M., *J. Comp. Pathol. Ther.* **73,** 208 (1963).
17. Buttner, W., *Arch. Oral Biol., Suppl.* **6,** 40 (1961); *J. Dent. Res.* **42,** 453 (1963).
18. Cjerniejewski, C.P., Shank, C.W., Bechtel, W.G., and Bradley, W.B., *Cereal Chem.* **41,** 65 (1964).
19. Cohen, H.J., Fridovich, I., and Rajagopalan, K.V., *J. Biol. Chem.* **246,** 374 (1971).
20. Cunningham, I.J., *in* "Symposium on Copper Metabolism" (W.D. McElroy and B.Glass, eds.), p. 246. Johns Hopkins Press, Baltimore, Maryland, 1950.
21. Cunningham, I.J., and Hogan, K.G., *N. Z. J. Agric. Res.* **1,** 841 (1958).
22. Cunningham, I.J., and Hogan, K.G., *N. Z. J. Agric. Res.* **2,** 134 (1959); Cunningham, I.J., Hogan, K.G., and Lawson, B.M., *ibid.* p. 145.
23. Curzon, M.E.J., Kubota, J., and Bibby, B.G., *J. Dent. Res.* **50,** 74 (1971).
24. Dale, S.E., Ewan, R.C., Speers, V.C., and Zimmerman, D.R., *J. Anim. Sci.* **37,** 913 (1973).
25. Davies, R.E., Reid, B.L., Kurnich, A.A., and Couch, J.R., *J. Nutr.* **70,** 193 (1960).
26. Davis, G.K., *in* "Symposium on Copper Metabolism" (W.D. McElroy and B. Glass, eds.), p. 246. Johns Hopkins Press, Baltimore, Maryland, 1950.
27. Doesthale, Y.G., and Gopalan, C., *Br. J. Nutr.* **31,** 351 (1974).
28. Dick, A.T., *Aust. Vet. J.* **28,** 30 (1952); **29,** 18 and 233 (1953); **30,** 196 (1954).
29. Dick, A.T., *in* "Inorganic Nitrogen Metabolism" (W.D. McElroy and B. Glass, eds.), p. 445. Johns Hopkins Press, Baltimore, Maryland, 1956.
30. Dick, A.T., Doctoral Thesis, University of Melbourne, Australia (1954).
31. Dick, A.T., *Aust. J. Agric. Res.* **5,** 511 (1954).
32. Dick, A.T., *Soil Sci.* **81,** 229 (1956).
33. Dick, A.T., Dewey, D.W., and Gawthorne, J.M., *J. Agric. Sci.* **85,** 567 (1975).
34. Dinu, I., Boghianu, L., and Sporn, A., *Rev. Roum. Biochim.* **9,** 215 (1972); *Nutr. Abstr. Rev.* **44,** 115 (1974).
35. Dowdy, R.P., and Matrone, G., *J. Nutr.* **95,** 191 and 197 (1968).
36. Dye, W.B., and O'Hara, J.L., *Nevada, Agric. Exp. Stn., Bull.* **208** (1959).
37. Ellis, W.C., Pfander, W.H., Muhrer, M.E., and Pickett, E.E., *J. Anim. Sci.* **17,** 180 (1958).
38. Ellis, W.C., and Pfander, W.H., *J. Anim. Sci.* **19,** 1260 (1960).
39. Ericsson, Y., *Acta Odontol. Scand.* **24,** 405 (1966).
40. Fairhall, L.T., Dunn, R.D., Sharpless, N.E., and Pritchard, E.A., *U.S., Public Health Serv. Bull.* **293** (1945).

41. Fell, B.F., Mills, C.F., and Boyne, R., *Res. Vet. Sci.* **6,** 170 (1965).
42. Ferguson, W.S., Lewis, A.H., and Watson,S.J., *Nature (London)* **141,** 553 (1938).
43. Ferguson, W.S., Lewis, A.H., and Watson, S.J., *J. Agric. Sci.* **33,** 44 (1943).
44. Fletcher, W.K., and Brink, V.C., *Can. J. Plant Sci.* **49,** 517 (1969).
45. Gawthorne, J.M., and Nader, C.J., *Br. J. Nutr.* **35,** 11 (1976).
46. Glenn, J.L., and Crane, F.L., *Biochim. Biophys. Acta* **22,** 111 (1956).
47. Goodman, F., *J. Dent. Res.* **44,** 565 (1965).
48. Gray, L.F., and Daniel, L.J., *J. Nutr.* **53,** 43 (1954).
49. Gray, L.F., and Daniel, L.J., *J. Nutr.* **84,** 31 (1964).
50. Gray, L.F., and Ellis, G.H., *J. Nutr.* **40,** 441 (1950).
51. Hadjimarkos, D.M., *Adv. Oral Biol.* **3,** 253 (1968).
52. Hadjimarkos, D.M., *Trace Subst. Environ. Health–7, Proc. Univ. Mo. Annu. Conf., 7th, 1973* (1973).
53. Halverson, A.W., Phifer, J.H., and Monty, K.J., *J. Nutr.* **71,** 95 (1960).
54. Hamilton, E.I., Minski, M.J., and Cleary, J.J., *Sci. Total Environ.* **1,** 341 (1972/1973).
55. Hamilton, E.I., and Minski, M.J., *Sci. Total Environ.* **1,** 375 (1972/1973).
56. Hart, L.I., and Bray, R.C., *Biochim. Biophys. Acta* **146,** 611 (1967).
57. Hart, L.I., McGartoll, M.A., Chapman, H.R., and Bray, R.C., *Biochem. J.* **116,** 851 (1970).
58. Hart, L.I., Owen, E.C., and Proudfoot, R., *Br. J. Nutr.* **21,** 617 (1967).
59. Healy, W.B., Bate, L.C., and Ludwig, T.G., *N. Z. J. Agric. Res.* **7,** 603 (1964).
60. Higgins, E.S., Richert, D., and Westerfeld, W.W., *J. Nutr.* **59,** 539 (1956).
61. Hogan, K.G., and Hutchinson, A.J., *N. Z. J. Agric. Res.* **8,** 625 (1965).
62. Hogan, K.G., Money, D.F.L., White, D.A., and Walker, R., *N. Z. J. Agric. Res.* **14,** 687 (1971).
63. Hogan, K.G., Money, D.J.L., and Blayney, A., *N. Z. J. Agric. Res.* **11,** 435, (1968).
64. Hogan, K.G., Ris, D.R., and Hutchinson, A.J., *N. Z. J. Agric. Res.* **9,** 691 (1966).
65. Huisingh, J., Gomez, G.G., and Matrone, G., *Fed. Proc., Fed. Am. Soc. Exp. Biol.* **32,** 1921 (1973).
66. Huisingh, J., Milholland, D.C., and Matrone, G., *J. Nutr.* **105,** 1199 (1975).
66a. Irreverre, F., Mudd, S.H., Heizer, W.D., and Laster, L., *Biochem. Med.* **1,** 187 (1967).
67. Jaulmes, P. and Hamelle, G., *Ann. Nutr. Aliment.* **25,** B133 (1971).
68. Jeter, M.A., and Davis, G.K., *J. Nutr.* **54,** 215 (1954).
69. Johnson, J.L., Rajagopalan, K.V., and Cohen, H.J., *J. Biol. Chem.* **249,** 859 (1974); Johnson, J.L., and Rajagopalan, K.V., *in* "Molybdenum in the Environment" (W. Chappell and K. Peterson, eds.), p. 201. Dekker, New York, 1976.
70. Kienholz, E., *in* "Transport and the Biological Effects of Molybdenum in the Environment" (W.R. Chappell, ed.), Prog. Rep., p. 148. University of Colorado, Denver, 1974.
71. Kiermeier, F., and Capellari, K., *Biochem. Z.* **330,** 160 (1958).
72. Kline, R.D., Corzo, M.A., Hays, V.W., and Cromwell, G.L., *J. Anim. Sci.* **37,** 936 (1973).
73. Kovalskii, V.V., Jarovaja, C.E., and Shnavonjan, D.M., *Zk. Obshch. Biol.* **22,** 179 (1961).
74. Kovalskii, V.V., Vorotnitskaya, I.E., and Tsoi, G.G., *in* "Trace Element Metabolism in Animals" (W.G. Hoekstra *et al.,* eds.), Vol. 2, p. 161. Univ. Park Press, Baltimore, Maryland, 1974.
75. Kratzer, F.H., *Proc. Soc. Exp. Biol. Med.* **80,** 483 (1952).
76. Krenitsky, T.A., *Adv. Exp. Med. Biol.* **41,** 57 (1973).

77. Leach, R.M., and Norris, L.C., *Poult. Sci.* **36**, 1136 (1957).
78. Lewis, A.H., *J. Agric. Sci.* **33**, 58 (1943).
79. Losee, F., Cutress, T.W., and Brown, R., *Trace Subst. Environ. Health–7, Proc. Univ. Mo. Annu. Conf. 7th, 1973,* p. 19 (1973).
80. Mackler, B., Mahler, H.R., and Green, D.E., *J. Biol. Chem.* **210**, 149 (1954).
81. Mahler, H.R., Mackler, B., Green, D.E., and Bock, R.M., *J. Biol. Chem.* **210**, 465 (1954).
82. Malthus, R.S., Ludwig, T.G., and Healy, W.B., *N. Z. Dent. J.* **60**, 291 (1964).
83. Marcilese, N.A., Ammerman, C.B., Valsecchi, R.M., Dunavant, B.G., and Davis, G.K., *J. Nutr.* **99**, 177 (1969).
84. Miller, R.F., and Engel, R.W., *Fed. Proc., Fed. Am. Soc. Exp. Biol.* **19**, 666 (1960).
85. Miller, R.F., Price, N.O., and Engel, R.W., *J. Nutr.* **60**, 539 (1956).
86. Mills, C.F., *Proc. Nutr. Soc.* **19**, 162 (1960).
87. Mills, C.F., *Rowett Res. Inst., Collect. Pap.* **17**, 71 (1961).
88. Miltmore, J.E., and Mason, J.L., *Can. J. Anim. Sci.* **51**, 193 (1971).
89. Monty, K.J., and Click, E.M., *J. Nutr.* **75**, 303 (1961).
90. Neenan, M., Walsh, T., and Moore, L.B., *Proc. Int. Grassl. Congr., 7th, 1956* N.Z. Paper 31A (1957).
91. Nielands, J.B., Strong, F.M., and Elvehjem, C.A., *J. Biol. Chem.* **172**, 431 (1948).
92. Owen, E.C., and Proudfoot, R., *Br. J. Nutr.* **22**, 331 (1968).
93. Renzo, E.C. de, Kaleita, E., Heytler, P., Oleson, J.J., Hutchings, B.L., and Williams, J.H., *Arch. Biochem. Biophys.* **45**, 247 (1953).
94. Richert, D.A., and Westerfeld, W.W., *J. Biol. Chem.* **203**, 915 (1953).
95. Richert, D.A., Bloom, R.J., and Westerfeld, W.W., *J. Biol. Chem.* **227**, 523 (1971).
96. Robinson, M.F., McKenzie, J.M., Thomson, C.D., and Van Rij, A.L., *Br. J. Nutr.* **30**, 195 (1973).
97. Rosoff, B., and Spencer, H., *Nature (London)* **202**, 410 (1964).
98. Scaife, J.F., *N. Z. J. Sci. Technol., Sect. A* **38**, 285 and 293 (1963).
99. Schroeder, H.A., Balassa, J.J., and Tipton, I.H., *J. Chronic. Dis.* **23**, 481 (1970).
100. Shirley, R.L., Jeter, M.A., Feaster, J.P., McCall, J.T., Cutler, J.C., and Davis, G.K., *J. Nutr.* **54**, 59 (1954).
101. Siegel, L.M., and Monty, K.J., *J. Nutr.* **74**, 167 (1961).
101a. Smith, B.S.W., and Wright, H., *J. Comp. Pathol.* **85**, 299 (1975); *Clin. Chim. Acta* **62**, 55 (1975).
102. Spais, A.G., Lazaridis, T.K., and Agiannidis, A.G., *Res. Vet. Sci.* **9**, 337 (1968).
103. Standish, J.F., Ammerman, C.B., Wallace, H.D., and Combs. G., *J. Anim. Sci.* **40**, 509 (1975).
104. Stiefel, E., *Proc. Natl. Acad. Sci. U.S.A.* **70**, 988 (1973).
105. Suttle, N.F., *Trace Subst. Environ. Health–7, Proc. Univ. Mo. Annu. Conf., 7th, 1973,* p. 245 (1974).
106. Suttle, N.F., *in* "Trace Elements in Soil-Plant-Animal Systems" (D.J.D. Nicholas and A.R. Egan, eds.), p. 271. Academic Press, New York, 1975.
107. Suttle, N.F., *in* "Trace Element Metabolism in Animals" (W.G. Hoekstra *et al.*, eds.), Vol. 2, p. 612. Univ. Park Press, Baltimore, Maryland, 1974.
108. Suttle, N.F., and Field, A.C., *Vet. Rec.* **95**, 166 (1974).
109. Thomas, J.W., and Moss, S., *J. Dairy Sci.* **34**; 929 (1951).
110. Tipton, I.H., and Cook, M.J., *Health Phys.* **9**, 103 (1963).
111. Tipton, I.H., Stewart, P.L., and Dickson, J., *Health Phys.* **16**, 455 (1969).
112. Tipton, I.H., Stewart, P.L., and Martin, P.G., *Health Phys.* **12**, 1683 (1966).
113. Todd, J.R., *J. Agric. Sci.* **79**, 191 (1972).

114. Van Reen, R., *Arch. Biochem. Biophys.* **53,** 77 (1954).
115. Van Reen, R., *J. Nutr.* **68,** 243 (1959).
116. Van Reen, R., and Williams, M.A., *Arch. Biochem. Biophys.* **63,** 1 (1956).
117. Ward, G.M., *in* "Transport and the Biological Effects of Molybdenum in the Environment" (W.R. Chappell, ed.), Prog. Rep., p. 179. University of Colorado, Denver, 1974.
118. Warren, H.V., *J. R. Coll. Gen. Pract.* **22,** 56 (1972).
119. Warren, H.V., Delavault, R.E., Fletcher, K., and Wilks, E., *Trace Subst. Environ. Health–4, Proc. Univ. Mo. Annu. Conf., 4th, 1970,* p. 94 (1971).
120. Westerfeld, W.W., and Richert, D.A., *J. Nutr.* **51,** 85 (1953).
121. Whitehead, D.C., *Commonw. Bur. Pastures Field Crops, (G.B.)* (1966).
122. Widjajakusuma, M.C.R., Basrur, P.K., and Robinson, G.A., *J. Endocrinol.* **57,** 419 (1973).
123. Williams, M.A., and Van Reen, R., *Proc. Soc. Exp. Biol. Med.* **91,** 638 (1956).
124. Winston, P.W., Hoffman, L., and Smith, W., *Trace Subst. Environ. Health–7, Proc. Univ. Mo. Annu. Conf., 7th,* p. 241 (1973).
125. Wynne, K.N., and McClymont, G.L., *Aust. J. Agric. Res.* **7,** 45 (1956).

5

Cobalt

I. COBALT IN ANIMAL TISSUES AND FLUIDS

The total cobalt content of the body of adult man has been reported to average 1.1 mg, with about 43% of this total stored in the muscles, 14% in the bones, and the remainder distributed among other tissues (156). Excessive accumulation does not occur in any particular organ or tissue, but the liver, kidneys, and bones usually carry the highest concentrations of this element. Distribution of retained cobalt in this manner has been demonstrated in mice, rats, rabbits, pigs, dogs, chicks, sheep, and cattle, with relatively small differences among species (23, 26, 77, 122). The Co concentrations reported for normal human tissues are similar to those of other species (142, 156), with the exception of those of Butt *et al.* (25) and Leddicotte (76) which are appreciably higher. Such differences probably reflect analytical variations, although regional differences in cobalt intakes may also be a factor. On the basis of rather limited evidence it seems that cobalt does not accumulate in human tissues with age (125, 142). Representative values for cobalt levels in human and ovine tissues are given in Table 18.

The concentrations of cobalt in the tissues are below normal in Co deficiency (17) and can be increased above normal by cobalt injections or oral supplements. Levels in the liver have attracted special attention because of their possible value in diagnosing Co deficiency in ruminants in the field. Thus in two Australian studies the mean liver Co levels of groups of Co-deficient sheep were 0.06 and 0.09 ppm (d.b.), compared with 0.28 and 0.34 ppm for healthy sheep (94, 148).

TABLE 18
Cobalt Concentration in Tissues[a]

Species	Liver	Spleen	Kidney	Heart	Pancreas
Normal human[b]	0.18	0.09	0.23	0.10	0.06
Healthy sheep[c]	0.15	0.09	0.25	0.06	0.11
Co-deficient sheep[c]	0.02	0.03	0.05	0.01	0.02

[a] Concentration expressed in ppm cobalt on dry basis.
[b] From Tipton and Cook (142).
[c] From Askew and Watson (17).

Similar values for the livers of Co-deficient and healthy cattle have been reported elsewhere (51, 98). McNaught (98) suggests that 0.04–0.06 ppm Co (d.b.) or less in the livers of sheep and cattle indicate Co deficiency and that 0.08–0.12 ppm Co indicate a satisfactory Co status. This worker showed that cobalt, unlike iron and copper, does not normally accumulate in the fetal liver. However, the cobalt (and vitamin B_{12}) content of the liver of the newborn lamb and calf is reduced below normal when the mother has been on a Co-deficient diet and can be raised to normal levels by prepartum cobalt administration (105).

Liver (and kidney) concentrations can be increased well above normal levels by cobalt injections or massive oral doses without this cobalt being available for vitamin B_{12} synthesis (4, 21, 66, 93, 113). Further, ruminants on Co-deficient diets can be maintained in health by vitamin B_{12} injections without raising their liver cobalt levels to normal. It is therefore apparent that freedom from signs of Co deficiency is compatible with a relatively low liver cobalt level and that a deficiency can occur in the presence of a high liver Co level. For these reasons liver Co concentration, although of some practical value, is not an entirely reliable criterion of the cobalt–vitamin B_{12} status of ruminants. The proportion of the liver cobalt that occurs as vitamin B_{12} varies with the Co status of the animal. In cobalt sufficiency most of the liver Co can be accounted for as vitamin B_{12}, whereas in cobalt deficiency only about one-fifth of the liver Co exists in this form (10). This indicates that in cobalt deficiency there is a greater reduction of vitamin B_{12}–cobalt than of nonvitamin B_{12}–cobalt. Conversely, oral dosing of Co-deficient lambs can increase B_{12} levels to normal, but the increase in total liver cobalt tends to be even greater.

Little is yet known of the forms in which cobalt exists in the tissues, other than as vitamin B_{12}. The existence of other bound forms of this metal has been demonstrated in the tissues of sheep (100), the blood plasma of the dog, and the intestinal wall of the chick (77).

Extremely variable Co concentrations in normal human blood have been reported as follows: 0.007–0.036 (mean 0.018) (139); 0.17–1.5 (155); and

0.35–6.3 (mean 4.3) μg Co/100 ml (72). Variations of this magnitude almost certainly reflect methodological errors rather than true variations of environmental origin. The red cells contain appreciably higher concentrations of cobalt than the plasma (72, 139).

Normal cow's milk analyzed by microcolorimetric methods is very low in cobalt. Concentrations ranging from 0.4 to 1.1 μg/liter, with a mean close to 0.5 μg/liter, have been reported (15, 36, 69, 110). Further studies of the level of cobalt in milk (and in blood) using modern physicochemical techniques are clearly necessary. Cow's colostrum is four–ten times higher in cobalt than milk (69) and can be further increased by prepartum supplementation of the cow's ration (140). Supplementing normal diets with cobalt salts can also increase the level in milk (15), but this cobalt cannot be used by young animals until they have developed a ruminal microbial population capable of incorporating it into vitamin B_{12}. Supplementary cobalt is similarly ineffective in raising the vitamin B_{12} activity of the milk above the normal range if the ration already contains adequate cobalt (60). However, cobalt supplementation of Co-deficient diets significantly increases the vitamin B_{12} content of milk and particularly colostrum (59, 105).

II. COBALT METABOLISM

1. Absorption

In several early studies with farm and laboratory animals dietary cobalt or cobalt administered orally as soluble salts was reported to be poorly absorbed (27). More recent investigations indicate that cobalt is surprisingly well absorbed by small laboratory animals and man. Thus Toskes *et al.* (144) found that normal mice on a normal diet absorbed 26.2% of an oral dose of cobalt labeled with ^{57}Co. Balance studies in man suggest that a variable amount, ranging from 20 to 95%, of dietary cobalt is absorbed from the intestine (37, 58, 65). The absorption of radioactive cobaltous chloride diminishes when it is administered after a meal or pretagged to protein (109). Cobalt absorption, as well as iron absorption, is significantly increased in iron deficiency in the rat (114) and man (150). Cobalt (and iron) absorption is also significantly increased in patients with portal cirrhosis, iron overload, and in patients with idiopathic hemochromatosis (150) (see Fig. 5). These findings have led Valberg and coworkers (141, 149) to postulate that cobalt shares a common intestinal mucosal transport pathway with iron, in which acceleration of transport of both elements is governed by the same mechanism. The concept that the intestinal transport of cobalt and iron shares a common pathway involving

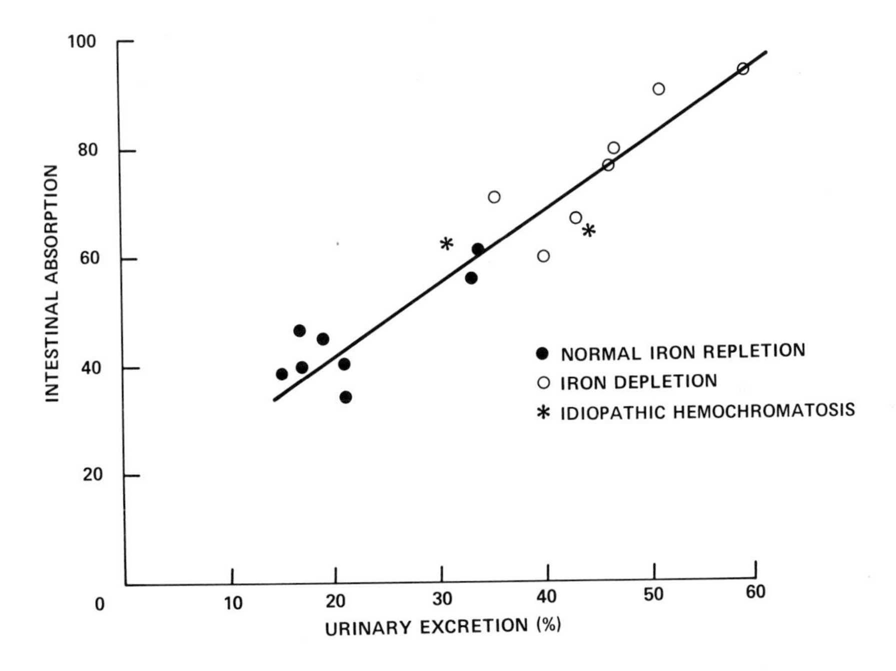

Fig. 5 Relationship between the absorption of cobalt from the gastrointestinal tract and the excretion of cobalt in the urine in man. From Valberg (149).

competition for absorptive sites is compatible with the demonstration of a mutual antagonism between the two elements at the absorptive level. Forth and Rummel (42) showed that the absorption of ^{59}Fe from jejunal loops of iron-deficient rats is reduced by almost two-thirds in the presence of a tenfold higher cobalt concentration. A further increase of 100-fold excess of cobalt was found to suppress absorption of ^{59}Fe nearly completely. In the reverse experiment a tenfold excess of iron decreased ^{58}Co absorption to about one-half the control value and a 100-fold excess to less than one-fifth the control value.

In dogs and chicks, injected radiocobalt rapidly reaches an equilibrium, with a standard and uniform partition among the body fluids, tissues, and intestinal contents and with little accumulation in any part of the body (77). In ruminants the time taken to reach uniform labeling is longer following injection (26). Following oral administration ^{60}Co is more slowly and less completely absorbed, indicating some immobilization in the digestive tract, probably due to binding of cobalt as vitamin B_{12} and related compounds by the rumen microorganisms (27, 122). A similar binding of cobalt by the bacteria of the cecum apparently occurs in chicks (77).

2. Excretion

The major route of excretion of cobalt in man is the urine (67, 125, 150), and a direct relationship has been observed between the proportion of an oral dose of cobalt that is absorbed from the intestine and the proportion that is excreted in the urine (149) (see Fig. 5). Small amounts of cobalt are also lost by way of the feces (149), sweat (29), and hair (125). A proportion of the fecal cobalt represents absorbed cobalt that has been reexcreted via the bile and to a small extent through the intestinal wall. The pancreatic juice is not a significant route of excretion of cobalt (101). Although control of the metabolic balance of cobalt is largely achieved by excretion in the urine, the question of whether some control is also exercised by the intestine in response to tissue levels of cobalt remains unanswered.

III. COBALT IN RUMINANT NUTRITION

1. The Discovery of the Need for Cobalt

Restricted areas in several parts of the world were long known to be unsatisfactory for the raising of sheep and cattle, in spite of apparently satisfactory pastures. Horses and other nonruminants thrived in these areas, whereas sheep and cattle became weak and emaciated, progressively anemic, and usually died. The condition was essentially similar in each locality and could only be controlled by periodic removal of the animals to healthy areas for varying periods.

The first serious scientific work on any of these diseases began in New Zealand at the end of the last century. A series of investigations led by Aston (18) culminated in the claim that "bush sickness" in cattle was due to a deficiency of iron. This was based on the occurrence of anemia in affected animals, the low-iron content of "bush sick" pastures and soils compared with those of healthy areas, and the effectiveness of crude iron salts and ores in preventing or curing the disease. The iron deficiency theory received support from other parts of the world where iron compounds were also found effective in preventing or curing similar maladies (53, 107, 120). Subsequent work in New Zealand revealing insignificant differences in the iron content of healthy and "bush sick" pastures (120), and little correlation between the iron content of different compounds and their curative effects (54), cast some doubt on the iron deficiency hypothesis. Indisputable evidence against this hypothesis was obtained by Filmer and Underwood (39, 41, 145, 147) in their investigation of a wasting disease of sheep and cattle (enzootic marasmus) occurring in the south coastal areas of Western Australia. These workers could find little relation between the size of an effective dose and the amount of iron it supplied. Furthermore, they discovered that (*i*) the liver and spleen of affected animals contained large stores of iron, (*ii*) whole liver was curative in doses that supplied insignificant amounts of iron, and (*iii*) an *iron-free* extract of one of the curative compounds (limonite, $Fe_2O_3 \cdot H_2O$) was just as potent as whole limonite.

The hypothesis was advanced that enzootic marasmus was due to a deficiency in the soils and herbage of some trace element which occurred as a contaminant of the iron compounds successfully used. Underwood and Filmer (147) then chemically fractionated limonite and, after some misleading tests with nickel suggested by the large amounts of that element present, found that the potency of the limonite resided in the cobalt that it contained. Normal growth and health of sheep and cattle on the deficient pastures were then secured by the administration of small oral doses of a cobalt salt, and the soils, pastures, and livers of affected animals were shown to contain subnormal cobalt concentrations (148).

While the above investigations were proceeding, studies of "coast disease" of sheep occurring in the calcareous sandy dunes of South Australia were under way. The possibility that this disease was due to a deficiency of a mineral element was recognized early. Supplements of phosphorus and copper proved ineffective, and the relatively small doses of the iron compounds used produced only a transitory improvement in the condition of "coasty" sheep (95). A mineral mixture supplying small amounts of iron, copper, boron, manganese, cobalt, nickel, zinc, arsenic, bromine, fluorine, and aluminum, on the other hand, was highly effective (95). The fact that coast disease is accompanied by anemia, coupled with Waltner and Waltner's (152) earlier finding that cobalt stimulates hemopoiesis in rats, led R. G. Thomas to suggest that cobalt might be the element responsible for the beneficial effects of the mineral mixture. Experiments by Marston and by Lines (90) revealed a dramatic improvement in the condition of coasty sheep from the oral administration of 1 mg cobalt per day. Subsequent experiments disclosed that supplementation with copper, as well as with cobalt, was necessary for completely successful treatment of coast disease.

Within a few years of these discoveries, cobalt supplements were found to be equally effective in the cure and prevention of all the diseases previously shown to respond to massive doses of iron compounds, and the soils and herbage of the affected areas were shown to contain subnormal levels of cobalt (see Russell and Duncan, 124). Subsequently it became apparent that larger areas existed in many countries where the deficiency was less severe.*

Author's note:* Later it was shown that Co deficiency in ruminants is, in effect, a vitamin B_{12} deficiency, responsive to injections of this vitamin [Smith *et al., J. Nutr.* **44, 455 (1951)].

2. Manifestations of Cobalt Deficiency

The appearance of a severely Co-deficient animal is one of emaciation and listlessness, indistinguishable from that of a starved animal except that the visible mucous membranes are blanched and the skin is pale and fragile. The emaciation or wasting of the musculature (marasmus) results from the failure of appetite, which is an early and conspicuous feature of the disease, along with paleness of the skin and mucous membranes from the anemia which usually develops progressively with the severity of the deficiency.

The cobalt deficiency syndrome varies from this acute and fatal condition through a series of less acute stages to a mild, ill-defined, and often transient state of unthriftiness that is difficult to diagnose. When ruminants are confined to Co-deficient pastures or are fed Co-deficient rations, there is a characteristic response. At first they thrive and grow normally for a period of several weeks or months, depending on their age, previous history, and the degree of deficiency of the diet. During this time they are drawing on the vitamin B_{12} reserves in the liver and other tissues. This period is followed by a gradual loss of appetite and failure of growth or loss of body weight, succeeded by extreme inappetence, rapid wasting and anemia, culminating in death (39, 95). Loss of appetite and low plasma glucose levels have been reported to be the best indicators of developing cobalt deficiency in steers (86). In some areas the Co deficiency is less severe, so that the acute disease conditions do not arise. Milder manifestations characterized by unthriftiness in young stock, and in mature stock also by diminution of lactation and birth of weak lambs and calves that do not survive long, are the only evidence of cobalt deficiency in such areas. In these mild or marginal areas the unthriftiness can be apparent in some years and absent in others (78).

At autopsy the body of severely affected animals presents a picture of extreme emaciation, often with a total absence of body fat. Cerebrocortical necrosis has been observed in some cases (87). The liver is fatty, the spleen hemosiderized, and in some animals there is hypoplasia of the erythrogenic tissue in the bone marrow (39). The red cell numbers and blood hemoglobin levels are always subnormal and sometimes very low. The anemia was reported by Filmer (39) to be normocytic and hypochromic in lambs and microcytic and hypochromic in calves. The nature of the anemia in Co-deficient calves does not appear to have been re-investigated. Later studies characterized the anemia in lambs as normocytic and normochromic (49, 132). However, it is not the anemia that is responsible for the main signs of cobalt deficiency. Inappetence and marasmus invariably precede any considerable degree of anemia. The first discernible response to cobalt feeding, or vitamin B_{12} injections, is a rapid improvement in appetite and body weight. Improvement in the blood picture may be equally dramatic but is sometimes delayed.

TABLE 19
Vitamin B_{12} Activity of Blood, Liver, and Rumen Ingesta of Sheep[a]

	Co-deficient		Co-sufficient (full fed)		Co-sufficient (limited fed)	
Sample	No.	Mean ± SD	No.	Mean ± SD	No.	Mean ± SD
Whole blood (mμg/ml)	16	0.47 ± 0.11	6	2.3 ± 0.6	3	4.3 ± 1.5
Liver (μg/g wet wt)	9	0.05 ± 0.01	6	0.93 ± 0.26	3	1.24 ± 0.20
Rumen ingesta (μg/g dry wt)	4	0.09 ± 0.06	5	1.3 ± 0.4	3	1.3 ± 0.9

[a]From Hoekstra *et al.* (63).

These clinical and pathological manifestations of cobalt deficiency in ruminants are accompanied by characteristic biochemical changes in the tissues. The decline in Co concentrations in the liver and kidney has already been mentioned (Table 18). The Co level in the rumen fluid also falls, as would be expected from the subnormal dietary intake of the element. When this has fallen below a critical level, tentatively set at < 0.5 ng/ml (128), vitamin B_{12} synthesis by the rumen microorganisms is inhibited and the levels of the vitamin decline in the rumen, blood, liver, and other tissues (10, 32, 63, 68, 91) (see Table 19). The tissue vitamin B_{12} depletion is accompanied by a marked depression of appetite, impaired propionate utilization, as discussed in Section III,6, and by a metabolic inefficiency (128). The latter was reflected in a 30% faster loss of body weight, a higher excretion of fecal nitrogen, and a higher fasting energy expenditure than was observed in pair-fed sheep injected with vitamin B_{12}. On the other hand, retention of combustible energy from the diet by vitamin B_{12}-deficient animals was not significantly different from that of pair-fed animals treated with cobalt or vitamin B_{12} (128). Abnormally low plasma glucose and alkaline phosphatase and abnormally high glutamic oxaloacetic transaminase (GOT) and pyruvate concentrations occur, together with low plasma ascorbic acid and thiamine levels, in severely Co-deficient sheep (85, 87). The low plasma glucose presumably reflects the impairment of propionate metabolism, the high GOT and low ascorbic acid values the severe liver damage, and the high pyruvate levels the induced thiamine deficiency. The low plasma ascorbic acid levels (mean 1 mg/liter, compared with normal concentrations of 4–8 mg/liter) are believed to be a factor in the increased susceptibility to infection shown by Co-deficient sheep (85). Cobalt supplementation is effective in preventing the vitamin deficiencies and the incidence of cerebrocortical necrosis (85).

3. Cobalt and Vitamin B_{12} Requirements

Under grazing conditions, lambs are the most sensitive to Co deficiency, followed by mature sheep, calves, and mature cattle, in that order (5). Field experience suggests that species differences among ruminants in Co requirements are small. Early evidence from Australia (41, 88) and New Zealand (97) indicated that 0.07 or 0.08 ppm Co in the dry diet was just adequate for sheep and cattle. This level of dietary cobalt therefore became accepted as the minimum requirement for these species. Later studies placed the minimum level of "pasture associated" cobalt required by growing lambs appreciably higher, namely, 0.11 ppm on the dry basis (12). In a later study of a marginally Co-deficient area in New Zealand, Andrews (6) assessed the position as follows: "Mean (pasture) values of 0.11 ppm Co or more would probably exclude the likelihood of cobalt deficiency. Mean values approaching 0.08 ppm would suggest but not prove actual or potential existence of the disease." More precise estimates of minimum cobalt requirements applicable under all grazing conditions are difficult because of the influence of many variables such as seasonal changes in herbage Co concentrations, selective grazing habits, and soil contamination. Lee and Marston (82) provide evidence that for sheep grazing grossly Co-deficient pastures, the total intake of cobalt to ensure optimum growth and hemoglobin production is 0.08 mg/day when supplementary cobalt is given three times each week. In growing lambs the requirement was stated to be higher. These findings apply to sheep consuming pastures, i.e., high-roughage diets. On low-roughage diets there is a significant reduction in vitamin B_{12} production (137).

The results of pen-feeding experiments with sheep consuming purified diets carried out by Somers and Gawthorne (135) point to a minimum cobalt requirement closer to the earlier level quoted above, namely, 0.07–0.08 ppm, than to a level of 0.11 ppm. These figures were obtained with sheep that were 18 months of age at the beginning of an experiment designed for other purposes and lasting only 39 weeks. Clinical signs of Co deficiency would almost certainly have appeared later in the sheep consuming 0.06 ppm Co, since their serum vitamin B_{12} concentrations had already fallen to a critically low level. In a study designed specifically to determine the cobalt and vitamin B_{12} requirements of sheep, Marston (89) found that sheep confined to pens and given a Co-deficient ration which supplied about 30 μg Co/day required for maintenance of normal growth rate a cobalt supplement approaching 40 μg administered by mouth daily; for maintenance of what appeared to be the maximum vitamin B_{12} status of sheep, namely, 3 ng vitamin B_{12}/ml serum and 1.4 μg vitamin B_{12}/g liver tissue, a supplement of between 0.5 and 1.0 mg Co/day was necessary. Natural rations high in cereal grains do not necessarily supply adequate cobalt. Growth responses to supplementary cobalt have, in fact, been demonstrated in steers fed a fattening ration based on barley grain (118), sorghum grain, and silage (102).

The requirements of ruminants for parenterally administered or absorbed vitamin B_{12} are of the same order as those found for the rat, pig, and the chick but are higher than those for man (see Smith and Loosli, 133). The minimum total requirement of sheep for parenteral or absorbed cobalamin has been estimated to be 11 ± 2 μg/day (89, 128)—a level consistent with a more recent estimate (61). Injections of 150 μg once every 2 weeks were found to be adequate for lambs fed a Co-deficient diet (132), and lambs grazing a Co-deficient pasture grew as well with injections of 100 μg of vitamin B_{12} at weekly intervals as they did with ample oral cobalt. By contrast, the oral requirements of ruminants for vitamin B_{12} are substantially higher than those of other species. Dairy calves have been reported to require between 20 and 40 μg vitamin B_{12} per kg dry matter consumed (75). Andrews and Anderson (8) fed crystalline vitamin B_{12} to lambs grazing a Co-deficient pasture at the rate of 1000 μg/week for 16 weeks. The growth response was much smaller than that obtained from either oral cobalt or injected vitamin B_{12}. Marston (89) also found oral doses of 100 μg vitamin B_{12} /day to be quite inadequate for young sheep. Extrapolation from these experiments suggests an oral requirement for growing lambs of some 200 μg/day, which is about 10 times the reported oral requirement of other species per unit of food intake. The reasons for this high requirement are discussed in Section III,6.

4. Diagnosis of Cobalt–Vitamin B_{12} Deficiency

The milder forms of cobalt deficiency in ruminants are impossible to diagnose with certainty on the basis of clinical and pathological observations alone. The only evidence of the deficiency is a state of unthriftiness, and there is usually no sign of anemia. A secure diagnosis of cobalt–vitamin B_{12} deficiency can be achieved in these circumstances by measuring the response in temperament, appetite, and live weight that follows cobalt feeding or vitamin B_{12} injections. However, if the ration of grazing consistently contains less than 0.08 ppm Co, cobalt–vitamin B_{12} deficiency can be predicted with confidence.

The level of cobalt in the livers of sheep and cattle is sufficiently responsive to changes in cobalt intake to have some value in the diagnosis of cobalt deficiency, but must be used with caution for the reasons given earlier (Section I). Liver vitamin B_{12} concentration is a more sensitive and reliable criterion than liver cobalt concentration. The criteria for sheep tabulated below were suggested by Andrews *et al.* (10) and can tentatively be applied to cattle.

The diagnostic value of liver vitamin B_{12} levels, and still more of kidney vitamin B_{12} levels, may be reached if the cobalt deficiency is coexistent with other diseases or conditions resulting in loss of appetite (9). From an examination of numerous controlled trials with sheep, Andrews (6) emphasized the

Condition of animal	Vitamin B_{12} concentration ($\mu g/g$ wet wt)
Severe cobalt deficiency	< 0.07
Moderate cobalt deficiency	0.07–0.10
Mild cobalt deficiency	0.11–0.19
Cobalt sufficiency	> 0.19

limitations of liver vitamin B_{12} assays as an aid to diagnosis of mild or intermittent forms of cobalt deficiency, but stated that "values of 0.10 $\mu g/g$ or less for livers from individual sheep can be accepted as clearly diagnostic of cobalt deficiency disease." Marston (89) also found that loss of appetite occurred when the concentration of vitamin B_{12} in the liver was reduced to about 0.1 $\mu g/g$ wet weight.

Serum vitamin B_{12} assays have obvious advantages over liver or kidney determinations. Low plasma vitamin B_{12} levels, loss of appetite, and elevated blood pyruvate have been reported to provide the best indications of the development of the Co-deficient state in sheep (85). With the onset of Co deficiency, serum vitamin B_{12} levels fall markedly and the levels can be related to the amount of cobalt ingested (10, 32, 63, 68, 91). Dawbarn *et al.* (32) concluded that "signs of Co deficiency may be expected to supervene when the mean vitamin B_{12} activity in the plasma falls to about 0.2 ng/ml." Andrews and Stephenson (11) reached similar conclusions from experiments with grazing ewes and lambs, although considerable individual variability was observed. Incipient stages of Co deficiency were associated with mean serum B_{12} values of 0.26 ng/ml for ewes and 0.30 ng/ml for lambs. When the Co deficiency had become marked in ewes and acute in lambs, mean serum B_{12} concentrations fell in both cases to levels less than 0.20 ng/ml. Similar conclusions were reached by Somers and Gawthorne (135), although they observed plasma vitamin B_{12} concentrations of 0.2 ng/ml or less for a few weeks in some sheep without the appearance of any clinical signs of deficiency.

5. Prevention and Treatment of Cobalt Deficiency

Cobalt deficiency in ruminants can be controlled either by treatment of the pastures with Co-containing fertilizers, direct oral administration of cobalt to the animal, or injections of vitamin B_{12}. Treatment with vitamin B_{12} is of little significance for reasons of cost and convenience. Vitamin $B_{12(b)}$ (hydroxycobalamin) is as effective as vitamin B_{12} itself (cyanocobalamin) in correcting cobalt deficiency in lambs when injected at the rate of 100–150 μg/week (71), but folinic acid at levels of 71 μg, 5 mg, or 15 mg daily produces no such response (68). Large oral doses of dried or fresh whole liver were also shown to be highly curative (39), a finding which led Filmer and Underwood (41) as early as 1937 to propose that "the potency of liver may be due to the presence of a stored factor and that cobalt may function through the production of this factor within the body." Subsequent attempts to cure cobalt deficiency with liver were unsuccessful (68, 93), presumably because of differences in the doses or vitamin B_{12} or cobalt contents of the liver preparations employed.

The most economical and widely practiced means of ensuring continuous and adequate supplies of cobalt to sheep and cattle grazing Co-deficient areas is by treatment or "top-dressing" of the pastures with fertilizers to which a small proportion of Co salts or ores has been added. In this way the concentration of cobalt in the herbage can be raised for extended periods to levels adequate for the requirements of the animal. In some areas improvement in the growth of the legume component of the pasture can also be achieved by such treatment (108, 117), but this is not common. As little as 100–150 g of cobalt sulfate per acre applied annually or biennially suffice on most deficient pastures (16, 121). A single dressing of 500–600 g of cobalt sulfate per acre has maintained satisfactory pasture cobalt levels for 3–6 years in New Zealand (7). The most efficient quantity and frequency of top-dressing with cobalt depends on the degree of deficiency and the soil type and has to be determined for each area. Cobalt uptake by plants on highly calcareous soils, or on soils high in manganese which fixes the cobalt in unavailable forms (1), can be so low as to render treatment with"cobaltized" fertilizers inefficient or ineffective.

Direct administration of cobalt to stall-fed animals is usually achieved by incorporating cobalt oxide or salts into the mineral mixtures normally fed as supplements. The inclusion of cobalt in such mixtures is now common, even where there is no clear evidence that the unsupplemented rations are Co-deficient. Cobalt may also be supplied successfully to either stall-fed or grazing sheep and cattle in the form of cobalt-containing salt "licks," but variable consumption of the lick, and therefore of cobalt, is a disability inherent in this form of treatment. Oral dosing or "drenching" of animals with solutions of cobalt salts is practiced in many areas and can be completely successful if the dosing is frequent enough.

Dosing sheep with 2 mg Co twice weekly or 7 mg Co weekly (79) is completely adequate, even under acutely Co-deficient conditions. Since individual drenching is a tedious and labor-demanding operation the effect of larger doses at longer intervals has been investigated. Lee (79) found that 35 mg cobalt administered once in 5 weeks "merely delays the onset of symptoms." Filmer (40) observed that monthly doses of 140 mg cobalt kept ewes and lambs alive on Co-deficient land, but the sheep did not thrive as well as those treated twice weekly or weekly with the usual small doses. Andrews and co-workers (13) gave monthly doses of 300 mg Co, either as soluble cobalt sulfate or insoluble cobalt oxide, to lambs grazing Co-deficient pastures for 5 months. Their growth rates and vitamin B_{12} levels in blood and liver were compared with those of untreated lambs and lambs dosed with 7 mg Co weekly. The two forms of cobalt were equally effective but neither of the monthly treatments prevented cobalt deficiency entirely. Both permitted growth at suboptimal rates and greatly reduced mortality but were less effective than the weekly dosing. Lee and Marston (82) showed further that the equivalent of 1 mg Co/day is as effective for 2–3 years when given weekly as when given more frequently. Over longer periods the body weight response is slightly but significantly in favor of the more frequent administration.

The necessity for regular and frequent dosing arises from the fact that cobalt, unlike iron and copper, is not readily stored in the liver or elsewhere in the body. The cobalt which is present in the tissues does not easily pass into the rumenoreticular region of the digestive tract where it is needed for vitamin B_{12} synthesis. Injections of cobalt are therefore largely ineffective. Where large amounts are injected some improvement in the condition of Co-deficient sheep has been observed (68), presumably as a consequence of small amounts of cobalt reaching the rumen in the saliva or via the rumen wall. Studies with radioactive cobalt indicate that the element only reaches the rumen, reticulum, and omasum when large amounts are injected, and even then only in very small quantities (27). Cobalt injections are capable of increasing the vitamin B_{12} content of the cecum and large intestine of Co-deficient lambs, probably as a result of bacterial synthesis following the excretion of cobalt into the duodenum via the bile. The vitamin so produced is apparently not absorbed since blood and body tissue vitamin B_{12} levels are not improved (68). Phillipson and Mitchell (113) placed cobalt salts directly into the abomasum or duodenum by means of appropriate fistulas, and found them to be much less effective than cobalt given by mouth, despite evidence of some movement of cobalt into the rumen. It is apparent that any treatment with cobalt, if it is to be fully successful, must be capable of maintaining adequate Co concentrations in the rumen fluid.

Many of the disadvantages of other forms of treatment can be avoided by the use of cobalt pellets or "bullets," first devised by Dewey and co-workers (33) in 1958. The small dense pellets composed of cobalt oxide and finely divided iron (S.G. 5.0), when delivered into the esophagus with a balling gun, lodge in the reticulorumen where they usually remain to yield a steady supply of cobalt to the rumen liquor. The usefulness of cobalt pellet therapy in the prevention of cobalt deficiency has been established in sheep and cattle (9, 10, 82, 119, 127). Two problems have arisen with this form of treatment—rejection of the pellets by regurgitation in a proportion of animals, and surface coating of the pellets with calcium phosphate so that the rate of cobalt release into the rumen is reduced. Millar and Andrews (99) prepared radioactive cobaltic oxide pellets so that they could be detected in live sheep and followed their retention in grazing ewes and lambs over a 16-month period. By the end of the period about one-third of the ewes and lambs had rejected their pellets, and a proportion of those retained were coated with calcium phosphate. Very similar results were obtained by Poole and Connolly (115) in Ireland. Furthermore, both groups of workers reported that the administration of two pellets or one pellet plus a $\frac{1}{2}$-in. grub screw grinder was ineffective in reducing the calcium phosphate coating. This is contrary to Australian experience in which abrasion between the two objects has been quite successful in keeping the pellet surface clean and maintaining a steady supply of cobalt to the rumen of sheep for more than 5 years (82).

A cobaltic oxide pellet has been designed and tested by Connolly and Poole (28) which is claimed to eliminate these drawbacks. The pellet consists of a $1\frac{3}{4} \times \frac{1}{2}$-in. steel rod covered with layers of cotton gauze bandage impregnated with 5 g cobaltic oxide. The cobalt is released during the slow digestion of the cotton by the microorganisms of the rumen. In an experiment with lambs lasting 73 days none of the new type pellets was lost, there was no evidence of coating, and live-weight gains and serum vitamin B_{12} levels were satisfactory.

6. Mode of Action of Cobalt

The production of vitamin B_{12} in the rumen depends on the cobalt and the roughage content of the diet and the total dietary intake (61, 137). When the concentration of cobalt in the rumen fluid falls below a critical level, placed at $<$ 5 ng/ml (128), the rate of vitamin B_{12} synthesis by the rumen organisms is reduced below the animal's needs. When the stores of vitamin B_{12} laid down mainly in the liver become depleted below a critical level close to 0.10 μg/g wet weight (6, 89), and when the plasma vitamin B_{12} levels fall below a critical level of 0.2 ng/ml as a result of inadequate production, absorption, and inability to obtain supplies from body stores, characteristic signs of cobalt deficiency begin to appear in most lambs and calves. The length of time taken for clinical signs of deficiency to become apparent depends primarily on the degree of dietary Co deficiency, the magnitude of the vitamin B_{12} stores previously built up, and the age of the animal. There is no evidence that vitamin B_{12} synthesis is possible within the body tissues (68), or that the vitamin B_{12} synthesized by the bacteria of the cecum and large intestine can be absorbed. The ruminant is therefore ultimately dependent on the synthetic capacity of its rumen organisms.

In Co-deficient lambs the total concentrations of rumen bacteria and the principal types of those organisms are reduced below normal (42). No direct evidence has been obtained implicating particular organisms in vitamin B_{12} synthesis in the rumen, or demonstrating a preferential diminution of the numbers of any such organisms as a consequence of lack of dietary cobalt.

However, Dryden *et al.* (35) established that several pure strains of rumen bacteria each produced a mixture of vitamin B_{12} and its analogs when cultured in a Co-adequate medium.

The rumen organisms normally produce many Co-containing vitamin B_{12} -like compounds, with no physiological activity in the body tissues, in addition to true vitamin B_{12} (46, 62, 116). Comparative assays with *Escherichia coli* or *Lactobacillus leichmannii* and with *Ochromonas mahlemensis*, which responds only to vitamin B_{12} itself, have shown that a considerable proportion of the total activity of the rumen contents as measured by the former assay organisms is normally contributed by the inactive analogs. Under cobalt deficiency conditions both the total and the true vitamin B_{12} activity fall markedly. In most circumstances (46, 62), but not always (61), the total activity falls more than that of true vitamin B_{12}. Thus Gawthorne (46), using a technique capable of estimating nanogram amounts of the individual cobamides in rumen contents (45), found the total concentrations of these compounds to fall significantly, and the proportion present as true vitamin B_{12} (5,6-dimethylbenzimidazolylcobamide cyanide or DMBC) to rise significantly as dietary cobalt was reduced from 0.34 to 0.04 ppm. The proportion of the total due to the physiologically inactive 2-methyladenylcobamide cyanide (2MAC) remained unaltered at about 44%. It was further shown that the total vitamin B_{12} activity of the blood plasma of sheep can be almost entirely (99%) accounted for as DMBC, even when the rumen concentrations of its analogs are high. From these data it appears that the Co-deficient sheep converts at least 60% of its limited supply of dietary cobalt into compounds that it cannot absorb and use. At higher cobalt intakes a still larger proportion is so converted. In the experiments of Smith and Marston (128) an even more inefficient production of cobalamin from cobalt was observed. This was assessed at about 15% under Co deficiency and about 3% under Co sufficiency conditions.

A further factor contributing to the relatively high cobalt requirement of the ruminant is its limited ability to absorb the vitamin B_{12} that is produced. This is apparent from a study with sheep using vitamin B_{12} containing ^{60}Co (111) and can be deduced from the studies of Kercher and Smith (68). These workers found that the effective oral dose of the vitamin in Co-deficient lambs was about 35 times the parenteral dose and calculated that about 3% of the orally administered vitamin must have been absorbed. This calculation involves the assumption that cobalamin injected in relatively high doses is retained as effectively as cobalamin absorbed from the alimentary tract, but there is evidence that it is not. Thus Dawbarn and Hine (31) found that cobalamin injected into sheep at the level of 50 μg/day was 30–36% excreted in the urine. Subsequently it was found that injected labeled cobalamin was 70–90% retained by Co-deficient sheep, with excretion in the urine and feces and evidence of substantial secretion into the duodenum (68). About half the cobalamin produced in the rumen was lost during passage through the alimentary tract and only about 5%, or even less than 3%, was estimated to be absorbed (68, 77). At higher dietary cobalt levels appreciably higher apparent absorption values for vitamin B_{12} were obtained by Hedrich and co-workers (61). The ruminant clearly makes extremely inefficient use of its dietary cobalt, both in respect to production of cobalamin in the rumen and to absorption of the vitamin from the alimentary tract.

The marked failure of appetite that is such a conspicuous feature of Co–vitamin B_{12} deficiency in ruminants is not nearly so noticeable in vitamin B_{12} deficiency in man or other species. This difference can probably be related to the means by which the two types of animals derive their energy. The main source of energy to ruminants is not glucose but acetic and propionic acids, together with smaller quantities of butyric and other fatty acids produced by fermentation in the rumen. Following the discovery that methylmalonyl-CoA isomerase, a vitamin B_{12} requiring enzyme, catalyzes the conversion of methylmalonyl-CoA to succinyl-CoA (19), and that the activity of this isomerase is severely depressed in the livers of vitamin B_{12}-deficient rats (56, 129), Marston and associates (91) investigated the possibility that (a) a breakdown in propionate utilization at this point in the metabolic pathway might be the primary defect in Co–vitamin B_{12} deficiency in ruminants, and (b) the depression of appetite might be due to an increased level in the blood of propionic acid or some other metabolic product stemmed back by the reduced capacity of the B_{12}-deficient animal to metabolize propionate. The production and absorption of propionic and other fatty acids were found to proceed more or less normally in the Co-deficient sheep. As the deficient state progressed, the rate of disappearance from the blood of injected propionate fell. Examination of liver homogenates from vitamin B_{12}-deficient sheep revealed (a) a failure to convert propionate efficiently to succinate, (b) an accumulation of the intermediary methylmalonyl-CoA, and (c) prevention of this accumulation by the addition of 5-deoxyadenosylcobalamin.

Subsequently Somers (134) found that propionate and acetate clearance rates from the blood of sheep were increasingly adversely affected as vitamin B_{12} deficiency intensified. These effects were greater than the effects of depressed feed intake alone, and both variables had a more pronounced effect on the rate of clearance of propionate than of acetate. Marston and co-workers (92) attribute the impaired acetate clearance to the elevated levels of circulating propionate, since propionate was shown to have a marked inhibitory effect on acetate metabolism in deficient sheep. Failure to metabolize acetate is therefore probably not a primary consequence of vitamin B_{12} deficiency. Further evidence of depressed propionate metabolism in Co–vitamin B_{12}-deficient sheep was obtained by Gawthorne (47), who found that the excretion of methylmalonic acid (MMA) in the urine of severely vitamin B_{12}-deficient sheep was 5–12 times greater than that of pair-fed, B_{12}-injected controls and was restored to normal within 3 weeks by injections of the vitamin. A significant increase in urinary MMA excretion occurred only when the sheep were severely affected by the deficiency. In the early stages of the deficiency it seems that the potential accumulation and augmented excretion of MMA are offset by the reduced feed intake and therefore the amount of propionate presented for metabolism. The most significant evidence linking inappetence with impaired propionate metabolism is an inverse relationship between voluntary feed intake of deficient sheep and the half-time for propionate clearance (92) (Fig. 6). Propionate clearance was shown to be already fairly severely impaired at the mildly deficient feed intake of 700 g/day. Impairment of propionate clearance is thus clearly an early consequence of vitamin B_{12} deficiency.

Some of the effects of vitamin B_{12} deficiency in sheep can be attributed to depressed activity of a second vitamin B_{12}-containing enzyme, i.e., in addition to the effects of a critical depression in activity of methylmalonyl-CoA mutase, as just discussed. This second enzyme is 5-methyltetrahydrofolate:homocysteine methyltransferase, which catalyzes the reformation of methionine from homocysteine, thus permitting the recycling of methionine following loss of its labile methyl group, in accordance with the reaction

$$\text{5-Methyltetrahydrofolate} + \text{homocysteine} \rightleftharpoons \text{methionine} + \text{tetrahydrofolate}$$

The activity of this methyltransferase is depressed in the liver of vitamin B_{12}-deficient sheep and is accompanied by subnormal levels of *S*-adenosylmethionine (48) and a fatty liver with a depleted choline content (131). Both defects are corrected by injection of extra methionine. Since the rate of turnover of the methyl group of methionine in the liver is rapid (2), restriction of methyltransferase activity in vitamin B_{12} deficiency could lead to a deficiency of available methionine. This provides a possible basis for the impaired nitrogen retention found in vitamin B_{12}-deficient sheep (128) and could be a critical limiting factor in both wool and body growth.

A further consequence of vitamin B_{12} deficiency is a marked depletion of liver folate stores. This occurs in rats (70) and sheep (130), and in both species the levels of liver folate can be restored by treatment with L-methionine (50, 131). In fact methionine, like vitamin B_{12}, increases the concentrations of all the major classes of folate in the liver of vitamin B_{12}-deficient sheep (131). It appears, therefore, that the effect of vitamin B_{12} on folate metabolism, as well as on liver lipids, is exerted via methionine. Gawthorne and Smith (48) have provided evidence that methionine exerts its effects on folate retention by promoting transport of folate into the vitamin B_{12}-deficient liver cell. On this evidence the deficit of folate in vitamin B_{12}-deficient sheep is due to the deficit of methionine. Other consequences of a deficit of methionine brought about by

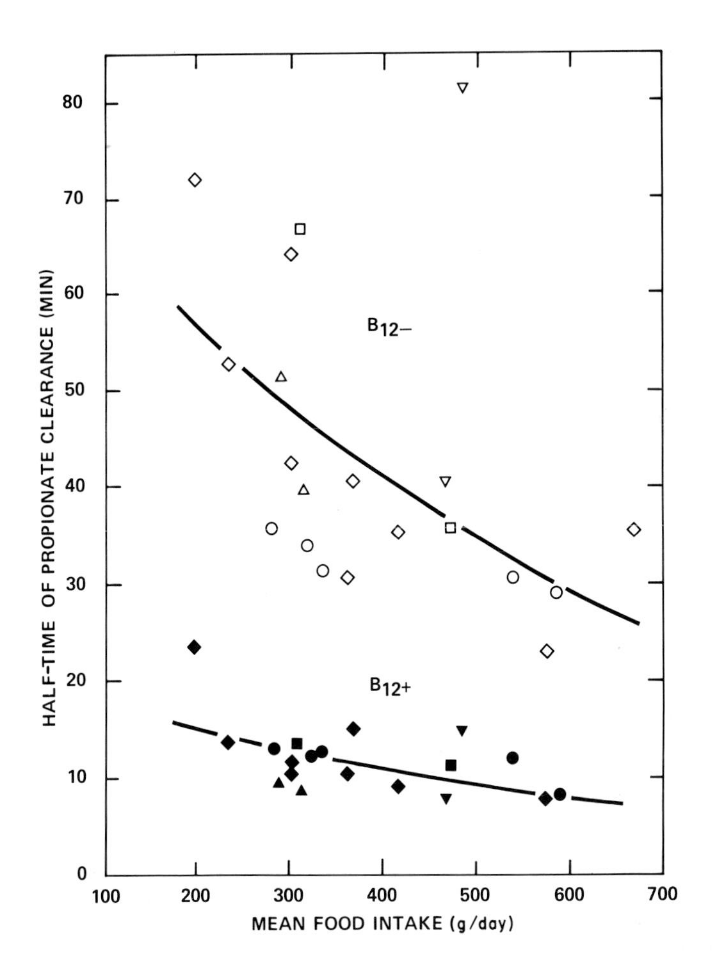

Fig. 6 Relationship between half-time of clearance of propionate and food intake for vitamin B_{12} -deficient (B_{12} –) and vitamin B_{12} -treated (B_{12} +) sheep. From Marston *et al.* (92).

the reduced activity of the methyltransferase in vitamin B_{12} deficiency may yet emerge.

7. Cobalt and *Phalaris* Staggers

In certain areas sheep, and to a lesser extent cattle, grazing pastures containing the perennial grass *Phalaris tuberosa* can develop either an acute form of a disease from which they quickly die, or a chronic form with nervous disorders characterized by a marked incoordination of gait (staggers), muscular tremors, and rapid breathing and pounding of the heart. *Phalaris* staggers, which refers particularly to the chronic condition, was first described by McDonald (96) in South Australia and has been observed elsewhere in Australia. An identical staggers syndrome occurs in sheep and cattle grazing *Phalaris minor* in South Africa (153), *Phalaris arundinacea* in New Zealand (126), and Ronpha grass (*P. tuberosa* × *P. arundinacea*) in Florida (123) and South Africa (151, 153).

Evidence has been obtained that the tryptamine alkaloids, closely related to serotonin, shown to be present in *P. tuberosa* (30), are responsible for acute and peracute poisoning syndromes (43). However, the acute toxicity is not always related to the tryptamine alkaloid concentration in the plant (106), and it is improbable that these alkaloids produce the persistent neurological disorders associated with chronic staggers.

Lee and Kuchel (80) first demonstrated the association of this condition with incipient cobalt deficiency and secured its complete prevention in sheep grazing affected pastures by regular oral dosing with Co salts. This finding has been confirmed (81, 136, 151) and extended to include the protective action of Co pellets. As cobalt does not appear to be effective against the acute forms of the disease, it seems that cobalt must in some way be concerned with preventing the development of degenerative, structural changes in the nervous system. In this respect the action of cobalt appears to be different from, and additional to, its action in preventing Co deficiency in ruminants. This is because (a) administration of vitamin B_{12} is ineffective against *Phalaris* staggers, and (b) a high toxic potential of *Phalaris* pastures increases the level of cobalt required for protection. Where chronic staggers does not develop in animals grazing *Phalaris* pastures it is contended that the soils maintain sufficient concentrations of Co in the pastures to meet the normal requirements of ruminants for this element, plus the extra requirements needed to prevent the degenerative changes in the nervous system.

IV. COBALT IN THE NUTRITION OF MAN AND OTHER NONRUMINANTS

1. Man

Cobalt must be supplied in the diet of man and other monogastric species entirely in its physiologically active form, vitamin B_{12}. The tissues of these animals are unable to synthesize the vitamin from dietary cobalt, and their intestinal microflora have an extremely limited capacity to effect this vital transformation at a point in the digestive tract where the vitamin can be absorbed. In these unique circumstances the cobalt status of human foods and dietaries is relatively unimportant–it is their vitamin B_{12} status that is crucial.

Reported data for the cobalt content of human foods and total diets are both meager and highly variable. Some of the variation undoubtedly stems from analytical errors or inadequacies, but some also reflects soil and climatic differences directly affecting the Co content of foods of plant origin and indirectly those of animal origin. Thus Murthy and co-workers (103), in a study of the diets of children from 28 widely separated institutions in the United States, found the Co concentration in the total diets to vary from 0.25 to 0.69 mg/kg and the total intakes to range from 0.30 to 1.77 mg Co/day, with a mean close to 1 mg Co/day. These levels are much higher than the 0.16–0.17 mg Co/day for adults consuming North American diets estimated by Tipton *et al.* (143), the 0.14–0.58 mg Co/day estimated by Schroeder *et al.* (125), the mean of 10 μg/Co day given for Japanese adults (157), or the mean intake of 31 μg Co/day reported for ten 17-year-old students in the Soviet Union (104). A positive Co balance was achieved on these intakes in eight of the 10 students. It is difficult to believe that differences of this magnitude are not mainly of analytical origin.

Among individual types of foods the green leafy vegetables are the richest and most variable in Co content, while dairy products, refined cereals, and sugar are the poorest. Typical values for the former group are 0.2–0.6 ppm (d.b.), and for the latter 0.01–0.03 ppm Co (d.b.). Plant products have been estimated to contribute up to 88% of the total cobalt of Japanese diets (157). Normal cow's milk is very low in cobalt, with most values lying close to 0.5 μg/liter, as discussed in Section I. The organ meats, liver and kidney, commonly contain 0.15–0.25 ppm (d.b.), and the muscle meats about half those levels. These foods contain much more cobalt than can be accounted for as vitamin B_{12}. Fruits, vegetables, and cereals contain none of their cobalt as vitamin B_{12}.

Cobalt has been used as a nonspecific erythropoietic stimulant in the treatment of the anemias of nephritis and infection (44, 154), and several reports of responses to cobalt, in addition to iron, in the treatment of iron deficiency anemia in children (138) and pregnant women (57) have appeared. The amounts of cobalt required to elicit the hemopoietic response are large (20–30 mg/day), so that serious toxic manifestations can occur, including thyroid hyperplasia, myxedema, and congestive heart failure in infants (74, 120a). Cobalt therefore occupies a very restricted place in the treatment of human anemias. The value of cobalt therapy in the treatment of human hypertension has also been investigated. Since vasodilatation, with flushing, has been observed following injection of cobalt salts (83), Perry and Schroeder (112) treated nine hypertensive patients with 50 mg oral cobalt chloride per day for 10–65 days. Marked lowering of blood pressure to normal was observed in three of the patients, with a 14–72% reduction in requirements for antihypertensive drugs in five. No toxic effects were apparent in any of the treated patients.

Some evidence for an effect of varying intakes of cobalt on the thyroid gland has been obtained. Cobalt, and also manganese, have been reported to be necessary for the synthesis of the thyroid hormone in rats (22). The addition of physiological doses of Co to a diet which was not naturally high in iodine or cobalt did not affect the weight of the gland but caused definite histological changes, including a decrease in the size of the follicles and an increase in the height of the epithelial cells. It was concluded that "the appearance of endemic disturbances of thyroid gland function in people inhabiting biogeochemical provinces with a low-iodine and a low-cobalt content depends not only on the level of I and Co, but also on the ratio of these elements in the environment." Other workers in the Soviet Union have observed an inverse correlation between the cobalt levels in the foods, waters, and soils in certain areas and the incidence of goiter in man and farm animals (73). A possible relationship between the Co status of the environment and the incidence of goiter warrants investigation in other areas, and further examination of the suggestive Co–I interaction in animals is desirable.

2. Nonruminant Animals

Cobalt deficiency has never been demonstrated in laboratory animals. Rabbits have been successfully maintained on diets reported to supply only 0.1 μg Co/day (140), and rats were found to grow as well on diets supplying 0.6 μg Co/day (146), or 0.3μg Co/day (64), as when those diets were supplemented with cobalt. Horses thrive on pastures so low in cobalt that sheep and cattle dependent on such pastures soon waste and die.

Monogastric animals consuming all-plant rations, which directly supply little or no vitamin B_{12}, obviously need some dietary Co so that their intestinal flora can synthesize sufficient vitamin B_{12} to meet the metabolic needs of the tissues. The diets consumed generally contain sufficient cobalt for this purpose. However, Dinusson *et al.* (34) observed a small but significant increase from additions of cobalt in the rate of gain and efficiency of feed use of pigs fed a corn–soybean meal diet. In some experiments a similar improvement was observed from supplements of vitamin B_{12} or meat scraps, with a response from cobalt even when those rations contained 5% meat scraps. Filippov (38) has more recently reported a marked improvement in daily weight gains and feed efficiency and reduced mortality in young pigs given a supplement of 1 mg/day of cobalt chloride. Intestinal synthesis of vitamin B_{12} is not always adequate for the needs of growing pigs and poultry consuming all-plant rations. Additional supplies of this vitamin can be obtained by coprophagy, and by the consumption of litter and refuse in which bacterial fermentation by the bacteria voided in the feces or contaminated from the soil has occurred. Where these adventitious sources of vitamin B_{12} are denied to the animals, growth responses can be obtained from supplements rich in the vitamin.

The "animal protein factor" effect with pigs and poultry receiving all-plant diets is due partly and in some cases largely to the vitamin B_{12} supplied. Cobalt, of itself, is rarely important in this respect.

V. COBALT TOXICITY

Cobalt has a low order of toxicity in all species studied, including man. Daily doses of 3 mg Co/kg body weight, which approximate 150 ppm Co in the dry diet or some 1000 times normal levels, can be tolerated by sheep for many weeks without visible toxic effects (20). With doses of 4 or 10 mg Co/kg body weight, appetite and body weight are severely depressed, the animals become anemic, and some deaths occur at the higher level. The anemia probably arises from a depression in iron absorption by the very high intakes of cobalt, as discussed in Section II,1. Andrews (4) estimated that a single dose of 300 mg Co/kg body weight as a soluble salt would usually be lethal to sheep, and that single doses as small as 40–60 mg Co/kg body weight would occasionally be fatal. From the work of MacLaren *et al.* (84) it appears that cattle are less tolerant of high-cobalt intakes than sheep.

Rats and various species other than the adult ruminant fed large amounts of Co as cobalt salts develop a true polycythemia accompanied by hyperplasia of the bone marrow, reticulocytosis, and increased blood volume (52, 152). The oral intakes of cobalt necessary to induce significant polycythemia approximate 200–250 ppm of the total diet and are therefore clearly many times greater than those that could conceivably be obtained from normal foods and beverages.

Under certain circumstances, as yet unexplained, cobalt intakes substantially lower than the 20–30 mg/day mentioned earlier as producing toxic manifestations, when used as an erythropoietic stimulant, can be toxic to man. Cobalt has been incriminated as the precipitating factor in several outbreaks of severe cardiac failure in heavy beer drinkers. Cobalt was suspected because of the high incidence of polycythemia, thyroid epithelial hyperplasia, and colloid depletion noted in the fatalities, in addition to congestive heart failure (14). Cobalt salts had been added to the beer to improve its foaming qualities at concentrations of 1.2–1.5 ppm Co, a practice no longer in use. At such concentrations the reported formidable consumption of 24 pints (approximately 5 liters/day) of beer would supply about 8 mg of cobalt sulfate, an amount well below that which can be taken with impunity by normal individuals. It seems that high-Co and high-alcohol intakes are both necessary to induce the distinctive cardiomyopathy, plus a third factor which may be low-dietary protein (3) or thiamine deficiency (55). However, in a recent study with pigs on the effects of cobalt, beer, and thiamine deficiency, no comparable cardiomyopathy was apparent (24).

REFERENCES

1. Adams, S.N., Honeysett, J.L., Tiller, K.G., and Norrish, K., *Aust. J. Soil Res.* **7,** 29 (1969).
2. Aguilar, T.A., Benevenga, N.J., and Harper, A.E., *J. Nutr.* **104,** 761 (1974).
3. Alexander, C.S., *Ann. Intern. Med.* **70,** 411 (1969).
4. Andrews, E.D., *N. Z. Vet. J.* **13,** 101 (1965).
5. Andrews, E.D., *N. Z. Agric.* **92,** 239 (1956).
6. Andrews, E.D., *N. Z. Agric. Res.* **8,** 788 (1965).
7. Andrews, E.D., *N. Z. J. Sci. Technol., Sect. A* **35,** 301 (1953).
8. Andrews, E.D., and Anderson, J.P., *N. Z. J. Sci. Technol., Sect. A* **35,** 483 (1954).
9. Andrews, E.D., and Hart, L.I., *N. Z. J. Agric. Res.* **5,** 403 (1962).
10. Andrews, E.D., Hart, L., and Stephenson, B.J., *N. Z. J. Agric. Res.* **2,** 274 (1959); **3,** 364 (1960).
11. Andrews, E.D., and Stephenson, B.J., *N. Z. J. Agric. Res.* **9,** 491 (1966).
12. Andrews, E.D., Stephenson, B,J., Anderson, J.P., and Faithful, W.C., *N. Z. J. Agric. Res.* **1,** 125 (1958).
13. Andrews, E.D., Stephenson, B.J., Isaacs, C.E., and Register, R.H., *N. Z. Vet. J.* **14,** 191 (1966).
14. Anonymous, *Nutr. Rev.* **26,** 173 (1968).
15. Archibald, J.G., *J. Dairy Sci.* **30,** 293 (1947).
16. Askew, H.O., *N. Z. J. Sci. Technol., Sect. A* **28,** 37 (1946).
17. Askew, H.O., and Watson, J., *N. Z. J. Sci. Technol., Sect. A* **25,** 81 (1943).
18. Aston, B.C., *N. Z. J. Agric.* **28,** 38 and 301; **29,** 14 and 84 (1924).
19. Beck, W.S., Flavin, M., and Ochoa, S., *J. Biol. Chem.* **229,** 997 (1957); Beck, W.S., and Ochoa, S., *ibid.* **232,** 931 (1958).
20. Becker, D.E., and Smith, S.E., *J. Anim. Sci.* **10,** 226 (1951).
21. Becker, D.E., Smith, S.E., and Loosli. J.K., *Science* **110,** 71 (1949).
22. Blokhima, R.I., *in* "Trace Element Metabolism in Animals" (C.F. Mills, ed.), Vol. 1, p. 426. Livingstone, Edinburgh, 1970.
23. Braude, R., Free, A.A., Page, J.E., and Smith, E.L., *Br. J. Nutr.* **3,** 289 (1953).
24. Burch, R.E., Williams, R.V., and Sullivan, J.F., *Am. J. Clin. Nutr.* **26,** 403 (1973).
25. Butt, E.M., Nusbaum, R.E., Gilmour, T.C., and DiDio, S.L., *in* "Metal-Binding in Medicine" (M.J. Seven and L.A. Johnson, eds.), p. 43. Lippincott, Philadelphia, Pennsylvania, 1960.
26. Comar, C.L., *Nucleonics* **3,** 30 (1948).
27. Comar, C.L., Davis, G.K., and Taylor, R.F., *Arch. Biochem.* **9,** 149 (1946); Comar, C.L., and Davis, G.K., *ibid.* **12,** 257 (1947).
28. Connolly, J.F., and Poole, D.B.R., *Ir. J. Agric. Res.* **6,** 229 (1967).
29. Consolazio, C.F., Nelson, R.A., Matoush, L.O., Hughes, R.C., and Urone, P., *U.S. Army Med. Res. Nutr. Lab., Rep.* **284** (1964).
30. Culvenor, C.C., DalBon, R., and Smith, L.W., *Aust. J. Chem.* **17,** 1301 (1964).
31. Dawbarn, M.C., and Hine, D.C., *Aust. J. Exp. Biol. Med. Sci.* **33,** 335 (1955).

32. Dawbarn, M.C., Hine, D.C., and Smith, J., *Aust. J. Exp. Biol. Med. Sci.* **35**, 273 (1957).
33. Dewey, D.W., Lee, H.J., and Marston, H.R., *Nature (London)* **181**, 1367 (1958).
34. Dinusson, W.E., Klosterman, E.W., Lapsley, E.L., and Buchanan, M.J., *J. Anim. Sci.* **12**, 623 (1953).
35. Dryden, L.P., Hartman, A.M., Bryant, M.P., Robinson, J.M., and Moore, L.A., *Nature (London)* **195**, 201 (1962).
36. Ellis, G.H., and Thompson, J.F., *Ind. Eng. Chem., Anal. Ed.* **17**, 254 (1947).
37. Engel, R.W., Price, N.O., and Miller, R.F., *J. Nutr.* **92**, 197 (1967).
38. Filippov, A., *Svinovodstvo (Moscow)* No. 4, p. 23 (1973).
39. Filmer, J.F., *Aust. Vet. J.* **9**, 163 (1933).
40. Filmer, J.F., *N. Z. J. Agric.* **63**, 287 (1941).
41. Filmer, J.F., and Underwood, E.J., *Aust. Vet. J.* **10**, 84 (1934); **13**, 57 (1937).
42. Forth, W., and Rummel, R., *in* "Intestinal Absorption of Metal Ions, Trace Elements and Radionuclides" (S.C. Skoryna and D. Waldron-Edward, eds.), p. 179. Pergamon, Oxford, 1971.
43. Gallagher, C.H., Koch, J.H., Moore, R.M., and Steel, J.D., *Nature (London)* **204**, 542 (1964).
44. Gardiner, F.H., *J. Lab. Clin. Med.* **41**, 56 (1953).
45. Gawthorne, J.M., *Aust. J. Exp. Biol. Med. Sci.* **47**, 311 (1969).
46. Gawthorne, J.M., *Aust. J. Exp. Biol. Med. Sci.* **48**, 285 and 293 (1970).
47. Gawthorne, J.M., *Aust. J. Biol. Sci.* **21**, 789 (1968).
48. Gawthorne, J.M., and Smith, R.M., *Biochem. J.* **142**, 119 (1974).
49. Gawthorne, J.M., Somers, M., and Woodliff, H.J., *Aust. J. Exp. Biol. Med. Sci.* **44**, 585 (1966).
50. Gawthorne, J.M., and Stokstad, E.L.R., *Proc. Soc. Exp. Biol. Med.* **136**, 42 (1971).
51. Gessert, C.F., Berman, D.T., Kastelic, J., Bentley, O.G., and Phillips, P.H., *J. Dairy Sci.* **35**, 696 (1952).
52. Grant, W.C., and Root, W.S., *Physiol. Rev.* **32**, 449 (1952).
53. Greig, J.R., Dryerre, H., Godden, W., Crichton, A., and Ogg, W.G., *Vet. J.* **89**, 99 (1933).
54. Grimmett, R.E.R., and Shorland, F.B., *Trans. R. Soc. N.Z.*, **64**, 191 (1934).
55. Grinvalsky, H.T., and Fitch, D.M., *Ann. N.Y. Acad. Sci.* **156**, 544 (1969).
56. Gurnani, S., Mistry, S.P., and Johnson, B.C., *Biochim. Biophys. Acta* **38**, 187 (1960).
57. Hamilton, H.G., *South. Med. J.* **49**, 1056 (1956).
58. Harp, M.J., and Scoular, F.I., *J. Nutr.* **47**, 67 (1952).
59. Hart, L.I., and Andrews, E.D., *Nature (London)* **184**, 1242 (1959).
60. Hartman, A.M., and Dryden, L.P., *Arch. Biochem. Biophys.* **40**, 310 (1952).
61. Hedrich, M.J., Elliot, J.M., and Lowe, J.E., *J. Nutr.* **103**, 1646 (1973).
62. Hine, D.C., and Dawbarn, M.C., *Aust. J. Exp. Biol. Med. Sci.* **32**, 641 (1954).
63. Hoekstra, W.G., Pope, A.L., and Phillips, P.H., *J. Nutr.* **48**, 421 (1952).
64. Houk, A.E., Thomas, A.W., and Sherman, H.C., *J. Nutr.* **31**, 106 (1946).
65. Hubbard, D.M., Creech, F.M., and Cholak, J., *Arch. Environ. Health* **13**, 190 (1966).
66. Ibbotson, R.N., Allen, S.H., and Gurney, C.W., *Aust. J. Exp. Biol. Med. Sci.* **48**, 161 (1970).
67. Kent, N. L., and McCance, R.A., *Biochem. J.* **35**, 877 (1941).
68. Kercher, C.J., and Smith, S.E., *J. Anim. Sci.* **14**, 458, 878 (1955); **15**, 550 (1956).
69. Kirchgessner, M., *Z. Tierphysiol., Tierernaehr. Futtermittelkd.* **14**, 270 (1959).
70. Kitzbach, C., Galloway, E., and Stokstad, E.L.R., *Proc. Soc. Exp. Biol. Med.* **124**, 801 (1967).

71. Koch, B.A., and Smith, S.E., *J. Anim. Sci.* **10,** 1017 (1951).
72. Koch, H.J., Smith, E.R., Shimp, N.F., and Connor, J., *Cancer* **9,** 499 (1956).
73. Kovalsky, V.V., *in* "Trace Element Metabolism in Animals" (C.F. Mills, ed.), Vol. 1, p. 385. Livingstone, Edinburgh, 1970.
74. Kriss, J.P., Carnes, W.H., and Gross, R.T., *J. Am. Med. Assoc.* **157,** 117 (1955).
75. Lassiter, C.A., Ward, G.M., Huffman, C.F., Duncan, C.W., and Webster, H.D., *J. Dairy Sci.* **36,** 997 (1953).
76. Leddicotte, G.W., *Int. Comm. Radiol. Protect. Rep. Permiss. Doses Intern. Radiat. 1958.*
77. Lee, C.C., and Wolterink, L.F., *Am. J. Physiol.* **183,** 173 (1955); *Poult. Sci.* **34,** 764 (1955).
78. Lee, H.J., *Brit. Commonw. Off. Sci. Conf. Aust., 1949,* p. 262 (1951).
79. Lee, H.J., *Aust. Vet. J.* **26,** 152 (1950).
80. Lee. H.J., and Kuchel, R.E., *Aust. J. Agric. Res.* **4,** 88 (1953).
81. Lee, H.J., Kuchel, R.E., and Trowbridge, R.F., *Aust. J. Agric. Res.* **7,** 333 (1956).
82. Lee, H.J., and Marston, H.R., *Aust. J. Agric. Res.* **20,** 905 and 1109 (1969).
83. LeGoff, J.M., *J. Pharmacol. Exp. Ther.* **38,** 1 (1930).
84. MacLaren, A.P.C., Johnston, W.G., and Voss, R.C., *Vet. Rec.* **76,** 1148 (1964).
85. MacPherson, A., Moon, F.E., and Voss, R.C., *Br. Vet. J.* **132,** 294 (1976).
86. MacPherson, A., Moon, F.E., and Voss, R.C., *Br. Vet. J.* **129,** 414 (1973).
87. MacPherson, A., and Moon, F.E., *in* "Trace Element Metabolism in Animals" (G.W. Hoekstra *et al.*, eds.), Vol. 2, p. 624. Univ. Park Press, Baltimore, Maryland, 1974.
88. Marston, H.R., *Physiol. Rev.* **32,** 66 (1952).
89. Marston, H.R., *Br. J. Nutr.* **24,** 615 (1970).
90. Marston, H.R., *J. Counc. Sci. Ind. Res (Aust.)* **8,** 111 (1935); Lines, E.W., *ibid.* p. 117.
91. Marston, H.R., Allen, S.H., and Smith, R.M., *Nature (London)* **190,** 1085 (1961).
92. Marston, H.R., Allen, S.H., and Smith, R.M., *Br. J. Nutr.* **27,** 147 (1972).
93. Marston, H.R., and Lee, H.J., *Nature (London)* **170,** 791 (1952).
94. Marston, H.R., Lee, H.J., and McDonald, I.W., *J. Agric. Sci.* **38,** 222 (1948).
95. Marston, H.R., Thomas, R.G., Murnane, D., Lines, E.W., McDonald, I.W., and Bull, L.B., *Commonw. Counc. Sci. Ind. Res. (Aust.), Bull.* **113** (1938).
96. McDonald, I.W., *Aust. Vet. J.* **17,** 165 (1942); **22,** 91 (1946).
97. McNaught, K.J., *N. Z. J. Sci. Technol., Sect. A* **20,** 14 (1938).
98. McNaught, K.J., *N. Z. J. Sci. Technol., Sect. A* **30,** 26 (1948).
99. Millar, K.R., and Andrews, E.D., *N. Z. Vet. J.* **12,** 9 (1964).
100. Monroe, R.A., Sauberlich, H.E., Comar, C.L., and Hood, S.L., *Proc. Soc. Exp. Biol. Med.* **80,** 250 (1952).
101. Montgomery, M.L., Skeline, G.E., and Chaikoff, I.L., *J. Exp. Med.* **78,** 151 (1943).
102. Morris, J.G., and Gartner, R.J.W., *J. Agric. Sci.* **68,** 1 (1967).
103. Murthy, G.K., Rhea, U., and Peeler, J.T., *Environ. Sci. Technol.* **5,** 436 (1971).
104. Nodya, P.I., *Hyg. Sanit. (USSR)* **37,** 108 (1972).
105. O'Halloran, M.W. and Skerman, K.D., *Br. J. Nutr.* **15,** 99 (1961).
106. Oram, R.N., *Proc. Int. Grassl. Cong. 11th, 1969,* p. 785 (1970).
107. Orr, J.B., and Holm, A., "Mineral Content of Pastures," 6th Rep. Econ. Advis. Counc., Great Britain, 1931.
108. Ozanne, P.G., Greenwood, E.A., and Shaw, T.C., *Aust. J. Agric. Res.* **14,** 39 (1963).
109. Paley, K.R., and Sussman, E.S., *Metab., Clin. Exp.* **12,** 975 (1963).
110. Paulais, R., *Ann. Pharm. Fr.* **4,** 110 (1946).
111. Pearson, P.B., Struglia, L., and Lindahl, I.L., *J. Anim. Sci.* **12,** 213 (1935).

112. Perry, H.M., Jr., and Schroeder, H.A., *Am. J. Med. Sci.* **228,** 396 (1954).
113. Phillipson, A.T., and Mitchell, R.L., *Br. J. Nutr.* **6,** 176 (1952).
114. Pollack, S., George, J.N., Reba, R.C., Kaufman, R.M., and Crosby, W.H., *J. Clin. Invest.* **44,** 1470 (1965).
115. Poole, D.B.R., and Connolly, J.F., *Ir. J. Agric. Res.* **6,** 281 (1967).
116. Porter, J.W., *Proc. Nutr. Soc.* **12,** 106 (1953).
117. Powrie, J.K., *Aust. J. Sci.* **23,** 180 (1960).
118. Raum, N.S., Stables, G.L., Pope, L.S., Harper, O.F., Waller, G.R., Renbarger, R., and Tillman, A.D., *J. Anim. Sci.* **27,** 1695 (1968).
119. Richardson, D., Tsien, W.S., Koch, B.A., Ward, J.K., and Smith, E.F., *J. Anim. Sci.* **19,** 910 (1960).
120. Rigg, T., and Askew, H.O., *Emp. J. Exp. Agric.* **2,** 1 (1934).
120a. Robey, J.S., Veazey, P.M., and Crawford, J.D., *N. Engl. J. Med.* **255,** 955 (1956).
121. Rossiter, R.C., Curnow, D.H., and Underwood, E.J., *J. Aust. Inst. Agric. Sci.* **14,** 9 (1948).
122. Rothery, P., Bell, J.M., and Spinks, J.W.T., *J. Nutr.* **49,** 173 (1953).
123. Ruelke, O.C., and McCall, J.T., *Agron. J.* **53,** 406 (1961).
124. Russell, F.C., and Duncan, D., *Commonw. Bur. Anim. Nutr., Tech. Commun.* No. 15 (1956).
125. Schroeder, H.A., Nason, A.P., and Tipton, I.H., *J. Chronic. Dis.* **20,** 869 (1967).
126. Simpson, B.H., Jolly, R.D., and Thomas, S.H.M., *N. Z. Vet. J.* **17,** 240 (1969).
127. Skerman, K.D., Sutherland, A.K., O'Halloran, M.W., Bourke, J.M., and Munday, B.L., *Am. J. Vet. Res.* **20,** 977 (1959).
128. Smith, R.M., and Marston, H.R., *Br. J. Nutr.* **24,** 857 and 879 (1970).
129. Smith, R.M., and Monty, K.J., *Biochem. Biophys. Res. Commun.* **1,** 105 (1959).
130. Smith, R.M., and Osborne-White, W.S., *Biochem. J.* **136,** 279 (1973).
131. Smith, R.M., Osborne-White, W.S., and Gawthorne, J.M., *Biochem. J.* **142,** 105 (1974).
132. Smith, S.E., Becker, D.E., Loosli, J.K., and Beeson, K.C., *J. Anim. Sci.* **9,** 221 (1950).
133. Smith, S.E., and Loosli, J.K., *J. Dairy Sci.* **40,** 1215 (1957).
134. Somers, M., *Aust. J. Exp. Biol. Med. Sci.* **47,** 219 (1969).
135. Somers, M., and Gawthorne, J.M., *Aust. J. Exp. Biol. Med. Sci.* **47,** 227 (1969).
136. Southcott, W.H., *Aust. Vet. J.* **32,** 225 (1956).
137. Sutton, A.L., and Elliot, J.M., *J. Nutr.* **102,** 1341 (1972).
138. Tevetoglu, F., *J. Pediatr.* **49,** 46 (1956).
139. Thiers, R.E., Williams, J.F., and Yoe, J.H., *Anal. Chem.* **27,** 1725 (1955).
140. Thompson, J.F., and Ellis, G.H., *J. Nutr.* **34,** 121 (1947).
141. Thomson, A.B.R., Valberg, L.S., and Sinclair, D.G., *J. Clin. Invest.* **50,** 2384 (1971).
142. Tipton, I.H., and Cook, M.J., *Health Phys.* **9,** 103 (1963).
143. Tipton, I.H., Stewart, P.L., and Martin, P.G., *Health Phys.* **12,** 1683 (1966).
144. Toskes, P.P., Smith, G.W., and Conrad, M.E., *Am. J. Clin. Nutr.* **26,** 435 (1973).
145. Underwood, E.J., *Aust. Vet. J.* **10,** 87 (1934).
146. Underwood, E.J., and Elvehjem, C.A., *J. Biol. Chem.* **124,** 419 (1938).
147. Underwood, E.J., and Filmer, J.F., *Aust. Vet. J.* **11,** 84 (1935).
148. Underwood, E.J., and Harvey, R.J., *Aust. Vet. J.* **14,** 183 (1938).
149. Valberg, L.S., *in* "Intestinal Absorption of Metal Ions, Trace Elements and Radionuclides" (S.C. Skoryna and D. Waldron-Edward, eds.), p. 257. Pergamon, Oxford, 1971.
150. Valberg, L.S., Ludwig, J., and Olatunbosun, D., *Gastroenterology* **56,** 241 (1969).

151. Van Der Merwe, F.J., *Farming S. Afr.* **35,** 44 (1959).
152. Waltner, K., and Waltner, K., *Klin. Wochenschr.* **8,** 313 (1929).
153. Wessells, C.C., *J. S. Afr. Vet. Med. Assoc.* **32,** 289 (1961).
154. Wilbert, C., *Muench. Med. Wochenschr.* **92,** 1373 (1950).
155. Wolff, H., *Klin. Wochenschr.* **28,** 280 (1950).
156. Yamagata, N., Murata, S., and Torii, T., *J. Radiat. Res.* **3,** 4 (1962).
157. Yamagata, N., Kurioka, W., and Shimizu, T., *J. Radiat. Res.* **4,** 8 (1963).

6

Nickel

I. NICKEL IN ANIMAL TISSUES AND FLUIDS

Nickel occurs in low concentrations in all animal tissues and fluids that have been examined by sufficiently sensitive and reliable analytical methods. The nickel is distributed throughout the body without particular concentration in any known tissue or organ and does not accumulate with age in any human organ other than the lungs (41, 54). Very low normal levels have recently been found in human tissues by Sunderman *et al.* (49). The reported range of values for four individuals were as follows: lung, 0.03–1.46; liver, 0.02–0.05; and heart, 0.02–0.03 μg/g dry weight. Appreciably higher Ni concentrations have been obtained for these tissues and for the kidney and spleen in calves (31) and rats (42) (see Table 20). Human dental enamel has been reported to contain a mean Ni concentration of 0.64 (range 0.11–3.00) μg/g dry weight (18). Schroeder and Nason (43) found the washed fat-free hair of 25 women and 79 men to average 3.96 ± 1.055 and only 0.97 ± 0.147 μg/g Ni, respectively. Of the eight elements studied only nickel, copper, and cobalt showed a significant sex difference. A lower Ni concentration of 0.5 ± 0.1 μg/g has been reported for jack rabbit hair (19).

The level of nickel in the blood serum of healthy individuals varies greatly among species but comparatively little within species. Sunderman *et al.* (49) give the following mean concentrations and range of values for various species, based in some cases on a small number of samples: human, 2.6 (1.1–4.6); horses, 2.0 (1.3–2.5); cattle, 2.6 (1.7–4.4); dogs, 2.7 (1.8–4.2); goats, 3.5 (2.7–4.4); cats,

TABLE 20
Effect of Nickel in the Diet on the Concentrations of Nickel in the Tissues (μg Ni/g Dry Matter)

	Level of Ni supplement[a]			Ni in the drinking water[b]	
Tissue ppm:	0	250	1000	0	5
Liver	0.76	0.37	0.53	1.3	1.2
Kidney	2.08	2.26	22.83	1.7	3.0
Spleen	0.00	0.00	1.85	6.2	4.9
Lung	2.94	13.64	5.88	2.4	2.7
Heart	1.11	0.32	0.50	3.0	3.2

[a] From O'Dell *et al.* (31).
[b] From Schroeder *et al.* (42).

3.7 (1.5–6.4); guinea pigs, 4.1 (2.4–7.1); pigs, 5.0 (4.2–5.6); rabbits, 9.3 (6.5–14.0); and Maine lobsters, 12.4 (8.3–20.1) μg Ni/liter. A small part of the nickel in human and rabbit serum occurs in ultrafilterable form, with the remainder present in nearly equal proportions in albumin-bound form and as the α_2-macroglobulin named nickeloplasmin (6, 29). Nickeloplasmin is presumed to be identical with the Ni-containing metalloprotein first isolated from human serum by Himmelhoch and co-workers (10). Rabbit nickeloplasmin has a molecular weight of 7.0×10^5, a Ni content (g-atom/mole) of 0.9, and is probably a ternary complex of serum α_1-macroglobulin with an ultrafilterable nickel constituent of serum (6, 29).

Abnormal serum Ni concentrations occur in several disease states in man. Serum Ni is elevated to about twice normal in the 12–36 hr after the onset of symptoms in patients with acute myocardial infarction (5, 20, 30, 50). Thus McNeely *et al.* (20) found a mean of 5.2 ± 2.8 μg Ni/liter in 33 patients 13–36 hr after acute myocardial infarction, compared with a mean of 2.6 ± 0.8 μg/liter in 47 healthy adults. The source of this increase in nickel and its cause are unknown. Sunderman *et al.* (49) have shown that the nickel in cardiac muscle, even if completely released and distributed solely in the serum, is insufficient to account for the hypernickelemia. They suggest that it could possibly come from other organs such as the liver or lungs. McNeely *et al.* (20) have also demonstrated abnormally high serum Ni concentrations following acute stroke and severe burns. Significantly diminished mean concentrations were found in patients with hepatic cirrhosis (1.6 ± 0.8 μg Ni/liter) and chronic uremia (1.7 ± 0.7 μg Ni/liter). Patients with acute myocardial ischemia without infarction, acute trauma with fractured bones, acute delirium tremens, and muscular dystrophy exhibited normal mean serum Ni concentrations. The sera from 12 mothers immediately after normal delivery contained a mean Ni concentration of 3.0 ± 1.2 μg/liter, which was the same as that of the sera from the umbilical cords of their 12 full-term infants (20).

The average level of nickel in cow's milk and colostrum has been reported to be 0.03 and 0.10 ppm, respectively (14). In a more recent study similar concentrations for milk were obtained (32). Most of the samples contained 0.02–0.05 ppm Ni and all were below 0.1 ppm. These levels were not increased when the cow's rations were supplemented with nickel carbonate to give a daily consumption of 365 and 1835 mg of supplemental nickel.

The nickel content of sweat is surprisingly high. Horak and Sunderman (11) found a mean concentration of 49 μg/liter in the sweat from the arms of five healthy men during sauna bathing. This is some 20 times the Ni concentration of normal blood serum.

II. NICKEL METABOLISM

Dietary nickel is poorly absorbed, probably in the range of 1–10%, even at high intakes. Most of this nickel is excreted in the feces, with smaller amounts appearing in the urine and sweat. The high-Ni content of sweat, pointing to active Ni secretion by the sweat glands, was mentioned in the previous section. Consolazio *et al.* (4) determined that approximately 8 μg Ni is lost in the sweat daily. It is obvious that under conditions of excessive sweating, dermal losses of nickel could be high.

Tedeschi and Sunderman (52) found that normal dogs excreted 90% of ingested Ni in the feces and 10% in the urine, with no significant retention in the body. Poor absorption and predominant excretion in the feces are also apparent from several studies in man. Thus Perry and Perry (36) found the urine of 24 healthy adults to contain 10–70 μg Ni/liter (mean 20 ± 2.6), with an average total urinary excretion of about 30 μg Ni/day. Sunderman (48) obtained somewhat lower values, with a mean total urinary excretion close to 20 μg Ni/day. Feces collected from 10 healthy adults over 3 days contained 14.2 ± μg Ni/g dry basis, giving a mean elimination in the feces of 258 ± 126 μg Ni/day, or about 20 times that of the urine (11). In Ni balance studies carried out on 10 Russian males (28) ingesting 289 ± 23 μg Ni/day, fecal excretion averaged 258 ± 23 μg Ni/day. Some of this fecal Ni would be absorbed Ni excreted via the bile, since bile is known to contain nickel (35, 45).

Data for ruminants and other farm species are meager, but O'Dell and co-workers (31) reported that calves excreted more than 20 times as much nickel in the feces as in the urine when consuming a normal ration. When this ration was supplemented with 62.5, 250 and 1000 ppm Ni as nickelous carbonate the animals excreted only 2.7, 1.9, and 4.3%, respectively, of the total excretion in the urine. Only at the two highest levels of supplementation was significant retention in the tissues observed. Similarly, little or no accumulation of nickel in the tissues was observed in rats fed 5 ppm Ni in the drinking water for life (42) (Table 20). It is clear that the animal body does not readily absorb or retain nickel, presumably through a homeostatic control mechanism that only begins to break down when exposed to abnormally high-Ni intakes.

Studies with radioactive ^{63}Ni injected into mice, rats, and rabbits reveal a different pattern of excretion from that just described. The metal is widely distributed throughout the tissues in a manner directly related to the blood volume of the particular organ or tissue and is rapidly eliminated in the urine, with smaller amounts appearing in the feces via the bile (35, 45, 57). In rats and rabbits ^{63}Ni is rapidly cleared from the plasma or serum during the first 48 hr after injection and then disappears at a much slower rate over 3–8 days, with 68–78% of the dose appearing in the urine during the first day after injection (35).

III. NICKEL DEFICIENCY AND FUNCTIONS

A dietary deficiency of nickel, leading to various pathological manifestations, has now been produced in chicks, rats, and pigs. Nielsen (22) obtained the first evidence that nickel may be essential for the chick. Chicks fed a diet containing less than 40 ppb Ni, compared with controls given a supplement of 3–5 ppm Ni, revealed (a) pigmentation changes in the shank skin, (b) thicker legs with slightly swollen hocks, (c) dermatitis of the shank skin, (d) a less friable liver, and (e) an enhanced accumulation of a tracer dose of ^{63}Ni in the liver, bone, and aorta. Sunderman *et al.* (51) were unable to confirm these findings with chicks using a diet containing 44 ppb Ni, but they observed ultrastructural changes in the liver. These changes were later confirmed and extended by Nielsen and co-workers (25, 26) employing a diet still lower in nickel (2–15 ppb Ni) and with more rigid environmental control. Biochemical and pathological abnormalities appeared at 3–5 weeks which were absent from control chicks fed 3 ppm Ni or more. The principal abnormalities were ultrastructural changes in the liver, with the most obvious abnormality in the organization of the rough endoplasmic reticulum; altered gross appearance; reduced oxidative ability and decreased lipid phosphorus in the liver; altered shank pigmentation with a decrease in yellow lipochrome pigments; and lower hematocrits. A tendency toward increased leg thickness and size of hock and decreased plasma cholesterol was also observed in the Ni-deficient chicks. Some evidence was obtained that rhodium interacts metabolically with nickel, although not competitively since rhodium at 50 ppm as $ClRh(NH_3)_5SO_4$ did not increase the signs of Ni deficiency (25).

Rats fed diets containing 2–15 ppb Ni throughout fetal, neonatal, and adult life exhibited impaired growth and reproductive performance, evidenced by

TABLE 21
Oxygen Uptake of Rat Liver Homogenates[a]

Group	No. of rats	O_2 uptake (U liter/hr/mg protein)
Ni deficient (4 ppb)	13	$3.20^b \pm 0.08$
+ 3 ppm Ni	12	4.17 ± 0.21

[a]Taken from Nielsen and Ollerich (24); α-glycerophosphate as substrate.
[b]Significant different ($p < 0.0001$) from + 3 ppm Ni group.

increased fetal death rate and perinatal mortality, particularly in the second generation (24, 26, 27). The Ni-deficient rats were less active during the suckling period, had a rougher hair coat, and weighed less at weaning. Liver changes were also observed in Ni-deficient rats, as in chicks, including a reduced oxidative ability of homogenates in the presence of α-glycerophosphate (Table 21), decreased cholesterol, and ultrastructural changes. Rations low in nickel have also been developed by Anke and co-workers (1) on which pigs showed a slower rate of weight gain, delayed sexual maturity, and higher piglet mortality than controls receiving 10 ppm Ni. A feature of these experiments was the appearance of signs of parakeratosis in 30–50% of the young pigs and goats on the Ni-low rations and the significantly lower Zn levels in their liver, hair, rib, and brain compared with those of Ni-supplemented animals.

Metabolic aspects of Ni deficiency have been further illuminated by the recent experiments of Kirchgessner and his colleagues (16, 38–40). No effect on the growth of rats initially exposed to Ni deficiency at 21 days of age was observed, even at dietary levels down to 15 ppb, presumably due to mobilization of Ni reserves in the tissues. When the mothers of the experimental rats were maintained on a Ni-deficient diet the young, similarly maintained, exhibited a marked growth inhibition, especially in the second generation. With 15 ppb Ni in the diet the weight differences at 30 days of age averaged 16% in the F_1 generation and 26% in the F_2 generation (38). An adverse effect on iron absorption resulting in severe anemia was apparent in the Ni-deficient rats. Iron storage in the liver, spleen, and kidneys was reduced 87%, 77%, and 46%, respectively, compared with rats on an adequate Ni intake (39). Doubling the iron supply from 50 to 100 ppm was insufficient to restore normal iron levels in the tissues or to prevent some degree of anemia (39). The Ni-deficient rats receiving the "normal" 50-ppm Fe diet developed severe anemia at 30 days of age, with the erythrocyte count reduced by 36%, the hematocrit 37%, and the hemoglobin content 44% compared with Ni-sufficient controls (40). In addition, greatly reduced dehydrogenase activities for the malate (MDH) and glucose-6-phosphate dehydrogenases (G6PDH) were demonstrated in the livers of

TABLE 22
Effect of Ni Deficiency on the Activity of Dehydrogenases in the Liver of 30-Day-Old Rats[a]

Dietary Ni (ppm)	Malate dehydrogenase (U/mg protein)	Glucose-6-phosphate dehydrogenase (mU/mg protein)
20	9.12 ± 1.81	209.1 ± 72.3
0.015	6.44 ± 1.45	38.3 ± 22.2
P	< 0.02	< 0.001

[a]From Kirchgessner and Schnegg (16).

Ni-deficient rats as shown in Table 22 (16). This effect was apparent in the absence of any significant differences in the food consumption or feeding patterns of the Ni-deficient and the Ni-sufficient rats.

The significance of the above important findings in relation to the precise biochemical role or roles of nickel and its minimum requirements remains to be determined. Nielsen and Ollerich (24) have speculated that nickel has a role in the metabolism or structure of membranes and point out that the morphological abnormalities, impaired oxidation, and changes in the phospholipid level in the liver are consistent with such a hypothesis. Increased liver nuclear RNA polymerase activity and increased total and active alkaline RNase–protein have also been demonstrated (23). The authors suggest that nickel, because it can complex with macromolecules, has a role in the metabolism of membranes and in RNA metabolism. Significant concentrations of nickel are known to be present in DNA and RNA (48, 56). Nickel stabilizes RNA (8) and DNA (7) against thermal denaturation and activates numerous enzymes *in vitro*, but it is not known if these nonspecific *in vitro* effects are related to an *in vivo* function for nickel. Indirect but suggestive evidence that nickel and other metals may play a role in pigmentation has been obtained by Kikkawa and co-workers (13), and the possibility that nickel plays a role in lactation at the pituitary level is suggested by the work of LaBella (17).

The minimum amounts of nickel required by animals to maintain health, based on the addition of graded increments to a known deficient diet in the conventional manner, are not yet known. Nielsen and Sandstead (27) suggest that 50–80 ng Ni/g is probably adequate for the rat and chick, and Schnegg and Kirchgessner (38) indicate that 50 ng/g is close to the Ni requirement for growth in the rat. This estimate cannot yet be extrapolated to man with great confidence, but if it is, then the minimum daily Ni intake required by individuals consuming 400 g dry matter/day would be only 20 μg. On this tentative basis, adequate dietary Ni supplies are unlikely to pose a practical problem to man, except perhaps, as discussed in the following section, in individuals consuming diets exceptionally high in fats, refined carbohydrates, and dairy products, or those with conditions that impede nickel absorption or promote excessive loss.

IV. SOURCES OF NICKEL

1. Nickel in Human Foods and Dietaries

The nickel content of foods has attracted very little attention so far. Fifty years ago Bertrand and Mâcheboeuf (2) reported levels of 1.5–3.0 ppm Ni (d.b.) for the green leafy vegetables and much lower values, ranging from 0.15 to 0.35 ppm, for fruits, tubers, and grains. Concentrations of this order were confirmed in a much later study by Schroeder and associates (41). These workers also established the very low-Ni levels of refined foods and most foods of animal origin, notably muscle meats, milk, and other dairy products. Considerable losses of nickel occur in the process of milling wheat to white flour. Thus Zook *et al.* (60), in a study of North American wheats and the flours and other products made from them, obtained the following Ni concentrations: common hard wheat, 0.47 ± 0.08; common soft wheat, 0.31 ± 0.08; and patent flour from soft wheat, 0.18 ± 0.07 ppm Ni dry basis. White bread was shown to be appreciably richer in nickel than would be expected from the figures given above for white flour, namely, 0.65 ± 0.13 and 0.52 ± 0.04 ppm Ni (d.b.). These values suggest either that the materials added to the flour were contributing nickel or that there was Ni contamination during the mixing or baking processes. In a later study of samples of wheat seed from 12 locations in North America Welch and Carey (59) obtained generally lower Ni concentrations with a wide range. Their values ranged from 0.08 to 0.35, with an overall mean of 0.18 ppm Ni.

Total dietary Ni intakes vary greatly with the amounts and proportions of foods of animal (Ni-low) and plant origin (Ni-high) consumed and with the amounts of refined and processed foods such as white sugar and flour included in the diets. Schroeder *et al.* (41) prepared a diet high in fat and refined foods and low in vegetables which was calculated to supply only 3–10 μg Ni/day, whereas another diet of similar calorie and protein value but low in fat and high in whole grain foods, vegetables and oysters was estimated to supply 700–900 μg Ni/day. Typical mixed Western-style diets consumed by adults generally supply 300–500 μg Ni/day (9, 12, 28, 41).

Little is known about the chemical forms of nickel in foods and the relation of different chemical forms to availability. For example, Ni forms a stable complex with phytic acid (55), but whether the phytate in grains decreases Ni bio-availability, as occurs with iron and zinc, is unknown. The extent to which organic Ni compounds, such as the anionic amino acid complex in which Ni is translocated in plants (53), occurs in edible plant tissues and affects Ni absorption also remains to be determined.

2. Nickel in Animal Feeds and Pastures

Common pasture plants contain 0.5–3.5 ppm Ni (d.b.) (15, 21, 44), and wheat grains 0.3–0.6 ppm (59). Little is known of the Ni content of other common grains and forages. Milk products and meat meals used as protein supplements are very poor Ni sources, as would be expected from the Ni concentrations given earlier for milk and animal tissues. There appear to be no published values for the Ni content of leguminous seeds, oilseed meals, or fish meals. Although Ni deficiency seems unlikely to arise under natural conditions in farm animals, further studies of nickel in animal feeds are desirable.

V. NICKEL TOXICITY

Nickel is a relatively nontoxic element, so that Ni contamination of foods does not present a serious health hazard. Acid foods take up Ni from nickel vessels during cooking, but it is poorly absorbed and causes no detectable damage (37). A high incidence of respiratory tract neoplasia and dermatitis has been observed among exposed workers in nickel refineries (46), and nickel has been implicated as a pulmonary carcinogen in tobacco smoke (47). Oral administration of nickel has been shown to aggravate the hand eczema in 9 of 12 patients with contact allergy to nickel (3). The authors concluded that the ingestion of small amounts of nickel within the physiological range may be of greater importance in maintaining the hand eczema than external contacts with the metal.

In a study of Ni toxicity in rats levels of 250, 500, and 1000 ppm Ni in three different forms did not affect growth rate or reproduction, and no signs of toxicity were apparent even after 3–4 months of continuous feeding (37). Adult monkeys maintained their weights and were in perfect health after 6 months on the same high-Ni intakes (37). A similar low toxicity for nickel is evident from the recent long-term study of Schroeder *et al.* (42) with rats. Rats of both sexes exposed to 5 ppm Ni as a soluble salt in the drinking water exhibited some increased growth, but there was no effect on survival, longevity, incidence of tumors, or specific lesions.

Studies with mice given the very high levels of 1100 and 1600 ppm Ni in the diet as acetate revealed some toxic effects (58). These levels did not influence the body weight of adult mice or their litter size, but the numbers of pups weaned were reduced at the higher Ni intake. The feeding of 1600 ppm Ni to young growing mice lowered growth and food consumption in both males and females, and 1100 ppm Ni induced similar changes in the females only. Experiments with growing chicks indicate a lower Ni tolerance in this species (58). Growth of chicks to 4 weeks of age was significantly depressed at 700 ppm Ni and above, when added as the sulfate or the acetate. This was associated with a reduction in food intake, an impairment of energy metabolism, and a marked reduction in nitrogen retention. In an experiment designed to differentiate between the effects on food consumption and those of Ni toxicity per se, no differences in growth rate were observed, even at 1100 ppm Ni, compared with pair-fed control chicks, but nitrogen retention was depressed.

The health, feed consumption, milk production, and milk composition of dairy cows were unaffected by dietary supplements of nickel carbonate at 50 and 250 ppm Ni (32). Dairy calves fed 62.5 and 250 ppm Ni as the carbonate were similarly unaffected, but at 1000 ppm feed intake was greatly depressed and nitrogen retention significantly lowered (33). In a further experiment O'Dell and co-workers (34) observed a linear depression in feed palatability with cattle as the Ni in the feed, as nickelous chloride, was increased from 50 to 100 to 200 ppm Ni, and from 250 to 500 ppm when given as nickelous carbonate. Nickel as the chloride was five times as potent in reducing feed palatability as nickel in the carbonate form.

REFERENCES

1. Anke, M., Grun, M., Dittrich, G., Broppel, B., and Hennig, A., *in* "Trace Element Metabolism in Animals" (W.G. Hoekstra *et al.*, eds.), Vol. 2, p. 715. Univ. Park Press, Baltimore, Maryland, 1974.
2. Bertrand, G., and Mâcheboeuf, M., *C.R. Hebd. Seances Acad. Sci.* **180**, 1380 and 1993 (1925); **182**, 1504 (1926); **183**, 5 (1926).
3. Christensen, O.B., and Möller, H., *Contact Dermatitis* **1**, 136 (1975).
4. Consolazio, C.F., Nelson, R.A., Matousch, L.O., Hughes, R.C., and Urone, P., *U.S., Army Med. Res. Nutr. Lab., Rep.* **284**, 1 (1964).
5. D'Alonzo, C.A., and Pell, S., *Arch. Environ. Health* **6**, 381 (1963).
6. Decsy, M.I., and Sunderman, F.W.,, Jr., *Bioinorg. Chem.* **3**, 95 (1974).
7. Eichhorn, G.L., *Nature (London)* **194**, 474 (1962).
8. Fuwa, K., Wacker, W.E.C., Druyan, R., Bartholomay, A.F., and Vallee, B.L., *Proc. Natl. Acad. Sci. U.S.A.* **46**, 1298 (1960).
9. Hamilton, E.I., and Minski, M.J., *Sci. Total Environ.* **1**, 375 (1972/1973).
10. Himmelhoch, S.R., Sober, H.A., Vallee, B.L., Peterson, E.A., and Fuwa, K., *Biochemistry* **5**, 2523 (1966).
11. Horak, E., and Sunderman, F.W., Jr., *Clin. Chem.* **19**, 429 (1973).
12. Kent, N.L., and McCance, R.A., *Biochem. J.* **35**, 837 and 887 (1941).
13. Kikkawa, H., *J. Jpn. Biochem. Soc.* **27**, 427 (1955); Kikkawa, H., Ogita, Z., and Fujito, S., *Science* **121**, 43 (1955).
14. Kirchgessner, M., *Schriftenr. Mangelkr.* **6**, 61 and 105 (1960).
15. Kirchgessner, M., Merz, G., and Oelschlager, W., *Arch. Tierernaehr.* **10**, 414 (1960).

16. Kirchgessner, M., and Schnegg, A., *Bioinorg. Chem.* **6,** 155 (1976).
17. LaBella, F.S., Dular, R., Vivian, S., and Queen, G., *Biochem. Biophys. Res. Commun.* **52,** 786 (1973).
18. Losee, F., Cutress, T.W., and Brown, R., *Trace Subst. Environ. Health–7, Proc. Univ. Mo. Annu. Conf., 7th, 1973* (1974).
19. Mangelson, N.F., Allison, G.M., Eatough, D.J., Hill, M.W., Izatt, R.M., Christensen, J.J., Murdoch, J.R., Nielson, K.K., and Welch, S.L., *Trace Subst. Environ. Health–7, Proc. Univ. Mo. Annu. Conf., 7th, 1973,* p. 369 (1973).
20. McNeely, M.D., Sunderman, F.W., Jr., Nechay, M.W., and Levine, H., *Clin. Chem.* **17,** 1123 (1971).
21. Mitchell, R.L., *Soil Sci.* **60,** 63 (1945).
22. Nielsen, F.H., *in* "Newer Trace Elements in Nutrition" (W. Mertz and W.E. Cornatzer, eds.), p. 215. Dekker, New York, 1971; Nielsen, F.H., and Sauberlich, H.E., *Proc. Soc. Exp. Biol. Med.* **134,** 845 (1970).
23. Nielsen, F.H., Ollerich, D.A., Fosmire, G.J., and Sandstead, H.H., *in* "Protein-Metal Interactions" (M. Friedman, ed.), p. 389. Plenum, New York, 1974.
24. Nielsen, F.H., and Ollerich, D.A., *Fed. Proc., Fed. Am. Soc. Exp. Biol.* **33,** 1767 (1974).
25. Nielsen, F.H., Myron, D.R., Givand, S.H., and Ollerich, D.A., *J. Nutr.* **105,** 1607 (1975).
26. Nielsen, F.H., Myron, D.R., Givand, S.H., Zimmerman, T.J., and Ollerich, D.A., *J. Nutr.* **105,** 1620 (1975).
27. Nielsen, F.H., and Sandstead, H.H., *Am. J. Clin. Nutr.* **27,** 515 (1974).
28. Nodya, P.I., *Hyg. Sanit. (USSR)* **37,** 108 (1972).
29. Nomoto, S., McNeely, M.D., and Sunderman, F., Jr., *Biochemistry* **10,** 1647 (1971).
30. Nozdryukina, L.R., Grinkevich, N.I., and Gribovskaya, I.F., *Trace Subst. Environ. Health–7, Proc. Univ. Mo. Annu. Conf., 7th, 1973,* p. 353 (1973).
31. O'Dell, D.G., Miller, W.J., Moore, S.L., King, W.A., Ellers, J.C., and Jurecek, H., *J. Anim. Sci.* **32,** 769 (1971).
32. O'Dell, D.G., Miller, W.J., King, W.A., Ellers, J.C., and Jurecek, H.J., *J. Dairy Sci.* **53,** 1545 (1970).
33. O'Dell, D.G., Miller, W.J., King, W.A., Moore, S.L., and Blackmon, D.M., *J. Nutr.* **100,** 1447 (1970).
34. O'Dell, D.G., Miller, W.J., Moore, S.L., and King, W.A., *J. Dairy Sci.* **53,** 1266 (1970).
35. Onkelinx, C., Becker, J., and Sunderman, F.W., Jr., *in* Trace Element Metabolism in Animals" (W.G. Hoekstra *et al.*, eds.), Vol. 2, p. 560. Univ. Park Press, Baltimore, Maryland, 1974.
36. Perry, H.M., and Perry, E.F., *J. Clin. Invest.* **38,** 1452 (1959).
37. Phatak, S.S., and Patwardhan, V., *Indian J. Sci. Ind. Res., Sect. A* **9,** 70 (1950); **11,** 172 (1952).
38. Schnegg, A., and Kirchgessner, M., *Z. Tierphysiol., Tierernaehr. Futtermittelkd.* **36,** 63 (1975).
39. Schnegg, A., and Kirchgessner, M., *Arch. Tierernaehr.* **26,** 543 (1976).
40. Schnegg, A., and Kirchgessner, M., *Int. J. Vit. Nutr. Res.* **46,** 96 (1976). *Nutr. Metab.* **19,** 268 (1975).
41. Schroeder, H.A., Balassa, J.J., and Tipton, I.H., *J. Chronic. Dis.* **15,** 51 (1961).
42. Schroeder, H.A., Mitchener, M., and Nason, A.P., *J. Nutr.* **104,** 239 (1974).
43. Schroeder, H.A., and Nason, A.P., *J. Invest. Dermatol.* **53,** 71 (1969).
44. Seay, W.E., and Denumbrum, L.E., *Agron. J.* **50,** 237 (1958).
45. Smith, J.C., and Hackley, B., *J. Nutr.* **95,** 541 (1968).

46. Stephens, G.A., *Med. Press* **187,** 216 (1933); **194,** 283 (1934).
47. Sunderman, F.W., and Sunderman, F.W., Jr., *Am. J. Clin. Pathol.* **35,** 203 (1961).
48. Sunderman, F.W., Jr., *Am. J. Clin. Pathol.* **44,** 182 (1965).
49. Sunderman, F.W., Jr., Decsy, M.J., and McNeely, M.D., *Ann. N.Y. Acad. Sci.* **199,** 300 (1972).
50. Sunderman, F.W., Jr., Nomoto, S., Pradhan, A.M., Levine, H., Bernstein, S.H., and Hirsch, R., *N. Engl. J. Med.* **283,** 897 (1970).
51. Sunderman, F.W., Jr., Nomoto, S., Morang, M., Nechay, M.W., Burke, C.N., and Nielsen, S.W., *J. Nutr.* **102,** 259 (1972).
52. Tedeschi, R.E., and Sunderman, F.W., Jr., *AMA Arch. Ind. Health* **16,** 486 (1957).
53. Tiffin, L.O., *Plant Physiol.* **48,** 273 (1971).
54. Tipton, I.H., and Cook, M.J., *Health Phys.* **9,** 103 (1963).
55. Vohra, P., Gray, G.A., and Kratzer, F.H., *Proc. Soc. Exp. Biol. Med.* **120,** 447 (1965).
56. Wacker, W.E.C., and Vallee, B.L., *J. Biol. Chem.* **234,** 3257 (1959).
57. Wase, A.W., Goss, D.M., and Boyd, J.M., *Arch. Biochem. Biophys.* **51,** 1 (1954).
58. Weber, C.W., and Reid, B.L., *J. Nutr.* **95,** 612 (1958); *J. Anim. Sci.* **28,** 620 (1959).
59. Welch, R.M., and Carey, E.E., *Agric. Food Chem.* **23,** 497 (1975).
60. Zook, E.G., Greene, F.E., and Morris, E.R., *Cereal Chem.* **47,** 720 (1970).

7

Manganese

I. MANGANESE IN ANIMAL TISSUES AND FLUIDS

1. Total Content and Distribution

The body of a normal 70-kg man is estimated to contain a total of 12–20 mg Mn (31). This relatively small amount of manganese is distributed widely throughout the tissues and fluids, without notable concentration in any particular location and with comparatively little variation among organs or species, or with age (145). However, manganese tends to be higher in tissues rich in mitochondria and is more concentrated in the mitochondria than in the cytoplasm or other organelles of the cell (114, 155). The pigmented portions of the eye are not exceptionally rich in Mn as they are in Zn and Cu, although the retina is richer in this metal than most body tissues. The pigmented melanin-containing parts of the conjunctiva are higher in manganese than the nonpigmented parts (35).

The mean Mn concentrations in the tissues, as reported by Hamilton *et al.* (66) and Tipton and Cook (158) for normal humans and by Fore and Morton (51) for rabbits and a range of animal species, are presented in Table 23. It is apparent that the bones, liver, and kidney normally carry higher Mn concentrations than do other organs, and that the muscles are among the lowest in this element of the tissues of the body. The levels of manganese in the bones can be raised or lowered by substantially varying the Mn intakes of the animal. This has been demonstrated in the newborn calf (79), rat (108), rabbit (41), pig (91), and

TABLE 23
Concentrations of Manganese in Animal Tissue in ppm Mn on the Fresh Basis

Tissue	Man[a]	Man[b]	Rabbit[c]	Av. figures for a range of animal species[c]
Bones (long)	–	–	–	3.3
Adrenals	0.20	–	0.67	0.40
Aorta	0.19	–	–	–
Brain	0.34	0.2 ± 0.03	0.36	0.40
Heart	0.23	–	0.28	0.34
Kidney	0.93	1.3 ± 0.5	1.2	1.2
Liver	1.68	0.5 ± 0.8	2.1	2.5
Lung	0.34	0.2 ± 0.03	–	–
Muscle	0.09	0.04 ± 0.007	0.13	0.18
Ovaries	0.19	0.7 ± 0.3	0.60	0.55
Pancreas	1.21	–	1.6	1.9
Pituitary	–	–	2.4	2.5
Prostate	0.24	–	–	–
Lymph nodes	–	1.1 ± 0.6	–	–
Spleen	0.22	–	0.22	0.40
Testes	0.19	0.1 ± 0.04	0.36	0.50
Hair	–	–	0.99	0.80

[a]Tipton and Cook (158).
[b]Hamilton *et al.* (66).
[c]Fore and Morton (51).

chick (112, 166). Mathers and Hill (112) reported a skeletal Mn concentration of 2.5 ppm on the dry, fat-free basis for pullets fed a low-Mn diet from 18 weeks of age to after 6–7 months of egg production, compared with 7.8 ppm for similar birds fed a high-Mn diet for this period. The skeletal manganese amounted to about 25% of total body Mn but did not constitute an important mobilizeable store of the element. Howes and Dyer (79) doubled the Mn concentration in the marrow-free radius of calves at 1 week of age by supplementing their milk diet with Mn at 14.5 ppm from birth. This treatment had little influence on the level of manganese in the calve's muscle or hair but strikingly increased the levels in their liver. For example, the liver of 7-day-old unsupplemented calves fed a 13-ppm Mn diet contained 6.36 ± 0.6 μg Mn/g of dry matter, compared with 614.67 ± 195.89 for 7-day-old calves from similar dams that were Mn-supplemented during the first week of life. It was also apparent from this study that newborn calves preferentially store manganese in their livers when this element is added to the diet of their mothers. Higher liver manganese in the newborn of Mn-supplemented mothers has similarly been demonstrated in rats and mice

(23). On the other hand, reserve Mn stores do not normally occur in the livers of newborn rats, rabbits, guinea pigs, cattle, or man (25, 110, 153, 163).

Human livers from healthy individuals of all ages contain about 6–8 ppm Mn (d.b.) (25, 158), with appreciable individual variation but very little variation from one part of the liver to another (135). Typical manganese concentrations of normal sheep and cattle livers are 8–10 ppm Mn (d.b.) (160, 163), with similar levels in the livers of laying hens (112). Manganese does not accumulate significantly in the lungs with age (145, 158). Lower Mn levels in the tumor than in the uninvolved lung, and in the rest of the involved lung, have been observed in patients who had died of lung cancer (124).

2. Manganese in Hair, Wool, and Feathers

The concentration of manganese in mammalian hair varies with the species, individual, season, color, and less certainly with the Mn status of the diet. Van Koetsveld (163) found the hair of healthy adult cows to range from 8 to 15 ppm (mean 12), and those showing signs of Mn deficiency to fall below 8 ppm. At very high intakes of Mn, concentrations as high as 80 ppm Mn were observed. Meyer and Engelbartz (120) found the hair of 351 cattle to range from 3.9 to 49.9 ppm Mn (mean 15.8), and O'Mary *et al.* (130) observed a range of 6–104 ppm Mn in Hereford cattle and calves. Red and black cattle hair are consistently higher in Mn than white cattle hair (20, 130). Groppel and Anke (62) reported that the Mn level in the hair of mature goats receiving a low-Mn diet averaged 3.5 ppm, compared with 11.1 ppm in comparable goats receiving adequate manganese. These workers maintained that the Mn level in the hair reflects the Mn dietary supply better than any other part of the body studied. No such relationship was observed in two later studies with cattle. Thus Howes and Dyer (79) found no significant differences in the hair Mn levels of 7-day-old calves that had received a Mn supplement from birth and those that had not, and Hartmans (69) found no difference between identical twin cattle that had received either a diet containing 21 ppm Mn or one containing 130 ppm Mn. This worker did find a significantly higher hair manganese in these cattle when at pasture than when receiving hay. The mean levels were 6.0 and 1.8 ppm Mn and 6.7 and 2.7 ppm for the low-Mn and the high-Mn groups, respectively. Marked seasonal changes in hair Mn levels have also been reported in the Alaskan moose *Alces Gigas* (49). Remarkably low levels, below 1 ppm, were observed in the late winter–early spring samples, with levels greater than 8 ppm appearing in the autumn samples. The extent to which these changes are a reflection of varying dietary Mn intakes remains to be determined.

The manganese levels in wool and feathers varies significantly with dietary Mn intakes. Lassiter and Morton (103) reported a mean of 6.1 ppm Mn in the wool of lambs fed a low-Mn diet for 22 weeks, compared with 18.7 ppm in the wool of control lambs. Mathers and Hill (112) found the skin and feathers of pullets fed a low-Mn diet for several months to average 1.2 ppm Mn, as opposed to 11.4 ppm in comparable birds fed a high-Mn diet.

3. Manganese in Blood

Widely varying values have appeared for the Mn levels in normal human blood. In a recent study of 102 individuals in England a mean of $6.88 \pm 0.86 \times 10^{-2}$ μg Mn/ml was reported (66). This is very close to 69 μg/liter. Bowen (21) found normal human blood to average 24 ± 8 μg Mn/liter, divided fairly equally between cells and plasma. Cotzias and co-workers (34, 133), in two separate studies, obtained much lower values and higher concentrations in the red cells than in the serum. In the first study 8.44 ± 2.73 μg Mn/liter was reported for whole blood and only 0.59 ± 0.18 μg/liter for plasma. In the second the following mean concentrations were obtained: whole blood, 9.84 ± 0.4; serum, 1.42 ± 0.2; and red cells, 23.57 ± 1.2 μg/liter. Similar low values were reported by Fernandez *et al.* (48). The manganese in human serum is selectively and almost totally bound in the trivalent form by a β_1 globulin (50). In rat serum the globulin that binds manganese is transferrin (95). In erythrocytes a firmly bound Mn compound exists which is probably a Mn porphyrin (18, 67).

Following acute coronary occlusion blood Mn levels are elevated sharply to about twice the normal values (72, 148), and this elevation can persist for several weeks (128). Hedge *et al.* (72) found this to be a better criterion of myocardial infarction than serum glutamic oxaloacetic transaminase (SGOT) levels. Elevated Mn concentrations in the red cells, but not in the serum, also occur in rheumatoid arthritis patients (36).

Levels of 20 μg Mn/liter have been reported for the blood of calves (71), with lower values for bovine blood in other studies (14, 16). Systemic bovine blood plasma contains about 5 μg Mn/liter and maintains this level within narrow limits at widely varying Mn intakes (139a). Comparable low values have also been obtained for avian blood (17). The Mn concentration in the blood plasma of pullets increases markedly with the onset of egg-laying (75). At 19 weeks of age the levels were 30–48 μg Mn/liter and at 25 weeks the levels had risen to 85–91 μg/liter. The influence of estrogen activity on plasma Mn levels is further apparent from the studies of Panic and co-workers (132). The radioactivity in the plasma 4 hr after the intramuscular injection of ^{54}Mn was shown to be about 15 and 70 times higher in laying hens and estrogen-treated immature pullets than in control immature pullets.

4. Manganese in Milk

Investigations in several countries have shown normal cow's milk to contain 20–40 μg Mn/liter, with concentrations of 130–160 μg/liter in colostrum (5, 6, 90, 101, 140). The level in the milk responds rapidly to changes in dietary Mn intakes. Archibald and Lindquist (6) increased the milk Mn concentration 2- to 4-fold by feeding $MnSO_4$ to cows in amounts equivalent to 10 or 13 g Mn/day. The provision of high-Mn feeds or feeding supplements providing only 3 g Mn/day produced smaller increases (98). The manganese level in human milk does not appear to be so well established, but there seems little doubt that this level is significantly lower than it is in the milk of cows, sheep, or goats (24, 59, 118). McLeod and Robinson (118) give a mean level of 15 μg Mn/liter (12–20) in breast milk, compared with 40 (32–52) μg/liter in pasteurized cow's milk in New Zealand.

5. Manganese in the Avian Egg

The manganese content of eggs varies widely with the Mn level of the diet. In two early studies, raising the Mn level in the hen's diet from 13 to 1000 ppm increased the amount in the yolk from 4 to 33 μg (54), and supplementing a Mn-deficient diet with 40 ppm Mn increased the concentration in the whole egg from 0.5 to 0.9 ppm (d.b.) (111). Hill and Mather (76) reported that eggs from pullets fed a low-Mn diet contained 4–5 μg Mn, while those from pullets fed a normal-Mn diet contained 10–15 μg. In a recent study, hens fed a good diet containing 81 ppm Mn produced eggs containing 28.6 ± 0.78 μg Mn (39). The Mn concentration of the yolk is four–five times that of the white.

II. MANGANESE METABOLISM

The pioneer studies of Greenberg and co-workers (60) with radiomanganese indicated that only 3–4% of an orally administered dose is absorbed in rats, and that the absorbed Mn quickly appears in the bile and is excreted in the feces. In the bovine approximately 1% of dietary Mn is absorbed, irrespective of dietary concentration (139a). Absorbed manganese is almost totally excreted via the intestinal wall by several routes. These routes are interdependent and combine to provide the body with an efficient homeostatic mechanism regulating the level of Mn in the tissues (15, 133). Under ordinary conditions the bile flow is the main route of excretion. The Mn concentration of the bile fluid can be increased 10-fold or more by the administration of large amounts of Mn to the animal (138). Excretion also occurs via the pancreatic juice (27). When the hepatic (biliary) route is blocked, or when overloading with Mn occurs, pancreatic excretion increases (133). Manganese excretion also takes place in the duodenum, jejunum, and to a smaller extent into the terminal ileum (15). These must be regarded as auxiliary routes. Very little Mn is excreted in the urine, even when injected or added to the diet (100, 115). A marked rise in urinary Mn can be achieved by the administration of chelating agents (115). The remarkable capacity of the bovine liver to remove and presumably excrete excess absorbed manganese has been demonstrated by Sansom *et al.* (139a). Their findings indicate that manganous ions absorbed in the portal circulation may either remain free or rapidly become bound to α_2-macroglobulin before traversing the liver, where they are removed nearly quantitatively, although some of the Mn bound to α_2-macroglobulin may enter the systemic circulation, become oxidized to the manganic state, and bound to transferrin.

Injected ^{54}Mn is cleared rapidly from the bloodstream in three phases (19). The first and fastest of these is identical with a clearance rate of other small ions, suggesting the normal transcapillary movement; the second can be identified with the entrance of Mn into the mitochondria of the tissues; the third and slowest component could indicate the rate of nuclear accumulation of the metal. These interpretations are supported by studies demonstrating early and preferential accumulation of ^{54}Mn in the mitochondria-rich organs of the body (94), localization of Mn in the mitochondria of the cell, and high mitochondrial and low nuclear Mn turnover rates (31). The kinetic patterns for blood clearance and liver uptake of Mn are similar, indicating that the two Mn pools, blood Mn and liver mitochondrial Mn, rapidly enter into equilibrium.

Loading the body with stable Mn, but not with other elements, rapidly elutes Mn from the body and redistributes it within the tissues (23, 32). The turnover of parenterally administered ^{54}Mn has been directly related to the level of stable Mn in the diet of mice over a wide range (23). A linear relationship between the rate of excretion of the tracer and the level of Mn in the diet was observed and the concentration of ^{54}Mn in the tissues directly related to the level of stable Mn in the diet. An inverse relationship between dietary Mn and the percentage of ^{54}Mn taken up by the tissues of chicks has also been demonstrated (149). These findings support the concept that variable excretion rather than variable absorption is the regulator of manganese homeostasis. However, Lassiter and co-workers (102) have provided evidence that dietary Mn level has a greater effect on absorption than on endogenous excretion, and that both variable excretion and absorption play important roles in Mn homeostasis. For example, 3 days after a single oral dose, total body ^{54}Mn retention in low-Mn baby calves was nine times greater (18.2% vs 2.2%) than in Mn-supplemented calves. In an experiment with rats given a single oral ^{54}Mn dose, liver ^{54}Mn was 15 times higher after 4 hr in those fed a diet containing 4 ppm Mn than in those fed 1000 ppm Mn. Such a great difference indicates a major effect on absorption. Increased absorption of manganese under conditions of low-Mn intakes and decreased absorption at higher Mn intakes, in a manner reminiscent of iron absorption in iron deficiency, is equally apparent from experiments of Howes and Dyer (79) with calves.

Manganese is equally well absorbed throughout the length of the small intestine by a two-step mechanism involving initial uptake from the lumen and then transfer across the mucosal cells to the body. The two kinetic processes operate simultaneously, with Mn competing with Fe and Co for common binding sites in both processes (156, 157). In this way one of the metals exerts an inhibitory effect on the absorption of the others. Thomson *et al.* (156) showed that the addition of iron competitively inhibited Mn absorption in Fe-deficient rats. In patients with varying iron stores subjected to duodenal perfusion with Mn the rate of Mn absorption was found to be increased in iron deficiency and the enhanced Mn absorption to be inhibited by iron. Conversely, high-Mn intakes reduce iron absorption in several species, as discussed in Section VI. It is apparent that the absorption mechanisms of manganese and iron show many similarities–similarities which are not shared by their excretion processes.

Manganese availability is further affected by excess dietary calcium. In birds the effect of high dietary levels of calcium phosphate in aggravating Mn deficiency is believed to be due to a reduction in soluble manganese through adsorption by solid mineral (142, 169). Chemical forms as diverse and varying in solubility as oxide, carbonate, sulfate, and chloride were shown some years ago to be equally valuable as sources of Mn in poultry rations (142). Differences in Mn availability among various inorganic Mn sources have now been demonstrated using leg abnormality scores and bone Mn levels as the response criteria (166). The carbonate ore (rhodochrosite) and silicate ore (rhodomite) are relatively unavailable (54, 142). The fecal excretion of parenterally administered ^{54}Mn is higher and liver retention much lower in rats on a 1.0% than on a 0.6% Ca diet (104). It appears that calcium can influence Mn metabolism by affecting retention of absorbed Mn as well as by affecting its absorption.

Ethanol feeding has been shown to increase hepatic manganese (8). This is apparently due to a significant effect (at least twofold increase) on Mn absorption, mediated by ethanol metabolism in the gut (141).

Manganese metabolism is influenced by the estrogenic hormones of the ovary and by the adrenal cortical hormones. The profound effect of estrogen on plasma Mn levels in poultry has already been mentioned (132). The administration of glucocorticoid hormones markedly affects ^{54}Mn distribution in the mouse. There is a shift in the partition within the body from the liver to the carcass (80). Stimulation of the animal's own adrenal cortices with ACTH results in similar changes (31, 81). However, adrenalectomy did not alter the Mn concentrations in the liver and diaphragm, except in animals receiving high-Mn intakes when the levels increased.

III. MANGANESE DEFICIENCY AND FUNCTIONS

Manganese deficiency has been demonstrated in mice, rats, rabbits, guinea pigs, pigs, poultry, sheep, goats, and cattle, occurs naturally on certain diets composed of normal feeds fed to pigs and poultry, and has been observed in man in association with a vitamin K deficiency (40). The main manifestations of Mn deficiency, namely, impaired growth, skeletal abnormalities, disturbed or depressed reproductive function, ataxia of the newborn, and defects in lipid and carbohydrate metabolism, are displayed in all species studied, but their actual expression varies with the degree and duration of the deficiency and its rate of development and with the age and stage of growth of the animal. Many of these gross effects of Mn deficiency can now be explained in terms of its effect on mucopolysaccharide synthesis, as discussed in the following section.

Hemoglobin levels do not appear to be significantly affected by lack of manganese (154, 164). The growth inhibition of Mn deficiency results from both reduced food consumption and impaired efficiency of food use (22), but severe inappetence is not a conspicuous feature of Mn deficiency, as it is in Zn and Co deficiencies.

1. Manganese and Bone Growth

The skeletal abnormalities of Mn deficiency are particularly characterized in mice, rats, rabbits, and guinea pigs by retarded bone growth with shortening and bowing of the forelegs (9, 150, 154) and defective development of the skull and the otoliths of the inner ear during gestation (83, 88, 151). In pigs the skeletal abnormalities are characterized by lameness and enlarged hock joints with crooked and shortened legs (122, 126); in calves by difficulty in standing (79); in sheep by joint pains with poor locomotion and balance (103); and in goats by tarsal joint excrescences, leg deformities, and ataxia (62). Manganese deficiency in chicks, poults, and ducklings is manifested as the disease perosis or "slipped tendon" (168) and in chick embryos by chondrodystrophy (111). Perosis is characterized by enlargement and malformation of the tibiometatarsal joint, twisting and bending of the tibia, thickening and shortening of the long bones, and slipping of the gastrocnemius tendon from its condyles. With increasing severity of the condition the chicks are reluctant to move, walk on their hocks, and soon die. Nutritional chondrodystrophy is characterized by shortened and thickened legs and wings, "parrot beak" resulting from a disproportionate shortening of the lower mandible, globular contour of the head, and high mortality. A similar defect in bone development occurs in the offspring of Mn-deficient rats. A severe shortening of the radius, ulna, tibia, and fibula is evident in these young at birth (85). In addition, a marked epiphyseal dysplasia at the proximal end of the tibia and a shortening and doming of the skull with an amomalous ossification of the inner ear develops in such animals (86, 88). The skulls of deficient young are shorter, wider, and higher than those of controls.

Impairment of the calcification process per se is not a primary causal factor in the bone abnormalities of Mn deficiency just described. Neither the volume of the bone nor the gross composition of the bone ash is reduced, although the length, density, and breaking strength are lowered (1, 28, 154). Skeletal maturation is retarded due to an inhibition of endochondral osteogenesis at the epiphyseal cartilages (52, 126, 171). Deficient bones appear normal to X-ray examination and $AgNO_3$ staining (28, 53), despite some reduction in bone ash content, and differ distinctly from those of Ca-, P-, vitamin D-, or Cu-deficient animals (28, 105, 154).

In view of the above findings attention was turned to a possible involvement of manganese in the synthesis of the organic matrix of cartilage. Leach and Muenster (106) discovered that radiosulfate uptake is lowered in the cartilage of the Mn-deficient chick, and that the total concentration of hexosamines and hexuronic acid is reduced in this tissue. A less pronounced reduction was observed in the hexosamine content of other tissues. Chondroitin sulfate is the mucopolysaccharide most severely affected by Mn deficiency. Subsequently Everson and associates (45, 151, 159) demonstrated a reduction in the concentration of acid mucopolysaccharides (AMPS) in rib and epiphyseal cartilage and in the otoliths of Mn-deficient newborn guinea pigs, and Hurley *et al.* (87) showed that Mn deficiency markedly reduced the *in vitro* rate of incorporation of radiosulfate into cartilage matrix in fetal rat tibia. The same group then studied histologically mucopolysaccharide synthesis in the developing inner ear of Mn-deficient and pallid mutant mice (152). Manganese deficiency or the pallid gene caused reduced incorporation of ^{35}S in the macular cells, the formation of nonmetachromatic, variably PAS-positive matrix which did not contain ^{35}S, and failure of otolithic calcification.

The impairment in mucopolysaccharide synthesis associated with Mn deficiency has been related to the activation of glycosyltransferases by this element (105). These enzymes are important in polysaccharide and glycoprotein synthesis, and manganese is usually the most effective of the metal ions required for their activity. The critical sites of Mn function in chondroitin sulfate synthesis have been identified by Leach and co-workers (107) as the two enzyme systems (a) polymerase enzyme, which is responsible for the polymerization of UDP-*N*-acetylgalactosamine to UDP-glucuronic acid to form the polysaccharide, and (b) galactotransferase, an enzyme which incorporates galactose from UDP-galactose into the galactose-galactose-xylose trisaccharide which serves as the linkage between the polysaccharide and the protein associated with it. Since mucopolysaccharides are vital structural components of cartilage the above findings provide a likely biochemical explanation of the skeletal defects associated with Mn deficiency. The wider role of manganese in activating glycosyltransferases involved in the synthesis of other glycoproteins, e.g., prothrombin, is considered in Section III,5.

It is also probable that the effect of manganese on mucopolysaccharide synthesis explains some of the deficiency defects observed in laying hens, such as reduced egg production and poor shell formation (29, 64), and chondrodystrophy in chick embryos (111). A reduction in the hexosamine content of the shell matrix has been associated with poor shell formation in laying hens (109). Of further interest is the finding that several types of chondrodystrophy of genetic origin are characterized by a reduced limb mucopolysaccharide content (113).

2. Manganese of Neonatal Ataxia

Ataxia in the offspring of Mn-deficient animals was first observed in the chick by Caskey and Norris (30) and in rats by Shils and McCollum (150). Hurley and Everson and their co-workers (46, 83, 84) later showed that Mn deficiency during pregnancy in rats and guinea pigs produces an irreversible congenital defect in the young characterized by ataxia and loss of equilibrium, and often also by head retraction and tremors, increased susceptibility to stimuli, and delayed development of the body-righting reflexes. The defect responsible for these disturbances arises relatively late in gestation–in the rat between the fourteenth and eighteenth days. There is a critical need for manganese on the fifteenth and sixteenth days because Mn supplementation begun at that time results in an improvement of survival time and of the incidence of ataxia in the young, whereas supplementation begun on the eighteenth day is ineffective in preventing ataxia (84) (Table 24). A structural defect in the inner ear then emerged as the causative factor in the ataxia and postural defects of Mn-deficient rats and guinea pigs (83, 88, 151). These disorders were shown to arise from impaired vestibular function, itself a reflection of a specific effect of lack of manganese on cartilage mucopolysaccharide synthesis and hence bone development of the skull, particularly the otoliths. Deficient animals do not exhibit normal otolith development in the utricular and saccular maculae as a consequence of impaired mucopolysaccharide synthesis (152).

A specific congenital ataxia in mice resulting from defective development of the otoliths, and caused by the presence of a mutant gene affecting coat color (pallid), was shown by Erway *et al.* (42) to be completely rectified by high levels of Mn supplementation during pregnancy. The effect of the pallid gene on mucopolysaccharide synthesis, in contrast to the position in Mn-deficient mice, is highly localized to the otolith matrix and does not extend to the surrounding periotic cartilage (152). The genetic involvement of manganese in otolith development is complex and as yet little understood. Six pigment genes in four species are now known to result in impaired otolith development, including

TABLE 24
Effect of Manganese Supplementation at Various Times during Gestation[a]

Initiation of supplementation (day of gestation)[b]	No. of litters	Young born		Survival to 28 days	
		Total	Per litter	Live young	Ataxic
7–12	14	105	7.5	53	0
14	6	42	7.0	87	0
15	8	60	7.5	36	48[c]
16	8	54	6.8	44	46[c]
18	8	65	8.1	26	100

[a]From Hurley *et al.* (84).
[b]Day of finding sperm considered first day of gestation.
[c]Mild.

pastel mink in which Mn supplementation prevents the otolith defect, as well as the associated "screw neck" phenomenon. All of these pigment genes interfere with the development of melanocytes within the inner ear, pointing to a causal relationship between the presence of pigment cells in the inner ear and the availability of Mn for otolith development (43).

3. Manganese and Reproductive Function

Defective ovulation, testicular degeneration, and infant mortality were observed in the earliest studies demonstrating the essentiality of manganese in the diet of rats (99, 131, 165). In the female three stages of Mn deficiency can be recognized. In the least severe stage the animals give birth to viable young, some or all of which exhibit ataxia. In the second, more severe stage the young are born dead or die shortly after birth. In the third, acute stage of deficiency estrous cycles are absent or irregular, the animals will not mate, and sterility results. A delay in the opening of the vaginal orifice may also occur (22). A similar impairment of reproductive performance occurs in hens, since lowered egg production and decreased hatchability can occur even in the absence of perosis and chondrodystrophy (7, 143). The severely Mn-deficient male rat and rabbit exhibit sterility and absence of libido, associated with seminal tubular degeneration, lack of spermatozoa, and accumulation of degenerating cells in the epididymis (22, 154). In guinea pigs omission of Mn from the maternal diet increases the proportion of young born dead or delivered prematurely and reduces litter size (46).

The feeding of low-Mn rations to cows (3, 14, 79) and goats (3) causes depressed or delayed estrus and conception, as well as increased abortion and stillbirths and lowered birth weights. Several claims that Mn supplements improve the fertility of dairy cows in parts of Europe have also been made (74, 167, 170). However, experiments by Hartmans (70) with identical cattle twins fed from the age of 1 or 2 months on low-Mn rations (16–21 ppm Mn on d.b.), i.e., low in comparison with most field pasture levels, revealed no differences in fertility compared with controls receiving supplementary manganese. It seems, therefore, that unidentified environmental factors that limit Mn absorption or utilization must be operating where Mn responses have been obtained.

The precise locus or mode of action of manganese in preventing the reproductive defects in both male and female described above has not yet been established. Doisey (40) has put forward the hypothesis that lack of manganese inhibits the synthesis of cholesterol and its precursors and this, in turn, limits synthesis of sex hormones, and possibily other steroids, with consequent infertility. The relation of manganese to cholesterol synthesis and to other lipids is considered in the next section.

4. Manganese and Lipid Metabolism

An association between manganese and choline in metabolism has been recognized for some years (2, 44, 93). Liver and bone fat in Mn-deficient rats are reduced by both Mn and choline supplements (2), and Mn supplementation of a Mn-deficient diet can reduce fat deposition and back fat thickness in pigs (137). Both nutrients are needed for complete protection against perosis in chicks (93). Rats placed on a choline-deficient diet for 25 days exhibited lower hepatic Mn levels than those of controls (96). The authors suggest that this is due to reduced intestinal transport of the metal. Furthermore, the changes in liver ultrastructure that arise in choline deficiency (26) are very similar to those that have recently been observed in Mn deficiency (11). Deficiencies of manganese and choline both appear to affect membrane order and integrity. These nutrients may therefore be, as Bell and Hurley (11) have suggested, "linked in a common pathway to establish the normal structure of the mitochondrial and cellular membranes, either directly through effects on membrane synthesis or indirectly through alterations of mitochondrial oxidations."

Manganese has a further role in the biogenesis of cholesterol. Curran (38) has shown that the addition of Mn to rat liver *in vitro* markedly increases cholesterol synthesis, and Olson (129) has defined two sites between acetate and mevalonate that require manganese. In addition, farnesylpyrophosphate synthetase is now known to require Mn^{2+}, and the two have been found to act together to add one 5-carbon unit to geranyl pyrophosphate to make farnesyl pyrophosphate (13). Lack of farnesyl pyrophosphate could inhibit production of squalene, so limiting the formation of important precursors of cholesterol. The effect of Mn deficiency on cholesterol in the living animal does not appear to have been directly investigated. However, a marked reduction in serum cholesterol level was observed by Doisey (40) in a patient inadvertently showing evidence of manganese (and vitamin K) deficiency. Serum phospholipids and triglycerides were also decreased. It is clear that the nature and extent of the involvement of manganese in lipid and sterol metabolism are ripe for further critical study.

5. Manganese, Vitamin K, and Prothrombin Formation

Prothrombin formation is known to be controlled by vitamin K. Since prothrombin is a glycoprotein its synthesis should also be influenced by manganese through the activating effect of the metal on glycosyltransferases, as discussed in Section III,1. Evidence for an effect of Mn on the prothrombin response to vitamin K has now appeared (40). Moderate Mn deficiency has been shown to reduce the clotting response from vitamin K in chicks, compared with that of chicks receiving ample dietary Mn (see Table 25). Moreover, a human patient suffering from a coincident deficiency of Mn and vitamin K was found unable to elevate his depressed clotting protein levels when given vitamin K, until manganese was restored to the diet. These data, together with recent evidence indicating that the addition and completion of the glycosyl portion of prothrombin is a major function of vitamin K (92, 134), led Doisey (40) to put forward a working hypothesis explaining the Mn–vitamin K nexus as follows: "since the presence of Mn^{++} appears to supersede and control the action of vitamin K in completing the biosynthesis of the glycoproteins involved in clotting, Mn^{++} might by necessary to initiate the addition of the proximal portion of the carbohydrate chain to the apoprotein 'preprothrombin,' in

TABLE 25
Effect of Available Mn^{2+} on Response to Vitamin K by the Chick[a]

	Diet content[b]			Results		
Group	Mn (ppm)	Ca (g%)	P (g%)	Body wt at 14 days (g)	Prothrombin (% normal)[c]	Incidence of perosis (%)
A	87	0.8	0.4	214	62	0
B	37	0.8	0.4	194	51	16
C	87	3.2	1.6	184	35	50
D	37	3.2	1.6	170	18	100

[a]From Doisey (40).

[b]Group B received basal, vitamin K-deficient diet; salt mix was replaced by 5.3 g% salts to reduce Mn^{2+} to 37 ppm, calculated.

[c]On and after day 9, all diets were supplemented with 0.15 μg vitamin K_1/g diet. Without vitamin K, survivors to day 14 had prothrombin fall to 8% of normal, and there was a high incidence of deaths from heart puncture.

addition to vitamin K. Mn^{++} would then act with the glycosyl transferases to complete the terminal carbohydrate units to convert preprothrombin to prothrombin. Equally likely, and more probably, Mn^{++} is a required intermediary to the action of vitamin K by being essential to the UDPG transferases which complete the glycosyl–glycosyl linkages of most of the glycoproteins yet examined."

6. Manganese and Carbohydrate Metabolism

Defects in carbohydrate metabolism, other than those known to arise from the impairment of the activities of glycosyltransferases already described, have been observed in Mn-deficient guinea pigs. Everson and Schrader (47) found that newborn guinea pigs severely affected with Mn deficiency exhibited aplasia, or marked hypoplasia, of all the cellular components of the pancreas. Smaller numbers of islet cells, containing fewer and less intensely granulated beta cells, were apparent than in the islets of control guinea pigs. Young, adult Mn-deficient guinea pigs also had subnormal numbers of pancreatic islets, with less intensely granulated beta cells and more alpha cells than Mn-supplemented controls. When glucose was administered orally or intravenously the Mn-deficient guinea pigs revealed a decreased capacity to utilize glucose and displayed a diabetic-like curve in response to glucose loading. Manganese supplementation completely reversed the reduced glucose utilization.

The mechanism of the impairment of glucose utilization in Mn deficiency is not yet known but it may be related, as with so many other metabolic disturbances in Mn deficiency, to the connective tissue defect that occurs in such animals. Decreased concentrations of stainable mucopolysaccharides have been demonstrated in the skin of young rats born to diabetic mothers (172), and it has been suggested that insulin may regulate the utilization of glucose in the synthesis of mucopolysaccharides (144). Manganese may therefore be involved in some way in insulin formation or insulin need.

7. Manganese and Enzyme Activities

The number of manganese metalloenzymes is very limited, whereas the enzymes that can be activated by Mn are numerous. They include hydrolases, kinases, decarboxylases, and transferases (162). Many of these metal activations are nonspecific, so that Mn may be partly replaced by other metal ions, particularly magnesium. Even the Mn metalloenzyme pyruvate carboxylase is subject to partial substitution of Mn by Mg, with only minor alteration in the catalytic properties of the enzyme. Pyruvate carboxylase isolated from chicken liver mitochondria (148) contains 4 moles Mn(II) per mole of enzyme. The following reaction has been proposed for this enzyme (121).

$$\text{Enzyme–biotin} + CO_2 + \text{pyruvate} \rightleftharpoons \text{enzyme–biotin} + \text{oxalacetate}$$

Scrutton *et al.* (147) found that Mg replaced Mn as the bound metal in pyruvate carboxylase from Mn-deficient chicks. The relative content of the two metals was related to the severity of the deficiency. Avimanganin, a further Mn metalloprotein obtained from avian liver containing 1 mole Mn(III) per mole of protein (146) and of unknown function, was reduced in amount in the liver of Mn deficiency as well as in its Mn content. It seems, therefore, that the animal can adapt to Mn deficiency with respect to pyruvate carboxylase by substituting Mg for Mn, whereas with avimanganin the production of the metalloprotein is limited by the Mn deficiency.

The superoxide dismutases, which catalyze the disproportionation of O_2^- to $H_2O_2 + O_2$, are metalloenzymes containing Cu, Zn, Fe, or Mn, or in some cases combinations of these (61). The superoxide dismutase isolated from *E. coli* has a molecular weight of 40,000, is composed of two subunits of equal size, and contains one Mn^{2+} per molecule (97). The manganoenzyme from chicken liver mitochondria is similar to the corresponding bacterial enzyme but is twice as large and contains four subunits instead of two (61). Nothing is yet known of the effect of Mn deficiency on the superoxide dismutases, although they apparently function as a defense against the deleterious reactivities of the free radical O_2^-.

Abnormalities in cell function and ultrastructure, particularly involving the mitochondria, occur in Mn deficiency. Oxidative phosphorylation was studied in isolated liver mitochondria of Mn-deficient mice and rats (89). In both species, ratios of ATP formed to oxygen consumed (P/O) were normal but oxygen uptake was reduced. Electron microscopy of these mitochondria revealed ultrastructure abnormalities including elongation and reorientation of cristae (89). Subsequently Bell and Hurley (11) confirmed and greatly extended these important findings. A dietary deficiency of manganese in mice was found to cause changes in the ultrastructural parameters of the cells of liver, pancreas, kidney, and heart tissues examined. All the Mn-deficient tissues observed revealed alterations in the integrity of their cell membranes. In addition, the endoplasmic reticulum was swollen and irregular mitochondria were found with elongated stacked cristae in liver, heart, and kidney cells, and there was an overabundance of lipid in liver parenchymal and kidney tubule cells. The lowered oxidation in liver mitochondria from Mn-deficient animals is believed by the authors to be possibly related to the morphological changes observed. The same workers (12) have also examined the histochemical enzyme changes in cephalic tissues of Mn-deficient fetal mice. No differences in enzyme activities were found in the inner-ear tissues between the Mn-deficient, pallid, and control fetuses, but in the cephalic epidermis of Mn-deficient fetuses, particularly the soft keratin layer, cytochrome oxidase and choline esterase activities and the false positive yellow reaction for alkaline phosphatase were reduced, and there was a reduced concentration of choline lipids. It is apparent from these studies that manganese is involved in a wide range of enzyme activities in a variety of tissues, and that mitochondrial structure and function are particularly affected in Mn deficiency.

IV. MANGANESE REQUIREMENTS

The minimum dietary requirements of manganese vary with the species and genetic strain of animal, the chemical form in which the element is ingested, the composition of the rest of the diet, and the criteria of adequacy employed.

1. Laboratory Species

Mice, rats, and rabbits are unable to grow normally on milk diets containing 0.1–0.2 ppm Mn (d.b.), or on purified diets containing 0.2–0.3 ppm Mn (150, 154, 164). A level of 1 ppm Mn appears to be adequate for growth but is inadequate for normal fetal development in mice or rats (11, 88). The genetic constitution of mice has been shown to affect their response to dietary deficiencies of manganese, suggesting that at low or borderline levels of intake of this element the responses of individuals may vary greatly depending in part on their genetic background (82). Holtkamp and Hill (77) concluded that 40 ppm Mn of the dry diet is optimum for rats. A level of 45 ppm Mn is found to be fully adequate for growth and normal reproduction in mice and rats in the studies of Hurley and co-workers (11, 88). High intakes of phosphorus relative to calcium increase the Mn requirements of rats (164).

2. Pigs

The requirements of pigs for satisfactory reproduction are substantially higher than those needed for body growth. In fact satisfactory growth in young pigs has been reported with diets supplying only 0.5 or 1.0–1.5 ppm Mn (91, 137). However, marked tissue Mn depletion was observed, and when such diets were fed throughout gestation and lactation skeletal abnormalities and impaired reproduction became apparent. All these manifestations of Mn deficiency were prevented by supplemental manganese at a level of 40 ppm Mn, but whether 40 ppm Mn is necessary or whether some lower level is adequate for all functional purposes still cannot be answered with certainty.

3. Sheep, Goats, and Cattle

The minimum dietary Mn requirements of sheep, goats, and cattle cannot be given with any precision. Hawkins *et al.* (71) maintain that the minimum requirement of young calves for growth is probably not more than 1 ppm, and that they are increased by high dietary intakes of calcium and phosphorus. Bentley and Phillips (14) state that 10 ppm Mn is adequate for growth in heifers but is marginal for optimal reproductive performance. Rojas and co-workers (139) concluded that the Mn requirements of Hereford cows for maximum fertility are in excess of 16 ppm of the dry diet. Howes and Dyer (79) found no differences in growth or weight gain in 2-year-old Hereford heifers fed diets containing 13 ppm Mn, 14 ppm Mn + 1.5% Ca, and 21 ppm Mn for several months, but calves born to heifers on the 13 and 14 ppm Mn diets were weak and had difficulty in standing. These findings suggested that a level of about 20 ppm Mn is adequate for growth and satisfactory reproductive performance. This finds support from the experiments of Hartmans (69), who fed practical rations containing 16 and 21 ppm Mn for $2\frac{1}{2}$ to $3\frac{1}{2}$ years without any clinical evidence of Mn deficiency or significant improvement from Mn supplementation of identical twin heifers.

Female goats fed rations containing 20 ppm Mn in the first year and 6 ppm Mn in the second were found to grow as well as those consuming the same rations supplemented with Mn at 100 ppm (4), but the former animals exhibited greatly impaired reproductive performance.

4. Poultry

The minimum dietary Mn requirements of poultry for the growth of chicks and for normal egg production and hatchability approximate 40 ppm under normal dietary conditions. A total intake of 50 ppm is recommended to provide a margin of safety and to cope with variations in calcium and phosphorus intakes. For optimal reproductive performance in turkey hens it has been shown that an ordinary diet containing 3.22% Ca and 0.78% inorganic P needs to be supplemented with between 54 and 108 ppm Mn (7).

Excess dietary Ca and P increases Mn requirements by reducing Mn availability (143). For example, 64% of the chicks fed a ration containing 3.2% Ca, 1.6% P, and 37 ppm Mn developed persosis, whereas no perosis was observed when a diet was fed which contained the same level of Mn but only 1.2% Ca and 0.9% P. Freedom from perosis was achieved either by omitting the bone meal, without additional Mn, or by retaining the bone meal and increasing the Mn levels to 62 and 145 ppm (143).

The heavier breeds of poultry have slightly higher Mn requirements than the lighter ones. Egg production does not impose particularly high extra Mn demands on the hen, so that the dietary Mn requirement does not increase greatly with a higher rate of egg production.

The higher Mn requirements of birds compared with mammals arise mainly from lower absorption from the gut. The special importance of manganese in poultry nutrition stems largely from this fact. A second factor is the low-Mn content of corn (maize) compared with other cereal grains, as discussed in Section V,2. Where corn is the main dietary component Mn supplementation becomes essential, even when Ca and P intakes are not unduly high.

V. SOURCES OF MANGANESE

1. Manganese in Human Foods and Dietaries

The common foods in human dietaries are highly variable in Mn concentration. Many years ago Peterson and Skinner (136) listed 12 major food groups in descending order of their Mn content on the fresh basis. These groups were nuts, whole cereals, and dried fruits; roots, tubers, and stalks; fruits, nonleafy vegetables, animal tissues, and fluids; poultry and poultry products; and fish and seafoods. The average concentrations ranged from 20–23 ppm for the first three groups to as low as 0.2–0.5 ppm for the last three groups. The leafy vegetables would, of course, rank much higher if expressed on a dry weight basis. Essentially similar but highly variable levels of manganese have been reported for fresh vegetables obtained from different parts of the United States (78), and in a later study of the mineral content of a very wide range of foods and beverages used in hospital menus there (57). Schroeder *et al.* (145) confirmed that the highest Mn levels occur normally in nuts and whole cereals, variable amounts in vegetables, and low concentrations in meat, fish, and dairy products. Tea was found to be exceptionally rich in manganese (127, 145).

The wide range for manganese in cereal grains and their products is due partly to plant species differences and partly to the efficiency with which milling separates the Mn-rich from the Mn-poor parts of the grain. In one U.S. study whole wheat containing 31 ppm Mn yielded 160 ppm in the germ, 119 ppm in the bran, and only 5 ppm in the white flour (143). In another such study (173) common hard wheat averaging 38 ppm Mn yielded patent flour containing 4.5 ppm Mn, and common soft wheat averaging 35 ppm Mn yielded patent flour containing 4.8 ppm Mn.

Daily Mn intakes from human dietaries vary greatly with the amounts of unrefined cereals, nuts, leafy vegetables, and tea consumed. The average daily intakes of two adults in which 40–50% of the calories came from white flour were 2.2–2.7 mg Mn, whereas the corresponding intakes for two individuals in which the same proportion of calories came from 92% extraction flour were 8.5–8.8 mg Mn (100). The latter figures are close to the 7 mg Mn/day calculated for adults on a typical English winter diet, although in this case no less than 3.3 mg of the total was estimated to come from tea (125). The total mean daily Mn intakes for nine college women in the United States was 3.7 mg (127), and four young adult women in New Zealand had a mean daily intake of 2.48–3.15 mg

Mn (117). Of these intakes 0.05–0.46 mg Mn/day were retained. Similar average total intakes from adult diets in the United Kingdom of 2.7 ± 0.85 mg Mn/day (65) and from Japanese adult diets of 2.7–2.9 mg Mn/day (125a) have been reported.

Dietary Mn supply during the first 6 months of life may vary from 2.5 to 75 μg/day/kg body weight, depending on the age of the infant, the type of milk given, and the quantity of solid foods consumed (118). McLeod and Robinson (118) showed that average cow's milk contains more than twice the Mn level of human milk, that full cream and skim milk powder levels are higher still, and that processed cereal foods and ready to serve vegetable infant foods may be rich sources of manganese. Nothing is known of the form or availability of the manganese present in these foods, nor of how much is absorbed and retained.

2. Manganese in Animal Feeds and Forages

The Mn levels in pastures and forages are extremely variable. This variation reflects substantial species differences and soil and fertilizer effects. Thus Beeson *et al.* (10), in an investigation of 17 grass species grown together on a sandy loam soil and sampled at the same time, found the Mn concentrations to range from 96 to 815 ppm (d.b.). Mitchell (123) reports mean levels of 58 and 140 ppm Mn for red clover and ryegrass, respectively, when grown on a granitic, unlimed soil. Heavy liming reduced the levels in the red clover to 40 ppm Mn and in the ryegrass to 120 ppm Mn. In a study of mixed pastures in New Zealand the Mn levels were found to range from 140 to 200 ppm (d.b.), but levels above 400 ppm Mn were reported to occur in some areas (58).

The cereal grains and their by-products also vary greatly in Mn concentration, mainly due to inherent species differences. Typical concentrations in Australian grown feeds are wheat, 40; bran, 120; barley, 25; oats, 50; maize (corn) 8; and sorghum, 16 ppm Mn dry basis (55, 116, 161). Soybean meal, an important protein supplement for poultry, contains 30–40 ppm Mn (116, 143). It is apparent that poultry rations based on corn, and to a lesser extent sorghum and barley, are deficient in Mn unless supplemented with this element or with Mn-rich feeds such as wheaten bran or middlings, in which the Mn of the wheat grain is concentrated. Rations based on wheat or oats, by contrast, are likely to be adequate in manganese unless the diet contains Ca and P in excess. Diets containing appreciable quantities of the high-protein seeds of Lupinus albus are even more likely to be adequate in manganese. Gladstones and Drover (56) obtained the remarkably high concentrations of 817–3397 ppm Mn for this species, or 10–15 times those of other lupin species growing on the same sites.

Protein supplements of animal origin are usually low or very low in manganese. Levels of 5–15 ppm Mn are common in such feeds as dried skim milk or buttermilk, fish meal, and meat meal (116).

VI. MANGANESE TOXICITY

Manganese is among the least toxic of the trace elements to mammals and birds. The growth of rats is unaffected by dietary Mn intakes as high as 1000–2000 ppm, although larger amounts interfere with phosphorus retention (54). Hens tolerate 1000 ppm without ill effects (54), but 4800 ppm is highly toxic to young chicks (73). Growing pigs are less tolerant, since 500 ppm Mn retards growth and depresses appetite (63). Depressed feed intake and lowered body weight were observed in calves fed a low-Mn ration supplemented with manganese sulfate at levels of 2460 and 4920 ppm Mn. When supplemented at a level of 820 ppm Mn no effect on growth or appetite was apparent (37). The adverse effects of excess manganese on growth were shown to be mainly a reflection of depressed appetite. Sheep grazing pastures containing 140–200 ppm Mn (d.b.) supplemented with 250 or 500 mg Mn/day as the sulfate exhibited a significant depression in growth rate from both levels of Mn compared with untreated controls, together with decreased heart and plasma iron levels in one experiment (58) (Table 26). These findings indicate that an overall intake of 400 ppm of dietary Mn can be toxic to growing sheep, despite the fact that the pastures grazed were reported to contain 1100–2200 ppm Fe.

A relationship between manganese, iron, and hemoglobin formation has been demonstrated in lambs (68, 113a), pigs, and rabbits (113a). Hemoglobin regeneration was retarded and serum iron depressed in anemic lambs fed diets containing 1000 or 2000 ppm Mn. In normal lambs higher Mn levels up to 5000 ppm produced similar effects and also brought about decreased concentrations of iron in the liver, kidney, and spleen. The depressing effects were overcome by

TABLE 26
Effect of High Dietary Manganese on Growth in Grazing Sheep[a]

Supplementary Mn[b]	Total gain (kg)	Daily gain (g)
1971 av initial wt 27.3 kg–treatment period 10 weeks		
0	5.21 ± 0.27	74
250 mg Mn/day	2.70 ± 0.52	38
500 mg Mn/day	3.31 ± 0.54	47
1972 av initial wt 32.3 kg–treatment period 14 weeks		
0	7.99 ± 0.29	81
250 mg Mn/day	5.36 ± 0.45	55
500 mg Mn/day	4.41 ± 0.39	45

[a]From Grace (58).

[b]The pastures contained 140–190 ppm Mn in 1971 and 140–200 ppm Mn (d.b.) in 1972.

a dietary supplement of 400 ppm Fe (113a) and undoubtedly arise as a consequence of a mutual antagonism between Mn and Fe at the absorptive level, as discussed in Section II. Matrone and co-workers (113a) estimate that the minimum level of dietary Mn capable of affecting hemoglobin formation is only 45 ppm for anemic lambs and lies between 50 and 125 ppm for mature rabbits and baby pigs. This suggests that the Mn:Fe dietary ratio should be given closer attention, especially as many pasture and stock feeds can be appreciably higher in manganese than the concentrations just cited.

Manganese toxicity in man arising from excessive intakes in foods and beverages has never been reported and is difficult to visualize ever arising, except where industrial contamination occurs. Chronic manganese poisoning occurs among miners following prolonged working with manganese ores. Excess manganese enters the body mainly as oxide dust via the lungs and also via the gastrointestinal tract from the contaminated environment (31). The lungs apparently act as a depot from which the Mn is continuously absorbed. Manganese poisoning is characterized by a severe psychiatric disorder (locura manganica) resembling schizophrenia, followed by a permanently crippling neurological disorder clinically similar to Parkinson's disease. Comparative studies of a population of "healthy" manganese miners and patients suffering from chronic Mn poisoning revealed faster losses of injected ^{54}Mn from the whole body and from an area representing the liver and higher tissue Mn concentration in the former group (33, 119) than in those suffering from chronic Mn poisoning. The presence of elevated tissue Mn levels is thus not necessary for the continuance of the neurological manifestations of the disease, and metal chelation therapy is unlikely to secure their remission.

REFERENCES

1. Amdur, M.O., Norris, L.C., and Heuser, G.F., *Proc. Soc. Exp. Biol. Med.* **59**, 254 (1945)
2. Amdur, M.O., Norris, L.C., and Heuser, G.F., *J. Biol. Chem.* **164**, 783 (1946).
3. Anke, M., Groppel, B., Reissig, W., Ludke, H., Grun, M., and Dittrich, G., *Arch. Tierernaehr.* **23**, 197 (1973).
4. Anke, M., Groppel, B., *in* "Trace Element Metabolism in Animals" (C.F. Mills, ed.), Vol. 1, p. 133. Livingstone, Edinburgh, 1970.
5. Antila, P., and Antila, V., *Suom. Kemistil. B* **44**, 161 (1971), cited by M. Heikonen, Publ. No. 1. Valio Lab., Helsinki, 1973.
6. Archibald, J.G., and Lindquist, J.G., *J. Dairy Sci.* **26**, 325 (1943).
7. Atkinson, R.L., Bradley, J.W., Couch, J.R., and Quinsenberry, J.H., *Poult. Sci.* **46**, 472 (1967).
8. Barak, A.J., Keefer, R.C., and Tuma, D.J., *Nutr. Rep. Int.* **3**, 243 (1971).
9. Barnes, L.L., Sperling, G., and Maynard, L.A., *Proc. Soc. Exp. Biol. Med.* **46**, 562 (1941).

10. Beeson, K.C., Gray, L.C., and Adams, M.G., *J. Am. Soc. Agron.* **39**, 356 (1947).
11. Bell, L.T., and Hurley, L.S., *Lab. Invest.* **29**, 723 (1973).
12. Bell, L.T., and Hurley, L.S., *Proc. Soc. Exp. Biol. Med.* **145**, 1321 (1974).
13. Benedict, C.R., Kett, J., and Porter, J.W., *Arch. Biochem. Biophys.* **110**, 611 (1965).
14. Bentley, O.G., and Phillips, P.H., *J. Dairy Sci.* **34**, 396 (1951).
15. Bertinchamps, A.J., Miller, S.T., and Cotzias, G.C., *Am. J. Physiol.* **211**, 217 (1966).
16. Blakemore, F., Nicholson, J.A., and Stewart, J., *Vet. Rec.* **49**, 415 (1937).
17. Bolton, W., *Br. J. Nutr.* **9**, 170 (1955).
18. Borg, D.C., and Cotzias, G.C., *Nature (London)* **182**, 1677 (1958).
19. Borg, D.C., and Cotzias, G.C., *J. Clin. Invest.* **37**, 1269 (1958).
20. Bosch, S., Van der Grift, J., and Hartmans, J., *Versl. Landouwk. Onderz.* **666** (1966).
21. Bowen, H.J.M., *J. Nucl. Energy* **3**, 18 (1956).
22. Boyer, P.D., Shaw, J.H., and Phillips, P.H., *J. Biol. Chem.* **143**, 417 (1942).
23. Britton, A.A., and Cotzias, G.C., *Am. J. Physiol.* **211**, 203 (1966).
24. Brock, A., and Wolff, L.K., *Acta Brevia Neerl. Physiol., Parmacol., Microbiol.* **5**, 80 (1935).
25. Bruckman, G., and Zondek, S.G., *Biochem. J.* **33**, 1845 (1933).
26. Bruni, C., and Hegsted, D.M., *Am. J. Pathol.* **61**, 413 (1970).
27. Burnett, W.T., Bigelon, R.R., Kimbol, A.W., and Sheppard, C.W., *Am. J. Physiol.* **168**, 520 (1952).
28. Caskey, C.D., Gallup, W.D., and Norris, L.C., *J. Nutr.* **17**, 407 (1939).
29. Caskey, C.D., and Norris, L.C., *Poult. Sci.* **17**, **433 (1938).**
30. Caskey, C.D., and Norris, L.C., *Proc. Soc. Exp. Biol. Med.* **44**, 332 (1940).
31. Cotzias, G.C., *Physiol Rev.* **38**, 503 (1958).
32. Cotzias, G.C., and Greenough, J.J., *J. Clin. Invest.* **37**, 1298 (1958).
33. Cotzias, G.C., Horiuchi, K., Fuenzalida, S., and Mena, I., *Neurology* **18**, 376 (1968).
34. Cotzias, G.C., Miller, S.T., and Edwards, J., *J. Lab. Clin. Med.* **67**, 836 (1966).
35. Cotzias, G.C., Papavasiliou, P.S., and Miller, S.T., *Nature (London)* **201**, 1228 (1964).
36. Cotzias, G.C., Papavasiliou, P.S., Hughes, E.R., Tang, L., and Borg, D.C., *J. Clin. Invest.* **47**, 992 (1968).
37. Cunningham, G.N., Wise, M.B., and Barrick, E.R., *J. Anim. Sci.* **25**, 532 (1966).
38. Curran, G.L., *J. Biol. Chem.* **210**, 765 (1954).
39. Dewar, W.A., Teague, P.W., and Downie, J.N., *Br. Poult. Sci.* **15**, 119 (1974).
40. Doisy, E.A., Jr., *Trace Subst. Environ. Health–6, Proc. Univ. Mo. Annu. Conf., 6th, 1972,* p. 193 (1972); *in* "Trace Element Metabolism in Animals" (W.G. Hoekstra, *et al.*, eds.), Vol. 2, p. 664. Univ. Park Press, Baltimore, Maryland, 1974.
41. Ellis, G.H., Smith, S.E., and Gates, E.M., *J. Nutr.* **34**, 21 (1947).
42. Erway, L., Hurley, L.S., and Fraser, A., *Science* **152**, 1766 (1966).
43. Erway, L.C., and Purichia, N.A., *in* "Trace Element Metabolism in Animals" (W.G. Hoekstra *et al.*, eds.), Vol. 2, p. 249. Univ. Park Press, Baltimore, Maryland, 1974.
44. Evans, R.J., Rhian, M., and Draper, C.I., *Poult. Sci.* **22**, 88 (1943).
45. Everson, G.J., DeRafols, W., and Hurley, L.S., *Fed. Proc., Fed. Am. Soc. Exp. Biol.* **23**, 448 (1964).
46. Everson, G.J., Hurley, L.S., and Geiger, J.F., *J. Nutr.* **68**, 49 (1959).
47. Everson, G.J., Shrader, R.E., *J. Nutr.* **94**, 89 (1968); Shrader, R.E., and Everson, G.J., *ibid.* p. 269.
48. Fernandez, A.A., Sobel, C., and Jacobs, S.L., *Anal. Chem.* **35**, 1721 (1962).
49. Flynn, A., and Franzmann, A.W., *in* "Trace Element Metabolism in Animals" (W.G. Hoekstra *et al.*, eds.), Vol. 2, p. 444. Univ. Park Press, Baltimore, Maryland, 1974.

50. Foradori, A.C., Bertinchamps, A., Gulebon, J.M., and Cotzias, G.C., *J. Gen. Physiol.* **50,** 2255 (1967).
51. Fore, H., and Morton, R.A., *Biochem. J.* **51,** 594, 598, 600, and 603 (1952).
52. Frost, G., Asling, C.W., and Nelson, M.M., *Anat. Rec.* **134,** 37 (1959).
53. Gallup, W.D., and Norris, L.C., *Science* **87,** 18 (1938).
54. Gallup, W.D., and Norris, L.C., *Poult. Sci.* **18,** 76 and 83 (1939).
55. Gartner, R.J.W., and Twist, J., *Aust. J. Exp. Agr. Anim. Husb.* **8,** 210 (1968).
56. Gladstones, J.S., and Drover, D.P., *Aust. J. Exp. Agric. Anim. Husb.* **2,** 46 (1962).
57. Gormican, A., *J. Am. Diet. Assoc.* **56,** 397 (1970).
58. Grace, N.D., *N. Z. J. Agric. Res.* **16,** 177 (1973).
59. Grebennikov, E.P., Soroka, V.R., and Sabadash, E.V., *Fed. Proc., Fed. Am. Soc. Exp. Biol.* **23,** Transl. Suppl. T461 (1964).
60. Greenberg, D.M., Copp, H.D., and Cuthbertson, E.M., *J. Biol. Chem.* **147,** 749 (1943).
61. Gregory, E.M., and Fridovich, I., *in* "Trace Element Metabolism in Animals" (W.G. Hoekstra *et al.,* eds.), Vol. 2, p. 486. Univ. Park Press, Baltimore, Maryland, 1974.
62. Groppel, B., and Anke, M., *Arch. Veterinaermed.* **25,** 779 (1971).
63. Grummer, R.H., Bentley, O.G., Phillips, P.H., and Bohstedt, G., *J. Anim. Sci.* **9,** 170 (1950).
64. Gutowska, M.S., and Parkhurst, R.T., *Poult. Sci.* **21,** 277 (1942).
65. Hamilton, E.I., and Minski, M.J., *Sci. Total Environ.* **1,** 375 (1972/1973).
66. Hamilton, E.I., Minski, M.J., and Cleary, J.J., *Sci. Total Environ.* **1,** 341 (1972/1973).
67. Hancock, R.G.V., and Fritze, K., *Bioinorg. Chem.* **3,** 77 (1973).
68. Hartman, R.H., Matrone, G., and Wise, G.H., *J. Nutr.* **57,** 429 (1955).
69. Hartmans, J., *in* "Trace Element Metabolism in Animals" (W.G. Hoekstra *et al.*, eds.), Vol. 2, p. 261. Univ. Park Press, Baltimore, Maryland, 1974.
70. Hartmans, J., *Landwirtsch. Forsch.* **27,** 1 (1972).
71. Hawkins, G.E., Wise, G.H., Matrone, G., and Waugh, R.K., *J. Dairy Sci.* **38,** 536 (1955).
72. Hedge, B., Griffith, G.C., and Butt, E.M., *Proc. Soc. Exp. Biol. Med.* **107,** 734 (1961).
73. Heller, V.H., and Penquite, R., *Poult. Sci.* **16,** 243 (1937).
74. Hignett, S.L., *Proc. Int. Congr. Anim. Reprod., 3rd,* p. 116 (1956).
75. Hill, R., *in* "Trace Element Metabolism in Animals" (W.G. Hoekstra *et al.*, eds.), Vol. 2, p. 632. Univ. Park Press, Baltimore, Maryland, 1974.
76. Hill, R., and Mather, J.S., *Br. J. Nutr.* **22,** 625 (1968).
77. Holtkamp, D.E., and Hill, R.M., *J. Nutr.* **41,** 307 (1950).
78. Hopkins, H., and Eisen, J., *J. Agric. Food. Chem.* **7,** 633 (1959).
79. Howes, A.D., and Dyer, I.A., *J. Anim. Sci.* **32,** 141 (1971).
80. Hughes, E.R., and Cotzias, G.C., *Am. J. Physiol.* **201,** 1061 (1961).
81. Hughes, E.R., Miller, S.T., and Cotzias, G.C., *Am. J. Physiol.* **211,** 207 (1966).
82. Hurley, L.S., and Bell, L.T., *J. Nutr.* **104,** 133 (1974).
83. Hurley, L.S., and Everson, G.J., *Proc. Soc. Exp. Biol. Med.* **102,** 360 (1959).
84. Hurley, L.S., Everson, G.J., and Geiger, J.F., *J. Nutr.* **66,** 309 (1958); **67,** 445 (1959).
85. Hurley, L.S., Everson, G.J., Wooten, E., and Asling, C.W., *J. Nutr.* **74,** 274 (1961).
86. Hurley, L.S., Wooten, E., and Everson, G.J., *J. Nutr.* **74,** 282 (1961).
87. Hurley, L.S., Gowan, J., and Shrader, R., *in* "Les tissues calcifíes. V. Symposium Européen," p. 101. Soc. d'Enseigment Supérieur, Paris, 1968.

88. Hurley, L.S., Wooten, E., Everson, G.J., and Asling, C.W., *J. Nutr.* **71,** 15 (1960).
89. Hurley, L.S., Thériault, L.T., and Dreosti, I.E., *Science* **170,** 1316 (1970).
90. Jaulmes, P., and Hamelle, G., *Ann. Nutr. Ailment.* **25,** B133 (1971).
91. Johnson, S.R., *J. Anim. Sci.* **2,** 14 (1943).
92. Joynson, H.V., Martinovic, J., and Johnson, B.C., *Biochem. Biophys. Res. Commun.* **43,** 1040, (1971).
93. Jukes, T.H., *J. Biol. Chem.* **134,** 789 (1940); *J. Nutr.* **20,** 445 (1940).
94. Kato, M., *Q. J. Exp. Physiol. Cogn. Med. Sci.* **48,** 355 (1963).
95. Keefer, R.C., Barak, A.J., and Boyett, J.D., *Biochim. Biophys. Acta* **221,** 390 (1970).
96. Keefer, R.C., Tuma, D.J., and Barak, A.J., *Am. J. Clin. Nutr.* **26,** 409 (1973).
97. Keele, B.B., McCord, J.M., and Fridovich, I., *J. Biol. Chem.* **245,** 6176 (1970).
98. Kemmerer, A.R., and Todd, W.R., *J. Biol. Chem.* **94,** 317 (1931).
99. Kemmerer, A.R., Elvehjem, C.A., and Hart, E.B., *J. Biol. Chem.* **92,** 623 (1931).
100. Kent, N.L., and McCance, R.A., *Biochem. J.* **35,** 877 (1941).
101. Kirchgessner, M., *Mangelkrankheiten* **6,** 61 and 105 (1955).
102. Lassiter, J.W., Miller, W.J., Neathery, M.W., Gentry, R.P., Abrams, E., Carter, J.C., Jr., and Stake, P.E., *in* "Trace Element Metabolism in Animals" (W.G. Hoekstra *et al.*, eds.), Vol. 2, p. 557. Univ. Park Press, Baltimore, Maryland, 1974.
103. Lassiter, J.W., and Morton, J.D., *J. Anim. Sci.* **27,** 776 (1968).
104. Lassiter, J.W., Morton, J.D., and Miller, W.J., *in* "Trace Element Metabolism in Animals" (C.F. Mills, ed.), Vol. 1, p. 130. Livingstone, Edinburgh, 1970.
105. Leach, R.M., Jr., *Fed. Proc., Fed. Am. Soc. Exp. Biol.* **26,** 118 (1967); **30,** 991 (1971).
106. Leach, R.M., Jr., and Muenster, A.M., *J. Nutr.* **78,** 51 (1962).
107. Leach, R.M., Jr., Muenster, A.M., and Wien, E.M., *Arch. Biochem. Biophys.* **133,** 22 (1969).
108. Lindow, C.W., Petersen, W.H., and Steenbock, H., *J. Biol. Chem.* **84,** 419 (1929).
109. Longstaff, M., and Hill, R., *Br. Poult. Sci.* **13,** 377 (1972).
110. Lorenzen, E.J., and Smith, S.E., *J. Nutr.* **33,** 143 (1947).
111. Lyons, M., and Insko, W.M., *Ky., Agric. Exp. Stn., Bull.* **371** (1937).
112. Mathers, J.R., and Hill, R., *Br. J. Nutr.* **22,** 635 (1968).
113. Mathews, M.B., *Nature (London)* **213,** 215 (1967).
113a. Matrone, G., Hartman, R.H., and Clawson, A.J., *J. Nutr.* **67,** 309 (1959).
114. Maynard, L.S., and Cotzias, G.C., *J. Biol. Chem.* **214,** 489 (1955).
115. Maynard, L.S., and Fink, S., *J. Clin. Invest.* **35,** 83 (1956).
116. McDonald, M.W., Humphries, C., Short, C.C., Smith, L., and Solvyns, A., *Proc. Aust. Poult. Sci. Conv.* p. 223 (1969).
117. McLeod, B.E., and Robinson, M.F., *Br. J. Nutr.* **27,** 221 (1972).
118. McLeod, B.E., and Robinson, M.F., *Br. J. Nutr.* **27,** 229 (1972).
119. Mena, I., Marin, O., Fuenzalida, S., and Cotzias, G.C., *Neurology* **17,** 128 (1967).
120. Meyer, H., and Engelbartz, T., *Dtsch. Tieraerztl. Wochenschr.* **67,** 124 (1960).
121. Mildvan, A.S., Scrutton, M.C., and Utter, M.F., *J. Biol. Chem.* **241,** 3488 (1966).
122. Miller, R.C., Keith, T.B., McCarty, M.A., and Thorp, W.T.S., *Proc. Soc. Exp. Biol. Med.* **45,** 50 (1940).
123. Mitchell, R.L., *Research (London)* **10,** 357 (1957).
124. Molokhia, M.M., and Smith, H., *Arch. Environ. Health.* **15,** 745 (1967).
125. Monier-Williams, G.W., "Trace Elements in Foods." Chapman & Hall, London, 1949.
125a. Murakami, Y., Suzuki, Y., and Yamagata, T., *J. Radiat. Res.* **6,** 105 (1965).
126. Neher, G.M., Doyle, L.P., Thrasher, D.M., and Plumlee, M.P., *Am. J. Vet. Res.* **17,** 121 (1956).

127. North, B.B., Leichsenring, J.M., and Norris, L.M., *J. Nutr.* **72,** 217 (1960).
128. Nozdryukina, L.R., Grinkevich, N.I., and Gribovskaya, I.F., *Trace Subst. Environ. Health–7, Proc. Univ. Mo. Annu. Conf., 7th, 1973,* p. 353 (1973).
129. Olson, J.A., *Rev. Physiol., Biochem. Exp. Pharmacol.* **56,** 173 (1965).
130. O'Mary, C.C., Butts, W.T., Reynolds, R.A., and Bell, M.C., *J. Anim. Sci.* **28,** 268 (1969).
131. Orent, E.R., and McCollum, E.V., *J. Biol. Chem.* **92,** 661 (1931).
132. Panic, B., Bezbradica, L.J., Nedeljkov, N., and Istwani, A.G., *in* "Trace Element Metabolism in Animals" (W.G. Hoekstra *et al.*, eds.), Vol. 2, p. 635. Univ. Park Press, Baltimore, Maryland, 1974.
133. Papavasiliou, P.S., Miller, S.T., and Cotzias, G.C., *Am. J. Physiol.* **211,** 211 (1966).
134. Pereira, M., and Couri, D., *Biochim. Biophys. Acta* **237,** 348 (1971).
135. Perry, H.M., Jr., Perry, E.F., Purifoy, J.E., and Erlanger, J.N., *Trace Subst. Environ. Health–7, Proc. Univ. Mo. Annu. Conf., 7th, 1973,* p. 281 (1973).
136. Peterson, W.H., and Skinner, J.T., *J. Nutr.* **4,** 419 (1931).
137. Plumlee, M.P., Thrasher, D.M., Beeson, W., Andrews, F.N., and Parker, H.E., *J. Anim. Sci.* **14,** 996 (1954); **15,** 352 (1956).
138. Reiman, C.K., and Minot, A.S., *J. Biol. Chem.* **45,** 133 (1920).
139. Rojas, M.A., Dyer, I.A., and Cassat, W.A., *J. Anim. Sci.* **24,** 664 (1965).
139a. Sansom, B.F., Gibbons, R.A., Dixon, S.N., Russell, A.M., and Symonds, H.W., *in* "Nuclear Techniques in Animal Production and Health." IAEA, Vienna, 1976.
140. Sato, M., and Murata, K., *J. Dairy Sci.* **15,** 461 (1932).
141. Schafer, D.F., Stephenson, D.V., Barak, A.J., and Sorrell, M.F., *J. Nutr.* **104,** 101 (1974).
142. Schaible, P.J., and Bandemer, S.L., *Poult. Sci.* **21,** 8 (1942).
143. Schaible, P.J., Bandemer, S.L., and Davison, J.A., *Mich., Agric. Exp. Stn., Tech. Bull.* **159** (1938).
144. Schiller, S., and Dorfman, A., *J. Biol. Chem.* **227,** 625 (1957).
145. Schroeder, H.A., Balassa, J.J., and Tipton, I.H., *J. Chronic. Dis.* **19,** 545 (1966).
146. Scrutton, M.C., *Biochemistry* **10,** 3897 (1971).
147. Scrutton, M.C., Griminger, P., and Wallace, J.C., *J. Biol. Chem.* **247,** 3305 (1972).
148. Scrutton, M.C., Utter, M.F., and Mildvan, A.S., *J. Biol. Chem.* **241,** 3480 (1966).
149. Settle, E.A., Mraz, F.R., Douglas, C.R., and Bletner, J.K., *J. Nutr.* **97,** 141 (1969).
150. Shils, M.E., and McCollum, E.V., *J. Nutr.* **26,** 1 (1943).
151. Shrader, R.E., and Everson, G.J., *J. Nutr.* **91,** 453 (1967).
152. Shrader, R.E., Erway, L.C., and Hurley, L.S., *Teratology* **8,** 257 (1973).
153. Smith, S.E., and Ellis, G.H., *J. Nutr.* **34,** 33 (1947).
154. Smith, S.E., Medlicott, M., and Ellis, G.H., *Arch. Biochem.* **4,** 281 (1944).
155. Thiers, R.E., and Vallee, B.L., *J. Biol. Chem.* **226,** 911 (1957).
156. Thomson, A.B.R., Olatunbosun, D., and Valberg, L.S., *J. Lab. Clin. Med.* 78, 642 (1971).
157. Thomson, A.B.R., and Valberg, L.S., *Am. J. Physiol.* **223,** 1327 (1971).
158. Tipton, I.H., and Cook, M.J., *Health Phys.* **9,** 103 (1956).
159. Tsai, H., and Everson, G.J., *J. Nutr.* **91,** 447 (1967).
160. Underwood, E.J., and Curnow, D.H., unpublished data (1944).
161. Underwood, E.J., Robinson, T.J., and Curnow, D.H., *J. Agric. West. Aust.* **24,** 259 (1947).
162. Vallee, B.L., and Coleman, J.E., *Compr. Biochem.* **12,** 165 (1964).
163. Van Koetsveld, E.E., *Tijdschr. Diergeneesk.* **83E,** 229 (1958).

164. Wachtel, L.W., Elvehjem, C.A., and Hart, E.B., *Am. J. Physiol.* **140,** 72 (1943).
165. Waddell, J., Steenbock, H., and Hart, E.B., *J. Nutr.* **4,** 53 (1931).
166. Watson, L.T., Ammerman, C.B., Miller, S.M., and Harms, R.H., *Poult. Sci.* **50,** 1693 (1971).
167. Werner, A., and Anke, M., *Arch. Tierernaehr.* **10,** 142 (1960).
168. Wilgus, H.S., Jr., Norris, L.C., and Heuser, G.F., *Science* **84,** 252 (1936); *J. Nutr.* **14,** 155 (1937).
169. Wilgus, H.R., Jr., and Patton, A.R., *J. Nutr.* **18,** 35 (1939).
170. Wilson, J.G., *Vet. Rec.* **64,** 621 (1952); **79,** 562 (1966).
171. Wolbach, S.B., and Hegsted, D.M., *Arch. Pathol.* **56,** 437 (1953).
172. Zhuk, V.P., *Bull. Exp. Biol. Med. (Engl. transl.)* **54,** 1271 (1964).
173. Zook, E.G., Greene, F.E., and Morris, E.R., *Cereal Chem.* **47,** 720 (1970).

8

Zinc

I. ZINC IN ANIMAL TISSUES AND FLUIDS

1. General Distribution

Typical normal levels of zinc in the principal soft tissues of several species are given in Table 27. Species differences are clearly small and zinc occurs widely in relatively high concentrations throughout the body (80, 97, 157, 159, 215, 309). The mammalian newborn does not consistently carry higher total body Zn concentrations than mature animals of the same species and there is little fetal Zn storage (291). During the suckling period whole body Zn concentration rises substantially from newborn levels in the rat and pig but not in the guinea pig (291). The whole body of adult man is estimated to contain 1.4–2.3 g of zinc (342), of which about 20% is present in the skin (235). The mean Zn concentrations of normal human epidermis and dermis have been reported to be 70.5 and 12.6 μg/g dry weight, respectively (189). In animals covered with hair or wool a substantial proportion of total body zinc is present in these tissues. In the rat and hedgehog no less than 38% of the whole body Zn is situated in the skin and hair, or the skin and bristles (291). Because of the large weight of the bones and teeth and their relatively high-Zn concentration (150–250 ppm) (30, 58, 152), an appreciable proportion of total body Zn resides in these structures. The mean Zn concentration in 56 samples of human dental enamel was recently reported to be 203 ± 12 μg/g dry weight, with a range of 42–510 (152).

The Zn concentration of muscles varies with their color and their functional

TABLE 27
Typical Zinc Concentrations of Normal Tissues[a]

Tissue	Man[b]	Man[c]	Monkey[d]	Rat[e]	Pig[f]
Adrenal	12	–	16	–	33
Brain	14	13	–	18	–
Heart	33	–	22	21	–
Kidney	55	37	29	23	40
Liver	55	76	51	30	40
Lung	15	14	19	22	–
Muscle	54	39	24	13	–
Pancreas	29	–	48	33	45
Prostate	102	–	–	223	–
Spleen	21	–	21	24	28
Testis	17	13	17	22	–

[a] μg/g fresh tissue.
[b] Tipton and Cook (309).
[c] Hamilton *et al.* (97).
[d] Macapinlac *et al.* (157).
[e] Gilbert and Taylor (80); Mawson and Fischer (159).
[f] W. G. Hoekstra, private communication.

activity. A 3-fold variation in the Zn concentration of eight bovine muscles (301) and a 4-fold variation in the same eight porcine muscles (40) have been reported. The mean Zn level of the *longissimus dorsi*, which is light-colored and shows little activity, was 69 ppm on the dry, fat-free basis, whereas that of the *serratus ventralis,* which is a dark, highly active muscle, was 247 ppm (40). Such differences are minimal in newborn pigs and develop with the use of the muscles, so that marked differences between red and white muscle are apparent by 8 weeks of age (41). The higher Zn content of the red muscle is situated entirely in the subcellular fraction, composed mainly of myofibrils and nuclei.

In liver and mammary cells the zinc is present in the nuclear, mitochondrial, and supernatant fractions, with the highest levels per unit of protein in the supernatant and microsomes (12, 306). Experiments with ^{65}Zn injections in mice indicate that about one-sixth of the ^{65}Zn in the tissues is firmly bound to protein and cannot be removed by dialysis or by ethylenediaminetetraacetic acid (EDTA) (12). This fraction probably consists of Zn metalloenzymes and the Zn bound to nucleic acids. The remainder, which is exchangeable with Zn ions or removable by EDTA, is bound to the imidazole or sulfhydryl groups of the proteins (335) and includes those Zn enzymes in which the Zn is less tightly bound and is removable by EDTA. The relation of zinc to the metallothionein present in liver, kidney, and pancreas and its formation and functions are considered in Section II,3. Becker and Hoekstra (13) found a pool of firmly bound Zn in rat liver cells, together with three further pools of cellular Zn bound with differing intensities. The binding pattern of liver Zn, like the total Zn concentration, was unaltered in Zn deficiency.

The Zn concentrations of the tissues, apart from the blood, milk, and hair as discussed later in this section, do not always reflect dietary Zn intakes. Under deficiency conditions the growth of the organs can be so inhibited that while the total amount of zinc is greatly reduced, the concentrations may not be significantly lowered. However, Kirchgessner and Pallauf (136) found the Zn concentrations of the liver, bones, tail, and whole body to be significantly reduced in young rats Zn-depleted over a 35-day period. In weaned rats Zn-depleted for 10 days and then fed diets ranging in Zn content from 2 to 500 ppm for 21 days, the levels of Zn in serum and liver increased almost linearly with increasing Zn intakes up to the optimal level (12 ppm dietary Zn). A further abrupt rise in these tissues occurred at 500 ppm Zn in the diet. The heart and the testes remain relatively insensitive to very high-Zn intakes, but large increases in the Zn concentrations of the plasma, liver, kidney, and spleen have been demonstrated in rats fed a normal diet plus 1000 and 2000 ppm of supplementary Zn for 15 days (43). Comparable large increases in tissue Zn, other than in the heart and muscles, have been observed in calves at high-Zn intakes (600 ppm of the diet) following the breakdown of Zn homeostasis (293).

2. Zinc in Eye Tissues

The highest Zn concentrations known to exist normally in living tissues occur in the choroid of the eye. Eye tissues other than the choroid and the iris carry more normal Zn concentrations. The iris and choroid of sheep's eyes have been reported to contain 436 and 277 ppm Zn and those of cattle 246 and 139 ppm (d.b.) (26). The Zn levels in the choroids of the dog, fox, and marten have been given as 14,600, 69,000, and 91,000 ppm (d.b.), respectively (336). In the iridescent layer (*tapetum lucidum*) of the choroid of the dog and fox levels up to 8.5% (85,000 ppm) and 13.8% (138,000 ppm) Zn, respectively, have been reported (337). The function of zinc in such high concentrations in eye tissues is unknown.

3. Zinc in Male Sex Organs and Secretions

The Zn concentration of the testes of rats ranging in age from 7 to 58 days remains fairly constant at about 120 ppm dry weight for the first 30–35 days and then increases to close to 200 ppm during the second month of life, when spermatids are transformed into spermatozoa (220). Values of 176 ± 12 and 132 ± 16 ppm Zn (d.b.) for the testes of normal and of Zn-deficient rats, respectively

(241), and 105 ± 4.4 and 74.0 ± 5.0 for the testes of normal and severely Zn-deficient rams (316) have been reported. Lower mean levels of 17 and 13 μg Zn/g wet weight have been found for the normal adult human testis (97, 309). These values are close to 85 and 65 ppm on the dry basis.

Higher Zn concentrations than those just cited occur in the prostate gland of the rat, rabbit, and man and in human seminal fluid and spermatozoa. The following average values may be given: human prostate, 859; dorsolateral rat prostate, 891; human semen, first fraction of ejaculate, 2930, second fraction, 1400, third fraction, 910; and human spermatozoa, 1990 ppm Zn (d.b.) (159). Sperm-rich boar semen is intermediate between bull semen (10 μg Zn/ml) (340) and human semen (50–200 μg Zn/ml) (19). The concentration in sperm and seminal plasma is approximately equal when expressed on a dry weight basis (159). In ram (158) and rat (168) sperm the zinc is concentrated in the tail and in the rat prostate in the luminar edge of the acinar cells (168).

The level of Zn in the prostate is reduced by castration, and the rate of Zn accumulation is greatly increased in young rats by testosterone or gonadotropin injections (84, 167). Chorionic gonadotropin increases the weight and ^{65}Zn uptake in all the other accessory reproductive structures in male rats, as well as the dorsolateral prostate, whereas FSH (follicle stimulating hormone) reduces this uptake per organ and gram of tissue (259). The administration of stilbestrol lowers the Zn content of human prostatic cancerous tissue, which even before treatment is lower in Zn concentration than that of normal, or hyperplastic but nonmalignant prostatic tissue. The Zn concentration of 28 normal prostates taken from accident victims ranged from 186 to 1340 ppm (mean 732 ppm Zn on the dry basis), while 33 hyperplastic but not malignant glands removed surgically ranged from 190 to 2900 ppm (mean 972 ppm Zn) (310). The range of concentration is clearly too wide to give this determination much diagnostic value.

4. Zinc in Nails, Hair, and Wool

The level of zinc in the nails of 18 normal human subjects was reported by Smith (282) to range from 93 to 292 ppm and to average 151 ppm. The head hair of 46 subjects ranged similarly from 92 to 255 ppm and averaged 173 ppm Zn. This is similar to the mean of 167 ± 5 ppm for healthy males and 172 ± 9 ppm for females obtained by Schroeder and Nason (274). A lower overall mean of 119.6 ± 4.6 ppm was obtained by Strain *et al.* (297) for the hair of North American males. The mean Zn level in the hair of normal Egyptians was 103.3 ± 4.4 ppm and of untreated "zinc-deficient" dwarfs only 54.1 ± 5.5 ppm. Oral zinc sulfate therapy increased hair Zn levels to 121.1 ± 4.8 ppm and alleviated the signs of Zn deficiency. The hair of Iranian villagers, in a region where Zn deficiency had been suspected, averaged 127 ppm, compared with 220 ppm for controls whose diet was diversified and abundant (253). Hambidge and co-workers (92) obtained the following mean Zn concentrations for the hair of 338 apparently normal subjects living in Denver with ages ranging from 0 to 40 years: neonates (25), 174 ± 8; 3 months to 4 years (93), 88 ± 5; 4–17 years (132), 153 ± 5; and 17–40 years (88), 180 ± 4 ppm. The lower levels in the young children are particularly noteworthy, as they included a number with hair Zn levels below 70 ppm and a history of poor growth, appetite, and taste acuity responsive to Zn supplementation. Amadar *et al.* (3) also found the hair Zn levels of 3- to 5-year-old healthy Cuban children to be lower than those of 12–14 year olds, but the mean value for the former group (209 ± 53 μg Zn/g) is high compared with most reported values.

In a group of 20 normal pregnant women a small fall in hair Zn levels in late pregnancy has been observed (93). The mean concentrations were close to 170 μg Zn/g dry weight at 16 weeks and 155 μg/g at 38 weeks gestation.

The level of zinc in hair reflects dietary Zn intakes in rats (252), although Pallauf and Kirchgessner (218) found that this level only rose with successive additions of dietary Zn when the suboptimal Zn level necessary for hair production was available. With levels of 2–8 ppm of dietary Zn stepwise additions of zinc resulted in an actual reduction in the level of zinc in the hair, because of the large increase in body weight and mass of hair induced by the supplementary zinc. Hair Zn concentrations also reflect dietary Zn intakes in pigs (148), cattle, and goats (175, 181). Individual variability is very high and there is some variation with age and part of the body (181). For these reasons hair analysis as a diagnostic criterion of Zn deficiency must be used with caution. O'Mary *et al.* (207) found that the Zn level in Hereford cattle hair did not vary with color or season. Most of their values fell within the range of 115–135 ppm. Hidiroglou and Spurr (112) observed lower levels in the winter hair than in the summer hair of cattle, probably due to the lower Zn levels in the winter herbage. The mean Zn concentration in 45 samples of wool from various areas has been reported as 115 ppm (32).

5. Zinc in Blood

a. Forms and Distribution. Zinc is present in the plasma, erythrocytes, leukocytes, and platelets, 30–40% of the plasma Zn is firmly bound to an α_2-macroglobulin, and 60–70% is loosely bound to the albumin (219). Almost all the Zn in erythrocyres occurs as carbonic anhydrase (117, 118), together with a small fraction associated with other Zn enzymes (318). Carbonic anhydrase cannot be detected in plasma or leukocytes, so that the Zn in erythrocytes must account for all of this enzyme in blood (147). Calculations from data obtained by Vallee and Gibson (317) indicate that 75–88% of the total Zn of normal human blood is contained in the red cells, 12–22% in the plasma, and 3% in the leukocytes.

The individual leukocyte contains 25 times as much Zn as the individual erythrocyte. Human blood platelets contain Zn in amounts ranging from 0.2 to 0.45 μg/10^9 platelets, with a mean of 7.1 μg of platelet Zn in 100 ml of whole blood (71). The Zn concentration of human serum is consistently higher than that of plasma by an average of 16%. Of this increase, 44% was shown to be derived from platelets disintegrating during clotting, 39% from a slightly greater dilution in plasma, and 4.0% from hemolysis (71). In newborn infants the Zn content of the erythrocytes is only one-quarter of the adult value, rising progressively over the first 12 years of life (16). This would be expected from the low levels of carbonic anhydrase present in the red cells of newborn and premature infants (295).

b. Normal Levels. The following normal levels for Zn in erythrocytes were reported by Smirnov (281): man, 13; rat, 10; dog and rabbit, 9; and goose, 6.5 μg/g. In the adult rabbit, average Zn levels are 2.5 μg/ml of whole blood, 2.7 μg/ml of plasma, and 9 μg/ml of erythrocytes (16). In young growing pigs 0.6 μg Zn/ml of plasma and 7 μg/ml of erythrocytes can be considered normal (116, 169). Plasma Zn levels of 100 μg/100 ml were found to be optimal for growth in sheep fed semisynthetic diets in one study (214), while Mills *et al.* (186) did not find it essential to maintain this level to satisfy the growth requirements of young sheep. Furthermore, Grace (81) obtained appreciably lower plasma Zn levels in healthy sheep grazing New Zealand pastures normal in Zn content. The mean levels for groups of mixed age ewes from 18 farms ranged from 52.8 to 88.8 μg Zn/100 ml, with an overall mean close to 70 μg/100 ml.

Vallee and Gibson (317), using a dithizone method, reported the following values for normal humans: 8.8 ± 0.2 for whole blood; 1.21 ± 0.19 for plasma; and 14.4 ± 2.7 μg Zn/ml for erythrocytes. These values were later confirmed by an atomic absorption method (240). Pekarek *et al.* (223) reported a mean serum Zn concentration for 99 healthy young men of 102 μg/100 ml (SD ± 17), with a calculated range of 68–136. On a day to day basis, 95% of values from an individual were within 18 μg/100 ml of his own mean value. Diurnal variations were slight, with a small circadian rythm related to the ingestion of food. Plasma Zn returns to the fasting level within 3 hr of taking food (33). A significant decrease in plasma Zn levels in Japanese quail receiving normal or excess Zn has been observed as a consequence of 24-hr fasting or when supplements containing soybean or egg white protein were given (99). These findings suggest that under conditions of rapid protein breakdown plasma Zn levels decline. Significant regional differences in the United States in the plasma Zn levels of normal males have also been reported (142), presumably reflecting differences in dietary Zn intakes.

Plasma Zn levels fall in women in late pregnancy and when taking oral contraceptives. For example, Halsted *et al.* (90) reported mean values of 0.96 ±

0.13 μg Zn/ml for control women, 0.60 ± 0.11 μg/ml for 25 women in the third trimester of pregnancy, and 0.65 ± 0.08 μg/ml for 10 women who had been taking oral contraceptives for 1 month to 4 years. Hambidge and Droegemueller (93) observed a similar fall in mean plasma Zn in 20 pregnant women to 68.2 ± 2.0 μg/100 ml at 16 weeks and 56.0 ± 2.1 μg/100 ml at 38 weeks gestation, but the levels in women taking oral contraceptives were within the normal range. Henkin *et al.* (106) obtained a slightly lower mean of 48 ± 3 μg Zn/100 ml serum for 15 women at normal delivery, compared with a mean of 83 ± 3 μg/100 ml for serum from the umbilical cord of their infants. A higher plasma Zn level in the fetus than in the mother also occurs in sheep and goats (150). The average plasma Zn levels of seven ewes and fetuses sampled in late pregnancy were 102 and 168 μg/100 ml, respectively. The corresponding values for 11 goats and their fetuses were 98 and 138 μg/100 ml. Plasma Zn levels do not change significantly during pregnancy in normal cattle (59, 245), but there is usually a marked fall during and immediately after parturition. This fall is particularly evident in cows exhibiting dystocia, presumably as a reflection of stress (59). Thus Dufty (59) obtained a mean maternal plasma Zn concentration of 0.68 ± 0.23 μg/ml for 31 cows with normal parturition 18–24 hr after calving, compared with 0.38 ± 0.14 μg/ml for eight cows with dystocia and stillborn calves.

In a study of 130 normal infants total plasma Zn concentrations were found to be at adult levels in the newborn (107). These values fell to just below adult levels within the first week of life, fell further at 2–3 months of age, returned toward adult values at 4 months, and except for a fall at about 1 year of age, remained at adult levels throughout the remainder of infancy.

c. Effect of Diet. Large oral doses of zinc greatly increase whole blood and plasma Zn concentrations in rats, rabbits, cats, pigs (16, 116, 156), sheep, and cattle (212). Cattle given incremental Zn supplements to provide total dietary Zn intakes from 18 to 189 ppm displayed mean serum Zn increases from 1.5 to 2.7 μg/ml (231). Most workers have observed a decline in plasma or serum Zn in Zn-deficient animals. Mills *et al.* (186) reported a decline from normal levels of 0.8–1.2 μg Zn/ml to below 0.4 μg/ml in the serum of severely Zn-deficient lambs or calves, and falls in whole blood and plasma Zn occur in less acutely Zn-deficient calves and goats (175). In Zn-deficient baby pigs a level of 0.22 μg Zn/ml serum was obtained, compared with 0.98 μg/ml in pair-fed controls (169). Wilkins *et al.* (343) found plasma Zn concentrations in rats to fall from 1.2 to 0.7 μg/ml after 1 day on a severely Zn-deficient (< 0.25 ppm) diet. Thereafter the fall became less marked and after 5 days the concentration usually varied between 0.4 and 0.6 μg/ml. Pallauf and Kirchgessner (217) found that the serum Zn level of weaned male rats receiving a Zn-deficient (1.8–1.9 ppm) diet fell by more than 30% within 2 days and by more than 50% in 4 days from the original value of 2.3 μg Zn/ml. Addition of zinc restored the serum Zn level to normal. Luecke *et al.* (154) provided incremental increases in supplementary Zn up to 12.5 ppm to young rats fed a severely Zn-deficient (0.8 ppm) diet. Serum Zn levels rose steadily from less than 0.4 μg/ml to a plateau close to 1.4 μg/ml. It was concluded that a serum Zn level of 1.3 μg/ml indicates dietary Zn adequacy for growth in this species. Subnormal plasma Zn levels have been recorded for Middle East males suffering from "conditioned" Zn deficiency (46).

d. Effect of Disease and Stress. Profound changes occur in the levels of zinc in the blood plasma and cellular elements in various disease states and under stress conditions. Subnormal plasma Zn levels have been reported in patients with malignant tumors (1, 326), atherosclerosis (328), postalcoholic cirrhosis of the liver and other liver diseases (91, 240, 320), tuberculosis and leprosy (209) (Table 28), untreated pernicious anemia (295, 303), chronic and acute infections (224, 225, 326), and after acute tissue injury, regardless of origin (25). Serum Zn concentrations were shown to decrease significantly in a dose-dependent response after endotoxin administration in the rat (222). These abnormalities in zinc metabolism manifested in depressed serum Zn levels are little understood. However, Beisel *et al.* (15) observed a redistribution of Zn within the body initiated by a hormonelike protein factor which is released from phagocytizing cells. This factor, leukocytic endogenous mediator (LEM), stimulates the liver to take up Zn (and iron). These actions of LEM have been demonstrated using ^{65}Zn in the rat (226) and by injecting a normal test animal with sterile serum obtained from an infected subject (15).

In patients with anemia, other than pernicious anemia, the Zn and carbonic anhydrase levels in the blood are lowered in parallel fashion with the decreases in hemoglobin and red cell counts, but in pernicious anemia the Zn and carbonic anhydrase per unit of RBC are significantly above normal, even when the

TABLE 28
Mean Serum Zinc Concentrations in Five Groups of Subjects[a]

Clinical groups	No. of subjects	Serum zinc (μg/100 ml)	p^b
Control	33	102.4 ± 3.0	–
Leprosy and ulceration	21	89.9 ± 4.0	< 0.02
Leprosy, no ulceration	18	89.8 ± 4.6	< 0.02
Pulmonary tuberculosis	16	85.0 ± 3.9	< 0.005
Dermatitis herpetiformis	5	86.9 ± 3.1	< 0.005

[a]From Oon *et al.* (209).
[b]Significance of difference from control value.

increased cell size is eliminated as a contributing factor (317). Gibson and co-workers (79) have demonstrated subnormal Zn concentrations in the peripheral leukocytes of patients with chronic leukemia and shown that they cannot be raised by injections of zinc gluconate. A rise to normal levels occurs in clinical remission and under therapy with X-rays or urethane, accompanying the falling leukocyte count. The Zn content of the leukocytes also decreases in patients with a variety of neoplastic diseases (302). The physiological significance of these changes is unknown.

The administration of large doses of corticosteroids to patients with burn and surgical stress presenting a low cardiac output syndrome produces a rapid and sustained depression in serum Zn (70). Wegner and co-workers (334) were unable to demonstrate consistent trends linking plasma Zn concentration directly to the adrenal response or stressful conditions in dairy cattle, but hyperthermal stress had a depressing effect on serum Zn with a correlation of −0.92 between serum and plasma corticoids. The depression in plasma Zn associated with the stress of difficult or prolonged parturition (dystocia) in cattle was mentioned in Section I,5,b.

6. Zinc in Milk

The zinc content of milk varies with the species, stage of lactation, and Zn status of the diet. In all species the colostrum is three–four times richer in Zn than true milk (16, 61, 135, 277), and the level declines with advancing lactation (83, 194, 277). Rat's milk is substantially richer than that of other species studied. Thus Mutch and Hurley (194) reported levels approximating 16–18 μg Zn/ml in the milk of normal rats at the first day of lactation and 11–13 μg/ml at day 18 of lactation. The corresponding values for the milk of Zn-deficient rats were close to 12 and 6 μg Zn/ml for days 1 and 12, respectively. A decline in milk Zn concentrations from mildly Zn-deficient dairy cows has been observed, even where there was no reduction in milk yield (183). In cows Zn depleted on a diet containing only 6 mg Zn/kg the milk declined from 7.0 to 3.7 ppm after 3 weeks to 3.2 ppm by the sixth week and 2.3 ppm Zn after 19 weeks (277).

A high proportion of reported values for cow's, goat's, and ewe's milk lies between 3 and 5 μg Zn ml (10, 83, 135, 221). In an extensive study of market milk in U.S. cities Murthy *et al.* (193) obtained a range of 2.3–5.10, with a national average of 3.28 mg Zn/liter. Little of this zinc is associated with the fat and most of it with the casein (221). A mean of 6.9 μg Zn/g has been reported for the milk of sows taken at the thirty-fifth day of lactation (61).

The Zn concentration of human milk is appreciably lower than that of cow's milk, although individual variability is high. In a recent study of 25 women a range of 0.4–2.68 mg/liter was reported (232). In another study of 22 women a mean of 1.34 ± 0.94 mg/liter was obtained (192). The possible significance of the fact that human milk contains less Zn and more Cu than cow's milk and therefore a much lower Zn:Cu ratio was discussed in Chapter 3. The effect of subnormal and high dietary Zn intakes on the Zn level of human milk does not appear to have been specifically studied, but there is no doubt from work with other species that the level of zinc in milk reflects both low (83, 194) and high (10, 16, 61, 180) dietary Zn intakes.

7. Zinc in the Avian Egg

The total Zn content of eggs from hens consuming a good laying ration has been reported to average 762 ± 11.4 μg (55). This conforms well with a much earlier estimate 0.7 to 1.0 mg Zn (20). Most of the zinc is present in the yolk associated with the lipoprotein lipovitellin (314). Lower levels of zinc than those just given occur in eggs from hens on Zn-deficient diets.

II. ZINC METABOLISM

1. Absorption

In rats zinc is absorbed mainly from the duodenum, ileum, and jejunum, with very little being absorbed from the stomach or colon (164, 264, 323). In cattle about one-third of an oral dose of ^{65}Zn was apparently absorbed from the abomasum, with further absorption occurring throughout the small intestine (171). In calves Zn absorption also occurs throughout the small intestine, with the amount absorbed per unit of length being as great in the distal as in the proximal ends (98). Substantial Zn absorption from the proventriculus, as well as from the small intestine, occurs in chicks (172).

Homeostatic control of body zinc in accordance with needs is achieved, in part, through regulation of Zn absorption (48, 64, 87, 138, 174, 276, 343). For example, Schwarz and Kirchgessner (276) measured ^{65}Zn uptake into the blood, liver, and kidneys in Zn-depleted, pair-fed adequate-Zn, and in ad libitum, adequate-Zn fed rats. The proportions of the administered dose that were absorbed were 13.4, 4.8, and 7.4%, respectively. Furthermore, the Zn uptake by the intestinal wall of the Zn-deficient rats was four times that of the pair-fed and twice that of the ad libitum fed controls.

The mechanism of Zn absorption and its control has recently been illuminated by the work of Evans and his associates. They have described a low molecular weight Zn-binding factor in the intestinal lumen (86), intestinal mucosa (64), and pancreas of rats and pancreatic secretions from a dog (66). The uptake of ^{65}Zn by epithelial cells from everted intestinal rat segments was found to be markedly increased in the presence of this Zn-binding ligand from pancreatic secretions, and 30% of the ^{65}Zn in the epithelial cell was associated with the partially purified basolateral plasma membrane. When these membrances were incubated in a medium that contained Zn-free albumin, some 96% of the ^{65}Zn was transferred to the medium, while less than 30% was released to media that contained either no albumin or a 3:1 Zn–albumin complex. These findings led Evans *et al.* (66) to propose the following sequence for Zn absorption: "(a) the pancreas secretes a Zn-binding ligand into the intestinal lumen; (b) in the lumen Zn binds to the ligand; (c) complexed with a ligand, Zn is transported through the intestinal microvillus and into the epithelial cell; (d) in the epithelial cell Zn is transferred to binding sites on the basolateral plasma membrane; (e) metal-free albumin interacts with the plasma membrane and removes Zn from the receptor sites." The quantity of metal-free albumin available at the basolateral plasma membrane, it is contended, determines the amount of Zn removed from the intestinal epithelial cell and thus regulates the quantity of Zn entering the body. Whether this can be regarded as a fully acceptable hypothesis for the mechanism controlling Zn absorption will only be known as research proceeds.

Dietary Zn is known to be sequestered into mucosal cell Zn-binding proteins formed in response to Zn, thus inhibiting its transfer to serum albumin and permitting the excretion of sequestered Zn via the desquamation of mucosal epithelium (255). Copper and other metals inhibiting Zn absorption, as discussed later, do so at least in part by competing with Zn for binding sites on the albumin molecule. Competition for binding sites on the ligand or ligands from the pancreatic secretions in the intestinal lumen provides a further opportunity for these metals to inhibit Zn absorption.

The extent of Zn absorption varies with the level of zinc and with a range of other dietary components. Methfessel and Spencer (165) observed a marked fall in Zn absorption from the duodenum with increasing age in the rat, but the evidence for an effect of age or of species on Zn absorption is conflicting (see Becker and Hoekstra, 14). In a study of ^{65}Zn absorption in bovines, no evidence was found of a decline with age in lactating cows, compared with 2- and 6-month-old calves (293a).

Many of the factors that influence Zn absorption have been inferred from the way they influence signs of Zn deficiency or toxicity in the animal. High-Ca intakes potentiate the Zn deficiency syndrome or affect Zn utilization in several species (14), and they exert a protective action against Zn toxicity in pigs (119). These effects result largely from an interference by Ca with Zn absorption and may be mediated by phytic acid (151, 206). Conversely, low-Ca intakes can alleviate the effects of Zn deficiency (304). This is achieved by the release of skeletal Zn that occurs during bone resorption. Skeletal zinc is only released under conditions bringing about bone breakdown, including functional parathyroid glands, but once released is readily utilized for metabolic purposes (130, 304). Harland *et al.* (100) have shown that the Zn stored in the bones of young

Japanese quail in excess of requirement may be available for utilization during a subsequent period of Zn deprivation in a growing animal that has rapidly remodeling bones.

High inorganic P as well as high-Ca intakes aggravate Zn deficiency in rats, and their effect appears to be additive and independent (34). An independent action of phosphorus on Zn absorption remains to be clearly established. The chemical form in which the zinc is ingested also influences absorption, so that diets of similar Zn contents may differ in the amounts available to the animal. Thus Edwards (62) found zinc to be equally well absorbed as the oxide, carbonate, sulfate, or metal, whereas the Zn in sphalerite (mainly zinc oxide) and franklinite (oxides of Zn, Fe, and Mn) was largely unabsorbed judging by the capacity of these compounds to promote growth in chicks fed a Zn-deficient diet.

Different protein sources vary in their effects on the Zn requirements of rats, pigs, chicks, and poults, with diets containing protein of plant seed origin requiring higher levels of dietary Zn than those containing protein from animal sources. O'Dell and Savage (205) attributed such differences to the presence or absence of phytates which bind the zinc in a form from which it is not readily released and absorbed. The addition of phytic acid to casein diets can reduce Zn retention to that in soybean protein diets (73, 201), and Zn absorption from combinations with phytic acid is increased by autoclaving or by treatment with EDTA (140, 205). The presence of phytate is only one of the factors affecting Zn availability from oilseed meals (145). Various chelating agents can improve the availability of zinc to chicks and poults consuming soybean protein diets (198, 327). Furthermore, certain natural feeds such as casein and liver extract contain chelates which improve Zn absorption and utilization (278).

Vitamin D has been reported to increase Zn absorption by some workers (133, 341) but not by others (333). From the work of Becker and Hoekstra (13) with rats it seems that the increased Zn absorption attributed to vitamin D is not a direct effect of the vitamin but results from a homeostatic response to the increased need for Zn which accompanies stimulated skeletal calcification and growth. These workers have shown that bulk is a further dietary factor affecting Zn absorption and metabolism (14). Weanling rats were fed semipurified casein–cerelose diets containing graded levels of cellulose. Added cellulose decreased ^{65}Zn absorption, increased the rate of passage of the isotope through the tract, and decreased body ^{65}Zn retention. It seems likely, as the authors suggest, that by adding bulk to the diet, the rate of passage through the digestive tract is enhanced and the ^{65}Zn is present at the sites of absorption for a shorter time, resulting in decreased absorption and a consequent slower turnover of that which is absorbed.

Mention has already been made of the depressing effect of copper on Zn absorption. This was shown directly by Van Campen (322) using isolated

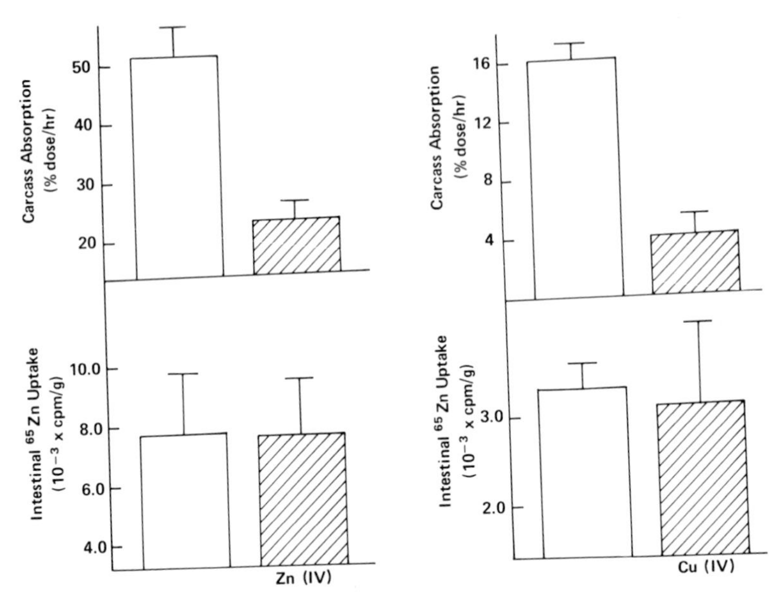

Fig. 7 Effect of intravenous Zn injections on ^{65}Zn uptake and absorption of Zn-deficient rats, and of intravenous Cu injections on ^{65}Zn uptake and absorption in normal rats. Each bar represents mean ± SD of four rats. (From Evans *et al.*, 66.)

duodenal segments of the rat. This worker had previously shown that high-Zn intakes depress Cu absorption (324). Conversely, Cu absorption is greatly increased in Zn deficiency (276). A mutual antagonism between Zn and Cu at the absorptive level is thus apparent. From the work of Evans *et al.* (66) it seems that the inhibition by copper of Zn absorption takes place at the site of transference from the mucosal cells to the plasma, since they demonstrated Cu-inhibited ^{65}Zn absorption in rats without affecting intestinal uptake of the isotope (Fig. 7). Metabolic interactions between zinc and cadmium, zinc and iron, and zinc and chromium also occur, as discussed in Chapters 2, 9, and 10. Such interactions are to be expected, as Hill and co-workers (113) have pointed out, among elements that share common chemical parameters and compete for common metabolic sites.

2. Excretion

Zinc leaves the body largely by way of the feces (63, 160, 177, 257, 288). Fecal zinc consists mostly of unabsorbed dietary Zn with a small amount of endogenous origin secreted into the small intestine. In the calf the pancreatic juice contributes about one-fourth of the total endogenous Zn loss (292). Small amounts are also secreted into the bile, cecum, and colon (164). Injected Zn is similarly excreted mostly in the feces (160, 177, 227). Radiozinc administered orally to steers was recovered 70% in the feces and 0.3% in the urine. When the zinc was given intravenously 20% appeared in the feces and 0.25% in the urine (67). This pattern of excretion was followed on normal diets and on diets high in zinc. Endogenous excretion of ^{65}Zn and stable Zn is significantly reduced in calves and goats on low-Zn diets, thus contributing to the homeostatic control of this element achieved primarily through increased absorption (175, 177).

The quantity of zinc excreted in the urine of healthy human adults is small (0.3–0.6 mg/day), compared with the 10–15 mg/day normally ingested (257, 288). The amounts so excreted do not vary greatly with dietary Zn levels and are not significantly increased following Zn injections. Cavell and Widdowson (42) found the urine of 10 one-week-old, breast-fed infants, nine of whom were in negative Zn balance, to average 1 μg Zn/ml, which is about five times the concentration of normal, adult urine.

Urinary Zn excretion is well above normal in nephrosis, postalcoholic hepatic cirrhosis, and hepatic porphyria (240, 298). Six patients with albuminuria averaged 2.1 mg Zn/day in their urine (range 1.0–3.8), whereas healthy individuals excreted only 0.3 mg Zn/day (160). These heavy losses can be explained by the inability of the nephrotic kidney to provide an effective barrier against the protein-bound Zn of the plasma. The pronounced zincuria of postalcoholic hepatic cirrhosis is less easily understood. Vallee and co-workers (321) found that sufferers from this disease excreted 1 mg Zn/day in the urine, compared with under 0.5 mg/day by noncirrhotic controls. Accumulative zincuria has also been observed following major operations and severe burns, its severity and duration being commensurate with the metabolic insult or injury incurred (110). Increased urinary Zn excretion occurs in total starvation in man (290) and during the administration of EDTA (230) and DTPA (diethylenetriaminepenta-acetic acid) (289).

Significant quantities of zinc can be lost in the sweat, especially in the tropics. Prasad *et al.* (242) found the sweat of normal individuals to average 1.15 ± 0.30 μg Zn/ml. Most of this was present in the aqueous phase, i.e., not associated with the cellular elements as occurs with iron. In Zn-deficient patients the mean Zn level in the sweat was reduced to 0.6 ± 0.27 μg/ml. On this basis a normal individual secreting 4 liters of sweat/day could lose 4 mg Zn/day and a Zn-deficient individual about 2 mg/day. In temperate climates the losses of zinc by this route for most individuals would be substantially smaller.

Menstrual losses of zinc are small and appear to be of little nutritional significance. If the Zn level of whole human blood is taken as 9 μg/ml and an average volume of menstrual flow as 50 ml, a loss of 450 μg Zn/period can be estimated. This represents only 15 μg Zn/day over a typical 30-day cycle, or one-thousandth of a normal 15 mg Zn/day dietary intake.

3. Intermediary Metabolism

The Zn absorbed from the intestine is carried to the liver in the portal plasma bound to transferrin (66a). In venous plasma the Zn is mostly bound to albumin and to a small extent to transferrin and α_2-macroglobulins. This Zn is incorporated at differing rates into different tissues which reveal varying rates of Zn turnover. Zinc uptake by the bones and the CNS is relatively slow and remains firmly bound for long periods. The Zn entering the hair is not available to the tissues and is only lost as the hair is shed (18, 80). The Zn in bones is also not normally readily available for metabolic use (130). The most rapid accumulation and turnover of retained zinc occurs in the pancreas, liver, kidney, and spleen (67, 101). In the rat radiozinc accumulates most rapidly in the pancreas and dorsolateral prostate (330). The zinc in these tissues, and the more slowly exchanging muscle and red cell Zn, constitute a soft tissue Zn pool composed of compartments of varying exchange rates. Further compartments, also with varying exchange rates, exist within the cells of particular tissues (12, 48).

Cotzias and co-workers (48, 49) showed that the tissues of mice, as well as various organelles of mouse liver cells, show distinct individuality relative to the time course of ^{65}Zn distribution after injection. The feeding of a Zn salt accelerated the loss of ^{65}Zn without affecting the characteristic partition of the isotope among either the organs or organelles. The specificity of the Zn pathway through the body was disturbed when stable Zn was injected and normal absorption bypassed. The sensitive homeostatic control mechanism of zinc metabolism acting at the sites of absorption and endogenous excretion therefore appears to operate only when the absorptive and excretory mechanisms are functioning together (49). Considerable homeostatic control, involving both a rapid increase in dietary Zn absorption and a decrease in endogenous fecal Zn excretion, has been demonstrated in young and lactating cows fed a moderately low-Zn, practical-type diet (174, 183).

The major organ involved in Zn metabolism is the liver. The liver cytosol of ruminants (27) and rats (43, 51) contains Zn-(and Cu)-binding components of differing molecular weight and lability, the amounts and proportions of which vary with the Zn status of the animal. Increases in liver Zn are associated with an increasing amount of Zn appearing in metallothionein or metallothionein-like forms. Zinc injections induce the synthesis *de novo* of these low molecular weight Zn-binding proteins (51, 255), and as the Zn content in the diet is increased to high levels almost the whole of the liver Zn increase can be accounted for as metallothionein (43). These findings have led Chen *et al.* (43) to suggest that the metabolic role of metallothionein, in addition to its role in cellular metal detoxication mechanisms, is to serve as a storage protein for zinc, analogous to ferritin for iron. Bremner and Davies (28) have similarly proposed that the labile metal-binding proteins may function as a temporary store of Zn (and/or Cu) prior to its utilization in essential functions. Incorporation of Zn into Zn-binding proteins in the mucosal cells in response to Zn loading of rats has also been demonstrated (255), thus preventing its transfer to serum albumin and constituting a further part of the mechanisms responsible for Zn homeostasis.

Injected radiozinc combines initially with the plasma proteins to form a plasma Zn pool, and with the cellular components of the blood more slowly. Clearance of plasma Zn, but not of erythrocyre Zn, is rapid to the soft tissues of the body and the feces via the small intestine (227). ^{65}Zn is also rapidly transported from the plasma through the placenta, so that 2 hr after injection into pregnant mice, a high content is present in fetal bones and liver (18).

III. ZINC DEFICIENCY AND FUNCTIONS

1. Effects on Growth, Appetite, and Taste

Growth retardation was observed in the original demonstration of Zn deficiency in rats (311) and has been a feature of this deficiency in all subsequent studies with all species. It probably arises primarily from a decreased activity of thymidine kinase and hence impaired DNA synthesis and cell division, as discussed in Section III,3. Lambs and calves fed a severely Zn-deficient diet (1.2 ppm Zn) ceased weight gain abruptly, and growth arrest occurred within 2 weeks (186). In pregnant rats a maternal dietary Zn deficiency severely impairs fetal growth (129), while such a diet fed during lactation impairs growth in the suckling pups. Nutritional dwarfism and hypogonadism are the main clinical manifestations of the conditioned Zn deficiency first demonstrated by Prasad and co-workers in young males in Iran (238) and Egypt (239). A clearly defined stimulus to growth is apparent when adequate Zn supplements are given to these individuals (258). Significant growth increments in male infants given Zn supplements to certain infant milk formulas have also been reported (331).

The growth inhibition of Zn deficiency results partly from impaired appetite, i.e., reduced food consumption, and partly from impaired food utilization (169, 180a, 241, 286). Chesters and Quarterman (45) found the voluntary food intake of Zn-deficient rats to fall to 70% of that of the controls, with marked day to day variation in intake and a cyclical pattern of food consumption. The Zn-deficient rats responded to a Zn-supplemented diet within 1–2 hr by an increased food intake. The apparent digestibility of the food is unaffected in Zn deficiency, at least in ruminants (182, 286). The fecal excretion of N and S is also unaffected but the urinary excretion of these elements was found to be greatly elevated in Zn-deficient lambs compared with pair-fed controls (286) (Table 29). Increased urinary excretion of total N, urea, and uric acid, but not of

TABLE 29
Effects of Zinc Deficiency on Feed, Digestibility, and Nitrogen and Sulfur Balances in Ram Lambs[a]

	Zn-deficient group[b]	Pair-fed controls[b]
Dry matter digestibility (%)	64.5 ± 1.2	66.8 ± 0.6
Nitrogen in feces (g/day)	4.6 ± 0.2	4.5 ± 0.2
Sulfur in feces (g/day)	0.4 ± 0.04	0.4 ± 0.04
Nitrogen in urine (g/day)	5.8 ± 0.30	3.7 ± 0.21[c]
Sulfur in urine (g/day)	0.63 ± 0.06	0.38 ± 0.02[c]
Nitrogen balance (g/day)	+2.0 ± 0.08	+4.2 ± 0.16[c]
Sulfur balance (g/day)	+0.25 ± 0.02	+0.49 ± 0.04[c]

[a]From Somers and Underwood (286).
[b]Each group consisted of four animals.
[c]Denotes that means were significantly different at the 0.01 level.

creatinine, has been demonstrated in Zn-deficient rats (121). These rats also had increased activities of liver tryptophane pyrrolase and arginase, indicating increased protein catabolism.

A physiological role for zinc in normal taste sensation was first established by Henkin and associates (103, 104, 108), together with the further important observation that the hypogeusia (loss of taste acuity) and dysgeusia (disordered taste or pica) that commonly occur in adults can respond to oral Zn therapy. Depletion of total body Zn by oral administration of histidine also produced anorexia and hypogeusia in man and was reversed by Zn administration (105). Hambidge *et al.* (92) discovered that poor growth and appetite, together with hypogeusia, in young children in Denver was associated with subnormal hair Zn levels (< 70 ppm). Supplementing their diet with small amounts of zinc normalized the taste acuity, increased the hair Zn levels, and improved growth. Subsequently Hambidge and Silverman (95) reported a case of a 2-year-old girl with a history of pica, manifested by "metal-eating." A diagnosis of Zn deficiency was suggested by poor appetite, subnormal growth, and low hair zinc. Three days after starting oral Zn therapy (10 mg $ZnSO_4 \cdot 7H_2O$/day) there was no further evidence of pica. Further evidence of an effect of Zn deficiency on taste has been obtained by McConnell and Henkin (162) with rats. Salt preference was shown to be significantly greater in Zn-deficient weanling rats than in pair-fed or ad libitum Zn-supplemented controls.

2. Zinc and Keratogenesis

Alopecia and gross skin lesions were observed in the early investigations of Zn deficiency in rats and mice. Histological studies disclosed a condition of parakeratosis, i.e., a thickening or hyperkeratinization, with failure of complete nuclear degeneration, of the epithelial cells of the skin and esophagus (72). In more severe Zn deficiency scaling and cracking of the paws with deep fissures develop, in addition to loss of hair and dermatitis (73). In pigs, parakeratosis mostly occurs around the eyes and mouth and on the scrotum and lower parts of the legs. A similar distribution occurs in Zn-deficient ruminants, with the legs becoming tender, easily injured, and often raw and bleeding (21, 316). In the Zn-deficient squirrel monkey (*Saimiri sciureus*) parakeratosis of the tongue has been observed in addition to alopecia (157). The healing response of the parakeratotic lesions to supplemental Zn is rapid and dramatic.

The influence of Zn deficiency on keratogenesis is particularly evident in sheep through changes in the wool and horns. In horned lambs the normal ring structure disappears from new horn growths, which are ultimately shed leaving soft spongy outgrowths that continually hemorrhage (186, 316). In breeds which show no horn growth, prolonged Zn depletion leads to the development of keratinous outgrowths or "buds" in the positions occupied by horns in other breeds (186). Changes in the structure of the hooves can also occur and the effects of Zn deficiency on wool growth in lambs are striking. The wool fibers lose their crimp and become thin and loose, and are readily shed. Sometimes the whole fleece is shed and no further wool growth occurs until additional zinc is supplied which results in immediate regrowth (316). Posthitis and vulvitis have also been observed in experimentally induced Zn deficiency in lambs, associated with enlargement of the sebaceous glands followed by an increased secretion of sebum (52).

In the Zn-deficient chick (203, 204), poult (141), and pheasant (279) feathering is poor and abnormal and dermatitis is usual. In the Japanese quail feathering is severely affected (75) due to a degeneration of the feather follicles resulting from hyperkeratosis (203). Involvement of zinc in keratogenesis is further evident from the gross disturbances of the integument, as well as the skeleton, observed in chick embryos from the eggs of severely Zn-deficient hens (134).

An association between chronic skin ulceration from different causes in man and subnormal plasma or serum Zn concentrations has been observed in several studies. For example Halsted and Smith (89) reported a mean serum Zn of 58 ± 15 μg/100 ml in eight patients with indolent ulcers, compared with 96 ± 12 in controls; Hallbook and Lanner (88) observed a more rapid healing of chronic leg ulcers in patients with serum Zn levels higher than 110 μg/100 ml initially or after oral zinc therapy than in those with a lower serum Zn level; and Burr (33) found fasting plasma Zn concentrations to be low in 20 out of 23 paraplegic patients with pressure sores.

Acrodermatitis enteropathica, a hereditary disease appearing in early infancy characterized by pustular and eczematoid skin lesions, alopecia, and diarrhea, is due to aberrant Zn metabolism and responds to Zn therapy (190, 196). Without treatment a relentless progression of severe malnutrition, poor growth and development, intercurrent infections, and death within 3 years is usual. With adequate oral Zn treatment clinical remission is rapid and complete. Plasma Zn, serum alkaline phosphatase, and urinary Zn excretion rise quickly to normal levels.

Facial eczema, a disease of sheep and cattle occurring in New Zealand on pastures infected with the fungus *Pithomyces chartarum* and producing the toxin sporidesmin, shows some clinical response to supplemental zinc at high doses (256). This important finding awaits confirmation and further biochemical investigation. The interdigital skin lesions of infectious pododermatitis in male cattle have also been shown to respond to zinc therapy (24, 53), following a suggestion by Bonomi (23) that a high incidence of infectious pododermatitis ("zoppina") in cattle in some parts of Italy might be related to a low-Zn status of the animals. Rapid and complete repair of severe interdigital lesions of infectious pododermatitis was obtained by Demertzis and Mills (53) in 11 of 12 young bulls given Zn supplements.

Zinc-related changes in the metabolism of skin acid mucopolysaccharides (AMP's) occur in pigs (308). No differences in total or sulfated AMP's in skin from Zn-deficient and pair-fed pigs were observed, but deficiency caused a significant increase in hyaluronic acid. Hsu and Anthony (120) similarly found no alteration in sulfated AMP's of Zn-deficient rat skin by ^{35}S incorporation. However, reduced collagen synthesis and apparent alterations in collagen cross-linking in Zn-deficient rat skin have been reported (161).

3. Zinc and Reproduction

Spermatogenesis and the development of the primary and secondary sex organs in the male, and all phases of the reproductive process in the female from estrus to parturition and lactation, can be adversely affected in Zn deficiency.

Atrophic seminiferous tubules were observed in the earliest study of the histopathology of Zn deficiency in the rat (72). Mawson and co-workers (159, 166) later reported retarded development of the testes, epididymis, prostate, and pituitary glands, with atrophy of the testicular germinal epithelium. Rapid growth and development of the dorsolateral prostate was induced in the rat by gonadotropin or testosterone treatment, without increasing the Zn concentration in this gland to its normal high level (167). The impaired development of the secondary sex glands appeared to be secondary to the inanition of Zn deficiency, which could result in a reduced gonadotropin output and consequent fall in androgen production. The possibility that cadmium played some part in these experiments cannot entirely be excluded since this element was included in a mineral supplement administered orally twice weekly. The testicular atrophy and failure of spermatogenesis, on the other hand, is due directly to lack of zinc.

The availability of sufficient Zn for incorporation of high amounts into sperm during the final stage of maturation is essential for the maintenance of spermatogenesis and survival of the germinal epithelium (167, 316).

Impaired development and functioning of the male gonads are apparent in other species. Hypogonadism, with suppression of the secondary sexual characteristics, is a conspicuous feature of the conditioned Zn deficiency observed in young men and young women in parts of Iran and Egypt (238, 239, 258). Hypogonadism also occurs in Zn-deficient bull calves (175, 233), kids (184), and ram lambs (316). In the experiment with lambs testicular growth was greatly impaired and spermatogenesis ceased within 20–24 weeks in a diet containing 2.4 ppm Zn, whereas no such effects were observed in comparable pair-fed lambs on an intake of 32.4 ppm Zn (316). The body growth and food consumption of the two groups of lambs receiving unrestricted diets containing 17.4 and 32.4 ppm Zn were similar, but testicular growth and sperm production were significantly greater in the rams receiving the larger Zn supplement (316). These differences were further reflected in the histological ratings given to the testes of the animals from the different treatments. The fructose concentration of the seminal plasma, which provides an index of testosterone production by the gonads (158), was similar for all dietary treatments. This suggests that testosterone output is not impaired in Zn-deficient ram lambs, as was suggested for rats (167). Remission of all signs of Zn deficiency and recovery of testicular size, structure, and sperm production were achieved during a Zn-repletion period lasting 20 weeks, indicating that the tissues had not been permanently damaged by the severe Zn deficiency imposed. Complete recovery was similarly obtained by Pitts and co-workers (233) with Zn-deficient calves.

The effects of Zn deficiency in the female depend on the severity, timing, and duration of the deficiency. When Hurley and Swenerton (129) fed a nearly Zn-free diet to female rats from weaning to maturity the animals made no growth and displayed severe disruption of the estrous cycles. In most cases no mating took place and the animals were infertile. With similar rats fed a marginally Zn-deficient diet (9 ppm) from weaning to maturity and then mated, the estrous cycles were not disrupted and mating took place normally. When these rats were given the severely Zn-deficient diet during pregnancy 54% of the implantation sites were resorbed and 99% of the total sites were affected. Of the young living at term, 98% exhibited gross congenital malformations involving numerous organ systems. Similar results were obtained when the rats had received a Zn-adequate diet prior to breeding (127). No congenital abnormalities and no reduction in the Zn content of the fetuses occurred in an earlier study in which the Zn deficiency was less severe (131). However, female rats exposed to only a transitory severe gestational Zn deficiency between days 6 and 14 of pregnancy gave birth to offspring reduced in weight and with a high incidence of malformations. Many of the young were stillborn and survival to weaning was poor (126). Dreosti and Hurley (56) have since provided evidence that the teratogenic effects of Zn deficiency in rats probably arise from impaired activity of fetal thymidine kinase during embryonic morphogenesis, and Hurley and Shrader (128) have reported that dietary Zn deficiency during the first few days of pregnancy in rats results in abnormal cleavage and blastulation in preimplantation eggs as early as day 3 of gestation. A study of the obstetric histories of patients with acrodermatitis enteropathica, due to the inherited defect in Zn metabolism discussed in the previous section, provides strong evidence that the human fetus is also susceptible to teratogenic effects from Zn deficiency (94).

The poor survival of young born to females fed a Zn-deficient diet for a short period during pregnancy could not be due to poor postnatal Zn nutriture because both the concentrations of zinc in the milk and plasma of mothers or young after birth were unaffected. The presence of congenital abnormalities and failure to suckle due to weakness at birth were thought to be the factors involved. Mutch and Hurley (194) subsequently showed that Zn deficiency imposed on rats during lactation rapidly reduced plasma Zn levels and caused an impairment in milk production which was specifically due to the lack of zinc rather than to inanition. The Zn level in the milk was also reduced so that the pups received only half the amount of zinc that the control pups received on the basis of body weight and became Zn deficient, as evidenced by reduced plasma Zn levels, impaired growth, and increased mortality.

Similar effects of severe Zn deficiency on mating and fetal development in rats have been obtained by Apgar (6), who showed that such a deficiency imposed on normal adult rats from the first day of gestation can markedly affect maternal behavior before and after parturition (5). The Zn-deficient females delivered their litters with extreme difficulty, suffered excessive bleeding, and failed to consume afterbirths or to prepare a nest site. Institution of a low-Zn diet as late as day 15 of pregnancy can also result in difficult parturition (7). Even more remarkably, administration of zinc by day 19 of pregnancy can protect a rat that has become Zn-deficient during pregnancy from severe stress at parturition. Administration as late as day 21 may protect some, though not all, of the females. The mechanism of this protective action and the role of zinc in parturition remain to be defined.

Comparable studies of severe Zn deficiency do not appear to have been carried out with female farm animals other than poultry. Decreased hatchability of eggs, grossly impaired development, and high mortality in chick embryos from the eggs of Zn-deficient hens have been reported (134). Suboptimal dietary levels of Zn fed to sows were shown to reduce the size of litters and the Zn content of some of the tissues of the young, but no abnormalities in fetal development or maternal behavior were observed (115).

4. Zinc and Skeletal Development

Skeletal abnormalities are a prominent feature of Zn deficiency in growing birds. This has been demonstrated in chicks, poults, pheasants, and quail (75, 141, 203, 204, 279, 346). The long bones are shortened and thickened in proportion to the degree of Zn deficiency (347). Changes and disproportions occur in other bones, giving rise to a perosis histologically similar to that of Mn deficiency (203, 204, 346). Gross skeletal and other deformities including agenesis of the limbs, dorsal curvature of the spine, and shortened and fused vertebrae develop in chick embryos from the eggs of hens fed severely Zn-deficient diets (22, 134).

Leg defects do not develop in Zn-deficient chicks on all types of diets, even when other signs of the deficiency are apparent (197). These "perosis-like" or "arthritis-like" abnormalities can be alleviated by histamine, histidine, and various antiarthritic agents without affecting the other manifestations of deficiency (198). Supplementary zinc prevents both the leg disorder and the other signs of Zn deficiency. The nature of the interaction between Zn and the various agents used remains obscure, although it is known that they do not act by increasing the availability of dietary zinc.

Bowing of the hind legs and stiffness of the joints in Zn-deficient calves, reversible by Zn feeding (173), and skeletal malformations in the fetuses from Zn-deficient rats (129) have been observed. A reduction in the size and strength of the femur in Zn-deficient baby pigs has been reported (169), but comparisons with pair-fed controls indicated that these changes were due to the reduced food intake. Bone growth was affected in direct proportion to the effect of Zn deficiency on body growth and maximum strength of the femur to bone size. In a later study with weanling pigs the reduced skeletal growth of Zn deficiency was most apparent in low activity of the epiphyseal growth plate and at other points of osteoblastic predominance (200).

The mechanism of action of zinc in bone formation is still not fully understood. Decreased osteoblastic activity in the bony collar of the long bones occurs in Zn-deficient chicks, together with a reduction in chondrogenesis associated with an increase in the amount of cartilage matrix (203, 346). Bone alkaline phosphatase activity is invariably reduced in Zn deficiency (294), but the significance of this is not clear. Zn-deficient chicks and rats develop abnormalities in the epiphyseal plate region of growing bones. In the former species the chondrocytes that are near blood vessels appear normal while those cells remote from blood vessels show cellular changes, are shaped differently, are surrounded by more extracellular matrix, and do not stain normally for alkaline phosphatase activity (339). As Westmoreland (339) has stated, "these changes may affect the normal maturation and degeneration of these cells which may in turn have an effect upon calcification. Other cells remote from blood vessels (epithelium of skin and oesophagus) also differentiate abnormally in Zn-deficient animals."

5. Zinc and Wound Healing

Numerous studies of the effect of Zn on wound healing have produced conflicting results, the reasons for which are not understood. The first indications of a role for Zn in wound healing came from the studies of Pories and associates, who demonstrated a significantly increased rate of wound healing, compared with unmedicated controls, when zinc sulfate, at the rate of 50 mg Zn three times daily, was added to the diet of young men following surgery for pilonoidal sinuses (236). Several workers subsequently demonstrated improved wound healing from oral zinc therapy in man following severe burns and major operations associated with high posttraumatic urinary Zn losses (37, 110, 271), and in patients with leg (88, 132) and gastric (76) ulcers and bedsores (47). Such responses can occur with or without the preexistence of low serum Zn levels. By contrast, other workers have observed no such beneficial effects on the rate of closure of granulating wounds (11, 195).

In Zn-deficient animals the evidence seems unequivocal that the rate of wound healing is impaired and accelerated by supplemental zinc. This has been demonstrated in calves (178) and rats (249, 268). The position with normally nourished animals is conflicting. Sandstead *et al.* (267) found that systemically administered Zn did not affect the strength of sutured incision wounds in normal rats, and Norman *et al.* (199) similarly were unable to demonstrate any beneficial effect from Zn in the development of tensile strength in incised wounds, or in the rate of healing of granulated wounds, in rats and guinea pigs fed standard laboratory diets. The supplementary Zn was ineffective whether administered orally, parenterally, or topically. By contrast, Lavy (144) reported a 35–36% increase in bursting strength of incision wounds 13 days postoperative from the oral administration of 40 mg zinc sulfate/rat/day to normal rats receiving a standard diet. This increase corresponded to a gain of about 13 days in healing time. On the sixth day after the operation and on the same daily Zn dose there was no significant increase of wound-bursting strength, suggesting that different results can be obtained depending on the time of observation after trauma. However, in the experiments of Norman *et al.* (199) no effect was observed even when the Zn was given for 14 days before wounding.

The enhancement of wound healing by Zn may stem from a heightened metabolic demand for this element for collagen synthesis in the process of tissue repair, with an increase in collagen synthesis and cross-linking explaining gains in wound tensile strength (161), but direct evidence for this is lacking. Furthermore, Zn responsive differences in tissue repair could be related to differences in the rate of cell division and DNA production in rapidly regenerating tissue, since Prasad and Oberleas (237) have shown that depressed activity of thymidine kinase, necessary for DNA synthesis and cell division, is an early metabolic defect of Zn deficiency.

6. Zinc and Atherosclerosis

Indications have been obtained that Zn therapy can be beneficial in some cases of atherosclerosis. Pories *et al.* (235) gave zinc sulfate orally to 13 patients with advanced vascular disease for 29 months. Twelve of these 13 showed marked clinical improvement, nine returned to normal activity, and seven had a return of previously absent pulses. In another study, 14 patients with inoperable symptomatic atherosclerosis, and two who had disabling vasospastic disease, were given zinc sulfate for 3–11 months (111). Six of these were excluded because of changes in their habits or diets, six of the remaining 10 experienced clinical improvement ascribed to the Zn therapy, and four experienced no improvement, two of whom had rapid progression of their disease.

The mode of action of zinc in atherosclerosis is unknown. Hair and plasma Zn levels are usually subnormal in atherosclerosis and myocardial infarctions (111, 328, 329), and the aortic wall has an active turnover of ^{65}Zn (296). This becomes even more active when the arterial wall is injured, in a manner similar to that seen in skin and muscle. Since atherosclerosis is thought to begin with some form of trauma it may be that it is, in part, an expression of inadequate arterial repair (235).

7. Brain Development and Behavior

Zinc deficiency during the critical period for brain growth permanently affects brain function. When this deficiency is imposed throughout the latter third of pregnancy, brain size is decreased, there is a reduced total brain cell number, and the cytoplasmic nuclear ratio is increased, implying an impairment of cell division in the brain during the critical period of macroneuronal proliferation (163). In adult life male rats so treated display impaired shock avoidance (266), and female rats are significantly more aggressive at a high level of shock than adult female rats whose dams were Zn sufficient during pregnancy (87a). When the Zn deficiency is imposed from birth to 21 days of age, brain size is diminished, brain DNA, RNA, and protein concentrations are reduced, and impaired maze acquisition ability is evident in such animals when adults (266). Retarded brain maturation, as indicated by reduced total cerebellar lipid concentration, is evident in Zn-deficient suckling rats (269), and a markedly lower rate of protein synthesis in the brain of Zn-deficient weanling rats has also been demonstrated (208). A delay in myelination is further implied from the observation that Zn deprivation during the gestational–lactational period produces a significant reduction in rat brain 2′,3′-cyclic nucleotide 3′-phosphohydrolase (244).

Prior to the above studies Caldwell and co-workers (35) observed a significantly inferior learning ability, as measured by water maze and platform avoidance conditioning tests, in the surviving offspring of mildly Zn-deficient mothers, compared with similar rats from Zn-supplemented mothers. These effects of Zn deficiency were subsequently confirmed and extended (36).

8. Protein and Nucleic Acid Metabolism

Zinc is involved primarily in nucleic acid and protein metabolism and hence in the fundamental processes of cell replication. Mention has already been made of the role of zinc in collagen synthesis and in DNA, RNA, and protein formation in the brain. Impaired DNA synthesis in the liver of Zn-deficient rats has been demonstrated in several studies (31, 77, 280). The total protein and RNA contents of the testes of Zn-deficient rats are reduced (157), and the testes of more severely Zn-deficient rats contain lower concentrations of Zn, RNA, DNA, and protein and higher nonprotein N levels and ribonuclease activity than either the restricted-fed or unrestricted-fed controls (287) (Table 30). Studies of the time course of RNA, DNA, and protein synthesis in phytohemaglutinin-stimulated lymphocytes deprived of Zn indicate an involvement of Zn in DNA synthesis, possibly at the time of gene activation (44). Zinc was found to be of prime importance in reversing the inhibition by EDTA of the expression of the genetic potential of these cells to synthesize the enzymes required for DNA synthesis and cell division. In other words the process of "gene activation" requires zinc. Evidence has also been obtained of a continuing Zn requirement for DNA synthesis in cultured chick embryo and mammalian cells (262), providing further evidence of a role for Zn in the regulation of cell multiplication.

The utilization of amino acids in the synthesis of protein is impaired in Zn deficiency (123). This could be due to an abnormality in the synthesis or the degradation of ribonucleic acid, or both. Support for this interpretation comes from the demonstration of increased RNase activity in the testes of Zn-deficient rats (287) and from the finding that liver DNA-dependent RNA polymerase activity is decreased in Zn-deficient suckling rats (305) (Table 31). Evidence had previously been obtained that this enzyme from *E. coli* is a "zinc metalloenzyme" (280). The earliest known metabolic defect in Zn deficiency is a decreased activity of thymidine kinase (TK) (237), the enzyme leading to thymidine triphosphate formation which is a prerequisite for DNA synthesis and thereby cell division.

Many years ago the proteolytic activity of the pancreas was found to be

TABLE 30
Composition of Testes from Zinc-Deficient Rats and Two Zinc-Supplemented Control Groups[a]

Group	Testes Zn (μg/g)	Testes RNA (mg/g)	Testes DNA (mg/g)	Testes total N (mg/g)	Testes protein N (mg/g)	Testes nonprotein N[b] (mg/g)	Testes RNase activity[c]
Zn-deficient	21.0	3.2	3.0	14.0	8.3	5.7	0.21
	±0.83	±0.16	±0.15	±0.40	±0.32	±0.13	±0.009
Pair-fed controls	32.5	5.0	4.5	14.5	10.8	3.8	0.12
	±1.08	±0.14	±0.24	±0.36	±0.28	±0.10	±0.005
Ad lib-fed controls	36.5	5.3	4.7	15.0	11.6	3.4	0.13
	±1.68	±0.19	±0.13	±0.41	±0.38	±0.21	±0.003
LSD 0.05	3.47	0.47	0.51	1.11	0.93	0.43	0.02
0.01	4.67	0.64	0.68	1.49	1.25	0.59	0.03

[a]From Somers and Underwood (287).
[b]Nonprotein nitrogen calculated by difference (total nonprotein N).
[c]Absorbancy at 260 mμ due to nucleotide production from RNA by 100 mg of testes.

TABLE 31
Effect of Zinc Deficiency on the Activity of Liver Nuclear, DNA-Dependent, RNA Polymerase of Suckling Rats[a]

	^{14}C per mg of deoxyribose (count mm^{-1} mg^{-1} ml^{-1})		
		Controls	
Age (days)	Zinc deficient	Pair-fed	Free access to food
2	83.0[b]	81.5	78.6
6	85.0	87.0	94.8
10	91.2	111.3[b]	–
12	80.4	114.4[b]	112.8
14	70.0	112.1	115.4[c]
16	60.4	107.8	107.4

[a]From Terhune and Sandstead (305).
[b]Average of two sets of pups.
[c]Average of three sets of pups.

reduced in the Zn-deficient rat (117). Pancreatic carboxypeptidase was later shown to be a zinc metalloenzyme (319). Carboxypeptidase activity was then found to be appreciably reduced in Zn-deficient rats and to return rapidly to normal with Zn therapy (122, 187, 261). No evidence was obtained that this reduction had limited protein digestion or absorption, so that this particular defect is unlikely to be directly responsible for the growth arrest and impaired food utilization of Zn deficiency.

9. Carbohydrate and Lipid Metabolism

Changes in carbohydrate metabolism and adipose tissue metabolism occur in Zn-deficient rats (247). Quarterman and Florence (246) have pointed out that these changes are the sort that could be affected by differences in feeding pattern between the Zn-deficient and the pair-fed control. Some effects of feeding pattern and previous day's feed intake on glucose tolerance, plasma free fatty acid (FFA) concentration, and plasma insulin levels were demonstrated (69). When these differences in feeding pattern were taken into account the glucose tolerance of Zn-deficient rats was no different from that of rats given a Zn-supplemented diet. The plasma insulin concentration after a dose of glucose tended to be lower in the Zn-deficient animals, and a transient rise in plasma FFA concentration was observed during the development of the Zn deficiency (246). It was suggested that some disturbance in adipose tissue metabolism is occurring, resulting in a degradation of the triglyceride reserves and a consequent increase in blood FFA concentrations.

Somewhat different findings have been reported by Huber and Gershoff (125) in their studies of rats at three dietary Zn levels (1, 20, and 1200 ppm), together with a pair-fed control group (20 ppm). The insulin content of the pancreas was not altered by high- or low-Zn diets but total serum insulinlike activity was reduced in the Zn-deficient animals compared with the pair-fed controls. *In vitro* studies of the release of pancreatic insulin during short incubation periods with glucose as a stimulant indicated that significantly less immunoreactive insulin, as well as insulinlike activity, was released by pancreata from Zn-deficient rats compared with both pair-fed and unrestricted-fed animals. The mechanism by which Zn affects the *in vitro* release of pancreatic insulin during glucose stimulation is unknown.

10. Changes in the Blood in Zinc Deficiency

In addition to the reduction in serum or plasma Zn level, already considered, many other changes in the plasma and formed elements of blood arise in Zn deficiency. The hematocrit values or erythrocyte numbers are above normal in the Zn-deficient rat (260), baby pig (169), and Japanese quail (75). In the Zn-deficient baby pig there is a relative and absolute reduction in lymphocytes (169). In male weanling rats fed a Zn-low diet for 14 days the percentage of polymorphonuclear leukocytes (P) markedly increased and of lymphocytes (L) decreased (57). Similar changes in the P:L ratio occurred in adult female rats fed a Zn-low diet during pregnancy but not in nonpregnant rats fed a similar diet for 3 weeks (57). The physiological significance of these white cell changes and their relationship to Zn deficiency remain obscure.

A decline in plasma protein levels has been observed in Zn deficiency in some investigations (117, 248), but not in others (75, 169). The electrophoretic patterns of these proteins are greatly altered in the Zn-deficient pig (169) and, after fasting, in quail (75). The plasma levels of the specific retinol-binding protein are significantly reduced in the Zn-deficient rat (283).*

Alkaline phosphatase activity is reduced in the blood serum of Zn-deficient rats (153, 260), pigs (169), and calves (180a). Comparisons with pair-fed controls indicate that this reduction is due to the Zn deficiency and not merely to reduced food intake (153, 169). Blood levels of malic and lactic dehydrogenases remain within normal levels in Zn deficiency (260). Subnormal levels of carbonic anhydrase have been reported in the blood of Zn-deficient calves (173) and rats (260), and in Middle East dwarfs suffering from conditioned Zn deficiency (270). Kirchgessner and his colleagues (260) found no difference between the carbonic anhydrase activity of the whole blood of severely Zn-

*Author's note: It is not clear whether the reduction in plasma vitamin A concentrations that occurs in Zn deficiency is a primary effect or a secondary effect from the reduced food intake [Smith *et al., J. Nutr.* **106,** 569 (1976)].

deficient rats and that of controls, but the carbonic anhydrase activity per unit of erythrocytes was greatly reduced. The increase in the erythrocyte numbers per unit of blood volume in Zn deficiency was sufficient to mask this effect.

11. Enzyme Activities in the Tissues

The levels and activities of zinc metalloenzymes and Zn-dependent enzymes in the tissues of Zn-deficient animals have been extensively studied. Histochemical determinations carried out by Prasad and co-workers (241) disclosed reduced activities of several enzymes, accompanied by reduced Zn levels, in the testes, bones, esophagus, and kidneys of Zn-deficient rats, compared with those of restricted-fed controls. In the testes, lactic dehydrogenase (LDH), malic dehydrogenase (MDH), alcohol dehydrogenase (ADH), and NADH (reduced nicotinamide adenine dinucleotide) diaphorase; in the bones, LDH, MDH, ADH, and alkaline phosphatase; in the esophagus, MDH, ADH, and NADH diaphorase; and in the kidneys, MDH and alkaline phosphatase were decreased. In a Zn-repleted group of rats the activities of these enzymes and the level of Zn increased in all the tissues examined. Similar results were obtained with baby pigs (243), although liver ADH and liver GDH (glutamate dehydrogenase) activities are unaffected in the Zn-deficient baby pig (169). The activities of LDH and ADH remained unaltered in the muscle tissue of Zn-deficient young rats, but MDH lost much of its activity in this tissue when an extreme state of Zn depletion was reached (261). A rapid and substantial decline also occurs in the alkaline phosphatase activity of the bones of young rats fed a Zn-deficient diet, with a restoration to nearly normal levels after 8 days of Zn repletion (261).

Huber and Gershoff (124) studied the effects of alterations in dietary Zn on the tissue levels of four zinc metalloenzymes (carbonic anhydrase, LDH, GDH, and alkaline phosphatase) in 4-week-old rats over a 2–4 week period. Zinc deficiency per se was found to significantly depress carbonic anhydrase activity in the stomach and pituitary; LDH in heart, kidney, and gastrocnemius muscle; and alkaline phosphatase in the duodenum, stomach, and serum. Mitochondrial GDH was not affected by high or low dietary Zn intakes in any tissues examined. In a further study with rats with gestational–lactational Zn deprivation the concentration of brain GDH was unchanged and significant reduction in brain 2′,3′-cyclic nucleotide 3′-phosphohydrolase (CNP) was found, implying a delay in myelination. Zinc restriction only in the lactational period resulted in smaller brain changes, with no differences in the levels of the cerebellar enzymes superoxide dismutase, GDH, or CNP between the deficient pups and those from dams pair fed a Zn-adequate diet (244). A reduction in the activity of the Zn-dependent enzyme thymidine kinase in rat connective tissue (237) and rat fetuses (56) as the earliest known and probably primary metabolic defect in Zn deficiency has already been mentioned.

The relationship of many of the changes in tissue enzyme activities to the clinical manifestations of Zn deficiency in the animal is far from clear. The activity of some enzymes will be affected more than others under conditions of limited Zn supply because of differences in binding affinities for the available metal. For example pancreatic carboxypeptidase, in which the Zn is not as firmly bound to the apoenzyme as it is in liver ADH, is reduced in activity under Zn deficiency conditions in which liver ADH activity remains largely unaffected. Furthermore, some enzymes are present at levels which appear to be much in excess of normal requirements. Under these circumstances a depletion of activity induced by a dietary deficiency would not necessarily limit or disrupt normal metabolic processes.

IV. ZINC REQUIREMENTS

Minimum zinc requirements vary with the age and functional activities of the animal and with the composition of the diet, particularly the amounts and proportions of the many factors, organic and inorganic, which effect zinc absorption and utilization. Zinc requirements are also influenced by ambient temperatures, where these cause profuse sweating and consequent large losses of Zn in the sweat, and by parasitic infestation with its attendant blood, and hence Zn, losses. The criteria of adequacy employed can also be important. For example, the requirements following trauma and disease are higher than "normal."

1. Laboratory Species

The minimum Zn requirements for growing rats have been given as 12 ppm on diets with casein or egg white as the protein source and 18 ppm on soybean protein diets, with their higher phytate content (74). The much higher long-term Zn requirements for males are presented in the next paragraph. Since raising the Ca level of the diets from 0.8 to 1.6% depressed weight gains, the requirements must be increased beyond the 12–18 ppm just stated as minimum, where Ca intakes are substantially above normal. In later experiments carried out by Luecke *et al.* (154) the minimum requirement of the rat for growth on an egg white diet was close to 11 ppm Zn, although higher serum Zn levels occurred at an intake of 13 ppm. Kirchgessner and Pallauf (136, 216) obtained maximum growth at 8 ppm dietary zinc and reported that at least 12 mg Zn/kg diet were necessary for optimal Zn contents in serum and liver of weanling rats on casein diets. The Zn requirements for maximal growth in young guinea pigs have been given as 12 ppm of Zn added to casein diets containing 2 ppm Zn and 20 ppm of Zn added to soybean protein diets containing 2 ppm Zn (2).

The minimum requirements for optimal reproductive performance in the male and female are higher than those just given for growth. In the male rat Swenerton and Hurley (300) have produced evidence that 60 ppm Zn is inadequate to prevent long-term testicular changes. They maintain that rat diets containing soybean protein should contain at least 100 ppm Zn, if extraneous sources of the element are minimal.

2. Pigs

On purified soybean protein diets 45 ppm Zn is adequate for growth in female baby pigs but not in males (170). Weanling pigs fed a soybean protein diet containing 0.66% Ca and 16 ppm Zn required Zn supplementation to give a total of 41 ppm Zn to achieve freedom from parakeratosis. A further increase to 46 ppm Zn improved the growth rate (284). Essentially similar results were obtained by Miller *et al.* (170), with no difference between male and female weanling pigs. Diets containing protein from animal sources, such as fish meal or meat meal, and therefore lower in phytate would reduce Zn requirements slightly below the 45–50 ppm just indicated. Where the diets contain Ca levels twice normal or higher these requirements would be increased, as judged by the increased incidence and severity of parakeratosis on Zn-low diets and the signs of Zn deficiency induced on diets otherwise marginal or adequate in zinc (114, 149, 313). High dietary intakes of copper also significantly increase Zn requirements above those supplied by most normal rations, as discussed in Chapter 3.

The reproductive performance of breeding sows was found to be satisfactory on a corn–soybean meal diet containing 35 ppm Zn and 1.4% Ca. No improvement was observed when this diet was supplemented with 50 ppm Zn (234). Hennig (109) also observed no effect on the reproductive performance of sows from Zn supplementation of barley–fish meal rations containing 36–44 ppm Zn, even when given excess Ca up to 1.5–2.2% of the diet. On the other hand, the addition of 100 ppm Zn to a corn–soybean diet containing 30–34 ppm Zn and 1.6% Ca has been shown in two trials to significantly increase the number of live pigs per litter without affecting birth or weaning weights (115). On this evidence, corn–soybean meal rations high in Ca must be considered marginal in zinc for reproduction in swine.

3. Poultry

The minimum Zn requirement of chicks for growth and health are given as 35 ppm when fed on soybean protein diets containing 1.6% Ca and 0.7% P (203, 204). Lowering the calcium to 1.1% slightly decreased this requirement but raising it to 2.1% had no effect. This estimate has been confirmed but where casein or egg white is the protein source total Zn requirements are lower (188, 229, 346). The minimum dietary level for Zn for growth in chicks can therefore be given as 35–40 ppm for soybean protein-type diets and 25–30 ppm on diets in which the protein comes mainly from animal sources. Furthermore, the chick is less vulnerable than the pig to excess calcium (229). There is ample evidence that the Zn intakes just given are also adequate to meet the requirements of egg production and hatchability.

4. Sheep and Cattle

Ott and co-workers (214) reported that 18 ppm Zn did not support maximal growth in lambs consuming a diet in which egg white was the N source and suggested a requirement between 18 and 33 ppm. Mills *et al.* (186) found 7 ppm Zn to be adequate for growth in lambs on a diet in which 60% of the N was supplied by urea, while 15 ppm were necessary to maintain normal plasma Zn levels. Underwood and Somers (316), also employing a diet in which 60–70% of the N came from urea, observed that ram lambs grew just as well when the diet supplied 17 ppm Zn as when it supplied 32 ppm. However, testicular growth and spermatogenesis were markedly improved at the higher Zn intake. It was concluded that 17 ppm Zn is adequate for body growth and appetite on such diets, but is quite inadequate for normal testicular growth and function.

A dietary Zn concentration of 8–9 ppm is adequate for the growth of calves (179, 186), and 10–14 ppm is necessary to maintain normal plasma Zn levels (186). These estimates of Zn requirements based on semipurified diets are lower than would be suggested from field observations. For example Perry *et al.* (231) obtained increases in daily weight gain from supplementary Zn in two out of four experiments with cattle fed practical fattening rations containing 18 and 29 ppm Zn. Raum and co-workers (250) similarly reported a small improvement in the growth of steers from Zn (and cobalt) supplementation of barley rations containing 29–33 ppm Zn, and Demertzis and Mills (53) observed lesions of infectious pododermatitis in young bulls, responsive to supplememtary Zn, on rations containing 30–56 mg Zn/kg. Signs of Zn deficiency responsive to zinc have been observed in cattle where the pastures or fodder contain 18–42 ppm (d.b.) (146), 19–83 ppm (60), and estimates of 19–28 ppm Zn (85). Since herbage Zn levels of this magnitude are common in areas where clinical Zn deficiency in cattle has not been reported, it seems that factors must be present that reduce Zn absorption or impair its utilization by the animal. Limited evidence with ruminants suggests that neither phytic acid nor calcium can be incriminated in this respect (186, 211).

5. Man

The minimum Zn requirements of humans compatible with satisfactory growth, health, and well-being vary with the type of diet consumed, climatic conditions, and the existence of stress imposed by trauma, parasitic infestations, and infections. The National Academy of Sciences (U.S.) has recommended the following daily dietary allowances for zinc (251): infants, 3–5 mg; children, 1–10 years, 10 mg; males, 10–51+ years, 15 mg; females, 10–51+ years, 15 mg; pregnant women, 20 mg; and lactating women, 25 mg. These allowances apply to Western-style mixed diets and are not necessarily adequate for diets consisting predominantly of unrefined cereals high in phytate. In fact such diets consisting mainly of unleavened whole wheat or corn bread and beans, and supplying approximately the same amount of zinc (15 mg/day) as a typical North American diet, are clearly inadequate in zinc for growth and sexual development in young males in parts of Middle East countries (238, 239, 258, 270). A similar Zn deficiency syndrome, responsive to large amounts of Zn, has recently been reported in females in Iran (258). The situation with respect to dietary Zn requirements in some of these areas is further complicated by the practice of clay-eating, by high-Zn losses in the blood and sweat, and from the effects of intestinal parasites (242). Even in the United States where such "conditioning" factors rarely prevail there are indications that the Zn requirements are not always met. In his critical assessment of zinc nutrition in that country, Sandstead (265) concludes that "If the estimate of the usual dietary availability of zinc is accurate, the findings suggest that some infants, pregnant women, teenage and college women, institutionalized individuals and some living on low income diets have a marginal to deficient intake of zinc. It seems reasonable to presume that some of these people may be adversely affected by their marginal zinc status, especially if they experience unusual stress as may occur with disease or trauma."

V. SOURCES OF ZINC

1. Human Foods and Dietaries

The richest sources of zinc are oysters (which may contain over 1000 ppm Zn) and to a lesser extent other seafoods and the muscle meats and nuts, which usually lie within the range 30–50 ppm Zn of the edible portion. The poorest sources are white sugar, pome and citrus fruits, nonleafy vegetables and tubers, and vegetable oils, which generally contain less than 1 ppm Zn (265, 275, 288). Whole cereal grains are relatively rich in total zinc, with small differences among species (54, 315). A considerable proportion of the zinc of cereals is lost in the milling process. For example Zook *et al.* (348), in a study of North American whests and their products, reported the following mean values in ppm on the dry basis: common hard wheat, 24.0 ± 4.5; common soft wheat, 21.6 ± 7.0; baker's patent flour, 6.3 ± 1.0; soft patent flour, 3.8 ± 0.8; and white bread, conventional dough, 8.9 ± 0.5. White flour from Australian soft wheats averaged only 5 ppm Zn, compared with 16 ppm in the wheats from which the flours were made (315). Whole kernel dent corn samples from 11 sites in North America ranged from 18.7 to 27.2 μg Zn/g (d.b.), with a mean of 22.9 ± 2.2 (78).

Variation in Zn content is high within types of foods, as well as among the different classes of foods, due to the effects of soil type and fertilizer treatment. This is well illustrated by the data obtained by Warren and Delavault (332) for vegetables obtained from a wide range of locations. Furthermore, Underwood (315) found the Zn concentration in wheat grown with the aid of Zn-containing fertilizers to be about twice that of wheat grown on the same soils, without Zn applications. Welch *et al.* (338) obtained marked increases in the Zn content of pea seeds from plants grown in solution culture when $ZnSO_4$ was added to the culture. They suggested that the nutritional value of legume seeds with respect to Zn content could be increased by applying zinc fertilizers "possibly in excess of requirements for optimal plant yields." Contamination with industrial sources of zinc provides a further source of variation.

Total dietary intakes are greatly influenced both by the choice of foods consumed and by their origin. The importance of choice is strikingly illustrated in a comparison of two types of hospital diets made by Osis *et al.* (210), one of which supplied only 4.7 mg Zn/day and the other 18.1 mg Zn/day. Most U.S. adult mixed diets supply 12–15 mg Zn/day (270, 275). The mean intakes by English (96) and Japanese adults (345) have been estimated to be close to 14 mg Zn/day. The daily Zn intakes by adults in nine different regions of India ranged from 14 to 24 mg (285). Young adult women consuming New Zealand-type diets ingested 16.1 to 20.9 mg Zn/day (257), while comparable young women in North America were earlier shown to have an average of 12 mg Zn/day (312). The daily Zn intakes of preadolescent girls were 4.6–9.3 mg (63), while healthy children in Norway 3–5 and 10–13 years of age had Zn intakes of 5–7 and 9–13 mg/day, respectively (272). The Zn intakes were mainly related to the protein content of the diets. A high positive correlation between Zn and protein, compared with Zn and energy content, was also found for U.S. school lunches (191).

Protein, particularly animal protein, is important to the Zn status of diets because of its general favorable effect on both total Zn intakes and Zn availability. Where the cereal portion of the diet is high and is consumed mainly in unleavened form the question of availability becomes highly significant. Reinhold and co-workers (254) showed that fermentation with yeast markedly increases the physiological availability of zinc in whole-meal bread, and that this effect is much greater than can be accounted for merely by the action of yeast in destroying phytate. Oberleas and Prasad (202) have emphasized a further aspect of zinc and protein nutrition. They have shown that the utilization of plant seed proteins for growth in rats is greatly influenced by the level of dietary Zn intake and contend that Zn supplementation of cereal diets may improve the growth and well-being of large segments of the human population in many areas.

Normal infants receiving breast milk were shown to be in negative Zn balance at 1 week of age, in some cases severely so, on daily Zn intakes ranging from 0.2 to 1.2 mg/kg body weight (42). The significance of such losses of body zinc depends on how long they continue into later infancy. The low-Zn and high-Cu content of human milk compared with cow's milk (see Klevay, 139) was considered in Chapter 3. Some infant milk formulas have a lower Zn content than the original cow's milk and provide insufficient zinc for infants (331). Supplementing with 4 mg Zn/liter was found to increase growth, raise plasma Zn levels, and decrease the incidence of disturbed gastrointestinal function in male infants and was not accompanied by any signs of toxicity (331). Evidence that adults (68) and infants (9) can develop signs of Zn deficiency when receiving total parenteral nutrition for several weeks has recently been presented.

2. Animal Feeds and Fodders

Typical Zn concentrations in pasture herbage remote from industrial areas have been given by Mills and Dalgarno (185) as 25–35 ppm (d.b.). Levels from 5 to 50 times higher were obtained for such herbage from agricultural land exposed to contamination from industrial sources. Grace (81) obtained values ranging from 23 to 70 (mean 38) ppm Zn (d.b.) from 10 improved mixed pastures in the north island of New Zealand, 17 to 27 (mean 22) ppm (d.b.) for such pastures in the south island of that country, and the very wide range of 8–48 ppm for tussock grassland from hill country. In northern Ontario, winter grazing contained 20–30 ppm and summer pasture 40–60 ppm Zn (d.b.) (112). The Zn concentration in plants usually falls with advancing maturity (137, 143), and leguminous plants invariably carry higher Zn levels than grasses grown and sampled under the same conditions (307). Heavy dressings with lime and to a lesser extent with superphosphate can greatly reduce pasture Zn levels (4).

The cereal grains used as the basis of pig and poultry rations typically contain 20–30 ppm Zn, with appreciably higher levels in most materials used as protein supplements. Typical values for soybean, peanut, and linseed meals may be given as 50–70 ppm Zn (315). The Zn contents of fish meal, whale meal, and meat meal are normally much higher than that of soybean meal. Levels of 90–100 ppm Zn or more are common (155, 315).

VI. ZINC TOXICITY

Zinc is relatively nontoxic to birds and mammals. Rats, pigs, poultry, sheep, cattle, and man exhibit considerable tolerance to high-Zn intakes, the extent of the tolerance depending greatly on the nature of the diet, particularly its content of calcium, copper, iron, and cadmium, with which it interacts in the process of absorption and utilization. For this reason studies of minimum toxic levels of dietary Zn are only meaningful when the status of the diet and the animal with respect to these interacting elements is known and defined.

In an early study with rats carried out before the above facts were known, dietary Zn intakes of 0.25% or 2500 ppm induced no discernible effects whether ingested as the metal, chloride, or carbonate. At 5000 ppm Zn growth was severely depressed and mortality was high in young animals when ingested as the chloride, with little mortality and only slight growth depression as the oxide (102). Subsequently Sutton and Nelson (299) confirmed that 5000 or 10,000 ppm Zn as the carbonate induced subnormal growth, anorexia, and at the higher rate heavy mortality in young rats. They also observed a severe anemia. Adult female rats fed a diet containing 2000 ppm Zn as the oxide maintained normal pregnancies with no malformations in the fetuses, but at 4000 ppm variable degrees of death and resorption of the fetuses were observed (273).

In the Zn-intoxicated rat an anemia of the microcytic hypochromic type develops, accompanied by high levels of Zn and subnormal levels of iron, copper, cytochrome oxidase, and catalase (50, 82, 325, 344). The anemia and the accompanying biochemical changes can largely be overcome by supplements of copper (82, 325) and completely overcome by supplements of copper plus iron (50). The anemia of Zn toxicity thus results from induced iron and copper deficiencies brought about by an interference with the absorption and utilization of these elements by the high-Zn intakes. In all the above studies the Cu contents of the diet were adequate for all purposes at normal intakes of zinc. Campbell and Mills (39) found that in weanling rats maintained on diets low or marginal in copper, supplementary Zn as low as 300 ppm of the diet can induce Cu deficiency as evidenced by reduced plasma ceruloplasmin activity, and also at 1000 ppm Zn by growth depression, hair depigmentation, and depressed liver Cu levels. The levels of dietary Zn at which toxic effects are evident thus depend markedly on the ratio of these levels to those of copper.

Weanling pigs fed for several weeks diets containing 1000 ppm Zn, either as the sulfate (148) or carbonate (29), suffered no obvious ill effects. At higher Zn levels depressed growth and appetite, arthritis, and internal hemorrhages were observed, and at 4000 and 8000 ppm mortality was high (29). Raising the dietary Ca level from 0.7 to 1.1% had a protective effect against the toxic effects of 4000 ppm Zn (as ZnO) in a recent experiment with weanling pigs. Growth and appetite were better in the pigs fed the high-Ca diet, and the Zn levels in their blood and tissues were significantly lower than in those ingesting the diet lower in Ca (119). Broilers and layer hens exhibit a tolerance to zinc similar to pigs at 1200–1400 ppm of the diet and a similar growth and appetite depression when the level is raised to 3000 ppm (228). The importance of diet composition on Zn toxicity is evident from the work of Berg and Martinson (17). Little or no evidence of toxicity was observed when 2000 ppm Zn as ZnO was added to corn–soybean, corn–fish meal, or sucrose–soybean diets fed for 2 weeks to baby chicks. The same amount of Zn added to a sucrose–fish meal diet reduced weight and bone ash, and as little as 800 ppm Zn was found to be toxic with this diet.

Sheep and cattle appear to be less tolerant of high Zn intakes than rats, pigs, and poultry. Consumption by lambs of diets containing 1000 ppm Zn as the oxide reduced weight gains and decreased feed efficiency, and diets containing more than 1500 ppm depressed feed consumption (213). In a recent study Campbell (38) showed that diets containing 700 ppm Zn offered to pregnant ewes resulted in high incidence of perinatal deaths in lambs. Steers and heifers have been shown to be unaffected by 500 ppm Zn or less, but 900 ppm caused reduced gains and lowered feed efficiency, and 1700 ppm induced, in addition, a depraved appetite characterized by excessive salt consumption and wood chewing (213). The tissue changes in sheep and cattle induced by high-Zn intakes differ from those described earlier for rats. Subnormal liver Cu levels and a mild anemia occur but liver Fe levels actually increase (212), and defective development and mineralization of bone are not apparent, as reported for Zn-toxic rats (263). At the higher levels of Zn intake changes in rumen metabolism evidenced by a reduction in the volatile fatty acid (VFA) concentration and acetic acid:propionic acid ratio occurred in lambs, probably through a toxic effect of the zinc on the rumen microorganisms (212).

The relatively low toxicity of Zn among the divalent cations, coupled with efficient homeostatic control mechanisms, make chronic Zn toxicity from dietary sources an unlikely hazard to man. Where Zn salts or compounds are given orally in large doses over prolonged periods, as in the treatment of chronic leg ulcers or the prophylaxis of cardiovascular disease, possibilities of toxic effects cannot be dismissed. Indeed doses of 150 mg Zn/day, which are equivalent to about 200–300 ppm of the total daily dry matter intake of an adult, are enough to interfere with copper and iron metabolism, since zinc is a metabolic antagonist of both these metals. Where the copper and iron intakes from the diet are ample this is unlikely to be serious, but where they are low or marginal the Cu and Fe status of the individual could decline and Zn-induced manifestations of Cu and Fe deficiencies ultimately arise. Hallbook and Lanner (88) found no "biochemical evidence" of toxicity during 4 months of oral administration of zinc sulfate at the rate of 200 mg three times a day to patients with venous leg ulcers, but iron and copper were not specifically investigated and the period is not long. On the other hand, zinc is also a metabolic antagonist of cadmium, as discussed in the next chapter, so that such high-Zn intakes would be expected to afford some protection against the potentially toxic effects of increasing Cd exposure from the environment.

REFERENCES

1. Addink, N.W.H., *Nature (London)* **186,** 253 (1960).
2. Alberts, J.C., Lang, J.A., and Briggs, S.M., *Fed. Proc., Fed. Am. Soc. Exp. Biol.* **34,** 906 (abstr.) (1975).
3. Amadar, M., Gonzalez, A., and Hermelo, M., *Rev. Cubana Pediatr.* **45,** 315 (1973).
4. Anonymous, *N. Z. Agriculturalist* **20,** 6 (1971).
5. Apgar, J., *Am. J. Physiol.* **215,** 160 and 1478 (1968).
6. Apgar, J., *J. Nutr.* **100,** 470 (1970).
7. Apgar, J., *J. Nutr.* **102,** 343 (1972).
8. Apgar, J., *J. Nutr.* **103,** 973 (1973).
9. Arakawa, T., Tumura, T., Igarashi, Y., Suzuki, H., and Sandstead, H.H., *Am. J. Clin. Nutr.* **29,** 197 (1976).
10. Archibald, J.G., *J. Dairy Sci.* **27,** 257 (1944).
11. Barcia, P.J., *Ann. Surg.* **172,** 1048 (1970).
12. Bartholemew, M.E., Tupper, R., and Wormall, A., *Biochem. J.* **73,** 256 (1959).
13. Becker, W.M., and Hoekstra, W.G., *J. Nutr.* **90,** 301 (1966); **94,** 455 (1968).
14. Becker, W.M., and Hoekstra, W.G., *in* "Intestinal Absorption of Metal Ions, Trace Elements and Radionuclides" (S.C.Skoryna and D. Waldron-Edward, eds.), p. 229. Pergamon, Oxford, 1971.
15. Beisel, W.R., Pekarek, R.S., and Wannemacher, R.W., Jr., *in* "Trace Element Metabolism in Animals" (W.G. Hoekstra *et al.*, eds.), Vol. 2, p. 217. Univ. Park Press, Baltimore, Maryland, 1974.
16. Berfenstam, R., *Acta Paediatr. (Stockholm)* **41,** Suppl. 87, 389 (1952).
17. Berg, L.R., and Martinson, R.D., *Poult. Sci.* **51,** 1690 (1972).
18. Bergman, R., and Soremark, R., *J. Nutr.* **94,** 6 (1968).
19. Bertrand, G., and Vladesco, R., *C.R. Hebd. Seances Acad. Sci.* **173,** 176 (1921).
20. Birckner, V., *J. Biol. Chem.* **38,** 191 (1919).
21. Blackmon, D.M., Miller, W.J., and Morton, J.D., *Vet. Med. & Small Anim. Clin.* **62,** 265 (1967).
22. Blamberg, D.L., Blackwood, U.B., Supplee, W.C., and Combs, G.F., *Proc. Soc. Exp. Biol. Med.* **104,** 217 (1960).
23. Bonomi, A., *18th Convegno. Soc. Ital. Sci. Vet. Prescara* (1964), cited by Demertzis and Mills (53).
24. Bosticco, A., and Bonomi, A., *Prog. Vet. Anno., 1965,* p. 1 (1965), cited by Demertzis and Mills (53).
25. Bottomley, R.G., Cornelison, R.L., Jacobs, L.A., and Lindeman, R.D., *J. Lab. Clin. Med.* **74,** 852 (abstr.) (1969).
26. Bowness, J.M., and Morton, R.A., *Biochem. J.* **51,** 530 (1950); **53,** 620 (1953).
27. Bremner, I., *in* "Trace Element Metabolism in Animals" (W.G. Hoekstra *et al.*, eds.), Vol. 2, p. 489. Univ. Park Press, Baltimore, Maryland, 1974.
28. Bremner, I., and Davies, N.T., *Rep. Rowett Inst.* **29,** 126 (1973).
29. Brink, M.F., Becker, D.E., Terril, S.W., and Jensen, A.H., *J. Anim. Sci.* **18,** 836 (1959).
30. Brudevold, F., Steadman, L.T., Spinelli, M.A., Amdur, B.H., and Gron, P., *Arch. Oral Biol.* **8,** 135 (1963).
31. Buchanan, P.J., and Hsu, J.M., *Fed. Proc., Fed. Am. Soc. Exp. Biol.* **27,** 483 (1968).
32. Burns, R.H., Johnston, A., Hamilton, J.W., McCullough, R.J., Duncan, W.E., and Fisk, H.G., *J. Anim. Sci.* **23,** 5 (1964).

33. Burr, R.G., *J. Clin. Pathol.* **26,** 773 (1973).
34. Cabell, C.A., and Earle, I.P., *J. Anim. Sci.* **24,** 800 (1965).
35. Caldwell, D.F., Oberleas, D., Clancy, J.J., and Prasad, A.S., *Proc. Soc. Exp. Biol. Med.* **133,** 1417 (1970).
36. Caldwell, D.F., Oberleas, D., and Prasad, A.S., *Nutr. Rep. Int.* **7,** 309 (1973).
37. Calhoun, N.R., and Smith, J.C., *Lancet* **2,** 682 (1968).
38. Campbell, J.K., personal communication (1975).
39. Campbell, J.K., and Mills, C.F., *Proc. Nutr. Soc.* **33,** 15A (1974).
40. Cassens, R.G., Briskey, E.J., and Hoekstra, W.G., *J. Sci. Food Agric.* **6,** 427 (1963).
41. Cassens, R.G., Hoekstra, W.G., Faltin, E.C., and Briskey, E.J., *Am. J. Physiol.* **212,** 688 (1967).
42. Cavell, P.A., and Widdowson, E.M., *Arch. Dis. Child.* **39,** 496 (1964).
43. Chen, R.W., Eakin, D.J., and Whanger, P.D., *Nutr. Rep. Int.* **10,** 195 (1974).
44. Chesters, R.K., *Biochem. J.* **130,** 133 (1972).
45. Chesters, R.K., and Quarterman, J., *Br. J. Nutr.* **24,** 1061 (1970).
46. Coble, Y.D., Van Reen, R., Schulert, A.R., Koshakji, R.P., Farid, Z., and Davis, J.T., *Am. J. Clin. Nutr.* **19,** 415 (1966).
47. Cohen, C., *Br. Med. J.* **2,** 561 (1968).
48. Cotzias, G.C., Borg, D.C., and Selleck, B., *Am. J. Physiol.* **202,** 359 (1962).
49. Cotzias, G.C., and Papavasiliou, P.S., *Am. J. Physiol.* **206,** 787 (1964).
50. Cox, D.H., and Harris, D.L., *J. Nutr.* **70,** 514 (1960).
51. Davies, N.T., Bremner, I., and Mills, C.F., *Biochem. Soc. Trans.* **1,** 985 (1973).
52. Demertzis, P.N., *Bull. Hellenic Vet. Med. Soc.* **23,** 256 (1972).
53. Demertzis, P.N., and Mills, C.F., *Vet. Rec.* **92,** 219 (1973).
54. Dewar, W.A., *J. Sci. Food Agric.* **18,** 68 (1967).
55. Dewar, W.A., Teague, P.W., and Downie, J.N., *Br. Poult. Sci.* **15,** 119 (1974).
56. Dreosti, I.E., and Hurley, L.S., *Proc. Soc. Exp. Biol. Med.* **150,** 161 (1975).
57. Dreosti, I.E., Tsao, S., and Hurley, L.S., *Proc. Soc. Exp. Biol. Med.* **128,** 169 (1968).
58. Drinker, R.E., and Collier, E.S., *J. Ind. Hyg.* **8,** 257 (1926).
59. Dufty, J., personal communication (1975).
60. Dynna, P., and Havre, G.N., *Acta Vet. Scand.* **4,** 197 (1963).
61. Earle, I.P., and Stevenson, J.W., *J. Anim. Sci.* **24,** 325 (1965).
62. Edwards, H.M., Jr., *J. Nutr.* **69,** 306 (1959).
63. Engel, R.W., Miller, R.F., and Price, N.O., *in* "Zinc Metabolism" (A.S. Prasad, ed.), p. 326. Thomas, Springfield, Illinois, 1966.
64. Evans, G.W., and Hahn, C., *in* "Protein-Metal Interactions" (M. Friedman, ed.), p. 285. Plenum, New York, 1974.
65. Evans, G.W., Grace, C.I., and Hahn, C., *Proc. Soc. Exp. Biol. Med.* **143,** 723 (1973).
66. Evans, G.W., Grace, C.I., and Votava, H.J., *Am. J. Physiol.* **228,** 501 (1975).
66a. Evans, G.W., and Winter, T.W., *Biochem. Biophys. Res. Commun.* **66,** 1218 (1975).
67. Feaster, J.P., Hansard, S., McCall, J.T., Skipper, F.H., and Davis, G.K., *J. Anim. Sci.* **13,** 781 (1954).
68. Fleming, C.R., Hodges, R.E., and Hurley, L.S., *Am. J. Clin. Nutr.* **29,** 70 (1976).
69. Florence, E., and Quarterman, J., *Br. J. Nutr.* **28,** 63 (1972).
70. Flynn, A., Pories, W.J., Strain, W.H., Hill, O.A., and Fratianne, R.B., *Lancet* **2,** 1169 (1971).
71. Foley, B., Johnson, S.A., Hackley, B., Smith, J.C., and Halsted, J.A., *Proc. Soc. Exp. Biol. Med.* **128,** 265 (1968).
72. Follis, R.H., Day, H.G., and McCollum, E.V., *J. Nutr.* **22,** 223 (1941).

73. Forbes, R.M., *Fed. Proc., Fed. Am. Soc. Exp. Biol.* **19,** 643 (1960).
74. Forbes, R.M., and Yohe, M., *J. Nutr.* **70,** 53 (1960).
75. Fox, M.R.S., and Harrison, B.N., *Proc. Soc. Exp. Biol. Med.* **116,** 256 (1964); *J. Nutr.* **86,** 89 (1965).
76. Frommer, D.J., *Med. J. Aust.* **2,** 793 (1975).
77. Fujioka, M., and Lieberman, I., *J. Biol. Chem.* **239,** 1164 (1964).
78. Garcia, W.J., Blessin, C.W., and Inglett, G.E., *Cereal Chem.* **51,** 788 (1974).
79. Gibson, J.G., Vallee, B.L., Fluharty, R.G., and Nelson, J.E., *Proc. Int. Congr. Cancer Res., 6th*, p. 1102 (1954).
80. Gilbert, I.G.F., and Taylor, D.M., *Biochim. Biophys. Acta* **21,** 546 (1956).
81. Grace, N.D., *N. Z. J. Agric. Res.* **15,** 284 (1972).
82. Grant-Frost, D.R., and Underwood, E.J., *Aust. J. Exp. Biol. Med. Sci.* **36,** 339 (1958).
83. Groppel, B., and Hennig, A., *Arch. Exp. Veterinaermed.* **25,** 817 (1971).
84. Gunn, S.A., and Gould, T.C., *Endocrinology* **58,** 443 (1956); *Am. J. Physiol.* **193,** 505 (1958).
85. Haarenen, S., *Nord. Veterinaermed.* **14,** 265 (1962); **15,** 536 (1963).
86. Hahn, C.J., and Evans, G.W., *Proc. Soc. Exp. Biol. Med.* **144,** 794 (1973).
87. Hahn, C.J., and Evans, G.W., *Am. J. Physiol.* **228,** 1020 (1975).
87a. Halas, E., Hanlon, M.J., and Sandstead, H.H., *Nature (London)* **257,** 221 (1975).
88. Hallbook, T., and Lanner, E., *Lancet* **2,** 786 (1972).
89. Halsted, J.A., and Smith, J.C., *Lancet* **1,** 322 (1970).
90. Halsted, J.A., Hackley, B.M., and Smith, J.C., *Lancet* **2,** 278 (1968).
91. Halsted, J.A., Hackley, B.M., Rudzki, C., and Smith, J.C., *Gastroenterology* **54,** 1098 (1968).
92. Hambidge, K.M., Hambidge, C., Jacobs, M., and Baum, J.D., *Pediatr. Res.* **6,** 868 (1972).
93. Hambidge, K.M., and Droegemueller, W., *Obstet. Gynecol.* **44,** 666 (1974).
94. Hambidge, K.M., Neldner, K.H., and Walravens, P.A., *Lancet* **1,** 577 (1975).
95. Hambidge, K.M., and Silverman, A., *Arch. Dis. Child.* **48,** 567 (1973).
96. Hamilton, E.I., and Minski, M.J., *Sci. Total Environ.* **1,** 375 (1972/1973).
97. Hamilton, E.I., Minski, M.J., and Cleary, J.J., *Sci. Total Environ.* **1,** 341 (1972/1973).
98. Hampton, D.L., Miller, W.J., Blackmon, D.M., Gentry, R.P., Neathery, M.W., and Stake, P.E., *Fed. Proc., Fed. Am. Soc. Exp. Biol.* **34,** 907 (abstr.) (1975).
99. Harland, B.F., Fox, M.R.S., and Fry, B.E., Jr., *Proc. Soc. Exp. Biol. Med.* **145,** 316 (1974).
100. Harland, B.F., Fox, M.R.S., and Fry, B.E., Jr., *J. Nutr.* **105,** 1509 (1975).
101. Heath, J.C., and Liquier-Milward, J., *Biochim. Biophys. Acta* **5,** 404 (1950).
102. Heller, V.G., and Burke, A.D., *J. Biol. Chem.* **74,** 85 (1927).
103. Henkin, R.I., and Bradley, D.F., *Life Sci.* **9,** 701 (1970).
104. Henkin, R.I., Graziadei, P.P.G., and Bradley, D.F., *Ann. Intern. Med.* **71,** 791 (1969).
105. Henkin, R.I., Keiser, H.R., and Bronzert, D., *J. Clin. Invest.* **51,** 44a (1972).
106. Henkin, R.I., Marshall, J.R., and Meret, S., *Am. J. Obstet. Gynecol.* **110,** 131 (1971).
107. Henkin, R.I., Schulman, J.D., Schulman, C.B., and Bronzert, D.A., *J. Pediatr.* **82,** 831 (1973).
108. Henkin, R.I., Schechter, P.J., Hoye, R., and Mattern, C.F.T., *J. Am. Med. Assoc.* **217,** 434 (1971).
109. Hennig, A., *Arch. Tierernaehr.* **15,** 331, 345, 353, 363, and 377 (1965).

110. Henzel, J.H., Deweese, M.S., and Lichti, E.L., *Arch. Surg. (Chicago)* **100**, 349 (1970).
111. Henzel, J.H., Holtman, B., Keitzer, F.W., Deweese, M.S., and Lichti, E.L., *Trace Subst. Environ. Health–2, Proc. Univ. Mo. Annu. Conf., 2nd, 1968* (1969).
112. Hidiroglou, M., and Spurr, D.T., *Can. J. Anim. Sci.* **55**, 31 (1975).
113. Hill, C.H., Matrone, G., Payne, W.L., and Barber, C.W., *J. Nutr.* **80**, 227 (1963).
114. Hoefer, J.A., Miller, E.R., Ullrey, D.E., Ritchie, H.D., and Luecke, R.W., *J. Anim. Sci.* **19**, 249 (1960).
115. Hoekstra, W.G., Faltin, E.C., Lin, C.W., Roberts, H.F., and Grummer, R.H., *J. Anim. Sci.* **26**, 1348 (1967).
116. Hoekstra, W.G., Lewis, P.K., Phillips. P.H., and Grummer, R.H., *J. Anim. Sci.* **15**, 752 (1956).
117. Hove, E., Elvehjem, C.A., and Hart, E.B., *Am. J. Physiol.* **119**, 768 (1937); **124**, 750 (1938).
118. Hove, E., Elvehjem, C.A., and Hart, E.B., *J. Biol. Chem.* **136**, 425 (1940).
119. Hsu, F.S., Krook, L., Pond, W.G., and Duncan, J.R., *J. Nutr.* **105**, 112 (1975).
120. Hsu, J.M., and Anthony, W.L., *J. Nutr.* **101**, 445 (1971).
121. Hsu, J.M., and Anthony, W.L., *J. Nutr.* **105**, 26 (1975).
122. Hsu, J.M., Anilane, J.K., and Scanlan, D.E., *Science* **153**, 882 (1966).
123. Hsu, J.M., Anthony, W.L., and Buchanan, P.J., *J. Nutr.* **99**, 425 (1969).
124. Huber, A.M., and Gershoff, S.N., *J. Nutr.* **103**, 1175 (1973).
125. Huber, A.M., and Gershoff, S.N., *J. Nutr.* **103**, 1739 (1973).
126. Hurley, L.S., and Mutch, P.B., *J. Nutr.* **103**, 649 (1973).
127. Hurley, L.S., Gowan, J., and Swenerton, H., *Teratology* **4**, 199 (1971).
128. Hurley, L.S., and Schrader, R.E., *Nature (London)* **254**, 427 (1975).
129. Hurley, L.S., and Swenerton, H., *Proc. Soc. Exp. Biol. Med.* **123**, 692 (1966).
130. Hurley, L.S., and Swenerton, H., *J. Nutr.* **101**, 597 (1971).
131. Hurley, L.S., Swenerton, H., and Eichner, J.T., *Fed. Proc., Fed. Am. Soc. Exp. Biol.* **23**, 292 (1964).
132. Husain, S.L., *Lancet* **1**, 1069 (1969).
133. Kienholz, E.W., Sunde, M.L., and Hoekstra, W.G., *Poult. Sci.* **43**, 667 (1964).
134. Kienholz, E.W., Turk, D.E., Sunde, M.L., and Hoekstra, W.G., *J. Nutr.* **75**, 211 (1961).
135. Kirchgessner, M., *Z. Tierphysiol., Tierernaehr. Futtermittelkd.* **14**, 270 (1959).
136. Kirchgessner, M., and Pallauf, J., *Z. Tierphysiol., Tierernaehr. Futtermittelkd.* **29**, 65 and 77 (1972).
137. Kirchgessner, M., Merz, G., and Oelschlager, W., *Arch. Tierernaehr.* **10**, 414 (1966).
138. Kirchgessner, M., Schwarz, F.J., and Grassman, E., *Bioinorg. Chem.* **2**, 255 (1973).
139. Klevay, L.M., *Nutr. Rep. Int.* **11**, 237 (1975).
140. Kratzer, F.H., Allred, J.B., Davis, P.N., Marshall, B.J., and Vohra, P., *J. Nutr.* **68**, 313 (1959).
141. Kratzer, F.H., Vohra, P., Allred, J.B., and Davis, P.N., *Proc. Soc. Exp. Biol. Med.* **98**, 205 (1958).
142. Kubota, J., Lazar, V.A., and Losee, F., *Arch. Environ. Health* **16**, 788 (1966).
143. Lang, V., Kirchgessner, M., and Voightländer, G., *Z. Acker- Pflanzenbau* **135**, 216 (1972).
144. Lavy, U.I., *Br. J. Surg.* **59**, 194 (1972).
145. Lease, J.G., and Williams, W.P., *Poult. Sci.* **46**, 233 and 242 (1967).
146. Legg, S.P., and Sears, L., *Nature (London)* **186**, 1061 (1960).
147. Lewis, H.D., and Altschule, M.D., *Blood* **4**, 442 (1949).
148. Lewis, P.K., Hoekstra, W.G., and Grummer, R.H., *J. Anim. Sci.* **16**, 578 (1957).

149. Lewis, P.K., Hoekstra, W.G., Grummer, R.H., and Phillips, P.H., *J. Anim. Sci.* **15**, 741 (1956).
150. Lichti, E.L., Almond, C.H., Henzel, J.H., and Deweese, M.S., *Am. J. Obstet. Gynecol.* **106**, 1242 (1970).
151. Likuski, H.J.A., and Forbes, R.M., *J. Nutr.* **85**, 230 (1965).
152. Losee, F., Cutress, T.W., and Brown, R., *Trace Subst. Environ. Health–7, Proc. Univ. Mo. Annu. Conf., 7th, 1973,* p. 19 (1974).
153. Luecke, R.W., Holman, M.E., and Baltzer, B.V., *J. Nutr.* **94**, 344 (1968).
154. Luecke, R.W., Rukson, B.E., and Baltzer, B.V., *in* "Trace Element Metabolism in Animals" (C.F. Mills, ed.), Vol. 1, p. 471. Livingstone, Edinburgh, 1970.
155. Lunde, G., *J. Sci. Food Agric.* **19**, 432 (1968).
156. Lutz, R.E., *J. Ind. Hyg.* **8**, 177 (1926).
157. Macapinlac, M.P., Pearson, W.N., Barney, G.H., and Darby, W.J., *J. Nutr.* **93**, 511 (1967); **95**, 569 (1968).
158. Mann, T., "The Biochemistry of Semen and the Male Reproductive Tract." Butler & Tanner, London, 1964.
159. Mawson, C.A., and Fischer, M.I., *Biochem. J.* **55**, 696 (1953); *Can. J. Med. Sci.* **30**, 336 (1952).
160. McCance,R.A., and Widdowson, E.M., *Biochem. J.* **36**, 696 (1942).
161. McClain, P.E., Wiley, E.R., Beecher, G.R., Anthony, W.L., and Hsu, J.M., *Biochim. Biophys. Acta* **304**, 457 (1973).
162. McConnell, D.P., and Henkin, R.I., *J. Nutr.* **104**, 1108 (1974).
163. McKenzie, J.M., Fosmire, G.J., and Sandstead, H.H., *J. Nutr.* **105**, 1466 (1975).
164. Methfessel, A.H., and Spencer, H., *Fed. Proc., Fed. Am. Soc. Exp. Biol.* **25**, 483 (abstr.) (1966).
165. Methfessel, A.H., and Spencer, H., *in* "Trace Element Metabolism in Animals" (W.G. Hoekstra *et al.*, eds.), Vol. 2, p. 541. Univ. Park Press, Baltimore, Maryland, 1974.
166. Millar, M.J., Elcoate, P.V., and Mawson, C.A., *Can. J. Biochem. Physiol.* **35**, 865 (1957).
167. Millar, M.J., Fischer, M.I., Elcoate, P.V., and Mawson, C.A., *Can J. Biochem. Physiol.* **36**, 557 (1958); **38**, 1457 (1960).
168. Millar, M.J., Vincent, N.R., and Mawson, C.A., *J. Histochem. Cytochem.* **9**, 111 (1961).
169. Miller, E.R., Luecke, R.W., Ullrey, D.E., Baltzer, B.V., Bradley, B.L., and Hoefer, J.A., *J. Nutr.* **95**, 278 (1968).
170. Miller, E.R., Liptrap, D.O., and Ullrey, D.E., *in* "Trace Element Metabolism in Animals" (C.F. Mills, ed.), Vol. 1, p. 377. Livingstone, Edinburgh, 1970.
171. Miller, J.K., and Cragle, R.G., *J. Dairy Sci.* **48**, 370 (1965).
172. Miller, J.K., and Jensen, L.S., *Poult. Sci.* **45**, 1051 (1966).
173. Miller, J.K., and Miller, W.J., *J. Dairy Sci.* **43**, 1854 (1960); *J. Nutr.* **76**, 467 (1962).
174. Miller, W.J., *Am. J. Clin. Nutr.* **22**, 1323 (1969).
175. Miller, W.J., Blackmon, D.M., Gentry, R.P., Powell, G.W., and Perkins, H.E., *J. Dairy Sci.* **49**, 1446 (1966).
176. Miller, W.J., Blackmon, D.M., Hiers, J.M., Fowler, P.R., Clifton, C.M., and Gentry, R.P., *J. Dairy Sci.* **50**, 715 (1967).
177. Miller, W.J., Blackmon, D.M., Powell, G.W., Gentry, H.P., and Hiers, J.M., *J. Nutr.* **90**, 335 (1966).
178. Miller, W.J., Morton, J.D., Pitts, W.J., and Clifton, C.M., *Proc. Soc. Exp. Biol. Med.* **118**, 427 (1965).
179. Miller, W.J., Clifton, C.M., and Cameron, N.W., *J. Dairy Sci.* **46**, 715 (1963).

180. Miller, W.J., Clifton, C.M., and Fowler, P.R., *J. Anim. Sci.* **23,** 885 (1964).
180a. Miller, W.J., Pitts, W.J., Clifton, C.M., and Morton, J.D., *J. Dairy Sci.* **48,** 1329 (1965).
181. Miller, W.J., Powell, G.W., Pitts, W.J., and Perkins, H., *J. Dairy Sci.* **48,** 1091 (1965).
182. Miller, W.J., Powell, G.W., and Hiers, J.M., *J. Dairy Sci.* **49,** 1012 (1966).
183. Miller, W.J., Neathery, M.W., Gentry, R.P., Blackmon, D.M., and Stake, P.E., *in* "Trace Element Metabolism in Animals" (W.G. Hoekstra *et al.,* eds.), Vol. 2, p. 550. Univ. Park Press, Baltimore, Maryland, 1974.
184. Miller, W.J., Pitts, W.J., Clifton, C.F., and Schmittle, S.C., *J. Dairy Sci.* **74,** 556 (1964).
185. Mills, C.F., and Dalgarno, A.C., *Nature (London)* **239,** 171 (1972).
186. Mills, C.F., Dalgarno, A.C., Williams, R.B., and Quarterman, J., *Br. J. Nutr.* **21,** 751 (1967).
187. Mills, C.F., Quarterman, J., Williams, R.B., Dalgarno, A.C., and Panic, B., *Biochem. J.* **102,** 712 (1967).
188. Moeller, M.W., and Scott, H.M., *Poult. Sci.* **37,** 1227 (1958).
189. Molokhia, M.M., and Portnoy, B., *Br. J. Dematol.* **81,** 759 (1969).
190. Moynahan, E.J., *Lancet* **1,** 399 (1974).
191. Murphy, E.W., Page, L., and Watt, B.K., *J. Am. Diet. Assoc.* **58,** 115 (1971).
192. Murthy, G.K., and Rhea, U.S., *J. Dairy Sci.* **54,** 1001 (1971).
193. Murthy, G.K., Rhea, U.S., and Peeler, J.T., *J. Dairy Sci.* **55,** 1666 (1972).
194. Mutch, P.B., and Hurley, L.S., *J. Nutr.* **104,** 828 (1974).
195. Myers, M.B., and Cherry, G., *Am. J. Surg.* **120,** 77 (1970).
196. Neldner, K.H., and Hambidge, K.M., *N. Engl. J. Med.* **292,** 879 (1975).
197. Nielsen, F.H., Sunde, M.L., and Hoekstra, W.G., *J. Nutr.* **89,** 24 and 35 (1966).
198. Nielsen, F.H., Sunde, M.L., and Hoekstra, W.G., *Proc. Soc. Exp. Biol. Med.* **116,** 256 (1964); *J. Nutr.* **86,** 89 (1965).
199. Norman, J.N., Rahmat, A., and Smith, G., *J. Nutr.* **105,** 815 and 822 (1975).
200. Norrdin, R.W., Krook, L., Pond, W.G., and Walker, E.F., *Cornell Vet.* **63,** 264 (1973).
201. Oberleas, D., Muhrer, M.E., and O'Dell, B.L., *J. Anim. Sci.* **19,** 1280 (1960).
202. Obesleas, D., and Prasad, A.S., *Am. J. Clin. Nutr.* **22,** 1304 (1969).
203. O'Dell, B.L., Newberne, F.M., and Savage, J.E., *J. Nutr.* **65,** 503 (1958).
204. O'Dell, B.L., and Savage, J.E., *Poult. Sci.* **36,** 489 (1975).
205. O'Dell, B.L., and Savage, J.E., *Proc. Soc. Exp. Biol. Med.* **103,** 304 (1960).
206. O'Dell, B.L., Yohe, M., and Savage, J.E., *Poult. Sci.* **43,** 415 (1964).
207. O'Mary, C.C., Butts, W.T., Reynolds, R.A., and Bell, M.C., *J. Anim. Sci.* **28,** 268 (1969).
208. O'Neal, R.M., Pla, G.W., Fox, M.R.S., Gibson, F.E., and Fry, B.E., Jr., *J. Nutr.* **100,** 491 (1970).
209. Oon, B.B., Khong, K.Y., Greaves, M.W., and Plummer, V.M., *Br. Med. J.* **2,** 5918 (1974).
210. Osis, D., Kramer, L., Wiatrowski, E., and Spencer, H., *Am. J. Clin. Nutr.* **25,** 582 (1972).
211. Ott, E.A., Smith, W.H., Stob, M., and Beeson, W.M., *J. Nutr.* **82,** 41 (1964).
212. Ott, E.A., Smith, W.H., Harrington, R.B., Stob, M., Parker, H.E., and Beeson, W.M., *J. Anim. Sci.* **25,** 432 (1966).
213. Ott, E.A., Smith, W.H., Harrington, R.B., and Beeson, W.M., *J. Anim. Sci.* **25,** 414 and 419 (1966).

214. Ott, E.A., Smith, W.H., Stob, M., Parker, H.E., Harrington, R.B., and Beeson, W.M., *J. Nutr.* 87, 459 (1965).
215. Pallauf, J., and Kirchgessner, M., *Z. Tierphysiol., Tierernaehr. Futtermittelkd.* **28,** 121 (1971).
216. Pallauf, J., and Kirchgessner, M., *Int. J. Vitam. Nutr. Res.* **41,** 543 (1971).
217. Pallauf, J., and Kirchgessner, M., *Zentralbl. Veterinaermed., Reihe A* **19,** 594 (1972).
218. Pallauf, J., and Kirchgessner, M., *Zentralbl. Veterinaermed., Reihe A* **20,** 100 (1973).
219. Parisi, A.F., and Vallee, B.L., *Biochemistry* **9,** 2421 (1970).
220. Parizek, J., Boursnell, J.C., Hay, M.F., Babicky, A., and Taylor, D.M., *J. Reprod. Fertil.* **12,** 501 (1966).
221. Parkash, S., and Jenness, R., *J. Dairy Sci.* **50,** 127 (1967).
222. Pekarek, R.S., and Beisel, W.R., *Appl. Microbiol.* **18,** 482 (1969).
223. Pekarek, R.S., Beisel, W.R., Bartelloni, P.J., and Bostian, K.A., *Am. J. Clin. Pathol.* **57,** 506 (1972).
224. Pekarek, R.S., Burghen, G.A., Bartelloni, P.J., Calia, F.M., Bostian, K.A., and Beisel, W.R., *J. Lab. Clin. Med.* **76,** 293 (1970).
225. Pekarek, R.S., Kluge, R.M., Dupont, H.L., Wannemacher, R.W., Hornick, R.B., Bostian, K.A., and Beisel, W.R., *Clin. Chem. (Winston-Salem, N.C.)* **21,** 528 (1975).
226. Pekarek, R.S., Wannemacher, R.W., and Beisel, W.R., *Proc. Soc. Exp. Biol. Med.* **140,** 685 (1972).
227. Pekas, J.C., *Am. J. Physiol.* **211,** 407 (1966); *J. Anim. Sci.* **27,** 1559 (1968).
228. Pensack. J.M., and Klussendorff, R.C., *Poult. Nutr. Conf., 1956.*
229. Pensack, J.M., Henson, J.N., and Pogdonorff, P.D., *Poult. Sci.* **37,** 1232 (1958).
230. Perry, H.M., Jr., and Schroeder, H.A., *Am. J. Med.* **22,** 168 (1957).
231. Perry, T.W., Beeson, W.M., Smith, W.H., and Mohler, M.T., *J. Anim. Sci.* **27,** 1674 (1968).
232. Picciano, M.F., and Guthrie, H.A., *Fed. Proc., Fed. Am. Soc. Exp. Biol.* **32,** 929 (1973).
233. Pitts, W.J., Miller, W.J., Fosgate, O.T., Morton, J.D., and Clifton, C.M., *J. Dairy Sci.* **49,** 455 (1966).
234. Pond, W.G., and Jones, J.R., *J. Anim. Sci.* **23,** 1057 (1964).
235. Pories, W.J., Henzel, J.H., and Hennessen, J.A., *Trace Subst. Environ. Health–1, Proc. Univ. Mo. Annu. Conf., 1st, 1967,* p. 114 (1968).
236. Pories, W.J., Henzel, J.H., Rob, C.G., and Strain, W.H., *Lancet* **1,** 121 (1967); *Ann. Surg.* **165,** 432 (1967).
237. Prasad, A.S., and Oberleas, D., *J. Lab. Clin. Med.* **83,** 634 (1974).
238. Prasad, A.S., Halsted, J.A., and Nadimi, M., *Am. J. Med.* **31,** 532 (1962).
239. Prasad, A.S., Miale, A., Farid, Z., Sandstead, H.H., Schulert, A.R., and Darby, W.J., *Arch. Intern. Med.* **111,** 407 (1963).
240. Prasad, A.S., Oberleas, D., and Halsted, J.A., *J. Lab. Clin. Med.* **66,** 508 (1965).
241. Prasad, A.S., Oberleas, D., Wolf, P., and Horwitz, J.P., *J. Clin. Invest.* **46,** 549 (1967); *J. Lab. Clin. Med.* **73,** 486 (1969).
242. Prasad, A.S., Schulert, A.R., Sandstead, H.H., Miale, A., and Farid, Z., *J. Lab. Clin. Med.* **62,** 84 (1963).
243. Prasad, A.S., Oberleas, D., Wolf, P., Horwitz, J., Miller, E.R., and Luecke, R.W., *Am. J. Clin. Nutr.* **22,** 628 (1969).
244. Prohaska, J.R., Luecke, R.W., and Jasinski, R., *J. Nutr.* **104,** 1525 (1974).
245. Pryor, W.J., private communication (1975).
246. Quarterman, J., and Florence, E., *Br. J. Nutr.* **28,** 75 (1972).
247. Quarterman, J., Mills, C.F., and Humphries, W.R., *Biochem. Biophys. Res. Commun.*

25, 354 (1966).

248. Rahman, M.M., Davies, R.E., Deyoc, C.W., Reid, B.L., and Couch, J.R., *Poult. Sci.* **40**, 195 (1960).
249. Rahmat, A., Norman, J.N., and Smith, G., *Br. J. Surg.* **61**, 271 (1974).
250. Raum, N.S., Stables, G.L., Pope, L.S., Harper, O.F., Waller, G.R., Renbarger, R., and Tillman, A.D., *J. Anim. Sci.* **27**, 1695 (1968).
251. "Recommended Dietary Allowances," 8th ed. Natl. Acad. Sci., Washington, D.C., 1974.
252. Reinhold, J.G., Kfoury, G.A., and Arslanian, M., *J. Nutr.* **96**, 519 (1968).
253. Reinhold, J.G., Kfoury, G.A., Chalambor, M.A., and Bennett, J.C., *Am. J. Clin. Nutr.* **18**, 294 (1966).
254. Reinhold, J.G., Parsa, A., Karimian, N., Hammick, J.W., and Ismael-Beigi, F., *J. Nutr.* **104**, 976 (1974).
255. Richards, M.P., and Cousins, R.J., *Bioinorg. Chem.* **4**, 215 (1975).
256. Rickard, B.F., *N. Z. Vet. J.* **23**, 41 (1975).
257. Robinson, M.F., McKenzie, J.M., Thomson, C.D., and Van Rij, A.L., *Br. J. Nutr.* **30**, 195 (1973).
258. Ronaghy, H.A., Reinhold, J.G., Maloudji, M., Ghavani, P., Fox, M.R.S., and Halsted, J.A., *Am. J. Clin. Nutr.* **27**, 112 (1974); Ronaghy, H.A., and Halsted, J.A., *ibid,* **28**, 831 (1975).
259. Rosoff, B., and Martin, C., *Fed. Proc., Fed. Am. Soc. Exp. Biol.* **25**, 316 (abstr.) (1966).
260. Roth, H.P., and Kirchgessner, M., *Z. Tierphysiol., Tierernaehr. Futtermittelkd.* **32**, 289 and 296 (1974); Kirchgessner, M., Stadler, A.E., and Roth, H.P., *Bioinorg. Chem.* **5**, 33 (1975).
261. Roth, H.P., and Kirchgessner, M., *Z. Tierphysiol., Tierernaehr. Futtermittelkd.* **33**, 57, 62, and 67 (1974).
262. Rubin, H., *Proc. Natl. Acad. Sci. U.S.A.* **69**, 712 (1972).
263. Sadavisan, V., *Biochem. J.* **48**, 527 (1951); **49**, 186 (1951); **52**, 452 (1952).
264. Sahagian, B.M., Harding-Barlow, I., and Perry, H.M., Jr., *J. Nutr.* **90**, 259 (1966).
265. Sandstead, H.H., *Am. J. Clin. Nutr.* **26**, 1251 (1973).
266. Sandstead, H.H., Fosmire, G.J., McKenzie, J.M., and Halas, E.Z., *Fed. Proc., Fed. Am. Soc. Exp. Biol.* **34**, 86 (1975).
267. Sandstead, H.H., Lanier, V.C., Shepard, G.H., and Gillespie, D.D., *Am. J. Clin. Nutr.* **23**, 514 (1970).
268. Sandstead, H.H., and Rinaldi, R.A., *Proc. Soc. Exp. Biol. Med.* **128**, 687 (1968).
269. Sandstead, H.H., Gillespie, D.D., and Brady, R.N., *Pediatr. Res.* **6**, 119 (1972).
270. Sandstead, H.H., Prasad, A.S., Schulert, A.S., Farid, Z., Miale, A., Bassily, S., and Darby, T.W., *Am. J. Clin. Nutr.* **20**, 422 (1967).
271. Savlov, E.D., Strain, W.H., and Hulgin, F., *J. Surg. Res.* **2**, 209 (1962).
272. Schlage, C., and Wortberg, B., *Acta Paediatr. Scand.* **61**, 425 (1972).
273. Schlicker, S.A., and Cox, D.H., *J. Nutr.* **95**, 287 (1968).
274. Schroeder, H.A., and Nason, A.P., *J. Invest. Dermatol.* **53**, 71 (1969).
275. Schroeder, H.A., Nason, A.P., Tipton, I.H., and Balassa, J.J., *J. Chronic Dis.* **20**, 179 (1967).
276. Schwarz, F.J., and Kirchgessner, M., *Int. J. Vitam. Nutrr. Res.* **44**, 258 (1974).
277. Schwarz, W.A., and Kirchgessner, M., *Z. Tierphysiol., Tierernaehr. Futtermittelkd.* **35**, 1 (1975).
278. Scott, M.L., and Zeigler, T.R., *J. Agric. Food. Chem.* **11**, 123 (1963).
279. Scott, M.L., Holm, E.R., and Reynolds, R.E., *Poult. Sci.* **38**, 1344 (1959).

280. Scrutton, M.C., Wu, C.W., and Goldthwait, D.A., *Proc. Natl. Acad. Sci. U.S.A.* **68**, 2497 (1971).
281. Smirnov, A.A., *Biokhimiya* **13**, 79 *Chem. Abstr.* **42**, 8302 (1948).
282. Smith, H., *J. Forensic Sci. Soc.* 7, 97 (1967).
283. Smith, J.E., Brown, E.D., and Smith, J.C., Jr., *J. Lab. Clin. Med.* **84**, 692 (1974).
284. Smith, W.H., Plumlee, M.P., and,Beeson, W.M., *Science* **128**, 1280 (1960).
285. Soman, S.D., Panday, V.K., Joseph, K.T., and Raut, S.J., *Health Phys.* **17**, 35 (1969).
286. Somers, M., and Underwood, E.J., *Aust. J. Agric. Res.* **20**, 899 (1969).
287. Somers, M., and Underwood, E.J., *Aust. J. Biol. Sci.* **22**, 1229 (1969).
288. Spencer, H., Osis, D., Kramer, L., and Norris, C., *Trace Subst. Environ. Health–5, Proc. Univ. Mo. Annu. Conf., 5th, 1971* p. 193 (1972).
289. Spencer, H., and Rostoff, B., *Health Phys.* **12**, 475 (1966).
290. Spencer, H., and Samachson, J., *in* "Trace Element Metabolism in Animals" (C.F. Mills, ed.), Vol. I, p. 312. Livingstone, Edinburgh, 1970.
291. Spray, C.M., and Widdowson, E.M., *Br. J. Nutr.* **4**, 361 (1951).
292. Stake, P.E., Miller, W.J., Blackmon, D.M., Gentry, R.P., and Neathery, M.W., *J. Nutr.* **104**, 1279 (1974).
293. Stake, P.E., Miller, W.J., Gentry, R.P., and Neathery, M.W., *J. Anim. Sci.* **40**, 132 (1975).
293a. Stake, P.E., Miller, W.J., Neathery, M.W., and Gentry, R.P., *J. Dairy Sci.* **58**, 78 (1975).
294. Starcher, B., and Kratzer, F.H., *J. Nutr.* **79**, 18 (1963).
295. Stevenson, S.S., *J. Clin. Invest.* **22**, 403 (1943).
296. Strain, W.H., Huegin, J., Lankau, C.A., Berliner, W.P., McEvoy, R.K., and Pories, W.J., *Int. J. Radiat. Isotop.* **15**, 231 (1964).
297. Strain, W.H., Steadman, L.T., Lankau, C.A., Berliner, W.P., and Pories, W.P., *J. Lab. Clin. Med.* **68**, 244 (1966).
298. Sullivan, J.F., and Lankford, H.E., *Am. J. Clin. Nutr.* **10**, 153 (1962).
299. Sutton, W.R., and Nelson, V.E., *Proc. Soc. Exp. Biol. Med.* **36**, 211 (1937).
300. Swenerton, H., and Hurley, L.S., *J. Nutr.* **95**, 8 (1968).
301. Swift, C.E., and Berman, M.D., *Food. Technol.* **13**, 365 (1969).
302. Szmigielski, S., and Litivin, J., *Cancer* **17**, 1381 (1964).
303. Talbot, T.R., and Ross, J.F., *Lab. Invest.* **9**, 174 (1964).
304. Tao, S., and Hurley, L.S., *J. Nutr.* **105**, 220 (1975).
305. Terhune, M.W., and Sandstead, H.H., *Science* **177**, 68 (1972).
306. Thiers, R.E., and Vallee, B.L., *J. Biol. Chem.* **226**, 911 (1957).
307. Thomas, B., Thompson, A., Oyenuga, V.A., and Armstrong, R.H., *Emp. J. Exp. Agric.* **20**, 10 (1952).
308. Thomson, R.W., Gilbreath, R.L., and Bielk, E., *J. Nutr.* **105**, 154 (1975).
309. Tipton, I.H., and Cook, M.J., *Health Phys.* **9**, 103 (1963).
310. Todd, J.R., private communication (1969).
311. Todd, W.R., Elvehjem, C.A., and Hart, E.B., *Am J. Physiol.* **107**, 146 (1934).
312. Tribble, H.M., and Scoular, F.I., *J. Nutr.* **52**, 209 (1954).
313. Tucker, H.F., and Salmon, W.D., *Proc. Soc. Exp. Biol. Med.* **88**, 613 (1955).
314. Tupper, R., Watts, R.W.E., and Wormall, A., *Biochem. J.* **57**, 254 (1954).
315. Underwood, E.J., *World's Poult. Congr., Proc., 12th, 1962,* p. 216 (1962).
316. Underwood, E.J., and Somers, M., *Aust. J. Agric. Res.* **20**, 889 (1969).
317. Vallee, B.L., and Gibson, J.G., *J. Biol. Chem.* **176**, 445 (1948); *Blood* **4**, 455 (1949).
318. Vallee, B.L., Hoch, F.L., Adelstein, S.J., and Wacker, W.E.C., *J. Am. Chem. Soc.* **78**, 5879 (1956).

319. Vallee, B.L., and Neurath, H., *J. Biol. Chem.* **217,** 253 (1955).
320. Vallee, B.L., Wacker, W.E.C., Batholomay, A.F., and Hoch, F.L., *Ann. Intern. Med.* **50,** 1077 (1959).
321. Vallee, B.L., Wacker, W.E.C., Bartholomay, A.F., and Robin, E.D., *N. Engl. J. Med.* **255,** 403 (1956); **257,** 1055 (1957).
322. Van Campen, D.R., *J. Nutr.* **97,** 104 (1969).
323. Van Campen, D.R., and Mitchell, E.A., *J. Nutr.* **86,** 120 (1965).
324. Van Campen, D.R., and Scaife, P.U., *J. Nutr.* **91,** 473 (1967).
325. Van Reen, R., *Arch. Biochem. Biophys.* **46,** 337 (1953).
326. Vikbladh, I., *Scand. J. Clin. Lab. Invest.* **2,** 143 (1950).
327. Vohra, P., and Kratzer, F.H., *J. Nutr.* **82,** 249 (1964).
328. Volkov, N.F., *Fed. Proc., Fed. Am. Soc. Exp. Biol.* **22,** Trans. Suppl., T897 (1963).
329. Wacker, W.E.C., Ulmer, D.D., and Vallee, B.L., *N. Engl. J. Med.* **255,** 449 (1956).
330. Wakely, J.C.N., Moffat, B., Crook, A., and Mallard, J.R., *Appl. Radiat.* **7,** 225 (1960).
331. Walravens, P.A., and Hambidge, K.M., *Am. J. Clin. Nutr.,* in press.
332. Warren, H.V., and Delavault, R.E., *Mem. Geol. Soc. Am.* **123,** (1971).
333. Wasserman, R.H., *J. Nutr.* **77,** 69 (1962).
334. Wegner, T.N., Ray, D.E., Lox, C.D., and Scott, G.H., *J. Dairy Sci.* **56,** 748 (1973).
335. Weitzel, G., *Angew. Chem.* **68,** 566 (1966).
336. Weitzel, G., and Fretzdorff, A.M., *Hoppe-Seyler's Z. Physiol. Chem.* **292,** 221 (1953).
337. Weitzel, G., Stecker, F.J., Roester, U., Buddecke, E., and Fretzdorff, A.M., *Hoppe-Seyler's Z. Physiol. Chem.* **296,** 19 (1954).
338. Welch, R.R., House, W.A., and Allaway, W.H., *J. Nutr.* **104,** 733 (1974).
339. Westmoreland, N., *Fed. Proc., Fed. Am. Soc. Exp. Biol.* **30,** 1001 (1971).
340. Westmoreland, N., First, N.L., and Hoekstra, W.G., *J. Reprod. Fertil.* **13,** 223 (1958).
341. Whiting, F., and Bezeau, L.M., *Can. J. Anim. Sci.* **38,** 109 (1958).
342. Widdowson, E.M., McCance, R.A., and Spray, C.M., *Clin. Sci.* **10,** 113 (1951).
343. Wilkins, P.J., Grey, P.C., and Dreosti, I.E., *Br. J. Nutr.* **27,** 113 (1972).
344. Witham, I.J., *Biochim. Biophys. Acta* **11,** 509 (1963).
345. Yamagata, N., and Iwashima, K., *Bull. Inst. Public Health, Tokyo* **11,** 131 (1962).
346. Young, R.J., Edwards, H.M., and Gillis, M.B., *Poult. Sci.* **37,** 1100 (1958).
347. Ziegler, T.R., Scott, M.L., McEvoy, R., Greenlaw, R.H., Huegin, F., and Strain, W.H., *Proc. Soc. Exp. Biol. Med.* **109,** 239 (1962).
348. Zook, E.G., Greene, F.E., and Morris, E.R., *Cereal Chem.* **51,** 788 (1974).

9

Cadmium

I. CADMIUM IN ANIMAL TISSUES AND FLUIDS

Cadmium is virtually absent from the human body at birth and accumulates with age up to about 50 years. At this age the average person not exposed to abnormal amounts of cadmium has a total body burden of 20–30 mg Cd, of which one-half to two-thirds occurs in the liver and kidneys (76, 84, 111). Anke and Schneider (1) found that cadmium increased in human tissues continuously with age, with kidney Cd concentrations of about 30 ppm in persons who had died from accident or cardiac infarction and 75–100 ppm in those who had died from diseases of the lungs and malignant neoplasms. Piscator (77) gives mean renal cortical Cd levels at age 50 of 24–50 ppm for some European countries and the United States and 60–120 ppm for three areas in Japan. The critical renal cortex concentration beyond which renal damage can be expected has been placed at 200 mg/kg wet weight (111). Substantial regional variation in kidney Cd levels has been reported (75). These are a reflection of differences in Cd intakes, mainly from the food, although heavy cigarette smoking can contribute significant amounts to the body's burden of cadmium (42).

The special concentration of cadmium in the liver and kidneys, particularly the kidney cortex, is apparent from many studies with several species. For example, Hamilton *et al.* (31) reported the following mean values for adult human tissues in England: kidney whole, 13.9 ± 0.7; kidney cortex, 14.3 ± 2.9; kidney medulla, 12.3 ± 2.8; liver, 4.3 ± 1.0; lung, 2.3 ± 0.8; brain, 0.30 ± 0.04; testis, 0.30 ± 0.09; ovary, 0.10 ± 0.03; lymph nodes, 0.06 ± 0.02; and muscle,

0.03 ± 0.01 μg Cd/g wet weight. A renal Cd gradient occurs in human kidney, with the concentration in the outer cortex being twice that of the inner medulla (44). The mean Cd concentrations in the tissues of young sheep fed a diet containing 0.2 ppm Cd for 191 days from 4 months of age were as follows: kidneys, 4.42; liver, 1.69; lungs, 0.13; testes, 0.12; heart, 0.06; spleen, 0.14; muscle, 0.025; and fat, 0.011 μg/g dry weight (15, 16). Cadmium does not accumulate markedly in bone as do Pb, Zn, and Sr (11, 31), except apparently in the teeth. Thus human dental enamel has been reported to contain from 0.03 to 6.70 (mean 0.99) μg Cd/g (45). The teeth of growing pigs receiving Cd supplements at levels of 50, 150, 450, and 1350 ppm of the diet contained 12.8, 32.4, 76.1, and 211.5 μg Cd/g, respectively (11).

The levels of cadmium in the liver and kidneys and to a lesser extent those in other tissues reflect dietary Cd intakes over a wide range. This is evident from studies with rats (12) and dogs (2) given cadmium in the drinking water. This is also apparent from two recent investigations with swine (11) and sheep (15, 16) in which the diets were supplemented with graded increments of Cd up to toxic levels. In the swine the Cd concentrations of the liver and kidneys were increased by all dietary Cd supplements, whereas those of the skin and muscles were only significantly elevated at the highest Cd intakes (450 and 1350 ppm of the diet). In the sheep the Cd supplements similarly increased liver and kidney Cd levels, but the levels in the muscle and fat, although increased by all levels of Cd, remained relatively low (14) (Table 32). In rats administered $CdCl_2$ by injection the accumulation of Cd in the liver is roughly proportional to dosage level (109). Control rats had 0.5 μg Cd/g of liver (wet weight) while rats injected with 0.5, 1.0, 2.0, 2.5, and 3.0 mg Cd/kg body weight had, respectively, 8.5, 27, 60, 60, and 73 μg Cd/g of liver. The appearance of pathological changes in the animal and the liver depended on the liver Cd level. It seems that at low liver Cd levels the metal is solely and safely bound to metallothionein, as discussed in the next paragraph and in Section IV,2, but as accumulation increases there is a spillage of Cd to other proteins and signs of toxicity appear.

The retention of Cd in the liver and kidneys is related to its selective storage or sequestration in the protein metallothionein, discovered by Margoshes and Vallee (47) in equine renal cortex and found to contain as high as 5.9% Cd and 2.2% Zn. Metallothionein from the livers of Cd-pretreated rats and chickens contains 7 g-atoms of metal ions per 12,000 g protein and has a Zn:Cd ratio of 1:1.4 ± 0.1 (106). The formation of this Cd-binding protein can be induced in rats by both oral Cd administration and subcutaneous Cd injections (92, 104). Reports of the occurrence of similar Cd-binding proteins in other tissues and fluids have appeared (see Bremner, 3). In fact cadmium is believed to occur throughout the body mainly bound to the low molecular weight (5000–10,000) protein metallothionein. This compound forms associative complexes with Zn, Cu, and Hg, and can bind up to 11% of total metal. It is believed to serve a

TABLE 32
Cadmium Concentrations in the Tissues of Lambs Fed Dietary Cadmium over a 191-Day Period[a]

Treatment	No. of animals	Cadmium concentration (μg/g) Liver	Kidney	Muscle	Fat
Control	6	1.69 ± 0.26	4.42 ± 0.50	0.025 ± 0.0005	0.011 ± 0.001
5 ppm Cd	6	14.92 ± 1.51	58.86 ± 3.50	0.047 ± 0.0014	0.010 ± 0.001
15 ppm Cd	6	51.72 ± 4.17	187.62 ± 19.5	0.091 ± 0.0014	0.012 ± 0.001
30 ppm Cd	6	62.73 ± 3.13	426.81 ± 30.3	0.170 ± 0.004	0.021 ± 0.002
60 ppm Cd	6	275.94 ± 38.69	768.84 ± 83.3	0.428 ± 0.012	0.113 ± 0.023

[a]From Doyle *et al.* (16).

detoxification function for Cd^{2+} (93, 109) and to reduce the toxicity of Hg^{2+} ions and possibly that of other divalent metals (94).

Normal human blood is low and variable in Cd content. Imbus *et al.* (33) reported a range of 0.3–5.4 μg Cd/100 ml with a median concentration of 0.7 μg/100 ml. Kubota and co-workers (41) determined the Cd content of the blood of 243 adults from 19 cities in the United States. Less than one-half had detectable amounts of Cd by the atomic absorption method used and no consistent geographical pattern was apparent. More than half the samples contained 0.5 μg Cd/100 ml or less, and 83% were below 1 μg/100 ml. The median concentration was stated to be close to 0.5 μg/100 ml. Very similar mean levels, with an equally wide variation (0.38–2.16 μg Cd/100 ml), have been reported for the blood of cattle (13).

The urinary Cd excretion of individuals with no known abnormal exposure to cadmium is low and variable, and according to Fassett (18) is usually less than 5 μg/day. However, McKenzie (49) found the mean urinary excretion of 42 male students in New Zealand to be 17.8 ± 11.6 μg Cd/day and that of 54 female students to be 16.3 ± 10.5 μg Cd/day. The overall mean urinary Cd concentration was 14 μg/liter, with a very wide range of 1–42 μg/liter. These concentrations are of the same order as those found in other investigations (72, 95). In a comparison of normal and 15 hypertensive patients an average of less than 1 μg Cd/liter in the urine of the former group and nearly 50 μg/liter in the latter were found (74). During treatment to control the blood pressure in the hypertensive group a fall in urinary Cd to about 5 μg/liter was observed. On the other hand, Szadowski *et al.* (99) found no correlation between the urinary Cd excretion and arterial pressure of 169 individuals not occupationally exposed to industrial sources of this element. Smith and Kench (95) reported values of 15–420, with most samples lying between 40 and 100 μg Cd/liter for workers exposed to cadmium oxide dust and 40–410 μg/liter for those exposed to cadmium fumes.

The Cd concentration of cow's milk varies among individuals and in different locations. Milk samples from 32 individual cows ranged from 20 to 37 μg Cd/liter, and the market milk from different cities in the United States averaged from 17 to 30 μg Cd/liter (55). The national weighted average was 26 μg Cd/liter, and there were significant variations from one sampling period and from one city to another. Regional, seasonal, and individual variations with lower average values for cow's milk are apparent from two other U.S. studies (13, 41). The level of cadmium in the milk of other species has been little studied, but Murthy and Rhea (55) found 22 samples of human milk to average 0.019 ± 0.027 ppm, or close to 19 μg Cd/liter.

The mean Cd concentration in human hair has been reported as 2.76 ± 0.48 μg/g in males and 1.77 ± 0.24 in females (88). These levels are slightly higher than the 1–2 ppm quoted for human hair by Friberg *et al.* (24) or for calf hair by Powell and co-workers (78). The Cd concentration of wool was found to range from 0.74 to 0.94 (mean 0.83) and from 0.55 to 1.22 (mean 0.90) μg/g, with no significant increase from dietary Cd supplementation (16).

II. CADMIUM METABOLISM

The salient features of Cd metabolism are (a) lack of an effective homeostatic control mechanism, (b) tenacious retention in the body with a long half-life, estimated as 16–33 years in man (38) and about 200 days in rats (52a), and (c) powerful interactions with other divalent metals, both at the absorptive level and in the tissues, particularly in the liver and kidney.* In studies with injected ^{109}Cd in the mouse almost no total body turnover of this element was observed, regardless of challenges with dietary Cd loads. Furthermore, absorption occurred irrespective of the body's Cd burden (10).

Cadmium uptake into the walls of intact strips of rat intestine occurs in the duodenum, jejunum, and ileum and is enhanced by Zn at 10^{-5} or 10^{-3} *M* and depressed at higher Zn concentrations (81). A well-defined competition for transmural transport, i.e., across the intestinal wall, was observed between Cd and Zn and between Cd and Hg. A severe depression by cadmium of ^{64}Cu uptake from ligated segments of rat intestine has been demonstrated (103). A significant reduction in ^{59}Fe absorption was similarly found in chicks fed 75 ppm of dietary Cd (23) and in mice fed a low-Fe diet and given Cd in the drinking water or as intragastric Cd doses (30). Cadmium had a similar effect on

*Author's note: Evidence that cadmium may be essential for growth in the rat has now appeared [Schwarz, K., and Spallholz, J., *Fed. Proc.* **35**, 255 Abstr. (1976)].

Co absorption. In the mice with low-Fe absorption Cd absorption was concentration dependent, whereas in mice with high-Fe absorption, Cd absorption appeared to be increased by the enhanced activity of a mucosal uptake step common to both metals. It seems, therefore, that cadmium shares a common or similar absorptive mechanism with several other metals and that competition within this mechanism provides at least a partial explanation of the mutual antagonism among them. It also explains some of the toxic manifestations of Cd in the animal, to be considered in Section IV. There is some evidence that Cd lowers the absorption of calcium and increases its excretion from the digestive tract. For example, Kobayashi (39) found that 30–100 ppm Cd mixed in a Ca-deficient diet resulted in a significant loss of calcium from rat bone, with much more fecal excretion of Ca than that gained from the diet. Simultaneously supplementing a diet with amounts of zinc, manganese, and copper above the requirement level of each results in decreased Cd concentrations in the liver and kidneys of Japanese quail when the dietary Cd levels are relatively high (10, 20, and 40 mg Cd/kg) and the dietary levels of zinc, manganese, and copper are high, and also when the dietary concentrations of Cd are lower and the excesses of the zinc, manganese, and copper are more modest (20). These effects on tissue accumulation of cadmium presumably reflect at least in part the antagonistic action of the divalent cations on Cd absorption.

Ingested cadmium is poorly absorbed from most diets, probably at about 3–8% in man (24). Only about 0.01% of the body burden of cadmium is excreted daily, i.e., excretion is very slow. Cadmium balance studies in animals are few and somewhat discordant with respect to the pathways of excretion. In two studies in human adults, Cd retention was found to be extremely low and considerable absorption and excretion in the urine were apparent (80, 101). In investigations with cows (51) and growing sheep (16) receiving dietary Cd supplements, by contrast, apparent Cd absorption ranged from 5 to 11% and excretion was mainly via the feces. Cadmium excretion in the urine and reexcretion in the bile were negligible. Several studies of Cd injections into laboratory animals further indicate that the feces constitute the major route of Cd excretion (see Schroeder *et al.,* 90). Inhaled cadmium is better absorbed (10–40%) than ingested cadmium, depending on its physical state (111). This cadmium is widely distributed from the lungs to the tissues, accumulates in the liver and kidneys, and is excreted in greatly increased concentrations in the urine (95).

The placenta and mammary gland effectively limit Cd transport into the fetus and milk, respectively (46). The Cd concentrations in normal human embryos and fetuses are extremely low (5).

Aspects of Cd metabolism are discussed further in Section IV and in the chapters on copper, zinc, and selenium, in relation to Cd interactions with those elements.

III. SOURCES OF CADMIUM

Cadmium enters the biosphere through its increasing use in electroplating, in plastics as stabilizers, in paints as pigments, in cadmium batteries, and as a contaminant in phosphate fertilizers and sewage sludges. In the absence of Cd-emitting factories, such as zinc refineries, the levels in the air approximate 0.001 μg Cd/m^3 which would lead to a maximum inhaled amount of 0.02 μg/person/day. In large cities higher levels approaching 0.03 μg/m^3 may be found (111). In the 28 U.S. cities studied by Carroll (4) the Cd level in the air ranged from "undetectable" to as high as 0.06 μg/m^3. The amount inhaled from the air in most circumstances is insignificant compared with that ingested with the food, with the exception of heavy smokers who could have an intake of 5 μg Cd/day or more from this source alone (24, 42). Moreover such inhaled Cd is much better absorbed than ingested Cd, as mentioned in the previous section. Most municipal waters contain less than 1–3 μg Cd/liter (19), which is well below the upper limit for drinking water of 10 μg/liter set by the World Health Organization (110). Even at this upper level the consumption of 2.5 liters would provide only 25 μg Cd/day. Food is thus normally the major source of cadmium to animals and nonsmoking humans.

1. Cadmium in Human Foods and Dietaries

Estimates of daily dietary Cd intakes by man are extremely variable. Robinson and co-workers (80) reported a range of 60–92 μg/day for young adult New Zealand women, and Murthy *et al.* (56) obtained average intakes of 27–64 μg/day for children in institutions. Duggan and Lipscomb (17) give a mean intake of 26 μg Cd/day from their "market basket" survey in the United States, and Hamilton and Minski (32) reported a mean of 64 ± 30 μg Cd/day for English adult diets. This is very close to the 67 μg/day estimated for Canadians, based on analysis of foods sampled during 1970–1971 (37). In another survey of British diets the average daily intake from food and beverages was found to be only 15–30 μg or to "amount to not more than 250 μg Cd/week" (98). It has recently been authoratively stated that human dietary Cd intake "probably varied from country to country from 50 μg/day, or less, up to 150 μg/day" (111).

In addition to regional differences Cd intakes by man also vary considerably with the amounts and types of food consumed. Oysters are exceptionally rich in cadmium with 3–4 ppm wet weight (90) or higher when grown in industrially contaminated waters. Limpets, winkles, and canned anchovies may also approach this level (90, 98), but all other foods are normally much lower. Milk and meat are poor sources of cadmium, except for the kidney. The edible portions of fruits, vegetables, and nuts generally contain 0.04–0.08 ppm Cd, with higher concentrations in vegetables grown on soils heavily fertilized with superphosphate or treated with $CdCl_2$ (107). A significant tendency toward increasing Cd concentration in Swedish wheat with time, between 1916 and 1972, has been demonstrated (38). How far this is due to the fertilizer used and how far to atmospheric Cd contamination is not known.

Losses of cadmium occur in the milling of wheat grain. Linman *et al.* (43a) reported levels of 0.05 ppm Cd in Swedish wheat, 0.033 ppm in flour, and 0.148 ppm in bran. Zook *et al.* (112) obtained the following mean values for wheat and flour: common hard wheat, 0.10 ± 0.02; common soft wheat, 0.07 ± 0.02; and baker's patent flour, 0.05 ± 0.01 ppm Cd dry basis. Their finding of 0.16 and 0.19 ppm Cd (dry basis) in white bread suggests that sources other than flour contribute significant amounts of Cd to the bread. A lower mean level of 0.022 ppm Cd has been reported for 16 Australian-grown wheats (107). Similar levels of Cd occur in rice (107) and maize (corn) from a range of geographical locations (102). Garcia *et al.* (25) report a mean of 0.055 ± 0.032 and the wide range of 0.035–0.148 μg Cd/g dry basis for whole kernel corn grown in different parts of the United States. Cadmium concentrations as high as 1 ppm have been produced in soybean seeds by the application of 144 metric tons/hectare of sludge solids to the soil (35).

2. Cadmium in Animal Feeds and Roughages

Data on the Cd content of animal feeds are sparse and fragmentary. It is evident from the preceding section that the cereal grains are generally low in Cd, although heavy fertilization with superphosphate or sewage sludge can significantly increase these concentrations. In a study of Australian superphosphates 38–48 ppm Cd was found. This cadmium was water soluble and appeared to be as readily available to plants as the cadmium in $CdCl_2$ (107). The cadmium contents of mixed feeds in Japan as reported by Tsuyagawa and Ohno (102) are as follows: for laying hens, 0.09–0.37 (mean 0.20); for dairy cattle, 0.07–0.27 (mean 0.14); and for growing swine, 0.07–0.19 (mean 0.09) mg Cd/kg.

From the limited data provided by Williams and David (107) it seems that clovers are normally richer in cadmium than grasses and respond in Cd content to superphosphate applications even more than grasses. For example, *Phalaris tuberosa* and *Trifolium subterraneum* from unfertilized pasture contained 0.020 and 0.048 ppm Cd, whereas on pastures that had received a total of 2500 or 2060 kg/hectare of superphosphate the Cd levels were found to be 0.060 and 0.072 in the *Phalaris* and 0.324 and 0.164 ppm in the subterranean clover. Ryegrass and subterranean clover from similarly fertilized pastures contained 0.153 and 0.411 ppm Cd, respectively. Mixed pasture herbage remote from industrial areas in Scotland is reported to contain 0.1–0.8 ppm Cd, compared with 1–10 and 4–41 ppm for similar summer- and winter-grown pastures exposed to contamination from industrial sources (52). Similarly, normal rice straw in Japan contained 0.1–0.3 ppm Cd, compared with 0.7–3.6 ppm for straw from "polluted" paddy fields (59).

IV. CADMIUM TOXICITY

1. General

Cadmium is toxic to virtually every system in the animal body, whether ingested, injected, or inhaled. Histological changes have been observed in the kidneys, liver, gastrointestinal tract, heart, testes, pancreas, bones, and blood vessels (33a, 85, 97, 108), and hepatic protein-bound Cd has been associated with emphysema and other chronic pulmonary diseases in a group of patients without unusual contact with cadmium (43). Anemia is a common manifestation of chronic Cd toxicity in all species, due at least in part to its metabolic antagonism to copper and iron. Changes in the integument characteristic of Zn deficiency can also be induced by Cd, acting similarly as a metabolic Zn antagonist. In rabbits fed 160 ppm Cd as $CdCl_2$ for 200 days, retarded growth, anemia, neutrophilia, lymphopenia, hypoalbuminemia, elevated α_1-, β_1-, and γ-globulins, splenomegaly, cardiomegaly, and renal enlargement were observed (97). The most striking morphological change produced by this long-term ingestion of toxic Cd levels was interlobar hepatic and interstitial renal fibrosis.

In order to produce adverse physiological effects with an adequate diet it is claimed that "5 mg Cd/kg diet are usually required" (20). However, minimum toxic levels or maximum safe dietary Cd levels cannot be given with any precision because Cd metabolism is so powerfully influenced by the dietary intakes of other elements with which it interacts, notably Zn, Cu, Fe, and Se. The toxicity of a particular intake is thus determined by the extent to which the interacting elements are present or absent from the diet. Furthermore, substances which modify the metabolism of the interacting elements can influence Cd toxicity. For example, ascorbic acid added to the diet of quail fed 75 mg Cd/kg of diet significantly alleviated or prevented almost all aspects of Cd toxicity at 4 or 6 weeks of age, due largely to improved utilization of dietary iron and/or zinc (21, 79). Increasing the availability of iron and zinc is also the most likely explanation of the reduction in toxicity of a 7.5 mg Cd/kg diet brought about in young Japanese quail by dried egg white compared with soybean protein or casein gelatine as the main protein source (22). By contrast Stowe *et al.* (96) have reported that high dietary pyridoxine levels increased the severity of the anemia produced by cadmium in rats.

Cadmium, as $CdCl_2$, added to a basal diet well supplied with iron and zinc at levels of 0, 50, 150, 450, and 1350 ppm over a 6-week period decreased the growth rate of young swine as a function of Cd level and stopped growth altogether in the 1350 ppm group. Hematocrit values, which were found to be the most sensitive indicator of toxicity, decreased in all Cd-fed animals. Bone ash content also decreased as a function of Cd intake (11). Male lambs provided with diets containing 0, 5, 15, 30, and 60 ppm Cd for 191 days exhibited a significant reduction in growth rate and feed intake at the two higher levels, but no effect on feed efficiency was apparent and only a slight reduction in hematocrit was observed (16). Dietary levels of 40 and 160 ppm Cd considerably reduced feed intake and weight gain in calves, but there was little clinical evidence of toxicity (78). Japanese quail given supplemental cadmium at the rate of 75 ppm of the diet developed a severe anemia, associated with increased plasma transferrin levels and transferrin to albumin ratios (34). Similar levels of cadmium fed to this species for 4–6 weeks induced testicular hypoplasia and growth retardation, severe anemia and bone marrow hyperplasia, hypertrophied heart ventricles, enteropathy of the small intestine, and decreased granules in the adrenal medullary cells (79).

For man a "provisional tolerable weekly intake" of 400–500 μg Cd per person has been proposed by the World Health Organization (111). This approximates 1 μg/kg body weight for most individuals or 55–70 μg Cd/day. Some normal diets would supply higher Cd intakes than these. However, no chronic toxic effects from the oral ingestion of cadmium have been reported in man not exposed to industrial sources of the element, with the possible exception of the incidence of emphysema and other bronchial disorders mentioned earlier in this section (43). The Cd poisoning observed in Japan (itai-itai or ouch-ouch disease) and manifested in renal and gastrointestinal lesions and osteomalacia resulted from industrial contamination of the food and water supply (40, 54). Furthermore, it occurred mostly in postmenopausal, multiparous women consuming poor diets low in protein and calcium. Itokawa *et al.* (33a) have shown that the body growth depression and severe bone changes brought about by cadmium are increased markedly when rats are given a low-protein, Ca-deficient diet.

2. Reproductive Disturbances

Reference to injurious effects on the kidneys was made in the early studies of Cd toxicity (91), but it was not until 1956 that Parizek and Zahor (66) showed that a single small injection of $CdCl_2$ induced selective hemorrhagic necrosis of the testes of rats. An analogous necrosis also occurs in the epididymis (27). Similar effects on the testis from injected Cd have been observed in the mouse (9, 50), rabbit, hamster, and guinea pig (61). The necrosis appears to be caused primarily by interference of Cd with testicular blood supply (9, 27, 57). Protection against these effects on the testes can be secured by concurrent injections of zinc (26, 48, 60), selenium (36, 48), estrogens, cysteine, and to a lesser extent glutathione (28). The drastic toxic effects of Cd on the testes are paralleled, in certain circumstances, by comparable effects on the female. Cadmium injections into young rats maintained in a permanent state of estrus induce massive hemorrhagic necrosis in their ovaries (62). Injection of Cd salts into pregnant rats results in complete destruction of the pars fetalis of the placenta and transformation of this organ into an extensive blood clot. In all cases pregnancy is interrupted with either resorption or delivery of the dead conceptus (63). This Cd-induced toxemia of pregnancy can be completely prevented by selenium, when injected as sodium selenite or selenate (65). Selenium compounds similarly protect against the toxic effects of bivalent mercury (64).

No general relationship exists between the Cd concentration in testicular tissue and the degree of testicular necrosis (58), and the protective effect of Se is associated with an increase in testicular Cd content (29). This increase is associated with a 2- to 3-fold *reduction* in the amount of Cd present in a fraction with a molecular weight of about 10,000 and an increase in Cd-binding to a high molecular weight protein in which Se also occurs (7). A novel Cd protein of around 30,000 mol. wt. has also been isolated which is very unstable and into which injected Cd rapidly accumulates (6). It therefore seems that the particular sensitivity of the testis to Cd may be due to the formation of this reactive Cd compound, and that the protection afforded by Se is due to its diversion of Cd to a higher molecular weight inactive compound. The diversion in binding of the Cd in the soluble fraction to higher molecular weight proteins has also been observed in the kidney and liver (8). This may be a second mechanism involved in the protection of these organs against Cd by Se, i.e., additional to the effect of Se in reducing their Cd concentrations.

The protective effect of prior administration of Zn against Cd toxicity has been ascribed to the more rapid accumulation of the hepatic and renal Cd as metallothionein, especially as such bound Cd is not mobilized during subsequent pregnancy to cause the toxemia and fetal malformations described earlier (105).

3. Cadmium and Hypertension

Cadmium induces hypertension in rats, rabbits and dogs, either when injected intraperitoneally (87) or intraarterially (76), or when administered orally (70, 82). Thus weanling rats given 1, 2.5, or 5 ppm Cd in the drinking water for a year had average increases in systolic pressure of 13–19 mm Hg (70). Comparable increases in systolic pressure were observed in rats fed drinking water containing 2.5 or 10 ppm Cd for a period of 6 months from weaning (68). Larger increases have been reported in male rabbits given 6–7 weekly intraperitoneal injections of cadmium acetate at 2 mg/kg body weight (100). The mean systolic ear blood pressure increased from 86 ± 8.0 mm Hg prior to treatment to

TABLE 33
Effect of Selenium on Weight and Cadmium-Induced Hypertension in Female Rats[a]

No. of rats	Cd in drinking water (mg/liter)	Se in drinking water (mg/liter)	Body weight at 6 months (g)	Systolic[b] pressure ± SD (mm Hg)
12	0	0	33	114 ± 14[c]
12	2.5	0	324	123 ± 14
12	10.0	0	320	135 ± 15[c]
7	0	0.9	343	125 ± 11
7	2.5	0.9	296	109 ± 11
7	0	3.5	334	120 ± 15
7	10	3.5	290	112 ± 9 [c]

[a]From Perry and Erlanger (68).
[b]Systolic pressure measured in tail.
[c]Systolic pressures in lines 1 and 3 and in lines 3 and 7 differ significantly ($p > 0.005$).

215 ± 10.0 mm Hg 1 week after the last injection was given. The corresponding values for the untreated rabbits were 95 ± 4.0 and 98 ± 4.0.

The hypertension resulting from cadmium can be reversed by injection of the chelate Na_2Zn–CDTA (cyclohexane diamine tetraacetic acid), which has a somewhat higher stability constant for cadmium than for zinc and would therefore deplete tissue Cd and replete tissue Zn (86). Administration of this chelate in the drinking water of Cd hypertensive rats reduced both Cd and Zn in the tissues and resulted in a regression of the hypertension (89). The increase in systolic pressure that follows low doses of cadmium to rats is entirely blocked by even lower doses of selenium (68). The results with female rats given drinking water containing 0, 2.5, or 10 mg Cd/liter, and/or 0, 0.9, or 3.5 mg Se/liter, while consuming a low-Cd diet for 6 months from the time of weanling are presented in Table 33.

The mechanism of the chronic Cd hypertension observed in animals is poorly understood, although some significant pointers have been obstained. It develops at dose levels insufficient to induce other signs of Cd toxicity. In fact at higher levels hypertension does not develop and systolic pressure in rats may actually decrease (70). Cadmium hypertensive rabbits carry higher concentrations of cadmium in the heart, pulmonary and mesenteric arteries, aorta, liver, and kidney than normal untreated rabbits, and their responsiveness to angiotension in the aortic strip is significantly lower (100). In addition the administration of Cd directly into the renal artery results in a dose-related reversible inhibition of the vasopressor renal constrictions brought about by injections of angiotensin, epinephrine, and norepinephrine (100). Cadmium is also known to influence vascular resistance (71) and salt and water metabolism (14, 73). For example, Doyle *et al.* (14) found that both Na retention and NaCl intake were significantly greater in rats given 5 μg Cd/ml in the drinking water for 23–42 weeks than in controls. Since sodium retention has been observed in some humans with hypertension it was suggested that such retention could precede the occurrence of hypertension in rats. Perry and Erlanger (69) have further shown that cadmium, in addition to increasing blood pressure, elevates the apparent renin activity in peripheral blood. In fact Cd-injected rats exhibited simultaneous and significant elevations in both blood pressure and renin activity, with a significant correlation ($p < 0.005$) between the two parameters.

The significance of the relationship, if any, between cadmium and human hypertension is uncertain. Hypertension is not a characteristic of Cd poisoning in man where other signs of toxicity are apparent, and the studies of Morgan showed that kidney and liver Cd concentrations are not related to the incidence of hypertensive or degenerative vascular disease in man (53). However, hypertensive patients have been reported to have significantly more renal cadmium than similar normotensive patients (83) and to excrete more cadmium in the urine (67). Furthermore, Thind (100) observed a significant elevation of plasma Cd levels in 32 patients with different forms of hypertension, compared with 15 normal subjects. It is clear that as Thind (100) has stated, "a program of detailed epidemiological studies and public health measures is needed for further definition of the role of cadmium in 'idiopathic' human hypertension."

REFERENCES

1. Anke, M., and Schneider, H.J., *Arch. Exp. Veterinaermed.* **25**, 805 (1971).
2. Anwar, R.E., Langham, R.J., Hoppert, C.A., Alfredson, B.V., and Byerrum, R., *Arch. Environ. Health* **3**, 456 (1961).
3. Bremner, I., *Q. Rev. Biophys.* **7**, 75 (1974).
4. Carroll, R.E., *J. Am. Med. Assoc.* **198**, 177 (1966).
5. Chaube, S., Nishimura, H., and Swigyard, C.A., *Arch. Environ. Health* **26**, 237 (1973).
6. Chen, R., Hoekstra, W.G., and Ganther, H.E., *Fed. Proc., Fed. Am. Soc. Exp. Biol.* **32**, Abstr. 3994 (1974).
7. Chen, R., Wagner, P., Ganther, H.E., and Hoekstra, W.G., *Fed. Proc., Fed. Am. Soc. Exp. Biol.* **31**, Abstr. 2725 (1972).
8. Chen, R.W., Whanger, P.D., and Weswig, P.H., *Bioinorg. Chem.* **4**, 125 (1975).
9. Chiquoine, A.D., *Anat. Rec.* **149**, 23 (1964).
10. Cotzias, G.C., Borg, D.C., and Selleck, B., *Am. J. Physiol.* **201**, 63 and 927 (1961).
11. Cousins, R.J., Barber, A.K., and Trout, J.R., *J. Nutr.* **103**, 964 (1973).
12. Decker, L.E., Byerrum, R.U., Decker, C.F., Hoppert, C.A., and Langham, R.J., *Arch. Ind. Health* **18**, 228 (1958).

13. Dorn, C.R., Pierce, J.O., Chase, G.R., and Phillips, P.E., *Trace Subst. Environ. Health–7, Proc. Univ. Mo. Annu. Conf., 7th, 1973,* p. 191 (1973).
14. Doyle, J.J., Bernhoft, R.A., and Sandstead, H.H., *J. Lab. Clin. Med.* **86,** 57 (1975).
15. Doyle, J.J., and Pfander, W.H., *J. Nutr.* **105,** 599 (1975).
16. Doyle, J.J., Pfander, W.H., Grebing, S.E., and Pierce, J.O., *J. Nutr.* **104,** 160 (1974).
17. Duggan, R.E., and Lipscomb, G.Q., *Pestic. Monit. J.* **2,** 153 (1969).
18. Fassett, D.W., *in* "Metallic Contaminants and Human Health" (D.H.K. Lee, ed.), p. 97. Academic Press, New York, 1972.
19. Fleischer, M., Sarofim, A.F., Fassett, D.W., Hammond, P., Shacklette, H.T., Nisbet, I.C.T., and Epstein, S., *Environ. Health Perspect.* p. 253 (1974).
20. Fox, M.R.S., *in* "Trace Elements and Human Disease" (A.S. Prasad, ed.), Vol. 2, pp. 401–416. Academic Press, New York, 1976.
21. Fox, M.R.S., Fry, B.E., Harland, B.F., Schertel, M.E., and Weeks, C.E., *J. Nutr.* **101,** 1295 (1971).
22. Fox, M.R.S., Jacobs, R.M., and Harland, B.F., *Fed. Proc., Fed. Am. Soc. Exp. Biol.* **32,** 924 (1973).
23. Freeland, J.H., and Cousins, R.J., *Nutr. Rep. Int.* **8,** 337 (1973).
24. Friberg, L., Piscator, M., and Nordberg, G., *in* "Cadmium in the Environment." Chem. Rubber Publ. Co., Cleveland, Ohio, 1971.
25. Garcia, W.J., Blessin, C.W., and Inglett, G.E., *Cereal Chem.* **51,** 788 (1974).
26. Gunn, S.A., Gould, T.C., and Anderson, W.A.D., *Arch. Pathol.* **71,** 274 (1961).
27. Gunn, S.A., Gould, T.C., and Anderson, W.A.D., *Am. J. Pathol.* **42,** 685 (1973).
28. Gunn, S.A., Gould, T.C., and Anderson, W.A.D., *Proc. Soc. Exp. Biol. Med.* **119,** 901 (1965); **122,** 1036 (1966).
29. Gunn, S.A., Gould, T.C., and Anderson, W.A.D., *J. Pathol. Bacteriol.* **96,** 89 (1968).
30. Hamilton, D.L., and Valberg, L.S., *Am. J. Physiol.* **227,** 1033 (1974).
31. Hamilton, E.I., Minski, M.J., and Cleary, J.J., *Sci. Total Environ.* **1,** 341 (1972/1973).
32. Hamilton, E.I., and Minski, M.J., *Sci. Total Environ.* **1,** 375 (1972/1973).
33. Imbus, H.R., Cholak, J., Miller, L.H., and Sterling, T., *Arch. Environ. Health* **6,** 286 (1963).
33a. Itokawa, Y., Tomoko, A., and Tanaka, S., *Arch. Environ. Health* **26,** 241 (1973).
34. Jacobs, R.M., Fox, M.R.S., and Aldridge, M.H., *J. Nutr.* **99,** 119 (1969).
35. Jones, R.L., Hinesly, T.D., and Ziegler, E.L., *J. Environ. Qual.* **2,** 351 (1973).
36. Kar, A.B., and Das, R.P., *Proc. Natl. Inst. Sci. India, Part B* **29,** (Suppl.), 297 (1963).
37. Kirkpatrick, D.C., and Coffin, D.E., *Can. Inst. Food Sci. Technol. J.* **7,** 56 (1974).
38. Kjellstrom, T., *Nord. Hyg. Tidskr.* **53,** 111 (1971); Kjellstrom, T., Lind, T.B., Linman, L., and Elinder, C.G., *Arch. Environ. Health* **30,** 321 (1975).
39. Kobayashi, J., *Trace Subst. Environ. Health–7, Proc. Univ. Mo. Annu. Conf., 7th, 1973,* p. 295 (1973).
40. Kobayashi, J., Morii, F., Muramoto, S., and Nakashima, S., *Jpn. J. Hyg.* **25,** 364 (1970).
41. Kubota, J., Lazar, A., and Losee, F., *Arch. Environ. Health* **16,** 788 (1968).
42. Lewis, G.P., Jusko, W.J., Coughlin, L.L., and Hartz, S., *Lancet* **1,** 291 (1972).
43. Lewis, G.P., Lyles, H., and Miller, S., *Lancet* **2,** 1330 (1969).
43a. Linman, L., Andersson, A., and Nilsson, K.O., *Arch. Environ. Health* **27,** 45 (1973).
44. Livingstone, H.D., *Clin. Chem. (Winston-Salem, N.C.)* **18,** 67 (1972).
45. Losee, F., Cutress, T.W., and Brown, R., *Trace Subst. Environ. Health–7, Proc. Univ. Mo. Annu. Conf., 7th, 1973,* p. 19 (1973).
46. Lucis, O.J., Lucis, R., and Shaikh, Z.A., *Arch. Environ. Health* **25,** 14 (1972).
47. Margoshes, M., and Vallee, B.L., *J. Am. Chem. Soc.* **79,** 4813 (1957).

48. Mason, K.E., Young, J.O., and Brown, J.E., *Anat. Rec.* **148**, 309 (1964).
49. McKenzie, J.M., *Proc. Univ. Otago Med. Sch.* **50**, 16 (1972).
50. Meek, E.S., *Br. J. Exp. Pathol.* **40**, 503 (1959).
51. Miller, W.J., Lampp, B., Powell, G.W., Salotti, C.A., and Blackmon, D.M., *J. Dairy Sci.* **50**, 1404 (1967).
52. Mills, C.F., and Dalgarno, A.C., *Nature (London)* **239**, 171 (1972).
52a. Moore, W., Jr., Stara, J.F., Crocker, W.C., Malanchuk, M., and Iltis, R., *Environ. Res.* **6**, 473 (1973).
53. Morgan, J.M., *Arch. Intern. Med.* **123**, 405 (1969).
54. Murata, I., Hirono, T., Saeki, Y., and Nakagawa, S., *Bull. Soc. Int. Chir.* **29**, 34 (1969).
55. Murthy, G.K., and Rhea, U., *J. Dairy Sci.* **51**, 610 (1968); **54**, 1001 (1971).
56. Murthy, G.K., Rhea, U., and Peeler, J.T., *Environ. Sci. Technol.* **5**, 436 (1971).
57. Niemi, M., and Kormano, M., *Acta Pathol. Microbiol. Scand.* **63**, 513 (1965).
58. Nordberg, G.F., *Environ. Physiol. & Biochem.* **2**, 7 (1972).
59. Ogura, Y., *JARQ* **6**, 117 (1971).
60. Parizek, J., *J. Endocrinol.* **15**, 56 (1957).
61. Parizek, J., *J. Reprod. Fertil.* **1**, 294 (1960).
62. Parizek, J., *Proc. Int. Congr. Endocrinol., 2nd, 1964* (1965).
63. Parizek, J., *J. Reprod. Fertil.* **7**, 263 (1964).
64. Parizek, J., Benes, I., Ostadalova, I., Babicky, A., Benes, J., and Pitha, J., *in* "Mineral Metabolism in Paediatrics" (D. Barltrop and W.L. Burland, eds.). Blackwell, Oxford, 1969.
65. Parizek, J., Ostadalova, I., Benes, I., and Babicky, A., *J. Reprod. Fertil.* **16**, 507 (1968).
66. Parizek, J., and Zahor, Z., *Nature (London)* **177**, 1036 (1956).
67. Perry, H.M., Jr., *Fed. Proc., Fed. Am. Soc. Exp. Biol.* **20**, 254 (1961).
68. Perry, H.M., Jr., and Erlanger, M.W., *Fed. Proc., Fed. Am. Soc. Exp. Biol.* **33**, 357 Abstr. (1974).
69. Perry, H.M., Jr., and Erlanger, M., *J. Lab. Clin. Med.* **82**, 399 (1973).
70. Perry, H.M., Jr., and Erlanger, M.W., *J. Lab. Clin. Med.* **83**, 541 (1974).
71. Perry, H.M., Jr., Erlanger, M., and Yunice, A., *J. Lab. Clin. Med.* **70**, 963 (1967).
72. Perry, H.M., Jr., and Perry, E.F., *J. Clin. Invest.* **38**, 1452 (1959).
73. Perry, H.M., Jr., Perry, E.F., and Purifoy, J.E., *Proc. Soc. Exp. Biol. Med.* **136**, 1240 (1971).
74. Perry, H.M., Jr., and Schroeder, H.A., *Circulation* **12**, 758 (1955); *J. Lab. Clin. Med.* **46**, 936 (1955).
75. Perry, H.M., Jr., Tipton, I.H., Schroeder, H.A., Steiner, R.L., and Cook, M.J., *J. Chronic Dis.* **14**, 259 (1961).
76. Perry, H.M., Jr., and Yunice, A., *Proc. Soc. Exp. Biol. Med.* **120**, 805 (1965).
77. Piscator, M., *Trace Subst. Environ. Health–7, Proc. Univ. Mo. Annu. Conf., 7th, 1973,* p. 31 (1973).
78. Powell, G.W., Miller, W.J., Morton, J.D., and Clifton, C.M., *J. Nutr.* **84**, 205 (1964).
79. Richardson, M.E., Fox, M.R.S., and Fry, B.E., *J. Nutr.* **104**, 323 (1974).
80. Robinson, M.F., McKenzie, J.M., Thomson, C.D., and Van Rij, A.L., *Br. J. Nutr.* **30**, 195 (1973).
81. Sahagian, B.M., Harding-Barlow, I., and Perry, H.M., Jr., *J. Nutr.* **90**, 259 (1966); **93**, 291 (1967).
82. Schroeder, H.A., *Am. J. Physiol.* **207**, 62 (1964).
83. Schroeder, H.A., *J. Chronic Dis.* **18**, 647 (1965).

84. Schroeder, H.A., and Balassa, J.J., *J. Chronic Dis.* **14**, 236 (1961).
85. Schroeder, H.A., Balassa, J.J., and Vinton, W.H., *J. Nutr.* **86**, 51 (1965).
86. Schroeder, H.A., and Bickman, J., *Arch. Environ. Health* **14**, 693 (1967).
87. Schroeder, H.A., Kroll, S.S., Little, J.W., Livingstone, P.O., and Myers, M.A.G., *Arch. Environ. Health* **13**, 788 (1966).
88. Schroeder, H.A., and Nason, A.P., *J. Invest. Dermatol.* **53**, 71 (1969).
89. Schroeder, H.A., Nason, A.P., and Mitchener, M., *Am. J. Physiol.* **214**, 796 (1968).
90. Schroeder, H.A., Nason, A.P., Tipton, I.H., and Balassa, J.J., *J. Chronic Dis.* **20**, 179 (1967).
91. Schwartze, E.W., and Alsberg, C.L., *J. Pharmacol. Exp. Ther.* **21**, 1 (1923).
92. Shaikh, Z., and Lucis, O.J., *Fed. Proc., Fed. Am. Soc. Exp. Biol.* **29**, Abstr. 301 (1970).
93. Shaikh, Z.A., and Lucis, O.J., *Arch. Environ. Health* **24**, 419 (1972).
94. Shaikh, Z.A., Coleman, R.J., and Lucis, O.J., *Trace Subst. Environ. Health–7, Proc. Univ. Mo. Annu. Conf., 7th, 1973,* p. 313 (1973).
95. Smith, J.C., and Kench, J.E., *Br. J. Ind. Med.* **14**, 270 (1957).
96. Stowe, H.D., Goyer, R.A., Medley, P., and Cates, M., *Arch. Environ. Health* **28**, 209 (1974).
97. Stowe, H.D., Wilson, M., and Goyer, R.A., *Arch. Pathol.* **94**, 389 (1972).
98. "Survey of Cadmium in Food." HM Stationery Office, London, 1973.
99. Szadowski, von D., Schaller, K.H., and Lehnert, G., *Z. Klin. Chem. Klin. Biochem.* **7**, 551 (1969).
100. Thind, G.S., *J. Air Pollut. Control Assoc.* **22**, 267 (1972).
101. Tipton, I.H., and Stewart, P.L., *Trace Subst. Environ. Health–3, Proc. Univ. Mo. Annu. Conf., 3rd, 1969,* p. 305 (1970).
102. Tsuyagawa, H., and Ohno, S., *Bull. Natl. Feeds & Fertil. Inspect. Off., Tokyo* No. 3, p. 128 (1973).
103. Van Campen, D.R., *J. Nutr.* **88**, 125 (1966).
104. Webb, M., *Biochem. Pharmacol.* **21**, 2751 (1972).
105. Webb, M., *Biochem. Pharmacol.* **21**, 2767 (1972); *J. Reprod. Fertil.* **30**, 99 (1972).
106. Weser, U., Rupp, H., Donay, F., Linnermann, F., Voelter, W., Voetsch, W., and Jung, G., *Eur. J. Biochem.* **39**, 127 (1973).
107. Williams, C.H., and David, D.J., *Aust. J. Soil Res.* **11**, 43 (1973).
108. Wilson, R.H., De Eds, F., and Cox, A.J., *J. Pharmacol. Exp. Ther.* **71**, 222 (1941).
109. Winge, D., Krasno, J., and Colucci, A.V., *in* "Trace Element Metabolism in Animals" (W.G. Hoekstra *et al.*, eds.), Vol. 2, p. 500. Univ. Park Press, Baltimore, Maryland, 1974.
110. World Health Organization, "International Standards for Drinking Water." W.H.O., Geneva, 1971.
111. World Health Organization, W.H.O., *Tech. Rep. Ser.* **505** (1972).
112. Zook, E.G., Greene, F.E., and Morris, E.R., *Cereal Chem.* **47**, 720 (1970).

10

Chromium

I. CHROMIUM IN ANIMAL TISSUES AND FLUIDS

Satisfactory methods of analysis (5, 9, 21, 57, 83) accompanied by low temperature processes of drying and ashing to avoid loss of chromium by volatilization have become available so recently that reliable data on the distribution of this element in the animal body are sparse and sporadic. There seems no doubt, however, that chromium is widely distributed throughout the human body in low concentrations without special concentration in any known tissue or organ, and that these levels decline with age, at least in U.S. populations, except in the lungs. Human stillborn and infant tissues carry higher Cr concentrations than those of adults. These levels decline rapidly in the first decade of life in the heart, lung, aorta, and spleen, while in the liver and kidney the neonatal concentrations are maintained until the second decade, when a decline occurs (70, 77). Substantial variations in human liver and kidney Cr levels have been observed in different geographical regions (70), presumably as a reflection of regional differences in environmental Cr intakes.

The reported levels of chromium in blood have declined markedly in recent years, but a reliable normal range for human blood can still not be given with complete confidence. Thus Hambidge (21) found the early morning serums of 36 normal subjects to range from 5–31 ng Cr/ml, with a mean of 13 ng/ml. Subsequently this worker obtained a mean of 7 ng Cr/ml (22), a level very close to that of Cary *et al.* (5). In two more recent studies lower normal serum levels, namely, 4.7 ± 0.15 (8) and 1.58 ng Cr/ml (57), were reported. Serum Cr levels decline in pregnancy (8) and under the stress of acute infectious illness (56). Since blood chromium is not in equilibrium with tissue stores, the level in the blood is not a good indicator of body chromium status. However, an increase in serum insulin concentration, whether by injection of insulin or by oral or parenteral administration of glucose, can lead to an acute increase of serum Cr within 30–120 min in normal healthy subjects (15). No such increase or a negligible increase has been observed in older people, except when receiving a Cr supplement, and in pregnant women (25). Furthermore, intravenous glucose given to individuals suffering from infection resulted in a still further decline in serum Cr concentration (56). Changes in serum Cr following glucose loading or insulin injection must therefore be treated with caution as indicators of Cr status.

The urinary Cr concentration of nine healthy young women consuming a typical U.S. diet ranged from 7.2 to 11.9 ng/ml (mean 9.4 ± 0.6), and the total excretion ranged from 5.9 to 10.0 (mean 7.2 ± 0.4) μg/day (53). Hambidge (22) gives the mean urinary Cr excretion of 20 normal young adults as 8.4 ± 5.2 μg/day and of 18 normal children (mean age 8 years) as 5.5 ± 2.9 μg/day. By contrast the Cr excretion of seven insulin-dependent diabetic children (mean age 11 years) was 19.2 ± 18.9 μg/day. Diabetics also excrete more intravenously administered ^{51}Cr than normal persons (10), and a strong negative correlation between Cr excretion and the efficiency of glucose utilization has been demonstrated in Turkish children (64). In this latter study there was no significant relation between urinary excretion and dietary Cr intakes, although increased urinary Cr levels occur at high intakes of the element (65). In chronically disabled subjects and drug addicts urinary Cr excretion is subnormal (19). However, urinary Cr excretion is closely related to glucose metabolism and can be a meaningful indicator of the Cr status of population groups.

The levels of chromium in hair provide a useful index of the Cr status of groups, although absolute levels indicative of Cr deficiency in the individual remain to be defined. The hair Cr concentration of normal newborn infants is significantly higher than that of older children (24), corresponding with the higher Cr levels known to be present in other tissues. Hambidge (22) reported a mean hair Cr concentration close to 900 ppb in 25 newborn infants, compared with approximately 440 ppb in 20 young children 24–36 months of age.

In premature newborn infants and those with evidence of intrauterine growth retardation the mean hair Cr levels are below normal (22, 23). The hair Cr levels of insulin-dependent diabetic children (27) and parous women (26) are also significantly lower than those of normal children and nulliparous women, respectively. For example, Hambidge and Rodgerson (26) obtained a mean of 0.75 μg Cr/g for the hair of 10 nulliparous women and 0.22 μg Cr/g for that of 11 parous women. These findings suggest a depletion of tissue Cr stores during pregnancy, which could be important where Cr intakes are low or marginal. In a later study, where the Cr nutrition was presumed to be more adequate, only a small, nonsignificant decline in hair Cr concentration was observed during pregnancy in women (25). The relation of hair Cr levels to nutritional Cr status may have further significance in light of the demonstration of substantial glucose tolerance factor (GTF) activity in ethanol extracts of hair follicles (46). This finding suggests that hair chromium is derived, at least in part, from a physiologically important body pool.

Some years ago mean levels of 57 and 13 ng Cr/g were reported for cow's colostrum and normal milk, respectively (34). The latter value compares well with the 10 ng/g given by Schroeder *et al.* (70) and the 8 ng/ml for undiluted cow's milk obtained more recently by Hambidge (22). This worker also examined 14 samples of breast milk from five women and reported a mean of 11.6 (range 6.4–18.5) ng Cr/ml.

Very few data on the chromium content of eggs are available. Kirkpatrick and Coffin (37) recently analyzed 100 shelled chicken egg samples from different parts of Canada. The Cr concentration reported for these samples ranged from less than 0.05 to 0.15, with a calculated mean of 0.06 μg/g fresh basis. Significant differences attributable to geographical origin were not apparent.

II. CHROMIUM METABOLISM

Inorganic Cr compounds are poorly absorbed in animals and man, to the extent of 1–3% or less, regardless of dose and dietary Cr status (11, 31, 48, 79). The absorbability of different Cr compounds at physiological levels does not appear to have been studied, but there is some evidence that the natural complexes in the diet are better available than simple Cr salts (44). In fact preliminary animal experiments suggest that 10–25% of the chromium in brewer's yeast is absorbed by rats (84).

Little is yet known of the site or mechanism of Cr absorption. In the rat the midsection of the small intestine appears to be the most diffusible segment for chromium, followed by the ileum and duodenum (6). Interaction with other cations and anions is also beginning to shed some light on the mechanism of absorption. Hahn and Evans (20) showed that in Zn-deficient rats ^{51}Cr absorption and intestinal contents were increased and that this increase is prevented by oral Zn administration. It was further shown that chromium inhibited ^{65}Zn absorption in Zn-deficient rats and decreased the intestinal content of this isotope. These findings suggest that Cr and Zn may be metabolized by a common pathway in the intestine—a suggestion that finds support from their demonstration of the presence of the two metals in the same mucosal supernatant fraction and their identical behavior on an anion exchange column. The authors conclude that "the ligand that binds these two metals may be one of the sites of antagonism between chromium and zinc." A mutual metabolic antagonism between chromium and vanadium, or more particularly between chromate and vanadate, has also been demonstrated (28), as discussed in the chapter on vanadium. The presence of other anions constitutes a further factor that can affect Cr absorption. For example Chen *et al.* (6) found that oxalate significantly increased and phytate significantly decreased trivalent chromium transport through the rat intestine, both *in vitro* and *in vivo*, whereas two other chelating agents, citrate and EDTA, showed no significant effect.

Hexavalent chromium is better absorbed than trivalent chromium. Mackenzie *et al.* (40) observed a 3- to 5-fold greater blood radioactivity following intestinal administration of hexavalent ^{51}Cr than after trivalent ^{51}Cr; Donaldson and Barreras (11) obtained greater intestinal Cr uptake from $Na_2{}^{51}CrO_4$ than from $^{51}CrCl_3$ in man and in rats; and Mullor *et al.* (54) found the uptake of Cr^{6+} to be double that of Cr^{3+} in the organs of rats given 400 μg $^{51}Cr^{6+}$ or $^{51}Cr^{3+}$/day orally for several months.

Absorbed anionic hexavalent Cr readily passes through the membrane of the red cells and becomes bound to the globin fraction of the hemoglobin. Cationic trivalent Cr cannot pass this membrane; it combines with the β-globulin of the plasma and, in physiological quantities, is transported to the tissues bound to siderophilin (transferrin) (31). Tissue uptake is rapid and the plasma is cleared of a dose of ^{51}Cr within a few days (29, 79). Whole body radioactivity disappears much more slowly and can be expressed by at least three components, with half-lives of 0.5, 6, and 83 days, respectively (48). Hopkins (29) injected ^{51}Cr trichloride into rats at levels of 0.01 and 0.10 μg Cr/100 g body weight and found little difference in blood clearance, tissue distribution, or excretion due to dose level, previous diet, or sex. The bones, spleen, testis, and epididymis retained more ^{51}Cr after 4 days than the heart, lung, pancreas, or brain. The fact that various tissues retain chromium much longer than the plasma suggests that there is no equilibrium between tissue stores and circulating chromium, and therefore that plasma Cr levels may not be good indicators of body Cr status. Tissue uptake of chromate is markedly affected by age in mice (80). When older mice were used, the concentration of intraperitoneally injected ^{51}Cr in liver, stomach, epididymal fat pad, thymus, kidney, and especially the testes declined to almost half the values observed in young animals. As Mertz (44) has pointed out, "these observations may offer one explanation for the declining tissue chromium levels with age, detected in a survey of the U.S. population." The chromium entering the tissues is distributed among the subcellular fractions in unusual proportions. Edwards and co-workers (12) found 49% to be concentrated in the nuclear fraction, 23% in the supernatant, and the remainder divided equally between the mitochondria and the microsomes. A high concentration of chromium in nucleic acids has long been known (81) and the hypothesis proposed that chromium and other transition metals may play a role in nucleic acid metabolism (38).

Chromium is excreted mainly in the urine, whether ingested or injected, although small amounts are lost in the feces via the bile and small intestine, and possibly through the skin (29, 42). Urinary output alone may therefore underestimate the amount actually absorbed. A challenge with insulin or glucose is accompanied by an acute increment of chromium in the urine (67, 83), and in diabetic patients urinary Cr is also increased after a glucose load (10, 22, 67), as discussed in Section I. A significant increase in Cr excretion in the urine has been demonstrated in infants after the first 3–4 days of life, both in those who received intravenous glucose and in those who did not (63). No such response to glucose was evident in the first 36 hr of life. The indirect method of assessing body stores, based on increases in urinary Cr excretion following glucose loading, is thus not a reliable test in the first days of life, as it appears to be later in life. It seems that at this very early age Cr excretion parallels insulin release.

Several studies with rats have failed to detect any significant transfer of ^{51}Cr from the mother into the fetus, regardless of chemical binding and valence state (49, 59, 79), and despite clear evidence of measurable amounts of stable chromium in the newborn (59, 70) and in human embryos (58). Furthermore, newborn rats from mothers fed an ordinary commercial diet contained significantly more chromium than those from mothers fed a *Torula* yeast diet, or this diet supplemented with chromium acetate in the drinking water in amounts at least equal to that supplied in the commercial diet (49). These results, and the further finding that trivalent ^{51}Cr extracted from brewer's yeast readily crossed the placenta, indicate that chromium in the form of a natural complex, but not as a simple salt, passes the placental barrier. Since the regulation of glucose tolerance responds fairly well to simple chromium salts, it appears that the rat has a more limited capacity to convert such salts into the biologically active form necessary for placental transport than it has to convert them into the glucose tolerance factor (GTF).

III. CHROMIUM DEFICIENCY AND FUNCTIONS

Chromium deficiency is characterized by impaired growth and longevity in experimental animals and by disturbances in glucose, lipid, and protein metabolism. Rats maintained on diets low in protein and chromium also develop a corneal lesion visible in one or both eyes and manifested as a pronounced opacity of the cornea and congestion of the iridal vessels (60). Cr supplementation prevents the appearance of the lesion but does not cure the fully developed defect. The biochemical mechanism underlying this pathological change has not been determined.

1. Glucose Metabolism

In 1957 Schwarz and Mertz (74) observed impaired glucose tolerance in rats fed certain diets and postulated that the condition was due to a deficiency of a new dietary agent, designated the glucose tolerance factor or GTF. The active component was subsequently shown to be trivalent chromium, and various trivalent Cr compounds administered at dose levels of 20–50 μg Cr/100 g body weight were found to be fully effective in restoring tolerance to injected glucose (75). Fractionation studies of brewer's yeast yielded fractions with much greater biological activity than chromium as $CrCl_3$, suggesting the existence of a Cr-containing complex GTF (3). Efforts to purify this factor have led to the detection of nicotinic acid, glycine, glutamic acid, and cysteine, as well as chromium, in purified fractions (47). Synthetic complexes of these ligands with chromium have a biological activity similar to those of purified yeast fractions and greatly superior to those of simple Cr compounds, but they do not meet all the criteria established for GTF. The exact structure remains to be determined.

Other systems involving carbohydrate metabolism, including glucose uptake by the isolated rat lens (13), glucose utilization for lipogenesis and CO_2 production (50) (Table 34), and glycogen formation from glucose (62) respond to chromium plus insulin, with little or no response in the absence of the hormone. Under more severe Cr deficiency conditions a syndrome resembling diabetes mellitus, with fasting hyperglycemia and glycosuria, has been observed in rats and mice and shown to be rapidly reversed when 2 or 5 ppm Cr was supplied in the drinking water (51, 66). These findings are consistent with the hypothesis that a decreased sensitivity of peripheral tissue to insulin is the primary biochemical lesion in Cr deficiency. In Cr-deficient systems the response to doses of insulin *in vitro* and *in vivo* is significantly inferior, and higher doses of the

TABLE 34
Production of $^{14}CO_2$ from [1-^{14}C]Glucose by Rat Adipose Tissue *In Vitro*[a]

Supplement	No. of rats	Insulin (μU/flask) 0	200	500	1000
None	5	10.7 ± 0.8	17.9 ± 2.5	15.4 ± 1.1	18.7 ± 2.0
5 ppm Cr	5	19.9 ± 1.3	19.9 ± 7.7	33.9 ± 8.0	38.9 ± 7.0

[a]From Mertz *et al.* (50). Values are expressed as mmoles CO_2/100 mg tissue.

hormone are required to elicit metabolic responses similar to those of Cr-sufficient controls, whereas in the absence of insulin these responses in Cr-deficient animals or tissues do not differ significantly from those of controls. Chromium thus emerges not as an insulinlike agent but, in the words of Mertz (45), as "a true potentiator of the action of the hormone."

A severe impairment of glucose tolerance is characteristic of kwashiorkor or protein–calorie malnutrition (PCM) of infants (1). In some areas (17, 32), but not in others (4), this condition can be rapidly and markedly improved by Cr treatment. For example, Hopkins and co-workers (32) treated six such malnourished infants from Jordan and six from Nigeria with 250 μg of chromium as $CrCl_3$. Within 18 hr the glucose removal rates of the Jordanian infants improved from an average of 0.6 to 2.9%/min, and those of the Nigerian infants from 1.2 to 2.9%/min. In nine out of 14 Turkish infants suffering from marasmic protein–calorie malnutrition the glucose removal rate showed a similar striking response to the same dose of $CrCl_3$, and the effect of the single dose of chromium continued during the period of observation lasting from 8 to 40 days (17).

Severe chromium deficiency in a human subject exhibiting weight loss, peripheral neuropathy, impaired glucose tolerance, and subnormal blood and hair Cr concentrations while on prolonged total parenteral nutrition, and responsive to Cr therapy, has recently been reported (33a). The results indicated that isolated Cr deficiency in man causes glucose intolerance, inability to utilize glucose for energy, neuropathy with normal insulin levels, and no impairment of insulin action on amino acid uptake and FFA release.

Several therapeutic trials with elderly and middle-age subjects with chemical evidence of impaired glucose tolerance have revealed responses to Cr supplementation, although these responses are slow and delayed compared with the rapid results obtained with infants. Thus Glinsmann and Mertz (14) obtained significant improvement in glucose tolerance in four out of six maturity onset diabetics from the daily administration of 180–1000 μg chromium as $CrCl_3$ for periods of 7–13 weeks. Shorter periods of Cr supplementation did not improve glucose utilization. In another study, 50% of a group of similar subjects treated with 150 μg Cr daily for 6 months displayed a marked improvement in glucose tolerance (30). Many elderly subjects over the age of 70 exhibit impaired glucose tolerance. Levine *et al.* (39) restored the glucose tolerance of four out of 10 of such elderly individuals treated with 150 μg Cr daily for 4 months. No improvement in glucose tolerance was observed in 10 adult diabetics treated with 150 μg Cr daily for 16 weeks in a double-blind crossover study conducted by Sherman and associates (76). It is apparent that many individuals are ingesting insufficient chromium to maintain normal glucose utilization. The fact that not every malnourished infant or every diabetic or old person responds to Cr supplementation indicates that the Cr status of the nonresponders is adequate for this function, or that one of the many other factors influencing carbohydrate metabolism is operating in these cases.

2. Lipid Metabolism

There is some evidence that chromium plays a role in serum cholesterol homeostasis. The addition of chromium to a low-Cr diet suppressed serum cholesterol levels in rats and in males inhibited the tendency of these levels to increase with age. Further, a depression in serum cholesterol was achieved in male rats by feeding 1 μg Cr/ml in the drinking water, with a similar effect in females fed 5 μg Cr/ml in this way (69, 72). This effect was not specific to chromium since it was also observed in the rats receiving lead and cadmium in the drinking water. Confirmatory evidence implicating chromium was obtained in later experiments by Schroeder (68). Serum cholesterol levels were relatively elevated and increased with age in the rats receiving white sugar, very low in Cr, whereas those receiving brown sugar or white sugar plus Cr were low. The effects were similar in both sexes and younger rats fed raw sugar had lower levels than those fed white sugar (68). A significant decline in serum cholesterol has been reported in some institutionalized patients fed 2 mg Cr as the acetate daily for a period of 5 months. Other patients given similar treatment showed no such response (67).

Examination of the aortas of rats at the end of their natural lives revealed a significantly lower (2%) incidence of spontaneous plaques in the Cr-fed animals than in the Cr-deficient animals (19%). There were also lower amounts of stainable lipids and fluorescent material in the aorta (69). A satisfactory explanation of this effect of chromium has not yet appeared, although it should be noted that diabetes in man is associated with an increased incidence of vascular lesions (16). Further evidence of chromium involvement in lipid metabolism comes from the experiments of Mertz *et al.* (50), who found that chromium plus insulin significantly increased glucose uptake and incorporation of glucose carbon into epididymal fat in Cr-deficient rats (Table 34). Previously Curran (7) had observed that trivalent Cr enhances the synthesis of cholesterol and fatty acids from acetate in the livers of rats fed a commercial chow.

3. Protein Synthesis

Rats fed diets deficient in Cr and protein have an impaired capacity to incorporate several amino acids into the protein of their hearts (61, 62). Slightly improved incorporation was achieved with insulin alone, which was significantly enhanced by Cr(III) supplementation. The amino acids affected by Cr were α-aminoisobutyric acid, glycine, serine, and methionine. No such effect of Cr was observed with lysine, phenylalanine, and a mixture of 10 other amino acids.

Insulin *in vivo* also stimulated the cell transport of an amino acid analog to a greater degree in rats fed a low-protein, Cr-supplemented diet than it did in Cr-deficient controls (62). The claim that Cr acts as cofactor for insulin can therefore also be applied to two insulin-responsive steps in amino acid metabolism which are independent of the action of insulin on glucose utilization.

4. Growth and Longevity

On a diet of rye, skim milk, and corn oil with added vitamins and Zn, Cu, Mn, Co, and Mo in the water, male mice and rats receiving 2 or 5 ppm Cr(III) in the drinking water grew significantly better than their controls (71, 73). This effect was associated with decreased mortality and greater longevity (14). The median age of male mice at death was 99 days longer when they were fed chromium than when they were not, and the mean age was 91 days longer (73). No such differences in longevity due to chromium were observed in female mice or rats. The Cr treatment had no effect on the incidence of tumors but appeared to protect female rats against lung infection. Mertz and Roginski (46a) have shown that raising rats in plastic cages on a low-protein, low-Cr diet results in a moderate depression of growth which can be alleviated by Cr supplementation. Subjecting the animals to controlled exercise or blood loss aggravated the low-Cr state. A beneficial effect of supplemental Cr on weight gain in infants suffering from marasmic protein–calorie malnutrition has also been demonstrated (18).

IV. CHROMIUM SOURCES AND REQUIREMENTS

A number of reports of the Cr content of human foods and dietaries have appeared but many of these are of little value for nutritional purposes because (a) analytical methodology and instrumentation in the past have been inadequate, and (b) little is known of the forms of Cr present and their relative absorbability and biological activity. Toepfer and co-workers (78) measured the total Cr in the edible portions of a wide range of foods and obtained the relative biological activity of extracts of these foods by measuring the CO_2 production from glucose oxidation using [1-^{14}C]glucose in the presence of rat epididymal tissue and 100 units of insulin, as described by Mertz (44). The total Cr concentration in μg/g dry weight ranged from 2.16 and 1.83 in oysters and egg yolk, respectively, to 0.02 and 0.06 in cane sugar and grits, respectively. More than half of the foods examined contained between 0.2 and 0.8 μg Cr/g. The calculated relative biological values of the edible portion of selected foods as purchased ranged from 44.88 for brewer's yeast and 10.21 for black pepper to 1.75 for chicken breast and 1.59 for skimmed milk. Most of the grain and cereal products were in the middle group between the above extremes. In a recent study of 190 Canadian ready-to-eat cured meats the mean chromium contents ranged from 0.02 to 0.14 ppm on the fresh weight basis. The overall calculated mean was 0.06 (36). The chromium concentrations reported for milk and eggs were given in Section I.

Appreciable losses of chromium occur in the refining and processing of certain foods. Thus Zook *et al.* (85) report a mean level of 0.38 ± 0.06 μg Cr/g for common hard wheat and 0.37 ± 0.06 for common soft wheat, compared with 0.22 ± 0.08 for baker's patent flour and 0.29 ± 0.15 for soft patent flour. The recovery of Cr in white flour derived from known wheats was calculated as 35–44%. In a recent study of 34 samples of different wheat seed varieties grown in 12 different U.S. locations, Welch and Cary (82) obtained appreciably lower values for chromium. The Cr concentrations ranged from 3 to 43 with a mean of 17 ppb. Masironi *et al.* (43) analyzed molasses and unrefined, brown, and highly refined sugar from several countries for Cr content by flameless atomic absorption, with low temperature ashing. The mean levels obtained were 266 ± 58 for the molasses, 162 ± 36 for the unrefined sugar, 64 ± 5 for the brown sugar, and 20 ± 3 ng Cr/g for the refined (white) sugar. It was pointed out that the high intake of refined sugar in typical U.S. diets (about 120 g/day/person) not only contributes virtually no chromium but could lead to a loss of body Cr through the Cr-depleting action of glucose (15).

Dietary Cr intakes by man are clearly greatly influenced by the amounts and porportions of refined carbohydrates consumed. An institutional diet provided about 80 μg Cr/person/day in one study (70), while in two studies with diabetics and old people in which some responses to Cr supplementation were obtained, the daily intake was estimated to be as low as 50 μg (14, 39). Chromium intakes in the United States have been stated to "vary from 5 μg to over 100 μg/day" (84). Higher levels than these were reported for well-balanced Japanese diets (55). Three such recommended diets, containing varying proportions of the different food groups, were found to supply 130, 230, and 253 μg Cr/day, and even higher levels were obtained when cooked servings were analyzed.

The minimum human Cr requirements compatible with satisfactory growth and long-term health and fertility cannot yet be given because of inadequate knowledge of the forms and availability of chromium in foods. It has been calculated that a daily intake varying from 20 to 500 μg Cr, depending on the chemical nature of chromium in individual foods, would be needed to compensate for a urinary loss of 5 μg Cr/day (84). As indicated earlier in this chapter many individuals excrete more than 5 μg/day and insulin-dependent diabetics much more.

Nothing is known of the Cr requirements of domestic livestock and very little about Cr intakes from animal feeds. In two separate spectographic studies of pasture grasses Cr levels of 0.1–0.5 μg/g were reported (35, 52).

V. CHROMIUM TOXICITY

Hexavalent chromium is much more toxic than trivalent. In fact trivalent chromium has such a low order of toxicity that a wide margin of safety exists between the amounts ordinarily ingested and those likely to induce deleterious effects. Cats tolerate 1000 mg/day and rats showed no adverse effects from 100 mg/kg diet (84). Lifetime exposure to 5 mg/liter of chromium(III) in the drinking water induced no toxic effects in rats and mice, and exposure of mice for three generations to chromium oxide at levels up to 20 ppm of the diet had no measurable effect on mortality, morbidity, growth, or fertility (33).

Chronic exposure to chromate dust has been correlated with increased incidence of lung cancer (2), and oral administration of 50 ppm of chromate has been associated with growth depression and liver and kidney damage in experimental animals (41).

REFERENCES

1. Baig, H.A., and Edozien, J.C., *Lancet* **2**, 662 (1965).
2. Brinton, H.P., Fraiser, E.S., and Koven, A.L., *Public Health Rep.* **67**, 835 (1952).
3. Burkeholder, J.N., and Mertz, W., *Fed. Proc., Fed. Am. Soc. Exp. Biol.* **25**, 759 (1966).
4. Carter, J.P., Kattab, A., Abd-al-Hadi, K.A., Davis, J.T., Gholmy, A.E., and Patwardhan, V.N., *Am. J. Clin. Nutr.* **21**, 195 (1968).
5. Cary, E.E., and Allaway, W.H., *J. Agric. Food Chem.* **19**, 1159 (1971).
6. Chen, N.S.C., Tsai, A., and Dyer, I.A., *J. Nutr.* **103**, 1182 (1973).
7. Curran, G.L., *J. Biol. Chem.* **210**, 765 (1954).
8. Davidson, I.W.F., and Burt, R.L., *Am. J. Obstet. Gynecol.* **116**, 601 (1973).
9. Davidson, I.W.F., and Secrest, W.L., *Anal. Chem.* **44**, 1808 (1974).
10. Doisy, R.J., Streeten, D.H.P., Souma, L.M., Kalafer, M.E., Rekant, S.I., and Dalakos, T.G., *in* "Newer Trace Elements in Nutrition" (W. Mertz and W.E. Cornatzer, eds.), p. 155. Dekker, New York, 1971.
11. Donaldson, R.M., and Barreras, R.F., *J. Lab. Clin. Med.* **68**, 484 (1966).
12. Edwards, C., Olson, K.B., Heggen, G., and Glenn, J., *Proc. Soc. Exp. Biol. Med.* **107**, 94 (1961).
13. Farkas, T.G., and Robertson, S.L., *Exp. Eye Res.* **4**, 124 (1965).
14. Glinsmann, W.H., and Mertz, W., *Metab., Clin. Exp.* **15**, 510 (1966).
15. Glinsmann, W.H., Feldman, F.J., and Mertz, W., *Science* **152**, 1243 (1966).
16. Goldenberg, S., and Blumenthal, H.T., *in* "Diabetes Mellitus; Diagnosis and Treatment," p. 177. Am. Diabetes Assoc., New York, 1964.
17. Gürson, C.T., and Saner, G., *Am. J. Clin. Nutr.* **24**, 1313 (1971).
18. Gürson, C.T., and Saner, G., *Am. J. Clin. Nutr.* **26**, 988 (1973).
19. Gürson, C.T., Saner, G., Mertz, W., Wolf, W.R., and Sokiicii, S., *Nutr. Rep. Int.* **12**, 9 (1975).
20. Hahn, C.J., and Evans, G.W., *Am. J. Physiol.* **228**, 1020 (1975).

21. Hambidge, K.M., *Anal. Chem.* **43,** 103 (1971).
22. Hambidge, K.M., *in* "Newer Trace Elements in Nutrition" (W. Mertz and W.E. Cornatzer, eds.), p. 169. Dekker, New York, 1971.
23. Hambidge, K.M., and Baum, J.D., *Clin. Res.* **19,** 220 (1971).
24. Hambidge, K.M., and Baum, J.D., *Am. J. Clin. Nutr.* **25,** 376 (1972).
25. Hambidge, K.M., and Droegemueller, W., *Obstet. Gynecol.* **44,** 666 (1974).
26. Hambidge, K.M., and Rodgerson, D.O., *Am. J. Obstet. Gynecol.* **103,** 320 (1969).
27. Hambidge, K.M., Rodgerson, D.O., and O'Brien, D., *Diabetes* **17,** 517 (1968).
28. Hill, C.H., *in* "Trace Elements in Human Disease" (A.S. Prasad, ed.), Vol. 2, pp. 281–300. Academic Press, New York, 1975.
29. Hopkins, L.L., Jr., *Am. J. Physiol.* **209,** 731 (1965).
30. Hopkins, L.L., Jr., and Price, M.G., *Proc. West. Hemisphere Nutr. Congr., 2nd, 1968* Vol. II, p. 40 (1968).
31. Hopkins, L.L., Jr., and Schwarz, K., *Biochim. Biophys. Acta* **90,** 484 (1964).
32. Hopkins, L.L., Jr., Ransome-Kuti, O., and Majaj, A.S., *Am. J. Clin. Nutr.* **21,** 203 (1968).
33. Hutcheson, D.P., Gray, D.H., Venugopal, B., and Luckey, T.D., *J. Nutr.* **105,** 670 (1975).
33a. Jeejeebhoy, K.N., Shu, R., Marliss, E.B., Greenburg, G.R., and Bruce-Robertson, A., *Clin. Res.* **23,** 636A (1975).
34. Kirchgessner, M., *Z. Tierphysiol., Tierernaehr. Futtermittelkd.* **14,** 270 and 278 (1959).
35. Kirchgessner, M., Merz, G., and Oelschlager, W., *Arch. Tierernaehr* **10,** 414 (1960).
36. Kirkpatrick, D.C., and Coffin, D.E., *J. Sci. Food Agric.* **26,** 43 (1975).
37. Kirkpatrick, D.C., and Coffin, D.E., *J. Sci. Food Agric.* **26,** 99 (1975).
38. Kornicker, W.A., and Vallee, B.L., *Ann. N.Y. Acad. Sci.* **153,** 689 (1969).
39. Levine, R.A., Streeten, D.H.P., and Doisy, R.J., *Metab. Clin. Exp.* **17,** 114 (1968).
40. Mackenzie, R.D., Anwar, R., Byerrum, R.U., and Hoppert, C., *Arch. Biochem.* **79,** 200 (1959).
41. Mackenzie, R.D., Byerrum, R., Decker, C.F., Hoppert, C.A., and Langham, R., *AMA Arch. Ind. Health* **18,** 232 (1958).
42. Manusco, T.J., and Hueper, W.C., *Ind. Med. Surg.* **20,** 358 (1951).
43. Masironi, R., Wolf, W., and Mertz, W., *Bull. W. H. O.* **49,** 322 (1973).
44. Mertz, W., *Physiol. Rev.* **49,** 163 (1969).
45. Mertz, W., *Proc. Nutr. Soc.* **33,** 307 (1974).
46. Mertz, W., and Hambidge, K.M., unpublished data (1974).
46a. Mertz, W., and Roginski, E.E., *J. Nutr.* **97,** 531 (1969).
47. Mertz, W., Toepfer, E.W., Roginski, E.E., and Polansky, M.M., *Fed. Proc., Fed. Am. Soc. Exp. Biol.* **33,** 2275 (1974).
48. Mertz, W., Roginski, E.E., and Reba, R.C., *Am. J. Physiol.* **209,** 489 (1965).
49. Mertz, W., Roginski, E.E., Feldman, F.J., and Thurman, D.E., *J. Nutr.* **99,** 363 (1969).
50. Mertz, W., Roginski, E.E., and Schwarz, K., *J. Biol. Chem.* **236,** 318 (1961).
51. Mertz, W., Roginski, E.E., and Schroeder, H.A., *J. Nutr.* **86,** 107 (1965).
52. Mitchell, R.L., *Research (London)* **10,** 357 (1957).
53. Mitman, F.W., Wolf, W.R., Kelsay, J.L., and Prather, E.S., *J. Nutr.* **105,** 64 (1975).
54. Mullor, J.B., Vigil, J., Bielsa, L.B., Imaz, F., and Prat, J.C., *Rev. Fac. Ing. Quim. Univ. Nac. Litoral* **39,** 73 (1972).
55. Murakami, Y., Suzuki, Y., Yamagata, T., and Yamagata, N., *J. Radiat. Res.* **6,** 104 (1965).

56. Pekarek, R.S., Hauer, E.C., Bayfield, E.J., Wannemacher, R.W., and Beisel, W.R., *Diabetes* **24**, 350 (1975).
57. Pekarek, R.S., Hauer, E.C., Wannemacher, R.W., and Beisel, W.R., *Anal. Biochem.* **59**, 283 (1974).
58. Pribluda, L.A., *Dokl. Acad. Nauk, B. SSR* **7**, 135 (1963); *Chem. Abstr.* **59**, 3142 (1963).
59. Roginski, E.E., Feldman, F.J., and Mertz, W., *Fed. Proc., Fed. Am. Soc. Exp. Biol.* **27**, 482 (1968).
60. Roginski, E.E., and Mertz, W., *J. Nutr.* **93**, 249 (1967).
61. Roginski, E.E., and Mertz, W., *Fed. Proc., Fed. Am. Soc. Exp. Biol.* **26**, 301 (1967).
62. Roginski, E.E., and Mertz, W., *J. Nutr.* **97**, 525 (1969).
63. Saner, G., *Nutr. Rep. Int.* **11**, 387 (1975).
64. Saner, G., Wolf, W.R., and Gürson, C.T., *Fed. Proc., Fed. Am. Soc. Exp. Biol.* **33**, 660 (1974).
65. Schaller, K.H., Essing, H.G., Valentin, H., and Schacke, G., *Z. Klin. Chem. Klin, Biochem.* **10**, 434 (1972).
66. Schroeder, H.A., *J. Nutr.* **88**, 439 (1966).
67. Schroeder, H.A., *Am. J. Clin. Nutr.* **21**, 230 (1968).
68. Schroeder, H.A., *J. Nutr.* **97**, 237 (1969).
69. Schroeder, H.A., and Balassa, J.J., *Am. J. Physiol.* **209**, 433 (1965).
70. Schroeder, H.A., Balassa, J.J., and Tipton, I.H., *J. Chronic Dis.* **15**, 941 (1962).
71. Schroeder, H.A., Balassa, J.J., and Vinton, W.H., *J. Nutr.* 83, 239 (1964); **86**, 51 (1965).
72. Schroeder, H.A., Vinton, W.H., and Balassa, J.J., *Proc. Soc. Exp. Biol. Med.* **109**, 859 (1962).
73. Schroeder, H.A., Vinton, W.H., and Balassa, J.J., *J. Nutr.* **80**, 39 and 48 (1963).
74. Schwarz, K., and Mertz, W., *Arch. Biochem. Biophys.* **72**, 515 (1957).
75. Schwarz, K., and Mertz, W., *Arch. Biochem. Biophys.* **85**, 292 (1959).
76. Sherman, L., Glennon, J.A., Brech, W.J., Klomberg, F.H., and Gordon, E.S., *Metab., Clin. Exp.* **17**, 439 (1968).
77. Tipton, I.H., *in* "Metal-Binding in Medicine" (M.J. Seven, ed.), p. 27. Lippincott, Philadelphia, Pennsylvania, 1960.
78. Toepfer, E.W., Mertz, W., Roginski, E.E., and Polansky, M.M., *Agric. Food Chem.* **21**, 69 (1973).
79. Visek, W.J., Whitney, I.B., Kuhn, U.S.G., and Comar, C.L., *Proc. Soc. Exp. Biol. Med.* **84**, 610 (1963).
80. Vittorio, P.V., and Wright, E.W., *Can. J. Biochem. Biophys.* **41**, 1349 (1963).
81. Wacker, W.E.C., and Vallee, B.L., *J. Biol. Chem.* **234**, 3257 (1959).
82. Welch, R.M., and Cary, E.E., *Agric. Food Chem.* **23**, 479 (1975).
83. Wolf, W.R., Greene, F.E., and Mitman, F.W., *Fed. Proc., Fed. Am. Soc. Exp. Biol.* **33**, 659 Abstr. (1974).
84. World Health Organization, *W.H.O., Tech. Rep. Ser.* **532**, (1973).
85. Zook, E.G., Greene, F.E., and Morris, E.R., *Cereal Chem.* **47**, 720 (1970).

11

Iodine

I. IODINE IN ANIMAL TISSUES AND FLUIDS

1. General Distribution

The healthy human adult body contains a total of 15–20 mg I, of which 70–80% is present in the thyroid gland. Since the normal thyroid weighs only 15–25 g, or about 0.03% of the whole body, this represents a unique degree of concentration of any trace element in a single organ. The I concentration in the skeletal muscles is less than one-thousandth of that of the thyroid, but because of their large mass they contain the next largest proportion of total body iodine. Hamilton *et al.* (75) give the following mean levels in a range of human tissues in μg I/g wet weight: muscle, 0.01 ± 0.001; brain, 0.02 ± 0.002; testis, 0.02 ± 0.003; lymph nodes, 0.03 ± 0.01; kidney, 0.04 ± 0.01; lung, 0.07 ± 0.03; ovary, 0.07 ± 0.03; and liver, 0.20 ± 0.06. Other workers have found the ovaries, pituitary gland, bile, and salivary glands to be appreciably higher in iodine than most other extrathyroidal tissues (29, 112, 170). Significant I concentrations also occur in parts of the eye, notably the orbitary fat and the orbicular muscle. In one study of a small range of samples this muscle averaged close to 0.25 μg I/g wet weight (64).

Iodine in the tissues occurs in both inorganic and organically bound forms. The former is normally present in extremely low concentrations, of the order of 0.01 μg/g (148). In the saliva the iodine is almost entirely in the inorganic form, even in conditions when organic I compounds are secreted in the urine (2, 132).

The salivary I concentration is proportional to the plasma inorganic I concentration at physiological levels (79) and at plasma concentrations up to 100 μg/100 ml, or about 500 times normal (78). Such increases can be achieved by the administration of iodide in the prophylaxis of simple goiter, and particularly at the much higher levels used in the therapy of exophthalmic goiter (28).

Most of the small amounts of organic iodine in the extrathyroidal tissues consists of thyroxine bound to protein, together with widely distributed low concentrations of other compounds, including triiodothyronine. The solubility of muscle iodine differs from that of thyroxine added to tissue extracts, and its distribution is not uniform between myosin and actin. However, muscle I levels decrease in hypothyroidism and increase in hyperthyroidism (149).

2. Iodine in the Thyroid Gland

The total concentration of I in the thyroid varies with the I intake and age of the animal and with the activity of the gland. Variation among species is small, except that the thyroids of sea fish are richer and those of rats slightly poorer than those of most mammalian species. The normal healthy thyroid of mammals contains 0.2–0.5% I (d.b.), giving a total of 8–12 mg in the adult human gland. This amount can be reduced to 1 mg or less in endemic goiter, with an even greater reduction in concentration because of the hyperplastic changes that characterize the disease. Many years ago Marine and co-workers (111) showed that hyperplastic changes are regularly found when the I concentration falls below 0.1%. This has been confirmed by later studies with several species. Thus sheep's thyroids with marked follicular hyperplasia contained 0.01% I, those with moderate hyperplasia 0.04%, and pig's thyroids showing very slight hyperplasia 0.11% (d.b.) (9). Further investigations of neonatal mortality in lambs associated with goiter also indicate that a thyroid I level of 0.1% or slightly higher is a critical level below which the gland cannot function properly (156, 160).

Iodine exists in the gland as inorganic I, mono- and diiodotyrosine, thyroxine, triiodothyronine, polypeptides containing thyroxine, thyroglobulin, and probably other iodinated compounds (147). The iodinated amino acids are bound with other amino acids in peptide linkage to form thyroglobulin, the unique iodinated protein of the thyroid. Thyroglobulin, the chief constituent of the colloid filling the follicular lumen, is a glycoprotein with a molecular weight of 650,000. It constitutes the storage form of the thyroid hormones and normally represents some 90% of the total iodine of the gland. The amounts and proportions of the various I-containing components of the thyroid vary with the supply of I to the gland, the presence of goitrogens that can inhibit the I-trapping mechanism or the process of hormonogenesis, as discussed in Section VI, and with the existence of certain disease states and metabolic defects of genetic origin. The thyroid glands of individuals living in areas of endemic goiter in the Himalayas show an interesting response in an attempt to maintain internal homeostasis and increase the efficiency of I use in the face of severe I deficiency (97). The nature of this adaptation was studied in the thyroids of goats. Thyroids of goats living in an area of severe I deficiency showed higher ratios of monoiodotyrosine (MIT) to diiodotyrosine (DIT) and of triiodothyronine (T_3) to thyroxine (T_4) than glands of those in an area of iodine abundance. There was also a higher incorporation in the I-deficient glands of ^{131}I into 27 S iodoproteins, which are more efficient in producing thyroid hormones than 19 S (thyroglobulin). These changes should be viewed more as a direct consequence of I deficiency than as an adaptive mechanism. A higher ratio of MIT to DIT has also been demonstrated in the thyroids of the offspring of rats maintained on a low-I diet, but the total labeled T_3 in the I-deprived young rats was lower than that in the deficient adults or normal newborn, indicating that the young rats did not adapt to I deficiency as did the adults (50).

3. Iodine in Blood

Iodine exists in blood in both inorganic and organic forms. The normal range of plasma inorganic iodide (PII) is stated by Wayne *et al.* (186) to be 0.08–0.60 μg/100 ml, with values below 0.08 suggesting iodine deficiency and values above 1 μg/100 ml pointing to exogenous I administration. Karmarkar *et al.* (97) give mean PII values of 0.096 ± 0.02, 0.088 ± 0.017, and 0.089 ± 0.013 for individuals from goitrous areas in India, Nepal, and Ceylon, respectively, compared with 0.137 ± 0.018 μg/100 ml for normal controls.

The organic iodine of the blood, which does not occur in the erythrocytes, is present mainly as thyroxine bound to the plasma proteins. Only a very small porportion, normally about 0.05%, is free in human serum (166). Up to 10% of the organic I of the plasma is made up of several iodinated substances, including tri- and diiodothyronine (68, 145). Thyroglobulin occurs only in pathological states involving damage to the thyroid gland, and the iodotyrosines do not normally appear in the peripheral circulation, or even in the thyroid vein when the thyroid contains 70% or more of its ^{131}I in the form of iodotyrosines (178).

The levels of several different I-containing components of blood have been estimated in attempts to develop convenient and satisfactory indices of thyroid function. The protein-bound iodine of the serum (PBI), or the butanol extractable iodine (BEI) of the serum, has had a considerable vogue and corresponds reasonably well with the level of thyroid activity in man (73, 186) and farm animals (93, 137). In adult man the limits of normality have been placed at 4–8 or 3–7.5 μg/100 ml with a "mean" close to 5–6 μg/100 ml (73, 186, 188) (see Table 35). Lower PBI norms (3–4 μg/100 ml) have been found for mice, rats and dogs (175), adult sheep (187), and beef cattle (102). Still lower mean serum

TABLE 35

Values (Mean and SD) for Thyroid Function Tests in Normal Subjects and in Patients with Untreated Primary Thyroid Disorders[a]

Group	PBI (μg/100 ml)	Resin uptake (% normal)	Free T_4 index $T_3/100 \times$ PBI)	Free T_4 concentration (ng/100 ml)	Free T_4 (%)
Hypothyroid (15)[b]	1.89 ± 0.69	72.5 ± 12.1	1.34 ± 0.51	1.71 ± 0.73	0.05 ± 0.01
Euthyroid (23)	5.33 ± 0.98	96.5 ± 10.2	5.09 ± 0.84	5.43 ± 0.99	0.07 ± 0.01
Hyperthyroid (15)	11.06 ± 2.40	131.4 ± 16.8	15.37 ± 3.90	19.01 ± 6.25	0.11 ± 0.02

[a]From Wellby and O'Halloran (188).
[b]Figures in parentheses indicate number of subjects in group.

PBI levels have been recorded in studies with the domestic fowl (197) and horses (93).

Estimation of total thyroxine (T_4) in serum, based on competitive protein-binding analysis as introduced by Ekins (49), correlates well with thyroid function (49) but is open to the objection that changes in T_4 binding proteins can invalidate the results. For example, as a consequence of raised thyroxine-binding globulin concentration, euthyroid women taking oral contraceptives or pregnant women may have false elevated serum T_4 levels (49, 62). The concept developed by Robbins and Rall (143) that the small amount of unbound or free T_4 is the factor determining the true thyroid status of the individual has received strong support from later studies (90, 125, 188). Free T_4 assays on plasma can sharply differentiate hypothyroid and thyrotoxic patients from euthyroid individuals (Table 35). Recently D'Haene *et al.* (46) compared four thyroid function indices in the sera of 181 patients: total thyroxine (T_4), the triiodothyronine resin uptake (T_3) BC index, the free thyroxine (T_4) index, and the effective thyroxine ratio (ETR). Their data gave no convincing reasons for using the ETR determination instead of the free T_4 determination as an index of thyroid function, although Thorson *et al.* (180) have stated that the ETR has a high (99%) diagnostic accuracy.

4. Iodine in Milk

The I concentration of milk is greatly influenced by dietary I intakes. Blom (26) increased the level in cow's milk for a "normal" 20–70 μg I/liter to 510–1070 μg I/liter by feeding a supplement of 100 mg KI/day, and Kirchgessner (100) demonstrated rising milk I concentrations from increments of dietary I. At an intake of 1.6 mg I/cow/day the milk contained 28 ± 6 μg I/liter, at 12.7 mg/day it was 78 ± 18, and at an intake of 20 mg/cow/day it was 267 ± 55 μg I/liter. Increases in milk I from a low 8 μg/liter to 81 and 694 μg/liter were obtained by Hemken *et al.* (82) from supplements of 6.8 and 68 mg I as KI/day, respectively. Such treatment has been suggested as a means of raising the I intakes of women and children in goitrous areas. In such areas the I level in milk is below normal and the determination of I in milk has been proposed as a convenient means of establishing the I status of an area (24). Binnerts (24) found the following mean differences between goitrous and nongoitrous regions in the Netherlands: summer milk, 9.7 and 21.1 μg I/liter, and winter milk, 20.6 and 83.5 μg I/liter, respectively.

Teat dipping with iodophors in the treatment of mastitis also increases milk I levels. Funke *et al.* (58a) reported levels of 89 and 94 μg I/liter for controls and 127 and 152 μg/liter for the milk of cows treated after each milking.

Cow's colostrum has much higher I levels than true milk, and there is a fall in concentration in late lactation. Kirchgessner (100) reports a mean value of 264 ± 100 μg for colostrum, compared with 98 ± 82 for true milk. Lewis and Ralston (106) found the colostrum of five cows to range from 200 to 350 μg I/liter compared with 72–136 in the later milk of the same animals. Salter (148) quotes values of 50–240 for the I content of human colostrum and 40–80 μg/liter for human milk once lactation is established.

In ruminants, iodine is present in the milk entirely in the form of iodide, since only iodide has been detected after ^{131}I administration and no thyroactive compounds can be found by chromatographic procedures (61, 193). This is in agreement with biological tests of bovine milk (25) and with the finding that the normal bovine mammary gland is impervious to thyroxine (141). In the milk of the rat, rabbit (30), and dog (117) an I-containing protein can be detected after ^{131}I administration.

II. IODINE METABOLISM

Iodine metabolism and thyroid function are closely linked, since the only known role of iodine is in the synthesis of the thyroid hormones–thyroxine and triiodothyronine. The activity of the thyroid is regulated by a negative feedback mechanism involving the adenohypophysis and the hypothalamus. The hypothalamus secretes the thyrotropin releasing Factor or TRF (69), a peptide which reaches the adenohypophysis via the portal vessels of the pituitary stalk and provokes the secretion of the thyroid stimulating hormone (TSH) by the b_2 cells. TSH is a glycoprotein with a molecular weight of about 25,000, which stimulates the gland to release its hormones and trap iodide. The thyroid hormones in turn inhibit the release of both TRF by the hypothalamus and TSH by the pituitary, in this way keeping the plasma level of the thyroid hormones normal. Triiodothyronine, which has some four times the potency of thyroxine, is also stronger in inhibiting TSH secretion. Iodine metabolism thus consists essentially of the synthesis and degradation of the thyroid hormones and the reutilization or excretion of the iodine so released.

1. Iodine Absorption and Excretion

Iodine occurs in foods largely as inorganic iodide and is absorbed in this form from all levels of the gastrointestinal tract. Other forms of iodine are reduced to iodide prior to absorption (38). Iodide administered orally is rapidly and almost completely absorbed from the tract, with little appearing in the feces (185). Iodinated amino acids are well absorbed as such, although more slowly and less completely than iodide. A proportion of their iodine may be lost in the feces in organic combination. The remainder is broken down and absorbed as iodide (98).

Iodine is excreted mainly in the urine, with smaller amounts appearing in the feces and sweat. In tropical areas of low dietary I status, losses in the sweat could impose a significant drain on limited I supplies. Vought and co-workers (185) found fecal I excretion in normal adults to range from 6.7 to 42.1 μg/day. Koutras (101) gives 5–20 μg as the normal range of fecal I excretion in man. The dominant role of the kidney in I excretion has led to efforts to relate thyroid states to (a) stable urinary I excretion and (b) radioiodide urinary excretion after a standard interval of time. The level of urinary I excretion correlates well with plasma iodide concentration (186) and ^{131}I thyroid uptakes (58, 97) (Table 36). Follis (58) has set a urinary level of 50 μg I/g creatinine as the "tentative lower limit of normal" for adolescents, with 32.5 μg/g as the corresponding figure for children 5–10 years of age and 75 μg/g for adult men. Koutras (101) considers that a urinary I excretion below 40 μg/day is suggestive of I deficiency in man, if renal clearance is normal. More recently Stanbury *et al.* (165) have suggested that a daily excretion of less than 50 μg I/24 hr or 50 μg I/g of creatinine in randomly obtained specimens in a fair sample of the population is indicative of the existence of endemic goiter in a community.

In the ruminant the rumen is the major site of absorption of iodide and the abomasum the major site of endogenous secretion, i.e., for the reentry of circulating iodide into the digestive tract. Net absorption also occurs from the small intestine and from the remainder of the tract (20, 120). It is apparent that considerable absorption and endogenous I secretion take place throughout the ruminant (bovine) digestive tract.

At high dose levels absorbed labeled thyroxine is partially deiodinated and the I excreted in the urine as iodide. Part of such thyroxine is also taken up by the liver and excreted through the bile into the feces, either unchanged or in

TABLE 36
Twenty-Four Hour ^{131}I Neck Uptake and Urinary I Excretion in Endemic Goiter of India, Nepal, and Ceylon[a]

Country	Place	% 24-hr neck uptake of ^{131}I	Urinary I excretion (μg/g creatinine)
India	Uttar Pradesh	68.7 ± 8.3 (70)[b]	30.2 ± 2.87 (46)[b]
	Bihar	67.1 ± 12.3 (43)	–
Nepal	Trishuli	71.1 ± 1.65 (41)	21.6 ± 1.59 (40)
	Jumba	84.7 ± 1.88 (17)	20.2 ± 3.04 (11)
Ceylon	Horana	77.6 (33)	20.15 ± 3.00 (6)
Control values	Delhi	42.4 ± 3.00 (15)	76.4 ± 10.2 (10)

[a]From Karmarkar *et al.* (97).
[b]Numbers in parentheses indicate number of samples analyzed.

conjugated form. Small amounts of thyroxine also pass directly into the stomach, jejunum, and colon, but the bile accounts for the major portion of intestinal thyroxine in the rat and dog (67, 174). At physiological doses of thyroxine or following biosynthetic labeling of plasma hormone, the liver plays a less prominent part in excretion and conjugation. A higher proportion is deiodinated so that only the resulting iodide which escapes the efficient thyroid trap reaches the urine and little appears in the feces (22). Most of the hormonal I in man is thus fated for metabolic degradation and return to the body iodide pool.

2. Intermediary Metabolism

Iodide ions resemble chloride ions in that they permeate all tissues. The total iodide pool therefore consists of the iodide present in the whole extracellular space, together with the red blood cells and certain areas of selective concentration, namely, the thyroid, salivary, and gastric glands. Equilibrium within the total pool is reached rapidly (21). In the rat about 52% of either absorbed or intraperitoneally injected ^{131}I is excreted, mainly in the urine, with a half-life of 6–7 hr (81). This half-life represents the turnover of the inorganic iodide phase. Metabolic equilibrium between the retained ^{131}I and the whole body I pool was achieved within 4 days, and the half-life of this iodine reached 9.5 days and remained constant at that level. This is considered to represent the turnover of the organic I.

Despite its high-I content and the efficiency with which it traps iodine, the thyroid gland contributes little to the iodide pool, because the binding into organic form is normally so rapid. Significant quantities of iodide are also trapped by the salivary glands (29), apparently by mechanisms similar to those of the thyroid (76, 77). Since salivary iodide is not converted into organic form and is normally all reabsorbed, this process represents little net loss to the iodide pool.

The iodide pool is replenished continuously, exogenously from the diet, and endogenously from the saliva, the gastric juice, and the breakdown of hormones produced by the thyroid. Gastric clearance of iodide exceeds that of chloride in dogs by 10 to 50 times (88), and in dairy calves net gastric (abomasal) secretion of I exceeds that of chloride by 15 times (118). It has been suggested that the I-concentrating action of the abomasum may promote I conservation by creating an extravascular I pool, thus preventing its excessive loss in the urine (120).

Iodine is continuously lost from the iodide pool by the activities of the thyroid, kidneys, salivary glands, and gastric glands which compete for the available iodine. Koutras (101) conceives of I metabolism as a metabolic cycle consisting of three principal pools. These are the plasma inorganic iodide, the intrathyroidal I, and the pool comprising the hormonal or protein-bound I of the plasma and tissues. The rates of removal of iodide from the first of these pools by the thyroid and kidneys are expressed as thyroid and renal clearances, calculated as organ accumulation of I per unit time divided by plasma iodide concentration. In normal man total clearance from the iodide pool occurs at the rate of about 50 ml/min, and renal iodide clearance is constant at about 35 ml/min over all ranges of plasma iodide examined (35). Thyroid clearance, by contrast, is sensitive to changes in plasma iodide concentration and varies with the activity of the gland. In normal individuals the thyroid clears an average of 10–20 ml/min, whereas in exophthalmic goiter or Graves' disease, a clearance of 100 ml/min is usual (148) and over 1000 ml is possible (21). The measurement of thyroid ^{131}I clearance is a valuable diagnostic aid in hyperthyroidism but is much less sensitive to myxedema (131).

The intrathyroidal I pool involves a series of consecutive steps in the synthesis of the thyroid hormones. First, the iodide brought to the gland in the plasma is trapped by an energy requiring active mechanism which can be blocked by antithyroid agents of the perchlorate and thiocyanate type. This iodide is oxidized to elemental iodine or to some similar reactive form as a preliminary to its incorporation into organic combination by a peroxidase (1a). The more reactive I combines with the tyrosine residues and thyroglobulin to form 3-monoiodotyrosine and 3,5-diiodotyrosine. This reaction takes place near the boundary between follicular cells and follicular lumen, with the same enzyme system probably acting as iodide peroxidase and as tyrosine iodinase (44). It can be blocked by a great number of antithyroid substances of the thiouracil type and even by iodide itself in high concentrations. The ability of the gland to trap iodide is not reduced by such agents. In fact the thiouracil-blocked gland can maintain an iodide concentration several hundred times that in the plasma (176). Two diiodotyrosine molecules combine to form one molecule of thyroxine, or one mono- and one diiodotyrosine combine to form one molecule of triiodothyronine. Taurog and Howells (177) have provided evidence that a peroxidase may be involved in this coupling reaction, as well as in the iodination reaction. These hormones are stored in the colloid bound to thyroglobulin. The iodine of the normal gland is present mainly in the form of the two iodotyrosines and the two iodothyronines bound in this way (142). In a study of pancreatin hydrolysates of normal human thyroids the distribution of the iodine was found to be 16.1% as iodide, 32.7% as monoiodotyrosine, 33.4% as diiodotyrosine, 16.2% as thyroxine, and 7.6% as triiodothyronine (52). In the goitrous gland the proportions change, as discussed earlier, in attempts to conserve iodine (97).

The thyroid hormones are released through proteolysis of thyroglobulin by a protease system which yields both iodotyrosines and thyroactive thyronines (114). The normal human thyroid has been estimated to secrete daily 51.6 μg of iodine as thyroxine and 11.9 μg as triiodothyronine (140). The iodotyrosines liberated from the proteolysis of thyroglobulin, unlike the iodothyronines, are not secreted into the circulation but are deiodinated by an enzyme called deiodinase or dehalogenase (144). The iodine so released is not lost from the gland but is reutilized for hormone synthesis. This leads to an economical use of iodine and ensures that virtually all the I entering the normal thyroid as iodide leaves it in the form of hormones after one or more entries into thyroglobulin (116).

In order to ensure an adequate supply of hormones, the human thyroid must trap about 60 μg of iodide daily. This is primarily achieved, irrespective of the plasma iodide level, by adjustment of the thyroidal iodide clearance rate, so that when the plasma iodide decreases the thyroidal clearance increases, with the actual iodide uptake remaining more or less constant. Adaptation to iodine deficiency thus occurs by increasing the thyroidal iodide clearance rate. Adaptation through an increase in the MIT:DIT and T_3:T_4 ratios in the gland and a higher incorporation of I into 27 S iodoproteins (97) were discussed earlier (Section I,2). Functional overactivity of the iodide-trapping mechanism is usually associated with an increase in the gland mass, or goiter, but in mild iodine deficiency the biochemical manifestations of the deficiency, namely, low plasma iodide and urinary iodine and high thyroidal iodide clearance and radioiodine uptake, have been demonstrated without any obvious goiter (186).

The circulating thyroid hormones, comprising the third metabolic iodine pool as stated by Koutras (101), occur mostly bound to a thyroxine-binding globulin, to prealbumin, and to albumin itself (63, 89), with only about 0.07% of the thyroxine normally present in the free state (188). According to Rall and co-workers (140) some 10% of the circulating thyroxine and 56% of the circulating triiodothyronine is metabolized daily in man. Once these hormones enter the tissues about 80% is broken down by several deiodinating enzymes, with the iodine so liberated returning to the iodine pool, thus completing the iodide cycle. The small remainder enters the enterohepatic cycle and is lost to the body either through excretion into the bile unchanged, or through excretion by this route in conjugated form following detoxication in the liver (174).

III. IODINE DEFICIENCY AND FUNCTIONS

So far as is known the entire functional significance of iodine is accounted for by its presence in the thyroid hormones. The manifestations of I deficiency are therefore those of a deficient supply of these hormones to the organism. However, the reverse is not necessarily true. Many factors can inhibit the capacity of the thyroid gland to accumulate iodine and convert it into thyroactive compounds. These factors can act independently of I supply, or may only become apparent in circumstances of borderline I deficiency. An enlargement of the thyroid, or goiter, must be regarded as a final common expression of a number of separate disease processes. An absolute dietary I deficiency is only one of those processes. An I deficiency "conditioned" by the presence of goitrogens is another. Metabolic defects in thyroid hormonogenesis, due either to the presence of particular types of goitrogens or to a constitutional disability, represent a further disease process resulting in goiter.

The thyroid hormones regulate a wide variety of physiological processes in virtually all tissues of higher organisms. Their effect on cellular oxidation is fundamental to many of these processes, as discussed in Section III,1. In addition, alimentary tract motility is reduced in hypothyroid monogastric (105) and ruminant (121) animals, resulting in prolonged feed retention. Administration of thyroxine or thyroprotein reduces feed retention to normal levels.

1. Cellular Oxidation

The rate of energy exchange and the quantity of heat liberated by an organism at relative rest is elevated in hyperthyroidism and reduced below normal in hypothyroidism. Where thyroid output is limited by lack of dietary I the basal metabolic rate (BMR) is lowered and can be restored to normal by I supplementation or thyroid hormone therapy. Tissues from I-deficient or thyroidless animals consume less oxygen than normal, whereas those from hyperthyroid animals consume more oxygen than normal (33). The calorigenic action of the thyroid hormones in homeothermic adult vertebrates is exerted on many tissues, including the heart, liver, kidney, and skeletal muscles but not the brain. Because of this effect on O_2 consumption, attention has been directed to the mitochondria as a likely primary locus of action of T_4 and T_3 (85). Administration of thyroid hormone leads to an increase in the size, number, and metabolic activity of mitochondria in mammalian skeletal muscle (72). For a sustained increase in mitochondrial respiration, as the basis for a sustained increase in thermogenesis, ATP utilization must be increased. Ismael-Beigi and Edelman (94) have proposed that activation of ATP utilization by transmembrane Na^+ pumping may be one of the primary mediators in the calorigenic response to T_4 and T_3. More than 90% of the increment in the oxygen consumption in the liver and muscle of euthyroid rats produced by injections of T_4 and T_3 could be attributed to increased energy use by the Na^+ pump. Furthermore, inhibition of Na^+ transport by ouabain eliminated the effect of T_3 on liver slices and reduced its effect on kidney preparations by about one-half. In this respect it is particularly pertinent that the brain shows neither a calorigenic nor a Na^+ pump effect. As Hochachka (86) has recently stated, "these studies allow a fairly specific modus operandi for thyroid calorigenesis."

2. Cell Differentiation and Growth

The thyroid hormones, and therefore iodine, are essential for growth during early life in all mammals and birds and for almost all developmental processes, including the metamorphosis of amphibian larvae (136). Total thyroidectomy induces severe dwarfism in rats and birds, while athyreosis can lead to a type of dwarfism in human infants that occurs in severely goitrous areas. Under these circumstances I administration during adolescence increases growth rates as well as reducing goiter incidence. In an early Swiss study assessed by Stocks (168), in which 2 mg NaI were administered weekly to schoolgirls for a 3-year period, large increases in growth rate above the mean growth curves were observed in those with pronounced goiters and smaller but significant increases in those with moderate to small goiters. Stunted growth of domestic animals in I-deficient areas responsive to I and not always associated with visible goiter has also been reported (53, 96), and growth rates have been correlated with thyroid secretion rates in chicks (138), lambs (161), and calves (137).

The fact that the thyroid hormones play an important role in amphibian metamorphosis, mammalian growth, and cell differentiation supports the concept that such diverse biological effects may result from a primary effect on the control of gene expression (173). Induction of RNA and protein synthesis had earlier been shown to be required for the action of the thyroid hormones (172). Recently Samuels and his associates (150–152) have made further encouraging progress. They have shown that (a) T_3 and T_4 at physiological levels can induce a 3-fold increase in the rate of growth of GH_1 cells, a rat pituitary tumor cell line in culture, (b) high affinity saturable binding sites for ^{125}I-labeled T_3 and T_4 are present in the cell nucleus and appear to function as receptors for the thyroid hormones, and (c) nuclear receptors for T_3 and T_4 can be isolated in a soluble and stable form from rat liver nuclei and GH_1 cells with no apparent change in hormonal affinity. The binding activity has characteristics of a non-histone protein and the association with T_3 and T_4, as well as with hormonal analogs, suggests that the binding activity determined *in vitro* is likely to function as a receptor for the thyroid hormones *in vivo*. These studies encourage optimism that the mechanisms of action of the thyroid hormones at the molecular level will before long be resolved.

3. Neuromuscular Functioning and Cretinism

Endemic goiter, when severe, is frequently associated with a condition known as endemic cretinism, which is characterized by a wide range of clinical abnormalities, including mental retardation, deafness and deaf-mutism, retarded growth, and neurological abnormalities, as well as hypothyroidism. Endemic cretinism has long been known to occur in areas of severe I deficiency in association with endemic goiter (see Trotter, 181). The condition has been described in South America (55, 56), the Himalayas (169), Africa (48), and New Guinea (83, 135). Among the cretins observed in highland Ecuador (56) 90% presented no evidence of myxedema. Thus the predominant form of the disorder in the Andean region corresponds to the so-called "nervous endemic cretinism" rather than to the "myxedematous cretinism" of central Africa. Those cretins in whom both neurological impairment and hypothyroid manifestations are evident would be examples of "mixed endemic cretinism." Pharaoh and co-workers (135), working in the highlands of eastern New Guinea, have produced evidence that I deficiency in the mother during the first trimester of pregnancy is the main factor in the causation of endemic cretinism, and that I repletion prior to conception can prevent the condition (see Table 37). However, the mechanism of action of the iodine so used is unknown.

4. Reproduction and Interactions with the Gonads and Other Glands

The delicate balance and interaction between the thyroid and the adenohypophysis and the hypothalamus and their hormones which control the activity of the thyroid were described in Section II. Thyrotropin produces hypertrophy of the thyroid cells when these are grown in blood serum (7), and the atrophy of the thyroid which follows hypophysectomy can be prevented by the administration of this hormone. Furthermore, thyroidectomy is followed by a reduction in the size of the adrenals in rats and administration of thyroxine induces adrenal cortical hypertrophy (87). Conversely, hyperfunction of the adrenal cortex may induce decreased thyroid activity, and the administration of adrenalin can induce hyperplastic changes in the thyroid (162).

A relationship between the thyroid and the gonads is apparent in all male and female mammals and birds (see Maqsood, 108). In man, colloid goiter often develops at puberty and hyperthyroidism is sometimes precipitated at the

TABLE 37
Children Born in Jimi River Subdistrict (New Guinea) Classified according to Treatment Received by Mother[a]

Treatment received by mother	Total no. of new births	No. children examined	No. deaths recorded	No. of endemic cretins
Iodized oil	498	412	66	7[b]
Untreated	534	406	97	26[c]

[a]From Pharaoh *et al.* (135).
[b]Six already pregnant when injected with oil.
[c]Five already pregnant when injected with saline solution.

menopause. Goitrous cretins are usually sterile and invariably fail to develop normal sexual vigor, with a delayed maturation of the genitalia. Thyroidectomy at an early age is followed, in all species, by a long period in which the gonads and the secondary sex organs remain in an infantile condition.

In some types of birds thyroid–gonadal interrelationships can be very conspicuous due to plumage changes. In the Brown Leghorn male, thyroidectomy is followed by a period in which the testes remain small and free from spermatozoa. The comb decreases in size, molting is inhibited, and the characteristic male plumage is lost. Administration of estrogen to such birds does not induce the female plumage pattern as it does in normal males, which suggests that this pattern results from a synergistic action of the thyroid and ovarian hormones (23). Thyroidectomy reduces egg production in hens (191), and there is some evidence that the seasonal cycle of egg production in poultry is related in part to seasonal variation in thyroid activity. Turner and co-workers (182) have achieved some success in preventing this seasonal decline, associated with high summer temperatures, by feeding thyroactive iodinated proteins, but this effect is not related to a lack of dietary iodine.

Reproductive failure is often the outstanding manifestation of I deficiency and consequent impairment of thyroid activity in farm animals. The birth of weak, dead, or hairless young in breeding stock has long been recognized in goitrous areas (53). Fetal development may be arrested at any stage, leading either to early death and resorption, abortion and stillbirth, or the live birth of weak young, often associated with prolonged gestation and parturition and retention of fetal membranes (4, 124). Allcroft *et al.* (4) demonstrated subnormal serum protein bound iodine (PBI) levels in herds showing a high incidence of aborted, stillborn, and weakly calves. Falconer (54) has shown that thyroidectomy of ewes some months before conception severely reduced both the prenatal and postnatal viability of the lambs, despite the presence of an apparently adequate thyroid in the lamb itself. Neonatal mortality in lambs from ewes fed goitrogenic kale, responsive to I administration during pregnancy, has frequently been observed (8, 157).

In addition to the reproductive disturbances just described, irregular or suppressed estrus in dairy cattle causing infertility has been associated with goiter and shown to respond to I therapy. Moberg (124), working with 190 herds totaling 1572 cows in goitrous areas in Finland, obtained a significant improvement from I therapy in first service conception rate and in the number of cows with irregular estrous incidence. McDonald *et al.* (113), working in an I-deficient area in Canada, also obtained a marked improvement in first service conception rate by feeding an organic iodine preparation, beginning 8–10 days before the cows came into estrus.

The thyroid gland plays an equally important part in the maintenance of male fertility (108). A decline in libido and a deterioration in semen quality have been associated with I deficiency in bulls and stallions, and a seasonal decline in semen quality in rams has been related to a mild hypothyroid state. This condition is responsive in part to doses of thyroactive proteins, but not to iodine except in known I-deficient areas.

5. Condition of the Integument

Changes in the skin and its appendages, hair, wool, fur, and feathers, are among the most constant features of iodine deficiency. Cretins exhibit a pale, gray skin lacking in mobility and their hair tends to be dry and scanty. In human myxedema the skin is dry, rough, and thickened and shedding of the hair is a characteristic symptom. In birds the relation of different levels of thyroid activity to the molting process and to the form, structure, and pigmentation of the feathers can be particularly striking. Calves and pigs born to I-deficient mothers are often hairless and have thick pulpy skins. Less severe deficiencies induce milder disorders of the pelt such as rough, dry skin, scanty wool, and hairiness of the fleece (9, 96).

6. The Nature and Potency of the Thyroid Hormones

Many different compounds possess thyroidal activity equal to or in excess of that of thyroxine. Thus, in comparison with thyroxine (3,5,3′,5′-tetraiodothyronine), 3,5,3′-triiodothyronine is three to five times as active, depending on the dose used; 3,3′-diiodothyronine is similar in potency; 3,5 diiodo-3′,5′-dibromothyronine is nearly as active; and 3,5,3′-triiodothyropropionic acid is 300 times more active in accelerating amphibian metamorphosis (116, 126, 146). On the other hand, 3,3′,5′-triiodothyronine has only one-twentieth the activity of thyroxine or one-hundredth that of its isomer (146). Studies of various thyroxine analogs have revealed positive correlations between their thyromimetic activities and the ability of various substituent groups to attract or release electrons and form hydrogen bonds (31). Although these findings have limited physiological meaning, they permit the following generalizations: the thyronine nucleus is essential to any activity; I substitution in the inner aromatic ring is required for substantial activity; partial replacement of I by Br results in little loss of activity; an alanine side chain is not essential; and, whatever the aliphatic side chain on the nucleus, a 3,5,3′ substitution in the rings ensures maximal activity. Present concepts of the mechanism of action of the thyroid hormones at tissue level were presented earlier in Sections III,1 and 2.

IV. IODINE REQUIREMENTS

Calculations based on average daily losses of iodine in the urine give an adult human requirement of 100–200 μg I/day (42), while the results of balance studies indicate that equilibrium or positive balance can be achieved at intakes ranging from 44 to 162 μg I/day (39, 153). From an assessment of many studies Elmer (51) placed the requirements of man at 100–200 μg I/day, and more recently, Wood (192) has put these requirements similarly at 130–199 μg/day. Wayne *et al.* (186) state that 160 μg I/day is "the minimum certainly safe amount of iodine which must be available in the individual's diet if I-deficiency goiter is to be avoided." They suggested that it mught be advisable to raise this figure to 200 μg/day for children and pregnant women.

The nutritional I requirement of the rat has been given as 0.7–0.9 μg/day on the basis of whole body I turnover rates (81). This would give a requirement of 20–40 μg I per 1000 cal of food consumed. If this reasoning is applied to human adults consuming 2800 cal daily, the I requirement would be 56–112 μg/day.

Mitchell and McClure (123) have calculated the minimum I requirements of different classes of farm animals on the basis of their heat production, rather than their energy production, as tabulated below.

Animal	Body weight (lbs)	Heat production (cal)	Iodine requirements (μg/day)
Poultry	5	225	5–9
Sheep	110	2,500	50–100
Pigs	150	4,000	80–160
Cow in milk (40 lb/day)	1000	20,000	400–800

The above estimates, except for the cow where they are much lower, compare well with the minimum intakes given by Orr and Leitch (130) for those species in nongoitrous areas. Hartmans (80) cites 0.6 mg/kg of ration as the I requirement of high-yielding cows, with a tolerance of 0.4 mg I/kg for short periods. A requirement of 0.8 mg I/kg for pregnant and lactating livestock and 0.12 mg I/kg for other animals in the feed dry matter is given by the Agricultural Research Council of Great Britain (128a). Normal growth in chickens has been reported on diets as low as 0.07 ppm I, although 0.3 ppm was required for completely normal thyroid structure (41). The dietary I requirements of pheasants and quail are no greater than 0.3 ppm, either for growth or the development of normal thyroid glands (155). The results of recent experiments with growing pigs suggest that their I requirement is not greater than 0.14 ppm of the diet, and on corn–soybean meal rations approximates 0.086–0.132 ppm (159a).

Iodine requirements are influenced by the presence of goitrogens and certain other elements in the diet. Thiocyanates, perchlorates, (6, 194), and rubidium salts (17) are known to interfere with I uptake by the thyroid, and high levels of arsenic can induce goiter in rats (158). A high incidence of goiter occurs in the Cordoba province of Argentina, where chronic arsenic poisoning is endemic (154). High fluoride intakes have also been proposed as contributing factors in the incidence of goiter in parts of South Africa (167) and in the Punjab of India (189). It is difficult to assess the significance of these claims, especially as experimental evidence on the effects of fluoride on the thyroid is confusing. The incidence of goiter in some areas has been associated with the presence of limestone formations yielding hard waters, but as with fluoride, the experimental evidence is equivocal. Malamos and Koutras (109) were unable to demonstrate an increase in renal iodide clearance by oral or intravenous administration of calcium to man. Interactions between iodine and cobalt are considered in Chapter 5.

V. SOURCES OF IODINE

1. Iodine in Water

The level of I in the drinking water reflects the I content of the rocks and soils of a region and hence of the locally grown foods or feeds. Four studies in widely separated areas illustrate this relationship. Kupzis (103) found the water supplies in goitrous areas in Latvia to range from 0.1 to 2.0 μg I/liter, compared with 2–15 μg/liter for nongoitrous areas. Young *et al.* (196) reported that the drinking water in English villages with a goiter incidence assessed at 56% averaged 2.9 μg I/liter, compared with 8.2 μg/liter in other villages where the goiter incidence was only 3%. Karmarkar *et al.* (97) found the water from goitrous areas in India, Nepal, and Ceylon to range from 0.1 to 1.2 μg I/liter, compared with a nongoitrous area in Delhi of 9.0 μg/liter. In a study of goiter incidence in Egyptian oases, the waters from goitrous villages ranged from 7 to 18 μg I/liter, compared with the very high levels of 44 and 100 μg/liter in two samples from villages with no goiter (37). These last figures indicate that in some circumstances the water supply can contribute significant amounts of iodine to the daily diet, although in most areas the proportion of the total intake from this source is very low.

2. Iodine in Human Foods and Dietaries

The I concentrations in human foods of all types are exceedingly variable, mainly due to differences in the content and availability of I in the soil and to the amount and nature of the fertilizers applied. Chilean nitrate of soda, the only mineral fertilizer naturally rich in iodine, can double or triple the I content of food crops (and pastures) when applied in the amounts required to meet their nitrogen needs. Gurevich (71) showed that the I level in a range of vegetables and cereals were increased 10–100 times or more by applications of seaweed and by-products of the fish, crab, and whale processing industries. Barakat *et al.* (19), in a study of iodine in the edible portion of 15 vegetables, found large variations between samples of the same vegetables, presumably reflecting differences in the availability of I in the soils. Cabbage, for example, was reported to range from 0 to 0.95 μg I/g.

Similarly, animal products such as milk or eggs are much richer in iodine when they come from animals that have consumed I-enriched rations than from animals not so treated. The magnitude of this effect on cow's milk was considered in Section I,4. On average rations hen's eggs contain 4–10 μg I, most of which is located in the yolk (110). Feeding the hens large amounts of iodine either as iodized salt, NaI, or seaweed can raise this amount 100-fold (70, 95), or 1000-fold (110). The significance to man of adventitious I sources, such as iodophor antiseptics and iodates in bread, is considered in Section VII.

Foods of marine origin are so much richer in iodine than any other class of normal foodstuff that differences among other foods are usually of relatively minor importance. The edible flesh of sea fish and shellfish may contain 300–3000 ppb I on the fresh basis, compared with 20–40 ppb for freshwater fish (91). The I content of composites of food categories in the United States taken from the work of Vought and London (184) is given in Table 38. Highly refined products such as sugar contain negligible amounts of iodine. Hamilton and Minski (74) found Barbados brown sugar to contain 30 ppb I, compared with less than 1 ppb in refined white sugar. These workers also reported a mean intake of 220 ± 51 μg I/day from English adult diets. In 1969 the estimated average daily intakes for the United States ranged from 238 μg in the Northeast to 738 μg in the Southwest (129). The daily dietary I intakes by Japanese people are suggested as about 300 μg with a basic diet containing a minimum of

TABLE 38
Iodine Content of Composites of Food Categories[a]

Food category	No. of samples	Iodine (μg/wet kg)	
		Mean ± SE	Median
Seafoods	7	660 ± 180	540
Vegetables	13	320 ± 100	280
Meat products	12	260 ± 70	175
Eggs	11	260 ± 80	145
Dairy products	18	130 ± 10	139
Bread and cereal	18	100 ± 20	105
Fruits	18	40 ± 20	18

[a]From Vought and London (184).

seaweed, with incidental consumption of seaweed leading to a maximum of 10 mg I/day (195).

Overall I intakes are determined more by the source of foods composing dietaries than by the choice or proportion of different foods, except for those of marine origin. Residents of an endemic goiter area can only obtain sufficient iodine for their needs by the consumption of substantial amounts of marine foods, or foods imported from I-rich areas elsewhere, or by the use of iodized salt, bread, or some other form of I-fortified material. The consumption of seaweeds, which may contain as high as 0.4–0.6% I on the dry basis (92), is believed to be one reason for the low incidence of goiter in Japan. The supply of sea salt thrown up in the spray into the country is suggested as a further factor contributing to a relatively high I content in Japanese grown foods (195).

3. Iodine in Animal Feeds

The I levels in pasture species were reported in earlier studies to range between 300 and 1500 ppb on the dry basis (95, 130). More recently Hartmans (80), in a study of the factors affecting herbage I content in the Netherlands, found that dicotyledonous species had up to 13 times higher I contents than grasses, while the I content of grass species varied over a 2-fold range. Samples of white clover (*Trifolium repens*) ranged from 160–180 ppb and pasture grasses from 60 to 140 ppb (d.b.). The mean levels in a range of Welsh pasture grasses ranged from 200 to 310 ppb (1). Roughages are usually substantially higher in iodine than cereal grains, or the oilseed meals commonly employed as protein supplements in farm rations (91). Protein concentrates of animal origin, other than fish meal, cannot be relied on as significant sources of dietary I, unless the animals from which they are obtained are ingesting exceptionally large amounts of the element.

4. Iodine Supplementation in Man

Methods that have been adopted for the control of endemic goiter in man are (a) the use of iodized salt, (b) the use of iodine compounds in bread, (c) the administration of iodized tablets or confections to individuals, (d) injections of iodized oil, and (e) the addition of iodine to municipal water supplies. The last of these procedures has mostly been abandoned because of the numbers of people who cannot participate and the small proportion of the iodized water actually used for drinking. The use of medicated tablets or confections requires the continuous cooperation of many people and authorities and has therefore rarely been completely satisfactory after the initial enthusiasm has waned. The incorporation of iodized salt, or sodium iodate in place of bromate as a dough conditioner into factory-baked bread, is a convenient and effective means of I supplementation in goitrous areas where bread is a staple foodstuff and home baking is not a common practice (84). Wide variations in the amount of bread consumed detract from its effectiveness.

Compulsory iodization of domestic salt is an economical and highly effective means of mass prophylaxis in goitrous areas and is now carried out as a public health measure in many countries. For example, in a 5-year study in the endemic goiter areas of the Himalayas, Sooch and Ramalingaswami (163) obtained a striking reduction in the prevalence of goiter from the use of salt fortified with either KI or KIO_4. During the same period goiter prevalence remained unchanged in the control zone which received plain, unfortified salt. Problems of storage, distribution, cost, and acceptance exist in many goitrous areas and the levels of supplementation vary widely from country to country (165), depending on the intensity of the goiter incidence and the average daily salt consumption. A "reasonable" level of I fortification of salt according to Stanbury *et al.* (165) might be 1/25,000–1/50,000.

Until an effective salt iodization program can be implemented, or in isolated communities, intramuscular injections of iodized oil can provide a cheap, long-acting, easily administered means of combating endemic goiter. Such treatment has been found to correct I deficiency in New Guinea people living in mountainous areas (32, 135) (Table 37). The preparation used (ethiodized oil) is the ethyl ester of the fatty acids of poppyseed oil containing 475 mg I/ml (i.e., 37% by weight). It should be administered to all females up to the age of 45 years and all males up to the age of 20 years. If the dosage schedule set out in Table 39 is adhered to the injection program needs to be repeated only once every 3

TABLE 39
Recommended Dosages of Ethiodized Oil Containing 35% I[a]

Age	Iodine (mg)	Dose (ml)
0–6 months	95.0–180.0	0.2–0.4
6–12 months	142.5–285.0	0.3–0.6
6 months–6 years	232.5–465.0	0.5–1.0
6–45 years	475.0–950.0	1.0–2.0

[a]The dosage should be reduced to 0.2 ml for all persons with nodular goiters or presenting single thyroid nodules without goiter. From Stanbury *et al.* (165).

years (165). Occasional cases of thyrotoxicosis occur following oil injection, mainly in individuals over 40 years of age, but the disease is mild and easily managed.

5. Iodine Supplementation of Farm Animals

With farm animals, the best I supplementation method to adopt depends on the conditions of husbandry. Stall-fed animals are usually provided with iodized salt licks or pellets or the iodine is incorporated into the mineral mixtures or concentrates provided to supply other nutritional needs. Salt licks containing KI lose iodine readily from volatilization or leaching if exposed for any length of time to hot or humid conditions (43). However, potassium iodate, which is more stable than iodide and is nontoxic at the levels required, can be used (160). Inclusion of this compound into salt licks or mineral mixtures for stock at a level of 0.01% I (107) is recommended to ensure adequate intakes in goitrous areas. In a comparison of KI, calcium iodate, and 3,5-diiodosalicylic acid as I sources for livestock, Shuman and Townsend (159) found the first two to be rapidly lost from the surface layer of salt blocks when exposed to outdoor weather conditions, whereas the diiodosalicylic acid remained constantly on the surface and was obtained by the animals in a normal manner. Aschbacher and co-workers (13, 14) found this compound to be an unsatisfactory source of supplemental I for calves and dairy cattle. Subsequently Aschbacher (12) showed that allowing ewes free access to salt containing 0.007% I as diiodosalicylic acid did not prevent I deficiency in their newborn lambs, whereas access to salt containing the same I concentration as KI successfully achieved this aim. These workers (119) then compared the nutritional availability of iodine from calcium iodate, pentacalcium orthoperiodate (PCOP), and sodium iodide to pregnant cows. All three forms were found to supply the fetal thyroid with equal efficiency. Since PCOP has greater physical stability than sodium iodide and calcium iodate under field conditions (115), this compound is clearly a valuable form of supplemental iodine for use in livestock salt blocks or mineral mixes.

With sheep and cattle under permanent grazing in goitrous areas different forms of treatment are necessary. Iodized fertilizers cannot be relied on to maintain satisfactory I levels in the herbage for long periods after application (80), and the provision of iodized salt licks is subject to the hazard, as with all trace elements, of spasmodic and uncertain consumption by the grazing animal. Unless such stable and insoluble compounds as pentacalcium orthoperiodate are used there is also the problem of physical loss by volatilization and leaching. Regular dosing or "drenching" with inorganic I solutions is effective but can be costly and time-consuming. This form of treatment, consisting of two oral doses of 280 mg KI or 360 KIO_3, given at the beginning of the fourth and the fifth months of pregnancy, is satisfactory for the prevention of the neonatal mortality and associated goiter in lambs that arise when ewes are grazed on goitrogenic kale (*Brassica oleracea*) (160). This condition can also be controlled by intramuscular injections of an iodized poppyseed oil preparation, containing some 40% I by weight. A single 1-ml injection of this preparation into ewes 2 months before lambing raised the I concentrations in the thyroid glands of lambs from kale-fed ewes to normal levels and prevented marked goiter and high neonatal mortality. Similar injections given 1 month later only partially prevented the thyroid enlargement, although the I concentrations in these glands were normal and the death rate was reduced (8, 160).

Recently a plastic capsule containing solid iodine has been developed by Laby (104) which provides a sustained release of iodine into the rumen of sheep and cattle over several years. Preliminary field data indicate that this could be a cheap and satisfactory means of providing supplemental I to grazing ruminants in goitrous areas, especially as the I release rate and the lifetime can be varied over a wide range by choice from a variety of capsule dimensions and types of plastic.

VI. GOITROGENIC SUBSTANCES

Goitrogens are substances capable of producing thyroid enlargement by interfering with thyroid hormone synthesis. The pituitary responds by increasing its output of TSH which induces hypertrophy in the gland in an effort to increase thyroid hormone production. The extent to which the increased thyroid tissue mass compensates for the inhibition or blocking of thyroid hormonogenesis depends on the dose of goitrogen and, in some circumstances, on the level of I intake by the animal.

The first clear evidence of a goitrogen in food was obtained by Chesney and co-workers in 1928 (34). Rabbits fed a diet of fresh cabbage developed goiters which could be prevented by a supplement of 7.5 mg I per rabbit per week. "Cabbage goiter" was subsequently demonstrated in other animal species (27), and goitrogenic activity was reported for a wide range of vegetable foods, including virtually all cruciferous plants (15, 27, 164). About this time workers in New Zealand showed that rapeseed goiter arises from interference with the process of thyroid hormonogenesis and, unlike cabbage goiter, is only partially controlled by supplemental iodine (66, 139). Astwood and co-workers (16) subsequently isolated and identified a new compound named goitrin (L-5-vinyl-2-thiooxalidone) from rutabagus and showed that the goitrogenic activity of *Brassica* seeds could largely be accounted for by the presence of this compound in combined form (progoitrin).

Goitrin occurs in the edible portions of most members of the *Brassica* family but is not responsible for the whole goitrogenicity of all foods. Other thioglycosides with antithyroid activity have been found in cruciferous plants (18), and similar activity has been associated with the presence of cyanogenetic glucosides in white clover (*Trifolium repens*) (57). The latter owe their potency to the conversion of the HCN into thiocyanate in the tissues. Thiocyanate is a goitrogenic agent which acts by inhibiting the selective concentration of I by the thyroid (191). Its action is reversible by iodine, whereas that of goitrin, which acts by limiting hormonogenesis in the gland, is either not reversible by such means or is only partly so. The number of goitrogens of this latter type is large. All of them act by inhibiting the iodination of tyrosine, presumably through inhibition of the thyroidal peroxidase. A possible third type of goitrogen, a sulfated unsaturated hydrocarbon, has been isolated from drinking water in Colombia, South America (59).

Goitrogens are widely employed in the treatment of thyrotoxicosis in man and have found some use in livestock and poultry husbandry through their favorable influence on the fattening process. Iodide itself, in large doses, can act as an antithyroid agent in thyrotoxicosis, and this form of therapy was common before goitrogenic drugs became available. Goitrogens are also of interest as etiological factors in the production of nontoxic goiter in man and animals. Where the diet is composed wholly or largely of goitrogenic foods such as cabbage or kale (especially when consumed raw), or cassava, the amounts of antithyroid material ingested can be sufficient to induce goiter, even when I intakes are normal. Reference was made in Section V,5 to the reports of goiter in farm stock intensively fed on kale. Where such foods are consumed by man in normal quantities and cooked, goiter is much less likely to develop because the cooking of vegetables destroys the enzyme which liberates this substance from its inactive precursor (16). However, Greer and associates (65) showed that fresh human and rat feces can hydrolyze pure progoitrin into goitrin, which suggests that goitrin may be formed in the gastrointestinal tract even from cooked vegetables.

The goitrogenicity of plants varies with the species and with the conditions under which they have grown (45), including the fertilizer treatment (3). This could be due to an effect on the level of iodine in the plant, as well as on the levels and types of goitrogen present, because the goitrogenicity of even *Brassica* species is due in part to the presence of thiocyanate and other goitrogens, the effects of which can be overcome by a sufficient increase in I intake. In fact, there is some evidence, at least with kale, that such goitrogens can contribute a major proportion of the goitrogenicity. These findings emphasize the necessity of maintaining adequate iodine intakes in all animal and human populations exposed to the hazards of goitrogenic foods, especially where environmental supplies of the element are otherwise low or marginal.

Milk from cows consuming certain cruciferous plants in the grazing in parts of Tasmania contains a potent goitrogen, the effect of which on children is not overcome by feeding 10 mg KI weekly (36, 60). This suggests that it is of the goitrin or thiouracil type. Goitrogens have also been found in the milk of cows fed cruciferous fodder in England (99) and Finland (133). Direct human consumption of cassava (*Manihot utilissima*) grown on goitrous areas in the Congo has been shown to inhibit I uptake by the thyroid, presumably due to thiocyanate production from the large amounts of cyanogenetic glucosides that the cassava contains (45).

VII. IODINE TOXICITY

Iodine toxicity has been critically studied in man, laboratory species, poultry, pigs, and cattle. Wolff (190) has defined four degrees of iodide excess in man as follows:

(i) Relatively low levels which lead to temporary increases in the absolute I uptake by the thyroid and the formation of organic I, until such time as the thyroid is required to reduce iodide clearances;

(ii) a larger amount which can inhibit I release from the thyrotoxic human thyroid or from thyroids in which I release has been accelerated by TSH;

(iii) a slightly greater intake which leads to inhibition of organic I formation and which probably causes iodide goiter (the so-called Wolff–Chaikoff effect); and

(iv) very high levels of iodide which saturate the active transport mechanism for this ion. The acute pharmacological effects of iodide can usually be demonstrated before saturation becomes significant.

This worker has also suggested that human intakes of 2000 μg I/day should be regarded as an excessive or potentially harmful level of intake. Normal diets composed of natural foods are unlikely to supply as much as 2000 μg I/day, and most would supply less than 1000 μg/day, except where the diets are exceptionally high in marine fish or seaweed, or where foods are contaminated with I from adventitious sources. Inhabitants of the coastal regions of Hokkaido, the northern island of Japan, whose diets contain large amounts of seaweed, have astonishingly high-I intakes amounting to 50,000–80,000 μg I/day (171). Urinary excretion in five patients exhibiting clinical signs of iodide goiter exceeded 20 mg I/day or about 100 times normal. Vought (183) has drawn attention to the upward trend in I consumption in the United States and to the cumulative additions of this element from adventitious sources, notably bread, salt, vitamin preparations, I-containing medications and antiseptics, and coloring matters. Connolly (40) showed that the use of iodophor antiseptics in milking machines, storage vats, and bulk milk tankers leads to a marked increase in the I content of milk, ice cream, and confections made from dairy products. The I concentration of milk from four areas where iodophors were not used ranged from 13 to 23 μg/liter, compared with 113–346 μg/liter in milk from five areas where iodophor bacteriocidal agents were used. The use of potassium iodate instead of bromate as a dough conditioner or "improver" in bread making represents a further abnormal source of I to man. Where these two sources of extra iodine were concurrently present, as in parts of Tasmania, a marked increase in the incidence of thyrotoxicosis occurred, mainly in women in the 40–80-year age group with preexisting goiter (40). The normal thyroid is tolerant of I intakes well above those provided by most diets, but it is clear that toxic or potentially toxic intakes can occur as just described, where abnormal amounts are included as a consequence of various technological developments involving industrial use of the element.

Large doses of stable I are known to reduce radioiodine uptake by the thyroid and decrease retention of this radionuclide, an effect of particular interest with respect to protection against radioactive fallout involving ^{131}I. Driever *et al.* (47) found that administration to calves of stable KI decreased whole body retention of ^{131}I given subsequently by about one-half in 23 days.

Significant species differences exist in tolerance to high-I intakes. In all species studied the tolerance is high, i.e., relative to normal dietary I intakes, pointing to a wide margin of safety for this element. Thus adult female rats fed 500, 1000, 1500, and 2000 ppm of iodine as KI from 0 to 35 days prepartum revealed increasing neonatal mortality of the young with increasing levels of iodine, but the effects of the lowest level of supplemental I fed (500 ppm) were slight when compared with those receiving no supplemental I (5). Examination of mammary gland tissue from females fed iodine indicated that milk secretion was absent or markedly reduced. The fertility of male rats fed 2500 ppm I from birth to 200 days of age appeared to be unimpaired (5). In subsequent studies with rats, rabbits, hamsters, and pigs (11), rabbits fed 250 ppm I or more for

2–5 days in late gestation showed significantly higher mortality of the young than controls receiving no supplemental I. Hamsters were unaffected except for a slightly reduced feed intake and a decreased weaning weight of the young.

In a series of experiments with poultry Arrington and co-workers (10, 110, 134) demonstrated profound effects on egg production and hatchability. When laying hens were fed 312–5000 ppm I as KI in a practical laying ration, egg production ceased within the first week at the highest level and was reduced at the lower levels (134). The fertility of the eggs produced was not affected but early embryonic death, reduced hatchability, and delayed hatching resulted. Within 7 days after cessation of I feeding the hens resumed egg production, indicating that the adverse effects of the excess iodine are only temporary. Similar results were obtained subsequently with pullets and hens fed supplementary I in amounts ranging from 625 to 5000 ppm for 6 weeks, but the effects were much smaller for sexually mature pullets than for the mature hens (10).

The mechanism by which excess iodine affects egg production and embryonic mortality, or reproduction in female rats and rabbits, is not understood. Preliminary experiments conducted by Marcilese *et al.* (110) indicate that thyroxine production is not impaired in hens fed high-I levels. However, the growing ova were shown to have a marked ability to concentrate I from the high doses administered, a finding in keeping with the earlier demonstration of the specific incorporation of orally administered radioiodine into hens' eggs and follicles (179). Thus iodine in the eggs of hens given 100 mg iodine daily as NaI increased linearly for 10 days and reached a plateau of 3 mg/egg at that time. When hens were given 500 mg I daily the level in the eggs increased rapidly to an average of 7 mg I/egg by 8 days, at which time most hens ceased production. Ova continued to develop in hens not laying and many ova were found to be regressing (110). It was suggested that when a threshold amount of iodine reaches the ova, development ceases and regression takes place.

The minimum toxic I intakes, as calcium iodate, for calves (80–112 kg body weight) were shown by Newton *et al.* (127) to be close to 50 ppm, although some experimental animals were adversely affected at lower levels. At 50, 100, and 200 ppm I weight gain and feed intake were depressed and signs of toxicity, including coughing and profuse nasal discharge, became evident. At the higher I levels heavier thyroid and adrenal glands were apparent at the end of the 104–112-day treatment periods, and blood hemoglobin levels were slightly depressed. With milking cows, no signs of toxicity were observed at I intakes of 50 ppm of the diet (as NaI, KI, or ethylenediaminedihydroiodide), although extremely high-I levels were present in the blood serum and in milk, urine, and feces (122).

Pigs are much more tolerant of excess iodine than cattle. Newton and Clawson (128), in experiments with growing–finishing swine fed graded increments of calcium iodate to give I additions from 10 to 1600 ppm of the diet, found the minimum toxic level to lie between 400 and 800 ppm. Growth rate, feed intake, hemoglobin levels, and liver iron concentrations were depressed at 800 and 1600 ppm I. Liver iron levels were also significantly depressed at 400 ppm, suggesting that the minimum toxic level over extended periods could be below 400 ppm I. The interaction between iron and iodine demonstrated in these studies with pigs is of particular interest. The effects of the elevated I intakes on growth, feed intake, and hemoglobin levels were found to be reduced by supplementary Fe, whether given orally or intramuscularly.

REFERENCES

1. Alderman, G., and Jones, D.I.H., *J. Sci. Food Agr.* **18**, 197 (1967).
1a. Alexander, N.M., *J. Biol. Chem.* **234**, 1530 (1959); *Endocrinology* **68**, 671 (1961).
2. Alexander, W.D., Papadopoulos, S., Harden, R. McG., MacFarlane, S., Mason, D.K., and Wayne, E., *J. Lab. Clin. Med.* **67**, 808 (1966).
3. Allcroft, R., and Salt, F.J., *Adv. Thyroid Res., Trans. Int. Goitre Conf., 4th, 1960,* p. 4 (1961).
4. Allcroft, R., Scarnell, J., and Hignett, S.L., *Vet. Rec.* **66**, 367 (1954).
5. Ammerman, C.B., Arrington, L.R., Warnick, A.C., Edwards, J.L., Shirley, R.L., and Davis, G.K., *J. Nutr.* **84**, 107 (1964).
6. Anbar, M., Guttman, S., and Lewitus, Z., *Nature (London)* **183**, 1517 (1959).
7. Anderson, R.K., and Alt, H.L., *Am. J. Physiol.* **119**, 67 (1937).
8. Andrews, E.D., and Sinclair, D.P., *Proc. N. Z. Soc. Anim. Prod.* **22**, 123 (1962).
9. Andrews, F.N., Shrewsbury, C.L., Harper, C., Vestal, C.M., and Doyle, L.P., *J. Anim. Sci.* **7**, 298 (1948).
10. Arrington, L.R., Santa Cruz, R.A., Harms, R.H., and Wilson, H.R., *J. Nutr.* **92**, 325 (1967).
11. Arrington, L.R., Taylor, R.N., Ammerman, C.B., and Shirley, R.L., *J. Nutr.* **87**, 394 (1965).
12. Aschbacher, P.W., *J. Anim. Sci.* **27**, 127 (1968).
13. Aschbacher, P.W., Cragle, R.G., Swanson, E.W., and Miller, J.K., *J. Dairy Sci.* **49**, 1042 (1966).
14. Aschbacher, P.W., Miller, J.K., and Cragle, R.G., *J. Dairy Sci.* **46**, 1114 (1963).
15. Astwood, E.B., *Ann. Intern. Med.* **30**, 1087 (1940).
16. Astwood, E.B., Greer, M.A., and Ettlinger, M.G., *J. Biol. Chem.* **181**, 121 (1949); Greer, M.A., Ettlinger, M.G., and Astwood, E.B., *J. Clin. Endocrinol.* **9**, 1069 (1949).
17. Bach, I., Braun, S., Gati, T., Kertai, P., Sos, J., and Udvardy, A., *Adv. Thyroid Res., Trans. Int. Goitre Conf., 4th, 1960,* p. 505 (1961).
18. Bachelard, H.S., and Trikojus, V.M., *Nature (London)* **185**, 80 (1960).
19. Barakat, M.Z., Bassiouni, M., and El-Wakil, M., *Bull. Acad. Pol. Sci.* **20**, 531 (1972).
20. Barua, J., Cragle, R.G., and Miller, J.K., *J. Dairy Sci.* **47**, 539 (1964).
21. Berson, S.A., *Am. J. Med.* **20**, 653 (1956).
22. Berson, S.A., and Yalow, R.S., *J. Clin. Invest.* **33**, 1533 (1954).
23. Bilvais, B.B., *Physiol. Zool.* **20**, 67 (1947).
24. Binnerts, W.T., *Nature (London)* **174**, 973 (1954).

25. Blaxter, K.L., *Vitam. Horm. (N.Y.)* **10,** 217 (1952).
26. Blom, I.J.B., *Onderstepoort J. Vet. Sci. Anim. Ind.* **2,** 139 (1934).
27. Blum, F., *Schweiz. Med. Wochenschr. [N.S.]* **70,** 1301 (1971).
28. Bogard, R., and Mayer, D.T., *Am. J. Physiol.* **147,** 320 (1946).
29. Brown-Grant, K., *Physiol. Rev.* **41,** 189 (1961).
30. Brown-Grant, K., and Galton, V.A., *Biochim. Biophys. Acta* **27,** 423 (1958).
31. Bruice, T.C., Kharasch, N., and Winzler, R.J., *Arch. Biochem. Biophys.* **62,** 305 (1962).
32. Buttfield, I.H., and Hetzel, B.S., *Bull. W. H. O.* **36,** 243 (1967).
33. Canzanelli, A., Guild, R., and Rapport, D., *Endocrinology* **25,** 707 (1939).
34. Chesney, A.M., Clawson, T.A., and Webster, B., *Bull. Johns Hopkins Hosp.* **43,** 261 (1928).
35. Childs, D.S., Keating, F.R., Rall, J.E., Williams, M.M., and Power, M.H., *J. Clin. Invest.* **29,** 726 (1950).
36. Clements, F.W., *Br. Med. Bull* **16,** 133 (1960).
37. Coble, Y., Davis, J., Schulert, A., Heta, F., and Awad, A.Y., *Am. J. Clin. Nutr.* **21,** 277 (1968).
38. Cohn, B.N., *Arch. Intern. Med.* **49,** 950 (1932).
39. Cole, V.V., and Curtis, G.M., *J. Nutr.* **10,** 493 (1935).
40. Connolly, R.J., *Med. J. Aust.* **1,** 1268 (1971); Connolly, R.J., Vidor, G.I., and Stewart, J.C., *Lancet* **1,** 500 (1970).
41. Creek, R.D., Parker, H.E., Hauge, S.M., Andrews, E.N., and Carrick, C.W., *Poultry Sci.* **33,** 1052 (1954).
42. Curtis, G.M., Puppel, I.D., Cole, V.V., and Matthews, N.L., *J. Lab. Clin. Med.* **22,** 1014 (1937).
43. Davidson, W.M., and Watson, C.J., *Sci. Agric.* **28,** 1 (1948).
44. De Groot, L.J., and Davis, A.M., *Endocrinology* **70,** 492 (1962).
45. Delange, F., and Ermans, A.M., *Am. J. Clin. Nutr.* **24,** 1354 (1971).
46. D'Haene, E.G.M., Crombag, F.J.L., and Tertoolen, J.F.W., *Br. Med. J.* **3,** 708 (1974).
47. Driever, C.W., Christian, J.E., Bousquet, W.F., Plumlee, M.P., and Andrews, F.N., *J. Dairy Sci.* **48,** 1088 (1965).
48. Dumont, J.E., Ermans, A.M., and Bastenie, P.A., *J. Clin. Endocrinol. Metab.* **23,** 847 (1963).
49. Ekins, R.P., *Clin. Chim. Acta* **5,** 453 (1960).
50. Ekpechi, O.L.V., and Van Middlesworth, L., *Endocrinology* **92,** 1376 (1973).
51. Elmer, A.W., "Iodine Metabolism and Thyroid Function." Oxford Univ. Press, London and New York, 1938.
52. Ermans, A.M., Kinthaert, J., Delcroix, C., and Collard, J., *J. Clin. Endocrinol. Metab.* **28,** 69 (1968).
53. Evvard, J.M., *Endocrinology* **12,** 539 (1928).
54. Falconer, I.R., *Nature (London)* **205,** 703 (1965).
55. Fierro-Benitez, R., Penafiel, W., de Groot, W., and Ramirez, I., *N. Engl. J. Med.* **280,** 296 (1969).
56. Fierro-Benitez, R., Ramirez, I., Garces, J., Jaramillo, C., Moncayo, F., and Stanbury, J.B., *Am. J. Clin. Nutr.* **27,** 531 (1974).
57. Flux, D.S., Butler, G.W., Johnson, J.M., Glenday, A.C., and Peterson, G.B., *N. Z. J. Sci. Technol. Sect. A* **38,** 88 (1956).
58. Follis, R.H., Jr., *Am. J. Clin. Nutr.* **14,** 253 (1963).
58a. Funke, H., Iwarsson, K., Olsson, S.O., Salomonsson, P., and Strandberg, P., *Nord. Veterinaermed.* **27,** 270 (1975).

59. Gaiton, E., MacLennon, R., Island, D.P., and Liddle, G.W., *Trace Subst. Environ. Health–5, Proc. Univ. Mo. Annu. Conf., 5th, 1971,* p. 55 (1971).
60. Gibson, H.B., Howeler, J.F., and Clements, F.W., *Med. J. Aust.* **5**, 875 (1960).
61. Glascock, R.F., *J. Dairy Res.* **21**, 318 (1954).
62. Goolden, A.W.G., Gartside, J.M., and Sanderson, G., *Lancet* **1**, 12 (1967).
63. Gordon, A.H., Gross, J., O'Connor, D., and Pitt-Rivers, R., *Nature (London)* **169**, 19 (1952).
64. Gorge, F.B. de, and Jose, N.K., *Nature (London)* **214**, 491 (1967).
65. Greer, M.A., Iiono, S., Barr, S., and Whallon, H., *Adv. Thyroid Res., Trans. Int. Goitre Conf., 4th, 1960,* p. 1 (1961).
66. Griesbach, W.E., Kennedy, T.H., and Purves, H.D., *Br. J. Exp. Pathol.* **22**, 249 (1941); Griesbach, W.E., and Purves, H.D., *ibid.* **24**, 174 (1943).
67. Gross, J., and Leblond, C.P., *J. Biol. Chem.* **171**, 309 (1947).
68. Gross, J., and Pitt-Rivers, R., *Lancet* **1**, 439 (1952).
69. Guillemin, R., Yamazaki, E., Gard, D.A., Jutisz, M., and Sakiz, E., *Endocrinology* **73**, 564 (1963).
70. Gurevich, G.P., *Vopr. Pitan.* **18**, 65 (1959); *Nutr. Abstr. Rev.* **30**, 697 (1960).
71. Gurevich, G.P., *Fed. Proc., Fed. Am. Soc. Exp. Biol.* **23**, Trans. Suppl., T511 (1964).
72. Gustafsson, R., Tata, J.R., Lindberg, O., and Ernster, L., *J. Cell Biol.* **26**, 555 (1965).
73. Hallman, B.L., Bondy, P.K., and Hagewood, M.A., *Arch. Intern. Med.* **87**, 817 (1951).
74. Hamilton, E.I., and Minski, M.J., *Sci. Total Environ.* **1**, 375 (1972/1973).
75. Hamilton, E., Minski, M.J., and Cleary, J.J., *Sci. Total Environ.* **1**, 341 (1972/1973).
76. Harden, R. McG., Alexander, W.D., Shimmins, J., and Robertson, J., *Q. J. Exp. Physiol. Cogn. Med. Sci.* **53**, 227 (1968).
77. Harden, R. McG., Alexander, W.D., Shimmins, J., Kostalas, N., and Mason, D.K., *J. Lab. Clin. Med.* **71**, 92 (1968).
78. Harden R. McG., Hilditch, T., Kennedy, I., Mason, D.K., Papadopoulos, S., and Alexander, W.D., *Clin. Sci.* **32**, 49 (1967).
79. Harden R. McG., Mason, D.K., and Alexander, W.D., *Q. J. Exp. Physiol. Cogn. Med. Sci.* **51**, 130 (1966).
80. Hartmans, J., *Neth. J. Agric. Sci.* **22**, 195 (1974).
81. Heinrich, H.C., Gabbe, E.E., and Whang, D.H., *Atomkernenergie* **9**, 279 (1964).
82. Hemken, R.W., Vandersall, J.H., Oskarsson, M.A., and Fryman, L.R., *J. Dairy Sci.* **55**, 931 (1972).
83. Hennessy, W.B., *Med. J. Aust.* **1**, 505 (1964).
84. Hipsley, E.H., *Med. J. Aust.* **1**, 532 (1956).
85. Hoch, F.L., *Physiol. Rev.* **42**, 605 (1962).
86. Hochachka, P.W., *Fed. Proc., Fed. Am. Soc. Exp. Biol.* **33**, 2164 (1974).
87. Hoskins, R.G., *J. Am. Med. Assoc.* **55**, 1724 (1910).
88. Howell, G.L., and Van Middlesworth, L., *Proc. Soc. Exp. Biol. Med.* **93**, 602 (1956).
89. Ingbar, S.H., *Endocrinology* **63**, 256 (1958).
90. Ingbar, S.H., Braverman, L.E., Dawber, N.A., and Lee, G.Y., *J. Clin. Invest.* **44**, 1679 (1965).
91. "Iodine Content of Foods." Chilean Iodine Educational Bureau, London, 1952.
92. "Iodine and Plant Life." Chilean Iodine Educational Bureau, London, 1950.
93. Irvine, C.H.G., *Am. J. Vet. Res.* **28**, 1687 (1967).
94. Ismael-Beigi, F., and Edelman, I.S., *J. Gen. Physiol.* **57**, 710 (1971).

95. Johnson, J.M., and Butler, G.W., *Physiol. Plant.* **10,** 100 (1957).
96. Jovanović, M., Pantić, V., and Marković, B., *Acta Vet. (Belgrade)* **3,** 31 (1953).
97. Karmarkar, M.G., Deo, M.G., Kochupillai, N., and Ramalingaswami, V., *Am. J. Clin. Nutr.* **27,** 96 (1974).
98. Keating, F.R., Jr., and Albert, A., *Recent Progr. Horm. Res.* **4,** 429 (1949).
99. Kilpatrick, R., Broadhead, G.D., Edmonds, C.J., Munro, D.S., and Wilson, G.M., *Adv. Thyroid Res., Trans. Int. Goitre Conf., 4th, 1960,* p. 273 (1961).
100. Kirchgessner, M., *Z. Tierphysiol., Tierernahr. Futtermittelkd.* **14,** 270 and 278 (1959).
101. Koutras, D.A., *in* "Activation Analysis in the Study of Mineral Metabolism in Man." IAEA, Vienna, 1968.
102. Kunkel, H.O., Colby, R.W., and Lyman, C.M., *J. Anim. Sci.* **12,** 3 (1953).
103. Kupzis, J., *Z. Hyg. Infektionskr.* **113,** 551 (1932).
104. Laby, R.H., private communication (1975).
105. Levin, R.J., *J. Endocrinol.* **45,** 315 (1969).
106. Lewis, R.C., and Ralston, N.P., *J. Dairy Sci.* **36,** 33 and 363 (1951).
107. Loosli, J.K., Becker, R.B., Huffman, C.F., Phillips, P.H., and Shaw, J.C.,, "Nutrient Requirements of Dairy Cattle." Natl. Res. Counc., Washington, D.C., 1956.
108. Maqsood, M., *Biol. Rev. Cambridge Philos. Soc.* **27,** 281 (1952).
109. Malamos, B., and Koutras, D.A., *Proc. Int. Congr. Intern. Med., 7th, 1960,* Vol. 2, p. 678 (1962).
110. Marcilese, N.A., Harms, R.H., Valsechhi, R.M., and Arrington, L.R., *J. Nutr.* **94,** 117 (1968).
111. Marine, D., and Williams, W., *Arch. Intern. Med.* **1,** 349 (1908); Marine, D., and Lenhart, C.H., *ibid.* **3,** 66 (1909).
112. Maurer, E., and Dugrue, H., *Biochem. Z.* **193,** 356 (1928).
113. McDonald, R.J., McKay, G.W., and Thomson, J.D., *Proc. Int. Congr. Anim. Reprod., 4th, 1961,* Vol. 3, p. 679 (1962).
114. McQuillan, M.T., Stanley, P.G., and Trikojus, V.M., *Aust. J. Exp. Biol. Med. Sci.* **6,** 617 (1953).
115. Meyer, R.J., Internal Report, Morton Salt Co., Chicago (unpublished).
116. Michel, R., *Am. J. Med.* **20,** 670 (1956).
117. Middlesworth, L. Van., *J. Clin. Endocrinol. Metab.* **16,** 989 (1956).
118. Miller, J.K., *Proc. Soc. Exp. Biol. Med.* **121,** 291 (1966).
119. Miller, J.K., Moss, B.R., Swanson, E.W., Aschbacher, P.W., and Cragle, R.G., *J. Dairy Sci.* **51,** 1831 (1967).
120. Miller, J.K., Swanson, E.W., Spalding, G.E., Lyke, W.A., and Hall, R.F., *in* "Trace Element Metabolism in Animals" (W.G. Hoekstra *et al.,* eds.), Vol. 2, p. 638. Univ. Park Press, Baltimore, Maryland, 1974.
121. Miller, J.K., Swanson, E.W., Lyke, W.A., Moss, B.R., and Byrne, W.F., *J. Dairy Sci.* **57,** 193 (1974).
122. Miller, J.K., and Swanson, E.W., *J. Dairy Sci.* **56,** 378 (1973).
123. Mitchell, H.H., and McClure, F.J., *Bull. Natl. Res. Counc. (U.S.)* **99,** 1937.
124. Moberg, R., *Proc. World Congr. Fertil. Steril., 3rd,* p. 71 (1959).
125. Murphy, B.E.P., Pattee, C.J., and Gold, A., *J. Clin. Endocrinol. Metab.* **26,** 247 (1966).
126. Musset, M.V., and Pitt-Rivers, R., *Lancet* **2,** 1212 (1954).
127. Newton, G.L., Barrick, E.R., Harvey, R.W., and Wise, M.B., *J. Anim. Sci.* **38,** 449 (1974).

128. Newton, G.L., and Clawson, T.A., *J. Anim. Sci.* **39,** 879 (1974).
128a. "Nutrient Requirements of Farm Livestock," No. 2, p. 104. HM Stationery Office, London, 1966.
129. Oddie, T.H., Fisher, D.A., McConahey, W.M., and Thompson, C.S., *J. Clin. Endocrinol. Metab.* **30,** 659 (1970).
130. Orr, J.B., and Leitch, I., *Med. Res. Counc. (G.B.), Spec. Rep. Ser.* **SRS-123** (1929).
131. Paley, K.R., Sobel, E.S., and Yalow, R.S., *J. Clin. Endocrinol. Metab.* **15,** 995 (1955); **18,** 850 (1958).
132. Papadopoulos, S., MacFarlane, S., Harden, R. McG., Mason, D.K., and Alexander, W.D., *J. Endocrinol.* **36,** 341 (1966).
133. Peltola, P., *Adv. Thyroid Res., Trans. Int. Goitre Conf., 4th, 1960,* p. 10 (1961).
134. Perdomo, J.T., Harms, R.H., and Arrington, L.R., *Proc. Soc. Exp. Biol. Med.* **122,** 758 (1966).
135. Pharaoh, P.O.D., Buttfield, I.H., and Hetzel, B.S., *Lancet* **1,** 308 (1971).
136. Pitt-Rivers, R., and Tata, J.R., "The Thyroid Hormones." Pergamon, Oxford, 1959.
137. Post, T.B., and Mixner, J.P., *J. Dairy Sci.* **44,** 2265 (1961).
138. Premachandra, B.N., Pipes, G.W., and Turner, C.W., *J. Anim. Sci.* **17,** 1237 (1958).
139. Purves, H.D., *Br. J. Exp. Pathol.* **24,** 171 (1943).
140. Rall, J.E., Robbins, J., and Lewallen, C.G., *Hormones* **5,** 159 (1964).
141. Reineke, R.P., and Turner, C.W., *J. Dairy Sci.* **27,** 793 (1944).
142. Riggs, D.S., *Pharmacol. Rev.* **4,** 282 (1952).
143. Robbins, J., and Rall, J.E., *Physiol. Rev.* **40,** 415 (1960).
144. Roche, J., Michel, R., Michel, O., and Lissitzky, S., *Biochim. Biophys. Acta* **9,** 161 (1952).
145. Roche, J., Michel, R., Nuñez, J., and Wolff, W., *C.R. Seances Soc. Biol. Ses Fil.* **149,** 885 (1955).
146. Roche, J., Michel, R., Truchot, R., and Wolff, W., *C.R. Seances Soc. Biol. Ses Fil.* **149,** 1219 (1955).
147. Roche, J., Michel, R., and Wolf, W., *C.R. Hebd. Seances Acad. Sci.* **240,** 251 and 921 (1955).
148. Salter, W.T., *Hormones* **2,** 181 (1950).
149. Salter, W.T., and Johnson, McA. W., *J. Clin. Endocrinol.* **8,** 924 (1948).
150. Samuels, H.H., and Tsai, J.S., *Proc. Natl. Acad. Sci. U.S.A.* **70,** 3488 (1973).
151. Samuels, H.H., Tsai, J.S., Casanova, J., and Stanley, F., *J. Clin. Invest.* **54,** 853 (1974).
152. Samuels, H.H., Tsai, J.S., and Cintron, R., *Science* **181,** 1253 (1973).
153. Sceffer, L., *Biochem. Z.* **259,** 11 (1933).
154. Scott, M., *Trans. Int. Goitre Conf., 3rd,* p. 34 (1958).
155. Scott, M.L., van Tienhoven, A., Holm, E.R., and Reynolds, R.E., *J. Nutr.* **71,** 282 (1960).
156. Setchell, B.P., Dickinson, D.A., Lascelles, A.K., and Bonner, R.B., *Aust. Vet. J.* **36,** 159 (1960).
157. Shand, A., *Br. Vet. Assoc. Publ.* **23,** (1952).
158. Sharpless, G.R., and Metzger, M., *J. Nutr.* **21,** 341 (1941).
159. Shuman, A.C., and Townsend, D.P., *J. Anim. Sci.* **22,** 72 (1963).
159a. Sihombing, D.T.H., Cromwell, G.L., and Hays, V.W., *J. Anim. Sci.* **39,** 1106 (1974); Cromwell, G.L., Sihombing, D.T.H., Hays, V.W., *ibid.* **41,** 813 (1975).
160. Sinclair, D.P., and Andrews, E.D., *N. Z. Vet. J.* **6,** 87 (1958); **7,** 39 (1959); **9,** 96 (1961).
161. Singh, O.N., Henneman, H.A., and Reineke, R.P., *J. Anim. Sci.* **15,** 625 (1956).

162. Soffer, L.J., Gabrilove, J.L., and Jailer, J.W., *Proc. Soc. Exp. Biol. Med.* **71,** 117 (1949).
163. Sooch, S.S., and Ramalingaswami, V., *Bull. W. H. O.* **32,** 299 (1965).
164. Srinovasan, V., Moudgal, N.R., and Sarma, P.S., *J. Nutr.* **61,** 87 (1957).
165. Stanbury, J.B., Ermans, A.M., Hetzel, B.S., Pretell, E.A., and Querido, A., *WHO Chron.* **28,** 220 (1974).
166. Sterling, K., and Bremner, M.A., *J. Clin. Invest.* **45,** 153 (1966).
167. Steyn, D.G., *Rep. S. Afr. Med. Assoc. (Onderstepoort)* (1938).
168. Stocks, P., *Biometrika* **19,** 272 (1927); *Ann. Eugen.* **2,** 382 (1927).
169. Stott, H., and Gupta, S.P., *Indian J. Med. Res.* **21,** 649 (1934).
170. Sturm, A., and Bucholz, B., *Arch. Klin. Med.* **161,** 227 (1928).
171. Suzuki, H., Higuchi, T., Sawa, K., Ohtaki, S., and Horiuchi, Y., *Jpn. Acta Endocrinol.* **50,** 161 (1965).
172. Tata, J.R., Ernster, L., Lindberg, O., Arrhenius, E., Pederson, S., and Hedman, R., *Biochem. J.* **86,** 408 (1963).
173. Tata, J.R., and Widnell, C.C., *Biochem. J.* **98,** 604 (1966).
174. Taurog, A., Briggs, F.N., and Chaikoff, I.L., *J. Biol. Chem.* **191,** 29 (1951).
175. Taurog, A., and Chaikoff, I.L., *J. Biol. Chem.* **163,** 313 (1946).
176. Taurog, A., Chaikoff, I.L., and Feller, D.D., *J. Biol. Chem.* **171,** 189 (1947).
177. Taurog, A., and Howells, E.M., *J. Biol. Chem.* **241,** 1329 (1966).
178. Taurog, A., Wheat, J.D., and Chaikoff, I.L., *Endocrinology* **58,** 121 (1956).
179. Thorell, C.B., *Acta Vet. Scand.* **5,** 224 (1964).
180. Thorson, S.C., Muncey, E.K., McIntosh, H.W., and Morrison, R.T., *Br. Med. J.* **2,** 67 (1972).
181. Trotter, W.R., *Br. Med. Bull.* **16,** 92 (1960).
182. Turner, C.W., Irwin, M.R., and Reineke, R.P., *Poultry Sci.* **24,** 171 (1945); Turner, C.W., and Kempster, H.L., *ibid.* **27,** 453 (1948).
183. Vought, R.L., *Trace Subst. Environ. Health–5, Proc. Univ. Mo. Annu. Conf., 1971,* p. 303 (1971).
184. Vought, R.L., and London, W.T., *Am. J. Clin. Nutr.* **14,** 186 (1964).
185. Vought, R.L., London, W.T., Lutwak, L., and Dublin, T.P., *J. Clin. Endocrinol. Metab.* **23,** 1218 (1963).
186. Wayne, E.J., Koutras, D.A., and Alexander, W.D., "Clinical Aspects of Iodine Metabolism." Blackwell, Oxford, 1964.
187. Weeks, N.H., Katz, J., and Farnham, N.C., *Endocrinology* **50,** 511 (1952).
188. Welby, M., and O'Halloran, M.W., *Br. Med. J.* **2,** 668 (1966).
189. Wilson, D.C., *Lancet* **1,** 211 (1941).
190. Wolff, J., *Am. J. Med.* **47,** 101 (1969).
191. Wolff, J., Chaikoff, I.L., Taurog, A., and Rubin, L., *Endocrinology* **39,** 140 (1946).
192. Wood, F.O., *in* "Iodine Nutriture in the United States," p. 30. Natl. Acad. Sci.–Natl. Res. Counc., Washington, D.C., 1970.
193. Wright, W.E., Christian, J.E., and Andrews, F.N., *J. Dairy Sci.* **38,** 31 (1955).
194. Wyngaarden, J.B., Wright, B.M., and Ways, P., *Endocrinology* **50,** 1537 (1952).
195. Yamagata, N., and Yamagata, T., *J. Radiat. Res.* **13,** 81 (1972).
196. Young, M., Crabtree, M.G., and Mason, I.M., *Med. Res. Counc. (G. B.), Spec. Rep. Ser.* **SES-217** (1936).
197. Zyl, A. Van., and Kerrich, J.E., *S. Afr. J. Med. Sci.* **20,** 9 (1955).

12

Selenium

I. SELENIUM IN ANIMAL TISSUES AND FLUIDS

1. General Distribution

Selenium occurs in all the cells and tissues of the animal body in concentrations that vary with the tissue and the level and chemical form of Se in the diet. The liver and kidney usually carry the highest Se concentrations, with much lower levels in the muscles, bones, and blood and very low levels in adipose tissue. Cardiac muscle is consistently higher in Se than skeletal muscle (48, 91, 134). Selenium concentrations in the tissues reflect the level of dietary Se over a wide range. In a study of the Se level in the longissimus muscle of pigs from 13 locations in the United States with known differing natural dietary Se intakes from 0.027 to 0.493 ppm Se, a highly significant ($P < 0.01$) linear correlation of 0.95 between dietary Se and tissue Se concentration was established (123). Selenium deposition in the blood, muscle, liver, kidney, and skin of chicks and poults has similarly been shown to bear a direct relationship to the inorganic Se of the diet up to dietary levels of 0.2–0.3 ppm (213). Increasing the dietary inorganic Se further up to 0.8 ppm resulted in higher Se levels in the liver and kidney, but there was no appreciable increase in blood or muscle Se. Increasing the dietary Se to 0.67 ppm by addition of organic Se in soybean meal, fish meal, and wheat induced higher Se levels in muscle and blood than equivalent levels of dietary Se as selenite. A similar superiority of natural Se over selenite Se in increasing the Se levels in muscle and liver has been demonstrated in laying hens (127).

TABLE 40
Mean Selenium Concentrations in the Tissues of Normal Pigs and Pigs with Nutritional Muscular Dystrophy (NMD)[a]

	Kidney	Liver	Skeletal muscle	Heart	Pancreas	Spleen	Lung
Healthy pigs[b] (6)	11.47 ± 1.18	1.82 ± 0.16	0.52 ± 0.06	1.05 ± 0.10	1.42 ± 0.14	1.26 ± 0.09	1.13 ± 0.13
NMD pigs[c] (6)	2.48 ± 0.29	0.20 ± 0.05	0.16 ± 0.08	0.19 ± 0.09	0.24 ± 0.13	0.40 ± 0.40	0.25 ± 0.11

[a] Se concentration in ppm on the dry basis. From Lindberg (134).
[b] Food averaged 0.126 ppm Se.
[c] Food averaged 0.021 ppm Se.

In rats fed a *Torula* yeast (low-Se) diet for 4 weeks the Se levels in the kidneys fell from 1.0 to 0.3 ppm and in the liver from 0.7 to 0.1 ppm (fresh basis) (26). Similar levels occur in these tissues in normal and Se-deficient sheep (37, 97, 104, 186), cattle (13), and pigs (134) (see Table 40). Andrews and co-workers (7) consider that Se levels greater than 1.0 ppm in the kidney cortex and 0.1 in the liver are normal for sheep and that half these levels are indicative of marginal Se deficiency. Levels below 0.25 ppm in the kidney cortex and 0.02 ppm in the liver (fresh basis) are regarded as indicative of marked Se deficiency.

At toxic intakes of selenium, i.e., 10–100 times or more greater than those normally ingested, much higher tissue Se concentrations are usual than those just given. Levels as high as 5–7 ppm in liver and kidneys and 1–2 ppm in the muscles may be reached. Beyond these tissue levels excretion begins to keep pace with absorption (37, 157, 186, 224). Even higher Se levels may be reached in the hair and hoofs of severely affected animals, as discussed in Section I,4. The tissue Se levels cited for selenotic animals are well above those that occur in animals treated at recommended rates to prevent Se deficiency. Food products from such treated animals do not contain undesirably or dangerously high-Se concentrations (7, 37, 45, 91, 126, 179).

Comparatively few data have appeared on the Se levels in normal human tissues, other than blood. Dickson and Tomlinson (40) reported the following values for autopsy specimens from 10 Canadian adults: liver, 0.18–0.66 (mean 0.44); skin, 0.12–0.62 (mean 0.27); and muscle, 0.26–0.59 (mean 0.37) μg Se/g of whole tissue. In a later investigation carried out in England (90) the mean Se concentrations in μg/g whole tissue were reported as follows: liver, 0.30 ± 0.10; kidney, 0.10 ± 0.02; muscle, 0.11 ± 0.01; lung, 0.10 ± 0.02; brain, 0.09 ± 0.02; testis, 0.20 ± 0.04; and ovary, 0.09 ± 0.03. In another study (136) the Se level in 45 samples of human dental enamel ranged from 0.12 to 0.90 (mean 0.43 ± 0.03) μg/g dry weight. The high-Se content of human finger nails compared with other normal tissues has been pointed out by Hadjimarkos and Shearer (84). The range reported for 16 individuals was 0.70–1.69 and the mean 1.14 ± 0.06 ppm Se. By contrast these workers found the Se concentration of the saliva of a group of children to be very low (range 1.1–5.2, mean 3.0 ± 0.3 ppb) (83).

In a recent study of human tissues, higher than normal tissue Se concentrations were observed in synovia from patients with rheumatoid arthritis and in pancreatic tissues associated with histopathological changes (148). The basis for these changes has not been established.

2. Selenium in Blood

The concentration of selenium in blood is highly responsive to changes in the Se level in the diet over a wide range. For example, the blood of sheep fed highly toxic amounts of selenium can contain as much as 1.34–3.1 μg/ml Se (186,

193). In sheep suffering from various Se-responsive diseases Se levels as low as 0.01–0.02 μg/ml occur (37, 97, 104). In two experiments reported by Kuchel and Buckley (125) the Se concentration of the whole blood of sheep grazing pastures of normal Se status ranged from 0.06 to 0.20 (mean 0.10) μg/ml and from 0.04 to 0.08 (mean 0.06) μg/ml. The administration of Se pellets induced a rapid rise to levels as high as 0.15–0.25 μg Se/ml depending on the amount of Se in the pellets. Hartley (97) considers that a level of 0.05 μg Se/ml is "satisfactory" for sheep. The mean whole blood Se level of 10 normal healthy cows has been reported as 0.08 μg/ml (13).

The above values appear low compared with those reported as normal for other species. Thus Burk *et al.* (26) found the blood of rats fed a *Torula* yeast diet to fall from a mean of 0.3 μg Se/ml to approximately 0.05 μg/ml in 4 weeks. The latter value is close to the levels obtained by Scott and Thompson (213) for chicks and poults after 4 weeks on a low-Se (0.07 ppm) diet. At 2.7 ppm dietary Se the blood Se levels were approximately doubled at 4 weeks to 0.188 ppm in the chicks and 0.106 ppm in the poults, without appreciable increase beyond these levels with further dietary Se increments up to 0.67 ppm Se.

The Se levels in whole human blood from 210 male donors in 19 sites in the United States were reported to range from 0.10 to 0.34 μg/ml (4). Some evidence was obtained of a geographic pattern reflecting established regional differences in the Se levels in crops. These values are close to those of another study of U.S. residents (25), in which the levels obtained for samples from the northeastern states are similar to those reported by Dickson and Tomlinson (40) for an adjacent area in Canada. Two studies of the blood of English residents have yielded disparate results. Bowen and Cawse (16) reported somewhat higher Se levels than those just cited, while Hamilton and co-workers (90) obtained a lower level of 0.08 μg/g wet weight for a "U.K. Master Mix" in which equal aliquots from each of the 2500 blood samples collected were combined. A level of 0.12 μg Se/g was reported for the blood of a small group of normal Swedish individuals (20). There is little doubt that these variations primarily reflect real regional differences in Se status. Such a contention receives strong support from data obtained in New Zealand–a country with extensive areas of low-Se status soils and Se-responsive diseases in livestock, as described in Sections III,5 and 7. In two separate studies the mean whole blood Se levels of normal New Zealand people were reported as 0.068 ± 0.013 (74) and 0.069 ± 0.010 μg/ml, respectively (239). Furthermore, the Se levels in the blood of visitors to that country have been shown to decline to within the low range of concentration of New Zealand residents within 3–9 months (74).

McConnell and co-workers (148) recently reported a study of Se in human blood sera in health and disease. Their mean normal healthy value of 0.118 μg Se/ml is very close to the mean values obtained in other studies (40, 87).

Significantly lower Se values were observed in the sera of cancer patients, especially those with gastrointestinal cancer. The sera of patients with primary neoplasms of the reticuloendothelial system revealed generally higher than normal Se levels, while patients with other medical and surgical disorders mostly displayed normal Se values.

In two groups of children suffering from kwashiorkor mean levels of 0.08 and 0.11 μg Se/ml were obtained, compared with 0.14 and 0.22 μg/ml in control and recovered children, respectively (25). Sudden infant death (SID) victims exhibited lower whole blood and plasma Se levels than normal adults, but these levels did not differ significantly from those of control infants (190).

Red blood cells contain higher Se concentrations than plasma (25, 40, 190). The values reported for the control children and for children recovered from kwashiorkor were, respectively, 0.36 and 0.23 μg Se/ml for red cells and 0.15 and 0.10 μg Se/ml for plasma (25). The levels obtained for the SID victims and normal adults were whole blood, 0.100 ± 0.036 and 0.130 ± 0.014; and plasma, 0.069 ± 0.022 and 0.102 ± 0.018 ppm Se (190). In chicks, levels of 0.25–0.35 μg Se/g in red cells and 0.1–0.2 μg/g in serum have been reported (230). The distribution of glutathione peroxidase in cells and plasma is considered in the next section.

3. Forms of Selenium in Blood and Tissues

Selenium occurs in animal tissues partly bound to proteins in a manner incompletely understood, partly incorporated into proteins as Se analogs of the S-containing amino acids, and as the main functioning form of Se, glutathione peroxidase (GSH-Px). The presence of a Se-containing cytochrome in ovine muscle has also been suggested (244). The ability of the animal to convert inorganic Se into selenoamino acids, as occurs in plants, has been questioned (38, 110) since McConnell first reported that selenomethionine and selenocystine were present in dog liver protein (146) and that ^{75}Se is incorporated into the α- and β-lipoproteins of rat and dog serum (147). Jenkins (110) proposed that the Se is bound between the S atoms of disulfide to form S–Se–S bonds. Fuss and Godwin (55) have demonstrated the incorporation of small though significant amounts of selenite Se, as selenoamino acids, into proteins of liver, kidney, and pancreas, as well as into the proteins of the milk and plasma of sheep. Uptake of ^{75}Se into the protein components of the enamel and dentine of the teeth of rats also occurs (218), although it is not unequivocally evident that it is present as selenomethionine or selenocystine.

Glutathione peroxidase (GSH-Px) was identified as a selenoprotein by Rotruck *et al.* (196) in 1973, and in the same year Flohe and co-workers (52) showed that the Se in the enzyme was present in stoichiometric amounts with 4 g-atoms Se/mole. Glutathione peroxidase catalyzes the removal of H_2O_2 according to the coupled reaction

$$2GSH + H_2O_2 \xrightarrow{\text{glutathione peroxidase}} GSSG + H_2O$$

$$GSSG + NADPH \xrightarrow{\text{glutathione reductase}} 2GSH + NADP^+$$

Glutathione peroxidase activity has been demonstrated in a wide range of body tissues, fluids, cells, and subcellular fractions at levels which vary greatly with the species, tissue, and Se status of the animal (see Ganther *et al.*, 57). The highest GSH-Px activity commonly occurs in the liver, moderately high activity in the erythrocytes, heart muscle, lung, and kidneys, and smaller activity in the intestinal tract and skeletal muscles.

The dramatic dependence of the GSH-Px activity of the tissues on dietary Se intakes is evident from numerous studies with several species (34, 67, 86, 178, 210). For example, Scott (210) found that the level of GSH-Px in the tissues of chicks made Se-deficient dropped to about 10% of the normal value in 5 days, and Omaye and Tappel (178) found the GSH-Px activity of the tissues of chicks to be dose related to the dietary Se level, with the specific activity increasing as a logarithmic function of this level. Chow and Tappel (34) found the GSH-Px activity of the tissues of rats fed a low-Se diet for 17 days to decrease in the following relative order: plasma > kidney > heart > liver > lung > erythrocytes > testes. Se supplementation of the depleted rats increased the GSH-Px activity of all tissues other than the testes linearly and significantly above that of unsupplemented controls. Hafeman and co-workers (86) observed a fall in liver GSH-Px activity to "undetectable" levels within 24 days of feeding an unsupplemented *Torula* yeast diet to weanling rats. The response of erythrocyte GSH-Px activity to the lack of dietary Se was smaller in magnitude and more gradual, but in both tissues rapid elevation of GSH-Px activity followed Se supplementation. Godwin and co-workers (67) examined the tissues of 2–3 week-old lambs from Se-deprived and Se-supplemented ewes. Even at this early stage significant reductions in GSH-Px activities in the erythrocyres and muscles and a small reduction in the plasma were evident in the lambs from the deficient ewes (Table 41). At 10 weeks of age GSH-Px activity in the erythrocytes was markedly different between the two groups of lambs, and there was a significant fall in the levels of this enzyme in the right and left ventricles of the lambs from the untreated ewes.

In ovine skeletal and cardiac muscle (67), rat liver (72), and chicken liver (170) the major proportion of the GSH-Px is localized in the soluble or cytosolic fraction. Most of the noncytosolic, particulate portion of liver GSH-Px is present in the mitochondria (51) or in the nuclear and mitochondrial fractions (170).

TABLE 41
Glutathione Peroxidase Activity in Muscle, Hemolyzed Erythrocytes, and Plasma of 2- to 3-Week-Old Lambs Reared by Ewes Supplemented Variously with Selenium and Vitamin E[a]

Ewe treatment	No. of samples	Enzyme activity		
		Muscle biopsy	Erythrocytes	Plasma
No Se supplement[b]	6	1.46 ± 0.23	16.3 ± 2.8	0.29 ± 0.07
Se-supplemented[c]	6	2.33 ± 0.23	37.4 ± 2.8	0.48 ± 0.08
Significance		$p = 0.05$	$p < 0.001$	NS

[a]Enzyme activities are expressed as μmol NADPH oxidized per 100 mg protein per minute ± SEM. From Godwin *et al.* (67).
[b]Includes lambs from ewes treated with vitamin E.
[c]Both orally drenched and Se pellet-treated ewes.

Selenium and particularly selenide is strikingly localized in the mitochondria and smooth endoplasmic reticulum of the liver in adequately fed rats (33). Levander *et al.* (131) have since shown that most of the Se in rat liver mitochondria occurs in the form of glutathione peroxidase.

4. Selenium in Hair and Feathers

The Se level of the hair of cattle is a useful indicator of both Se deficiency and Se toxicity. In one study cows with hair Se levels between 0.06 and 0.23 ppm produced calves sick or dead from white muscle disease (WMD), whereas no WMD was observed in calves from dams with hair Se levels greater than 0.25 ppm (103). By contrast, the hair of yearling cattle on a seleniferous range averaged over 10 ppm Se, with values as high as 30 ppm, compared with 1–4 ppm Se for the hair of cattle from unaffected areas (175). It was concluded that hair values consistently below 5 ppm Se indicate that the diet is unlikely to contain sufficient Se to induce clinical signs of selenosis.

The feathers of chicks fed low-Se diets for 64 weeks contained approximately 0.3 ppm Se, compared with four–five times this level when the diets were supplemented with 2 ppm Se as selenite and 10 times this level, i.e., 3.3–3.4 ppm Se, when the diet was similarly supplemented at a level of 8 ppm Se (9).

5. Selenium in Milk

The concentration of Se in cow's milk varies greatly with the Se intake of the animal. In a study of pasteurized milk from different areas of New Zealand, a 3-fold variation from the highest to the lowest areas was found, reflecting the Se status of the soils and pastures of these areas (151). The actual range reported was 2.9 ± 0.7 to 9.7 ± 0.7 ng Se/ml, with a mean of samples from eight areas of 4.9 ng/ml. A similar mean concentration of 5 ng/ml for six areas in New Zealand was obtained in an earlier study (152) and also for a known low-Se area in Oregon (82). Allaway *et al.* (4) reported that cow's milk from a low-Se area in Oregon contained "less than" 20 ng Se/ml, compared with 50 ng/ml from a high-Se area in South Dakota. Higher levels, ranging between 160 and 1270 ng Se/ml, have been reported for cow's milk from high-Se rural areas in the United States (194).

The Se level in the milk is readily raised by Se supplementation of the diet or the animal. Grant and Wilson (69) obtained substantial milk Se increases over a period of 3–4 weeks from cows receiving a single oral or subcutaneous dose of 50 mg Se as selenate, while the levels in the untreated cows remained steady at 3–4 ng Se/ml. The Se concentration of the milk of ewes fed a low-Se diet was tripled by supplementing this diet with 2.25 mg Se/day (63), and that of sows was doubled by supplementing a low-Se diet (0.03 and 0.05 ppm Se) with sodium selenite at a level of 0.1 ppm Se (139). The actual levels in the milk were 13–15 ng Se/ml in the control sows and 25–29 ng/ml in the Se-supplemented sows. The levels of Se in the colostrum of these sows were 43 and 47 ng/ml and 80 and 106 ng/ml, respectively. The Se in milk is largely associated with the protein fractions (141, 152).

Human milk contains about twice as much Se as normal cow's milk, despite its lower protein content. Millar and Sheppard (152) obtained mean Se concentrations for human milk from five areas in New Zealand ranging from 11.5 to 14.5 ng/ml, compared with 5 ng/ml for cow's milk from six such areas. These values compare well with the 13–53 ng Se/ml reported by Hadjimarkos (75) for the milk of 15 healthy U.S. women, and with the mean of 20 ng/ml obtained by Hadjimarkos and Shearer (85) for 15 U.S. women and 24 Greek women. In a later investigation these workers reported a mean of 0.018 ppm Se for mature human milk from 241 subjects from 17 sites across the United States (219). Geographic variations were apparent in this study but most of the individual values fell within the relatively narrow range of 0.007–0.033 ppm Se.

6. Selenium in the Avian Egg

Under normal dietary conditions a hen's egg contains a total of 10–12 μg Se, most of which is present in the yolk (230). The total amount and the proportions present in the yolk and white are markedly and rapidly influenced by the Se status of the hen's diet and by the chemical form or forms in which the dietary Se is supplied. Thus Cantor and Scott (30) found the Se concentration in the eggs from groups of hens receiving 0.02, 0.04, 0.06, and 0.08 ppm dietary Se to be approximately equal to the Se level in the diet within 2 weeks of feeding the respective diets. Similarly Arnold *et al.* (8), employing much higher dietary Se levels, observed that the egg Se concentration increased from approximately 0.05 ppm for hens receiving a practical diet containing 0.5 ppm of naturally occurring Se to a maximum of 1.7 ppm Se in 12 days when this diet was supplemented with 8 ppm Se as selenite. Within 8 days of withdrawing the supplemental Se the egg Se levels returned to preexperimental values. The superiority of natural dietary Se over selenite Se with respect to transfer to the egg is apparent from the work of Latshaw (127). This worker fed hens for 180 days a diet that contained either 0.10 mg/kg natural Se, 0.10 mg/kg natural Se plus 0.32 mg/kg selenite Se, or 0.42 mg/kg natural Se. The Se concentrations in the liver, breast muscle, and eggs were significantly greater in those fed the 0.42 mg/kg natural Se than in those given the equivalent selenite Se diet. The whole egg Se concentrations for the three diets, as just given were 0.32, 0.74, and 1.23 μg/g. It was further found that (a) a higher proportion of the natural than the selenite Se passed into the white than the yolk, (b) more of the selenite Se was present in the yolk than in the white, and (c) the Se in the white resulting from selenite feeding could be removed by dialysis but not that in the yolk. A difference in the chemical forms of Se present in the two parts is apparent from these findings. The nature of these forms is not yet known.

At highly toxic Se intakes extremely high-Se levels can occur in hen's eggs, particularly in the white. Many years ago, Moxon and Poley (159) observed increases in yolk Se from 3.6 to 8.4 ppm and white Se from 11.3 to 41.3 ppm (dry basis) when the Se level in the hen's rations was raised from 2.5 to 10.0 mg/kg. More recently Arnold and co-workers (9) reported 5- to 9-fold increases in the Se concentrations in eggs up to nearly 2 ppm Se, when 8 ppm Se as selenite was added to various low-Se or moderate-Se diets of hens for 42–62 weeks. The further addition of 8 or 15 ppm As as sodium arsenite to these diets reduced but did not eliminate the increases in egg Se levels.

II. SELENIUM METABOLISM

Selenium absorption, retention, and distribution within the body and the amounts, forms, and routes of excretion vary with the chemical forms and amounts of the element ingested and with the dietary levels of several other elements, such as arsenic and mercury. There are also differences between ruminant and nonruminant species in these aspects of metabolism. The level of tocopherol in the diet does not affect significantly the pattern of Se absorption or retention (26, 48).

Studies with ^{75}Se at physiological levels indicate that the duodenum is the main site of Se absorption and that there is no absorption from the rumen or abomasum of sheep or the stomach of pigs (249). When rations containing 0.35

and 0.50 ppm Se were ingested total net absorption represented some 35% of the ingested isotope in sheep and 85% in pigs (249). Monogastric species have a higher intestinal Se absorption than ruminants, which may be related to a reduction of selenite to insoluble forms in the rumen (28, 37). In a study of the long-term fate of an oral dose of ^{75}Se selenite in three young women it was found that intestinal absorption was 70, 64, and 44% of the dose; urinary excretion accounted for 14–20% of the absorbed ^{75}Se in the first week, with only trace amounts in the expired air and dermal losses. Cumulative fecal Se excretion at day 14 was 33, 40, and 58% of the dose (234). After an initial phase in which radioactivity decreased rapidly, whole body ^{75}Se diminished exponentially with a half-time of 96–144 days. Radioactivity in the liver, heart, and plasma decreased more rapidly than that in the whole body, but radioactivity in skeletal muscle and bone decreased more slowly (28). In lambs fed varying levels of dietary Se, whole body loss of ^{75}Se 48–336 hr after administration of the isotope was described by a first order rate constant which was inversely proportional to the dietary Se level (135). The ^{75}Se concentration in various tissues was also inversely related to dietary Se levels. A similar inverse relationship in whole carcass retention of ^{75}Se was demonstrated in rats which had previously been fed Se-low and Se-high diets (105). Radioselenium is more efficiently retained by chicks from Se-deficient than from Se-supplemented diets (115).

Toxicity studies with rats indicate a higher absorption of Se from seleniferous grains than from selenites and selenates and a very low absorption from selenides and elemental Se (53, 224). Some organic compounds, including selenodiacetic and selenopropionic acids, are less toxic to rats per unit of Se than selenite, presumably as a consequence of lower absorption (158). Different chemical forms of Se also vary in their capacity to prevent signs of Se deficiency in animals (170, 205), although these variations do not necessarily reflect differences in absorbability. A similar reasoning can be applied to the fact that selenomethionine and selenocystine promote greater Se retention in the body tissues than equivalent intakes of selenite or selenate. This applies when the Se is injected (55, 153) and when it is ingested (111, 127, 213). Fuss and Godwin (55) found ^{75}Se to enter ewe's milk more rapidly and to a greater extent when given as selenomethionine than when given as selenite, but the proportion of unbound Se in milk was much lower after the administration of selenite Se than after selenomethionine administration. Jenkins and Hidiroglou (111) obtained data indicating that the ewe makes better use of orally administered selenite than of selenomethionine for secreting higher Se levels in the milk, but the forms of Se in the milk were more available to the preruminant lamb when the ewe was fed selenomethionine.

Absorbed Se is at first carried mainly in the plasma, from which it enters all tissues, including the bones, hair, and leukocytes (22, 37, 143, 147, 224). Selenite Se has to undergo a chemical transformation by the erythrocytes in order to be bound by the plasma proteins (198). The process of expulsion of Se from the erythrocytes depends on adequate glutathione levels in these cells (198). Most of this Se was initially observed to be transported by albumin, after which it moved to the globular fractions. More recently Sandholm (198) has shown that the Se that has been processed by the erythrocytes is taken up largely by β-lipoprotein and an unidentified fraction located electrophoretically between α_1- and α_2-globulin fractions. Enhanced *in vitro* uptake of ^{75}Se by erythrocytes occurs in Se-deficient sheep (135) and in children suffering from kwashiorkor (25). In dogs the rate of disappearance of ^{75}Se from whole blood, plasma, and red cells conforms to a multiple component rate function (145). Following the administration of subtoxic amounts of ^{75}Se the isotope was detected in various blood proteins for as long as 310 days. The greatest rate of disappearance of radioactivity from the red cells was at 100–120 days, suggesting that once Se enters these cells it remains there throughout their life-span. Se is also incorporated into myoglobulin, cytochrome c, the muscle enzymes, myosin, aldolase, and nucleoproteins (144). The exact nature of the Se binding in these tissue and compounds and the extent of its incorporation into seleno-amino acids in tissue proteins remain to be determined. The incorporation of Se into the specific enzyme glutathione peroxidase, and the widespread distribution of this Se compound in the cells and tissues, were considered earlier in Section I,3.

Most of the Se present in the tissues is highly labile. Following transfer of animals from seleniferous to nonseleniferous diets (224), or following injections of stabile or radioactive Se (135, 252), retained Se is lost from the tissues at first rapidly and then more slowly. Se is excreted in the feces, urine, and the expired air, the amounts and proportions of each depending on the level and form of the intake, the nature of the rest of the diet, and the species. Exhalation of Se is an important route of excretion at high intakes of the element but is much less so at low intakes (59, 91). Increasing the protein and methionine contents of the diet increases the pulmonary excretion of injected ^{75}Se (59). Injections of arsenic, mercury, thallium (118, 131, 181), and cadmium (56) also increase such excretion whereas lead (129, 131) and zinc (181) have no such effect on Se volatilization.

Fecal excretion of ingested Se is generally greater than urinary excretion in ruminants (28, 37), but not in monogastric species (142, 224). Most of the Se in the feces consists of unabsorbed dietary Se, together with small amounts excreted into the bowel in the biliary, pancreatic, and intestinal secretions (130). Levander and Baumann (130) found that Se excretion into the gastrointestinal tract via the bile fluid is markedly increased and retention in the carcass, liver, and blood greatly decreased, with no effect on urinary excretion, when subacute As injections are given with the Se. Injections of mercury, thallium, and lead had no such effect on biliary Se excretion (131). Arsenic injections given prior to the

Se reduce urinary Se excretion in rats and a new Se metabolite, identified as trimethylsenelonium ion, has been isolated from the urine of rats injected with [^{75}Se] selenite (180).

Selenium has a strong tendency to complex with heavy metals. Selenium metabolism is therefore influenced by the dietary levels of several such elements and radicles and the metabolism of these interacting elements is influenced, in turn, by Se. The interactions of Se with arsenic, already considered briefly, and with sulfate are discussed further in Section VI. The protective action of Se against Cd toxicity is dealt with in the chapter on cadmium. The mutual metabolic antagonism between Se and Hg is also discussed in Section VI and again in the chapter on mercury. A metabolic interaction between Se and silver has been apparent since Bunyan *et al.* (24) showed that the greenish exudate produced in chicks given 0.15% silver acetate in the drinking water was prevented by supplementary Se or vitamin E, and Peterson and Jensen (185) found that the growth depression and high mortality induced in chicks by adding 900 ppm Ag (as $AgNO_3$) to a diet marginal in Se and vitamin E was prevented by including either 1 ppm Se or 100 IU vitamin E. Rats fed diets low in Se and vitamin E with 100 ppm Ag as silver acetate in the drinking water developed a liver necrosis characteristic of Se–vitamin E deficiency (60), which did not occur when either cadmium or methylmercury was similarly administered in the drinking water. This implies a specific complexing of silver with some biologically active form of Se. Ganther and co-workers (60) suggest that glutathione peroxidase might be this molecular target of silver which binds to this enzyme in the liver to inhibit peroxide decomposition and thus bring about liver necrosis.

Selenium, in either inorganic or organic forms, is transmitted through the placenta to the fetus (92, 135, 145, 242, 248). Such transmission is also evident from reports that Se administration to the mother during pregnancy prevents white muscle disease in lambs and calves. The placenta nevertheless presents something of a barrier to the transfer of Se in inorganic forms. Jacobson and Oksanen (107) showed that the ^{75}Se in lambs was higher when their mothers were injected with [^{75}Se] selenomethionine or [^{75}Se] selenocystine than when [^{75}Se] selenite, was injected. Similar results have been obtained by others in mice (92) and sheep (111). Furthermore, several workers have shown that following injection of the ewe with [^{75}Se] selenite, the ^{75}Se concentrations in the blood and most organs of the fetus are lower than those in the mother (27, 248). It is apparent that these inorganic Se compounds pass the placental barrier less readily than the selenoamino acids. Whether the mammary barrier is similarly selective is not entirely clear. Thus Fuss and Godwin (55) found a greater entry of ^{75}Se into the milk of ewes when the isotope was given as selenomethionine than when it was given as selenite, whereas Jenkins and Hidiroglou (111) reported data indicating that the ewe made better use of selenite than of selenomethionine for secreting higher Se levels in the milk. The ease with which Se passes from the hen to the egg and the rapid equilibrium that exists between egg Se concentration and dietary Se concentration was considered in Section I,6.

III. SELENIUM DEFICIENCY AND FUNCTIONS

Selenium is necessary for growth and fertility in animals and for the prevention of various disease conditions which show a variable response to vitamin E. These are liver necrosis in rats and other species, exudative diathesis and pancreatic fibrosis in poultry, muscular dystrophy (white muscle disease) in lambs, calves, and other species, and hepatosis dietetica in pigs. The resorption sterility in rats (96) and encephalomalacia in chicks (39) that occur on vitamin E-deficient diets do not respond to selenium. Shortly after the original demonstration by Schwarz and Foltz (207) that Se prevents liver necrosis in rats and by Patterson and co-workers (182) that this element prevents exudative diathesis in chicks, growth and fertility responses to Se in farm animals, greater than could be achieved by tocopherol, were observed in Oregon and New Zealand. Unequivocal evidence that Se is a dietary essential independent of or additional to its function as a substitute for vitamin E was obtained by Thompson and Scott (231), who showed that chicks consuming a purified diet containing 0.005 ppm Se or less exhibited poor growth and high mortality even when 200 ppm of D-α-tocopherol was added. Higher tocopherol levels prevented mortality but even with 1000 ppm growth was inferior to that obtained with Se and no tocopherol. Comparable findings obtained with rats by McCoy and Weswig (149) and Wu and co-workers (250) clearly established that Se does not function merely as a substitute for vitamin E. Increased growth in children suffering from kwashiorkor from the administration of Se as selenite had been reported earlier (140, 206).

1. Muscular Dystrophy

Nutritional muscular dystrophy is a degenerative disease of the striated muscles that occurs, without neural involvement, in a wide range of animal species. It was described in calves in Europe in the 1880's and has been observed as field occurrences in many countries in lambs, calves, foals, rabbits, and even marsupials (see Anderson, 6). In some countries the incidence is low and sporadic. Less than 1% of the block or herd may be affected and in some seasons or years the disease may not appear at all. In other countries, notably in parts of New Zealand, Turkey, and Estonia (117), the incidence of muscular dystrophy or white muscle disease (WMD) is higher and more consistent unless appropriate Se treatment is undertaken.

White muscle disease rarely occurs in mature animals. In lambs it can occur at birth (congenital muscle dystrophy), or at any age up to 12 months. It is most common between 3 and 6 weeks of age. Lambs affected at birth usually die within a few days. The deep muscles overlying the cervical vertebrae are particularly affected with the typical chalky white striations. Lambs affected later in life (delayed muscular dystrophy) show a stiff and stilted gait and an arched back. They are disinclined to move about, lose condition, become prostrate, and die. Animals with severe heart involvement may die suddenly without showing any such signs. Clinical signs of WMD may not appear until the lambs are driven or moved about. Mildly affected animals may recover spontaneously.

These disabilities are associated with a noninflammatory degeneration or necrosis of varying severity of the skeletal or the cardiac musculature, or of both. A bilaterally symmetrical distribution of the skeletal muscle lesions is characteristic of WMD in lambs. The symmetry frequently extends beyond a simple bilateral involvement of paired muscles to the distribution of lesions within the muscles (251). The lesions are usually most readily discernible in the thigh and shoulder muscles. The lesions in the cardiac muscle are commonly confined to the right ventricle but may occur in other compartments. They are seen either as subendocardial grayish-white plaques or as more diffuse lesions of a similar color extending up to 1 mm into the myocardium (61). Alterations in the electrophoretic pattern of affected muscles are a consistent feature of the disease (251). Characteristic abnormalities in the electrocardiograms have also been demonstrated by Godwin (65) in lambs with WMD. The changes in the ECG pattern develop early and become very marked as death approaches. Similar changes in ECG pattern have been observed in rats fed a *Torula* yeast, low-Se diet (64) and in lambs fed similar diets (66).

White muscle disease is characterized biochemically by subnormal Se and glutathione peroxidase (GSH-Px) concentrations in the blood and tissues (see Tables 40 and 41) and by abnormally high levels of serum glutamic oxaloacetic transaminase (SGOT) and lactic dehydrogenase (184, 245). The dependence of tissue GSH-Px activity on the Se status of the animal and the diet was discussed earlier in Section I,3. In normal lambs and calves SGOT activity rarely exceeds 200 units/ml and is usually only about half that level. SGOT concentrations 5–10 times higher occur in animals with WMD, with the increase above normal being roughly proportional to the amount of muscle damage (15) (Table 42). SGOT determinations are therefore of some value in the diagnosis of WMD, although individual variability is high in both healthy and affected animals. Oksanen (172) has further pointed out that "the SGOT value may be only moderately increased in animals with marked clinical symptoms caused by extensive subacute or chronic degenerative processes" and "a degeneration in the myocardium, even an acute one, may also produce only a moderate increase in the SGOT value, although this often ends in sudden death."

White muscle disease has received most attention in lambs and calves because

TABLE 42
Serum Glutamic Oxaloacetic Transaminase Activity in Lambs and Calves[a]

	Normal			White Muscle Disease	
	Lambs		Calves	Lambs	Calves
Mean (units/ml)	128	56	57	1890	1313
SE	±18	±31	±17	±254	–
Range (units/ml)	97–191	22–160	19–99	687–3460	295–2360
No. of animals	20	21	69	17	4

[a]From Blincoe and Dye (15).

of its economic importance and natural occurrence but similar degenerative changes occur in foals (99), and in association with hepatosis dietetica in pigs (172) and exudative diathesis in chicks (168). Scott (209) states that muscular dystrophy in chicks is characterized by degeneration of the skeletal muscle fibers, especially the pectoral muscles. In all these species the disease can be prevented by Se supplementation of the diet at appropriate levels. In fact the first recognized field case of Se deficiency occurred in turkeys in Ohio. Myopathies of the heart and gizzard, resulting in severe mortality at 5–6 weeks of age, were demonstrated and shown to be Se responsive (212). Muscular dystrophy in rabbits is not similarly Se responsive (106).

2. Exudative Diathesis

The disease exudative diathesis first appears as an edema on the breast, wing, and neck and later has the appearance of massive subcutaneous hemorrhages, arising from abnormal permeability of the capillary walls and accumulation of fluid throughout the body. The greatest accumulation occurs under the ventral skin, giving it a greenish-blue discoloration. Plasma protein levels are low in affected chicks and an anemia develops probably as a consequence of the hemorrhages that occur (169). Growth rate is subnormal and mortality can be high. In the outbreaks of the disease that occur in commercial flocks as a result of consuming low-Se grain, chicks are most commonly affected between 3 and 6 weeks of age. They become dejected, lose condition, show leg weakness, and may become prostrate and die (7, 99).

Exudative diathesis (ED) can be completely prevented either by selenium or vitamin E (232). Noguchi and co-workers (170) have produced evidence indicating that vitamin E and Se prevent ED by two different mechanisms and that the effectiveness of vitamin E is not due to a simple oxidative action. For example, the synthetic antioxidant ethoxyquin, which will substitute for vitamin E in the

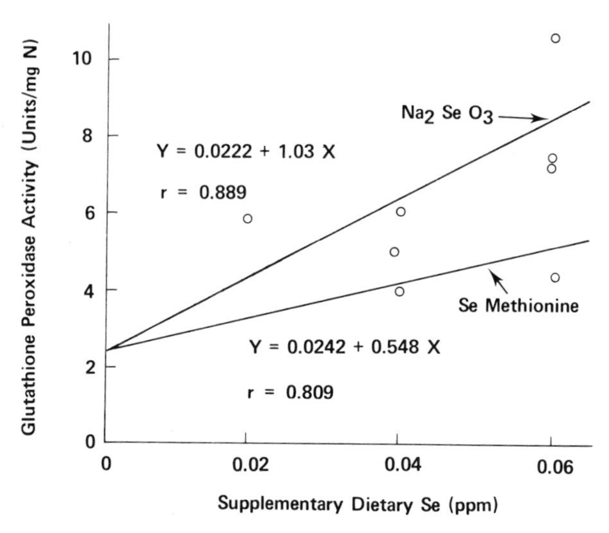

Fig. 8 Relationship of plasma glutathione peroxidase activity in chicks to dietary Se levels supplied by sodium selenite or DL-Se-methionine. From Noguchi *et al.* (170).

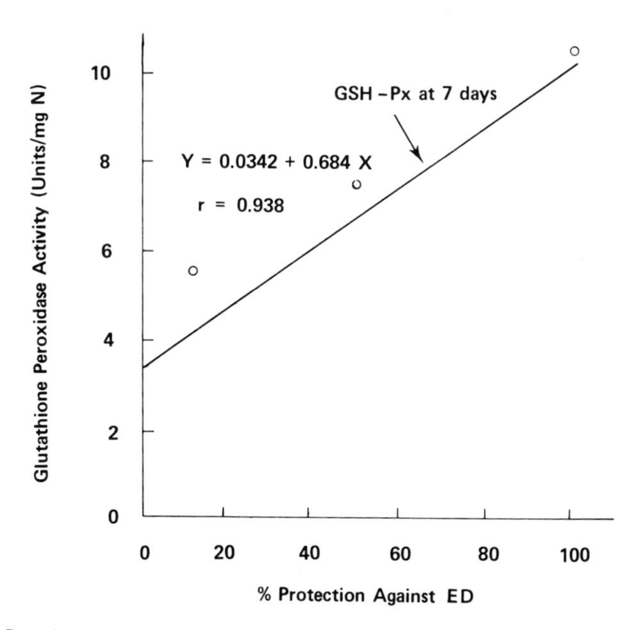

Fig. 9 Correlation between plasma glutathione peroxidase activities at 7 days of age and prevention of exudative diathesis (ED) measured at 13 days of age. GSH-Px, glutathione peroxidase. From Noguchi *et al.* (170).

prevention of encephalomalacia, has no preventive effect on the incidence or severity of ED in vitamin E- and Se-deficient chicks. These workers showed that the GSH-Px level of chick plasma is directly related to the Se level in the diet (Fig. 8) and to the effectiveness of Se in the prevention of ED (Fig. 9). Dietary vitamin E and Se were both found necessary for protection of hepatic mitochondrial and microsomal membranes from ascorbic acid-induced lipid peroxidation *in vitro.* On the basis of these results the hypothesis was put forward that "the plasma GSH-Px present when the diet contains adequate Se acts to prevent ED by destroying peroxides that may form in the plasma and/or cytosol of the capillary wall. Vitamin E appears to prevent ED by acting within the lipid membrane where it neutralizes free radicals, thereby preventing a chain-reactive autoxidation of the capillary membrane lipids." Subsequently the same group of workers (32) demonstrated marked differences in the biological availability of Se in different feedstuffs and Se compounds for the prevention of ED in chicks, and obtained evidence that these differences were related to the ability of the chick to utilize the various forms of Se for plasma GSH-Px activity. The whole question of Se availability and sources is considered in more detail in Section IV,2.

White muscle disease affecting particularly the breast muscles often develops concurrently with ED. A congenital myopathy, characterized by the hatching of dead chicks or chicks dying 3–4 days after hatching, is a further Se-responsive condition in this species (197). The chicks show extensive pale areas in the gizzard and sometimes in the hind limb skeletal musculature.

3. Pancreatic Fibrosis

Severe Se deficiency has been shown by Scott and his collaborators to result in atrophy of the pancreas of chicks, in addition to poor growth and feathering, even in the presence of high levels of dietary vitamin E (232). The initial deficiency lesions are apparent at 6 days of age and consist of vacuolization and hyaline body formation in the acinar cells, with loss of zonation of the acini. Fibroblasts are later observed in increasing numbers in the interacinar spaces while the acinar cytoplasm is shrinking basally, thus enlarging the central lumen of the acinus. Finally, the acini appear as rings of cells composed mainly of small dense staining nuclei, completely surrounded by the fibrotic tissue (73). Within 2 weeks of Se supplementation the pancreas returns to normal appearance (73). The degeneration of the Se-deficient pancreas has been further characterized by the detailed studies of Noguchi *et al.* (171), demonstrating a decrease in zymogen activity and an increase in the activities of the lysosomal enzymes cathepsin and acid phosphatase. The invasion of macrophages accounted for the increase in lysosomal enzymes, and no evidence of lysosomal disruption during the course of the pancreatic degeneration was obtained. It was concluded that the role of Se in preventing pancreatic fibrotic degeneration was not in the protection of lysosomal membranes of the acinar cells.

The pancreatic atrophy in the Se-deficient chick, as just described, is associated with an impairment of lipid and vitamin E absorption and other changes described by Thompson and Scott (232). Compared with Se-supplemented chicks, Se-supplemented animals revealed lower plasma tocopherol levels, higher fecal neutral lipids, higher bile weights, higher SGOT activities, and lower activities of pancreatic lipase and trypsin, in addition to decreased pancreas weights. These changes point to a decreased hydrolysis of fat, leading to an impairment of lipid bile micelle formation necessary for absorption of lipids and vitamin E. Supplementation with free fatty acids and monoolein increased vitamin E absorption and survival of chicks but did not prevent the pancreatic fibrosis.

Two features of pancreatic fibrosis in chicks are of particular interest. The first of these is the very much greater effectiveness of Se in selenomethionine and wheat than the Se in either selenite or selenocystine, both in preventing pancreatic degeneration and in increasing the relative weight and Se concentration of the pancreas (31). The high activity of selenomethionine contrasts with its low biological availability for the prevention of exudative diathesis (32). The second interesting feature of pancreatic fibrosis is that plasma and pancreatic glutathione peroxidase activities show no relationship between enzyme activity and the prevention of the disease (31). Again this contrasts with the position with respect to plasma GSH-Px activity and exudative diathesis. The failure to correlate protection against pancreatic fibrosis with GSH-Px activity led Cantor *et al.* (31) to suggest that "either (a) the addition of very low levels of Se (0.02 ppm) produces very small responses in GSH-Px activity that may go undetected or (b) the biochemical role of Se in protecting the pancreas is distinct from its effect on GSH-Px activity."

4. Hepatosis Dietetica in Pigs

Hepatosis dietetica has been produced on vitamin E-free diets based on *Torula* yeast (47) or soybean meal (70) and occurs spontaneously in New Zealand (7, 99) and Scandinavia (172) when pigs are ged grain rations naturally Se low. The disease is most common at 3–15 weeks of age and results in high mortality. Severe necrotic liver lesions are apparent at postmortem examination. There is also deposition of ceroid pigment in adipose tissue, giving a yellowish-brown color to the body fat and a generalized subcutaneous edema.

The mortality and liver lesions characteristic of hepatosis dietetica are associated with a marked depletion of tissue Se concentrations (see Table 40) and an increase in the level of the liver-specific enzyme ornithine carbamyltransferase (OCT) in the blood (172). Since the degenerative changes that may occur in the muscles do not affect the OCT level, blood OCT levels can be useful aids in the diagnosis of hepatosis dietetica. The mortality and liver lesions can be completely prevented by Se supplements, but vitamin E supplements appear to be more effective in preventing the muscle degeneration and pigment deposition (70, 172).

5. Selenium-Responsive Unthriftiness in Sheep and Cattle

In parts of New Zealand a serious condition known as "ill thrift" occurs in lambs at pasture and can occur in beef and dairy cattle of all ages (7, 97). The condition varies from a subclinical growth deficit (97, 150) to clinical unthriftiness with rapid loss of weight and sometimes mortality (46, 99). No characteristic microscopic lesions are apparent, there is no increase in SGOT levels, and the condition may or may not be associated with WMD and infertility. A further Se-responsive syndrome in lactating ewes, characterized by rapid weight loss and scouring in association with heavy parasitic infestation, has been described (150).

Ill thrift can be prevented by Se treatment, with striking increases in growth and wool yield in some instances (7, 46, 99, 116). Significant increases in wool yield from Se supplements have also been observed in Canadian (221) and Scottish (189) experiments, in the growth of lambs and calves in the western states of the United States (173), and in lambs from farms in northeast Scotland (14). In most parts of the world the responses to Se are less spectacular than those reported in New Zealand. In one series of New Zealand trials, lambs 5-months old were divided into two groups, the controls and those receiving 5 mg Se/lamb as selenite, given orally at commencement and again at 2 and 6 weeks. The mortality was reduced from 27 to 8% by the treatment and highly significant weight gains were observed in the Se-treated groups. Smaller oral doses, namely, 1 mg Se at docking, 1–5 mg at weaning, and 1–5 mg at 3-month intervals thereafter, were later found adequate. Neither vitamin E nor ethoxyquin had any effect on the unthriftiness (97).

6. Immunological Responses

Dietary Se at levels above those generally accepted as adequate (0.1 ppm) enhance the primary immune response in mice, as measured by the PFC (plaque-forming cell) test and by hemagglutination (227). In one experiment the PFC increased as the dietary level of selenite Se increased from 0 to 1.25 ppm and declined as this level exceeded 1.25 ppm. Almost a 4-fold increase in PFC was observed from the spleens of mice fed 1.25 ppm Se over that of mice fed a *Torula* yeast diet (0.009 ppm Se) or a chow diet (0.05 ppm Se). Sodium selenite administered to mice intraperitoneally at about 5μg Se was similarly shown to enhance the primary immune response to the sheep red cell blood antigen, the greatest enhancement occurring when the Se was administered with or prior to the antigen (227).

7. Reproductive Disorders

In all species studied Se deficiency results in impaired reproductive performance. In hens, Se deficiency reduces both hatchability and egg production (30), and in Japanese quail reduced hatchability of fertile eggs and viability of newly hatched chicks have been demonstrated (112). The fertility of rats fed a low-Se, *Torula* yeast diet with adequate vitamin E over successive generations was severely affected (149). The animals grew and reproduced normally but their offspring were almost hairless, grew slowly, and failed to reproduce. Sterility was determined by lack of breeding with fertile rats and, in the males, by examining for spermatozoa. In five of the eight untreated males immotile sperms, with separation of heads from tails, were observed, with no spermatozoa in the remaining untreated males. A supplement of 0.1 ppm Se as Na_2SO_3 restored hair coat, growth, and reproductive capabilities. Further investigation revealed degenerative changes in the epididymis, in addition to impaired testicular growth and function, in the rats fed the Se-deficient diets (250). Epididymal function, probably related to sperm maturation, appeared to be even more sensitive to Se deficiency than the development and functioning of the testis. The motility of spermatozoa improved almost linearly with increasing amounts of Se, from 0.01 to 0.08 ppm, added to the basal diet.

A high seasonal incidence of infertility in ewes occurs in parts of New Zealand in association with WMD and unthriftiness. In certain of these areas, 30% of ewes may be infertile and losses of lambs are high. The fertility of these ewes is dramatically increased by Se administration before mating, and further Se treatment reduces the lamb losses and incidence of WMD (98). The recommended doses are 5 mg Se as sodium selenite by mouth before mating and a similar dose 1 month before lambing. Hartley (98) has shown that estrus, ovulation, fertilization, and early embryonic development proceed normally in the affected flocks. The infertility results from a high embryonic mortality occurring between 3 and 4 weeks after conception, i.e., at about the time of implantation. This mortality can be prevented by Se but not by either vitamin E or by an antioxidant (Table 43).

In ewes fed a Se-deficient purified diet satisfactory reproductive performance was obtained only when Se and vitamin E were administered in combination (21). A combination of Se and vitamin E injected a month before mating was similarly shown to improve ewe prolificacy in Se-deficient areas of Scotland (163). A sodium selenite–vitamin E mixture injected into cows a month before calving completely prevented losses from the birth of premature, weak, or dead

TABLE 43
Effects of Selenium, Vitamin E, and Antioxidant on Ewe Fertility (200 Ewes/Group)[a]

	Untreated	Selenium[b]	Vitamin E	Antioxidant
Barren ewes (%)	45	8	50	43
Lambs born/100 ewes lambing	105	120	109	112
Lamb mortality (%)	26	15	16	14
Lambs marked/100 ewes lambing	43	93	46	54

[a] From Hartley (98).
[b] Treatment was confined to the mating period. Normal practice involves a further dose of Se 1 month before lambing to reduce lamb mortality.

calves in parts of California (138) and greatly reduced the incidence of retained placentas in a herd of cows in Scotland (235). In this last study selenate alone was less effective than the vitamin E–selenate combination. In a recent Australian study in commercial piggeries a significant response in farrowing percentage has been observed following a single subcutaneous injection of 2.5 mg Se as selenite given just prior to mating (100). The farrowing percentages were 82.5 and 77.0 (mean 79.3) in the two groups of treated gilts compared with 68 and 62% in two groups of untreated controls. The Se response appeared to be due very largely to decreased neonatal losses.

The remarkable protective action of Se injections against the testicular necrosis caused by cadmium is discussed in Chapter 9.

8. Mode of Action of Selenium

Two major steps forward have been made in our understanding of the mode of action of Se and its relationship to vitamin E, since liver necrosis in rats and exudative diathesis in chicks were first shown to be prevented by small amounts of selenite, as described in Sections III,1 and 2. The first of these was the demonstration of an essential requirement for dietary Se by chicks and rats, in the presence of adequate vitamin E (231). The second major advance was the discovery that GSH-Px is a Se-containing enzyme with an activity in the blood and tissues of rats and chicks directly related to the Se level in the diet. In fact Smith and co-workers (226) observed a close correlation between the specific activity of GSH-Px in seven tissues of the rat and the dietary Se level from 0.005 to 2.0 mg/kg. These important findings, also described earlier in this chapter, gave a great impetus to the antioxidant role for Se compounds fostered by Tappel (229) but open to some doubt through the researches of Green and Diplock and their group and others (see Green and Bunyan, 71).

Both Se and vitamin E protect biological membranes from oxidative degradation in the prevention of exudative diathesis (see Section II,2). Noguchi *et al.* (170) have put forward the hypothesis that vitamin E functions as a specific lipid-soluble antioxidant in the membrane and that Se functions as a component of cytosolic GSH-Px which reduces peroxides. In other words the GSH-Px is considered to be of primary importance, acting to destroy peroxides before they can attack the cellular membranes, while vitamin E acts within the membrane itself in preventing the chain reactive autoxidation of the membrane lipids.

A further hypothesis which can be regarded as complementary to that just mentioned has been put forward by Diplock (43, 44) as an outcome of studies of the intracellular forms and distribution of Se in rat liver. In this hypothesis Se, and particularly selenide, is presented as having a role in the electron transfer functions associated with mitochondria and smooth endoplasmic reticulum. Since mitochondria contain nonheme iron proteins (12), the active centers of which contain selenide, it has been tentatively suggested that the selenide might form part of a class of nonheme Fe selenide proteins.

Further elucidation of the precise mode of action of Se and confirmation or otherwise of the hypotheses already advanced will no doubt emerge as further research is undertaken. Meanwhile evidence is accumulating that Se may serve other functions in mammalian and avian tissues, unrelated to that of glutathione peroxidase. For example, Whanger *et al.* (244) have isolated a selenoprotein from the heart and muscle of lambs given adequate Se and have shown that this compound is not present in these tissues in the Se-deficient animal. The compound shows a marked resemblance to cytochrome c in several of its properties and may be a Se-containing cytochrome with a possible participation in oxidation–reduction reactions. This group has also produced evidence involving Se in the metabolism of sulfhydryl compounds (18). A significant decrease in total and protein sulfhydryl groups was observed in the muscle of WMD lambs compared with normals, while a significant increase in nonprotein sulfhydryl and reduced glutathione was found in WMD lamb muscle. Since there was no increase in disulfide bonds in the muscle of deficient lambs it was suggested that the sulfhydryls were not being oxidized to disulfides. Livers from Se-deficient rats had a higher content of total and nonprotein sulfhydryl groups compared with Se-supplemented rats. The opposite trend was apparent for the protein sulfhydryl groups. Further evidence that Se may be functioning in the metabolism of sulfhydryl groups has been produced by Sprinkler *et al.* (228). These workers found fewer sulfhydryl groups by histopathological staining methods in the skin, hair follicles, livers, testicles, and skeletal muscle of Se-deficient rats than in Se-supplemented rats. No evidence that Se is involved in creatine metabolism in muscle was obtained in a study with rats in which the Se deficiency was indicated by the low muscle Se level (0.21 ± 0.16 μg/g protein), compared with 1.45 ± 0.15 μg/g protein in the Se-supplemented controls (11).

A direct involvement of Se in the oxidative processes of the tricarboxylic cycle in rats is suggested by the work of Bull and Oldfield (23), showing that Se is associated with the oxidation of pyruvate by liver preparations and that Se and vitamin E are coinvolved in succinate metabolism. Subsequently Whanger (243) found that skeletal muscle slices from dystrophic lambs converted glutamate to CO_2 at a slower rate than such muscle from Se-supplemented lambs. No differences were apparent with pyruvate or succinate oxidation. These findings were taken further by Godwin *et al.* (68), who demonstrated lower respiratory rates with palmityl-DL-carnitine and acetyl-DL-carnitine as substrates in skeletal muscle mitochondria from dystrophic lambs than from normal lambs. Smaller but nonsignificant differences were found with succinate and pyruvate while no differences were apparent between heart muscle mitochondria isolated from normal and dystrophic lambs. The abnormalities in the oxidation rates of metabolic intermediates in the skeletal muscle mitochondria of dystrophic lambs were associated to some extent with abnormally high levels of calcium in their tissues and mitochondria.

IV. SELENIUM SOURCES AND REQUIREMENTS

1. Selenium Requirements

The minimum Se requirements of animals vary with the form of the Se ingested and the nature of the rest of the diet. A dietary intake of 0.1 ppm Se provides a satisfactory margin of safety against any dietary variables or environmental stresses likely to be encountered by grazing sheep and cattle (3). A minimum requirement of 0.06 ppm Se for the prevention of WMD disease in lambs is indicated by the work of Oldfield *et al.* (173), while New Zealand experience indicates that lambs can grow normally and remain free from clinical signs of Se deficiency on pastures containing 0.03–0.04 ppm Se (99). However, Gardiner and Gorman (62) showed that WMD can occur in lambs grazing Australian pastures estimated to contain 0.05 ppm Se.

Some of the above differences probably reflect analytical uncertainties, including variable Se losses in drying and storing of samples, while variations in the herbage levels of tocopherol and of other substances affecting Se metabolism can be important. In fact, a direct rather than an inverse relationship between the Se content of alfalfa and the occurrence of WMD in calves has been reported, despite evidence of the effectiveness of Se supplementation (204). This can probably be explained by the unusually high sulfate levels in the alfalfa. High-sulfate intakes are known to reduce Se availability to animals at high-Se intakes (88), so that Se requirements are likely to be greater when sulfate intakes are high than when they are low.

The minimum Se requirements of poultry have been greatly illuminated by the critical studies of Scott and his collaborators. Nesheim and Scott (168) found that a *Torula* yeast diet containing 0.056 ppm Se and 100 IU of vitamin E/lb did not sustain maximum growth rate in chicks unless it was supplemented with 0.04 ppm Se as selenite. Subsequently Thompson and Scott (231) demonstrated a Se requirement for growth close to 0.05 ppm when chicks were fed a purified diet with no vitamin E. With 10 ppm of added vitamin the requirement was more than 0.02 ppm and with 100 ppm of vitamin E it was less than 0.01 ppm Se. The dietary Se requirement of the vitamin E-adequate chick for the prevention of *in vitro* ascorbic acid stimulated hepatic microsomal peroxidation (36), and for complete protection of the chick pancreas from fibrotic degeneration (232), approximates 0.06 ppm when the Se is provided as sodium selenite. The Se requirements of turkey poults are higher than those of chicks on the basis of experiments carried out by Scott *et al.* (212). These workers reported a Se requirement of 0.17 ppm for the prevention of gizzard and heart myopathies when the diet was well supplied with vitamin E, and approximately 0.28 ppm in diets marginal in this vitamin and in sulfur amino acids.

The above requirements do not necessarily apply when the Se is provided in other forms or when the diets contain abnormal amounts of elements with which Se interacts metabolically. For example Cantor *et al.* (32) established a dose–response curve for the effectiveness of sodium selenite in the prevention of exudative diathesis (ED) and in the maintenance of plasma GSH-Px activity. The effectiveness for these purposes of the Se in various feedstuffs and Se compounds was then compared with the selenite at equivalent Se dose levels. The Se in most of the feeds of plant origin was highly available ranging from 60–90%, whereas the Se in the animal products tested was less than 25% available. It is clear that the Se requirements of chicks are higher when such animal products are included as protein sources than when plant products are similarly employed.

The Se requirements of chicks are also increased when the diet contains abnormally high dietary levels of silver, copper, and zinc. Jensen (114) observed high mortality and a high incidence of exudative diathesis and muscular dystrophy in chicks fed a diet containing 0.2 ppm Se supplemented with 800 or 1600 ppm Cu or with 2100–4100 ppm Zn. A supplement of 0.5 ppm Se completely prevented the deficiency signs and markedly reduced mortality. In other words the Se requirement of the chicks was raised from 0.2 ppm or less to 0.7 ppm by the treatments imposed. Similarly Peterson and Jensen (185) observed a marked growth depression and high mortality, mostly due to exudative diathesis, in chicks fed a diet marginal in vitamin E and containing 0.2 ppm Se, when 900 ppm Ag as silver nitrate was added. Including either 1 ppm Se as selenite or 100 IU per kg of vitamin E prevented these signs of Se deficiency.

Studies with weanling male rats fed *Torula* yeast based diets (0.01 ppm Se) supplemented with 0, 0.05, 0.1, 0.5, 1.0, and 5.0 ppm Se as Na_2SO_3 revealed optimal growth to 134 days with the 0.05 ppm supplement (i.e., 0.06 ppm total dietary Se). This level did not satisfy the GSH-Px requirements for Se. This requirement was apparently met by 0.1 ppm Se since increasing dietary Se from 0.1 to 0.5 or 1.0 ppm caused only comparatively small increases in GSH-Px activities in erythrocytes and liver (86). Supplemental Se at the rate of 0.1 ppm Se as selenite has similarly been found adequate to meet the Se requirements of sows throughout gestation and lactation through two reproductive cycles when corn–soybean meal diets containing 0.02 or 0.03 ppm Se were fed (139). The value of dietary Se levels, above those generally accepted as adequate, in enhancing the immunological response of mice (227) was mentioned in Section III,6.

The Se requirements of man are unknown. No pathological conditions unequivocally resulting from Se deficiency have been identified in human individuals, even in areas where this deficiency is severe and widespread in livestock. Increased susceptibility to cancer, as discussed in Section V, may conceivably provide an exception to this generalization. In the subhuman primate (*Saimiri sciureus*) Se deficiency characterized by loss of weight, listlessness, and alopecia has been produced by feeding a low-Se *Torula* yeast diet adequate in vitamin E for 9 months (166). Nontreated monkeys died and showed cardiac and skeletal muscle degeneration and hepatic necrosis. Monkeys treated with small (0.04 mg Se as sodium selenite) injections at 2 week intervals recovered rapidly.

2. The Potency of Different Forms of Selenium

Variations exist among different chemical forms of selenium in their capacity to meet the Se needs of animals and induce toxic effects. In the early studies of Schwarz, Se compounds were divided into three categories with respect to their potency against liver necrosis in rats (207). The first category includes elemental Se and certain compounds that are practically inactive due to poor absorption. Elemental Se was later shown to be almost completely unavailable for the prevention of exudative diathesis in chicks (32). Compounds in the second category, which included most inorganic salts such as selenites and selenates and the Se analogs of cystine and methionine, were found to be more or less equally protective against liver necrosis. Later studies have revealed significant differences within this group and even differences between the effectiveness of the same Se compound in protecting against different manifestations of Se deficiency. For example the Se from selenomethionine is less effective than Se from sodium selenite in preventing exudative diathesis in chicks (170), but the reverse is true for the prevention of pancreatic fibrosis (31). In fact selenomethione was four times as effective as either selenite or selenocystine for this latter purpose. The marked superiority of the Se in most feedstuffs of plant origin over the Se in animal products (various fish meals and meat and bone meal) for the prevention of exudative diathesis (32) was mentioned in the previous section. These differences must stem largely from differences in the chemical forms in which the Se occurs in the various feeds, but neither the nature of these forms nor the reason for their differing metabolic effectiveness is well understood.

The third category of Se compounds consists of organic forms more active per unit of Se than those of the second group. An organic component of hydrolysates of kidney and other potent sources was shown by Schwarz (205) to be three–four times as potent against liver necrosis as selenite Se. This fraction was designated factor 3 and established as organic in nature, strongly bound to proteins, and separable into two factors known as α- and β-factor 3. The isolation and identification of these potent preparations have been handicapped by low yields and chemical instability of the purified fractions, but Schwarz and Fredga (207a) demonstrated remarkable differences in the capacity of various aliphatic monoseleno- and diselenodicarboxylic acids to prevent liver necrosis in rats. In the course of further studies on the structure–activity interrelationship of other organic Se compounds in the prevention of dietary liver necrosis, comparable differences in biopotency were demonstrated by these workers (208). It is evident that potency is critically dependent on the nature of the organic moieties attached to the Se atom, but the basis of this dependence is not clear.

3. Selenium in Human Foods and Dietaries

The level of Se in individual foods of plant origin is highly variable depending mainly on the soil conditions under which they are grown. This variation is evident on a national basis from the work of Lindberg (134), who found samples of Swedish wheat to contain 0.007–0.022 ppm Se. These are similar to the levels found in wheat from Se-deficient areas in New Zealand. The corresponding values for wheat samples from other countries were Argentina, 0.05; the United States, hard winter, 0.37; U.S. durum, 0.70; and Canada, Manitoba, 1.30 ppm Se. Variation with locality within countries is equally apparent. Thus Scott and Thompson (213) reported median values of 0.05 and 0.80 ppm Se for wheats from the eastern United States and South Dakota, respectively. Hard wheat blends from different growing areas in the United States revealed a smaller range from 0.33 to 0.78 ppm Se (253). In this study there was little evidence of loss of Se in milling. The whole wheats averaged 0.50 ± 0.19 ppm Se and the white flour made from them 0.47 ± 0.08 ppm, both on the dry basis. However, Morris and Levander (155) reported considerably lower Se levels in white flour and bread than in whole wheat flour and bread.

Marked variation with locality is evident from other studies of human foods (80, 155). Hadjimarkos (80) observed a 10-fold difference in the Se concentrations in milk and eggs. The mean levels for a "Se-low" county in Oregon were

0.005 and 0.056 ppm for milk and eggs, respectively, compared with 0.05–0.07 and 0.4–0.5 ppm for these foods from two "Se-normal" counties. Similar differences in the Se content of milk, eggs, and muscle and organ meats from animals grown under Se-deficient and Se-adequate conditions were mentioned in Section I.

The Se content of human foods is further influenced by the class of food and by processing and cooking. Seafoods, kidney, liver, meat, and whole grains are generally good sources (> 0.2 ppm Se) and fruits and vegetables mostly poor sources (0.01 ppm Se or less wet weight) (155, 246). Fish flour for human consumption has been reported to contain 1.8 ppm Se (76) and tuna fish meal median levels as high as 5.1 and 6.2 ppm (dry basis) (213). High natural Se levels in tuna (58) and marine mammals (120) have been reported. The effect of cooking on the Se in meat and fish does not appear to have been studied, but much of the Se in vegetables can be discarded with the cooking water (10). Selenium is also lost in the refining of sugar. Morris and Levander (155) report a level of 0.012 for brown sugar and 0.003 μg Se/g for white sugar.

Total daily dietary Se intakes by man will clearly vary greatly with the source of the foods consumed and to some extent with the choice of foods, particularly the extent of the consumption of marine fish. Robinson *et al.* (192) found the dietary Se intake of four young women in New Zealand to range from 18 to 26 μg/day. As would be expected, this is much lower than the 62 μg/day for a U.S. hospital diet and the calculated Se intake of 60–150 μg/day for adults living in the northeastern part of the United States (201). Average Se intakes by Canadians are still higher. Thompson *et al.* (233) found that four composite diets from three cities, each representing the daily per capita consumption of foods in Canada, contained 191, 220, 113, and 150 μg Se. Cereals provided the most Se (62–112 μg), followed by meat, poultry, fish (25–90 μg), and dairy products (5–25 μg).

Water supplies do not normally constitute a significant source of Se to man, either in Se-deficient, "normal" (80), or seleniferous areas (174), although levels in water from some seleniferous areas as high as 50–300 μg/liter have been reported (246).

4. Selenium in Animal Feeds and Forages

The levels of selenium in animal feeds and forages have been extensively studied. These levels vary widely with the plant species and with the Se status of the soils in which they have grown. The effect of species is most apparent with the variously called accumulator, converter, or indicator plants that occur in seleniferous areas and which carry Se concentrations frequently lying between 1000 and 3000 ppm. One sample of *Astragalus racemosus* from Wyoming was reported to contain no less than 14,920 ppm Se, and over 4000 ppm Se on the dry basis was observed in the annual legume *Neptunia amplexicaulis* growing on a selenized soil in Queensland (121). The Se in these plants is present in organic combinations which are readily water extractable and which play an important part in the incidence of selenosis in grazing stock, as considered in Section VI,1. For all ordinary edible grasses and legumes the primary determinant of Se concentration is the level of available Se in the soil. This is apparent from the area studies of forages carried out by Allaway and associates (3, 165) and from numerous investigations in New Zealand and elsewhere. Pastures and forages free from Se-responsive diseases in animals generally contain 0.1 ppm Se or more, while in areas with a variable incidence of such diseases the levels are mostly below 0.05 ppm and sometimes as low as 0.02 ppm Se (dry basis). Forages and grains from British Columbia exhibited an exceedingly wide range of Se concentrations, with a considerable proportion of the samples containing between 0.12 and 0.32 ppm (154). The mean Se concentration of all forages was 0.19 ppm, with no significant differences between grasses and clovers, and of all grains 0.27 ppm, with wheat significantly higher (0.32 ppm) than barley and oats (0.20 ppm). Much lower Se levels in cereal grains from Se-deficient areas, down to 0.006–0.007 ppm, have been reported. The richest sources of Se observed in the Canadian study were salmon and herring meals (1.9 ppm). The relatively high-Se content of such marine products is also evident from the comprehensive study of feedstuffs carried out by Scott and Thompson (213).

The value of the different feedstuffs as a source of Se depends on the form in which the element is present, as well as on its concentration, at least for chicks and rats. Plant sources generally have a much higher Se availability than animal products (32). Much of the Se in wheat and probably other grains (177) and in alfalfa (1) is in the form of selenomethionine. Little of the Se in animal products is likely to be in this form, although it is largely associated with the protein fraction. The low availability of the Se in marine products such as tuna may be related to their high content of mercury or to interaction with other elements which are metabolic antagonists of selenium. These aspects are considered in the section on Se toxicity and in the chapter on mercury.

5. Selenium Supplementation

The methods available for providing Se supplements to animals include periodic injections or oral dosing with Se salts, provision of salt mixtures or licks containing small amounts of Se salts, use of feeds obtained from Se-rich areas, treatment of the soil with Se compounds, and administration of Se-containing heavy pellets to ruminants. The method of choice depends primarily on the conditions of husbandry and is influenced by the desirability of minimizing the amounts of Se introduced into the soil–plant–animal–man Se cycle. In the United States grains and alfalfa grown in areas where the Se levels in the soil are naturally high are often obtained and used in Se-low areas. As Allaway *et al.* (1) have pointed out, this system of blending imposes the need for controlled analyses of the feeds because the Se content of feeds grown in the so-called high-Se parts of the United States varies widely. In New Zealand, Se-containing supplements are commercially available for preventing Se-responsive conditions in pigs and poultry and are recommended for incorporation into rations to provide 0.15 ppm of added selenium (7).

Direct subcutaneous Se injections, usually as sodium selenite, constitute the best procedure in terms of the amounts of Se used and are widely practiced to control a range of Se-responsive diseases, particularly in cattle. New Zealand experience indicates that doses ranging from 10 mg Se for calves to 30 mg for adults, given at 3-month intervals, are satisfactory (7). Excessive Se concentrations in edible animal products from the use of Se injections or oral doses in the amounts prescribed have not been observed (37, 45, 126, 179). With sheep, oral dosing with selenite or selenate in doses from 1 to 5 mg Se at intervals as described previously is the most common means of preventing Se-responsive diseases in this species, because the drenching can often be carried out when the animals are yarded for routine management procedures. Both injections and oral dosing have the advantage of providing known amounts of Se to individual animals, but have the disadvantage of requiring individual handling and movement of stock.

The disadvantages attached to Se dosing or injecting led Kuchel and Buckley (125) to investigate the possibilities of Se-containing heavy pellets for grazing sheep, to be used in a similar manner to the heavy cobalt pellets. Such pellets, in which the Se sources were calcium selenate, barium selenate, and elemental Se, were compared when dosed singly to sheep grazing pastures of normal Se status. Blood and tissue Se levels were significantly enhanced by all three types of pellets for periods up to 12 months. In an experiment with pellets composed of finely divided metallic iron and elemental Se and containing 1.25%, 2.5%, 5%, or 10% Se, blood Se concentrations were increased significantly in all cases within 1 week. Within 10–12 weeks the extent of the rise was related to the proportion of Se in the pellets. The pellets were retained in the reticulorumen, and the Se concentrations in the tissues of the treated sheep after 6–12 months were no greater than those reported for untreated lambs in the United States (183) and New Zealand (37). Handreck and Godwin (91) have confirmed these findings with sheep, using dense pellets consisting of elemental iron and selenium (9:1 by weight) labeled with ^{75}Se. During the experimental period of 1 month the pellets released 0.5–1.3 mg Se/day, of which about 30% was excreted in the urine and about 1% in the expired air. Blood Se levels leveled out at 0.18–0.34 μg/ml, and there was no evidence of toxicity or excessive Se accumulation in the tissues. Heavy Se pellets must therefore be considered a safe, reliable, and economical means of preventing Se deficiency in grazing ruminants.

Treatment of the soil with various Se compounds, by spraying of herbage or by application alone or following incorporation of the Se into the normal fertilizers, offers further possibilities for raising the Se in pastures and forages to satisfactory levels. In terms of the amounts of Se required and introduced into the Se cycle this is an expensive and hazardous process, because of the inefficiency with which such added Se is taken up by most plants, especially from acid soils (1). Under experimental conditions Allaway *et al.* (5) showed that the addition of 2 lb Se/acre as Na_2SeO_3 increased the Se content of alfalfa from very low levels to concentrations that protect lambs from WMD. Furthermore, the effects of one application lasted for at least 3 years. A serious difficulty in supplying animal requirements by application to pasture is the high levels which occur immediately following application, whether these are due to foliar or root uptake or to initial surface contamination. The problem is to supply sufficient Se in a form which will maintain adequate concentrations of the element in the plants for a sufficiently long period to avoid the necessity of frequent application, without inducing potentially toxic levels following treatment.

Watkinson and Davies (240) maintain that, "with proper precautions to minimize pasture contamination," 1 oz/acre and possibly 2 oz/acre (as sodium selenite) should present no hazard, "at least for a few years." A further possibility has been proposed by Allaway *et al.* (2) involving the use of "selenized superphosphate." This material, prepared by incorporating the selenium into the superphosphate during the treatment of the rock phosphate with acid, seems to have promise on some soils since plants containing adequate Se levels can be produced at additions equivalent to 16 oz Se/acre, and five times this amount has not resulted in plants with toxic levels of selenium. On very sandy soils the apparent safe range of application of selenized superphosphate is much narrower.

V. SELENIUM AND CANCER

It is now evident that selenium, far from being the carcinogenic agent that some early experiments had appeared to suggest, can actually be a cancer-protecting element.

Selenium acquired its reputation as a carcinogen from the early studies of Nelson *et al.* (167) with rats. Later Tscherkes and co-workers (236) studied the effects of Se with rats on low- and high-protein diets. It is impossible to incriminate Se as a carcinogen from these latter observations because no controls, without added Se, were used. Harr *et al.* (93) fed Se at varying levels to rats and evaluated its carcinogenic potential in comparison with the known hepatocarcinogen *N*-2-fluorenylacetamide. Hyperplastic liver lesions occurred in the Se-fed rats which did not regress when the added Se was withdrawn.

Sixty-three neoplasms were observed, but none of these could be attributed to the addition of selenium. The experiments of Schroeder and Mitchener (202) in which rats were fed selenite in the drinking water at 2 ppm Se for a year and then selenite or selenate at 3 ppm Se for the remainder of their life-span are equally unsatisfactory from the point of view of demonstrating a carcinogenic effect from selenium. As Scott (211) has pointed out, the incidence of tumors was lower in the selenite-fed than in the control rats, and the higher incidence of tumors in the selenate-fed animals may have been due to the beneficial effects of the selenate on their longevity. In a later experiment by Schroeder *et al.* (203) with mice given selenite or selenate at 3 ppm Se in the drinking water for life no significant effect on the incidence of spontaneous tumors was observed.

Evidence that Se has an inhibitory effect on carcinogenesis comes from both experimental and epidemiological studies. Clayton and Baumann (35) found that the inclusion of 5 ppm Se as selenite in a purified diet reduced the incidence of liver tumors in rats which had previously received a carcinogenic azo dye. Injection of 1 mg of selenocystine/kg body weight for 14 days following injection of Murphy's lymphosarcoma was later shown to significantly reduce the average size of the tumors in rats (241), and the feeding of Se inhibited the cocarcinogenic effect of croton oil in mice (216). This finding was subsequently confirmed by Riley (191) using the carcinogenic principle from croton oil, compound A_1. In more recent studies with mice additional evidence of an anticarcinogenic action of Se was obtained. Shamberger (214) showed that the incidence of cancers from the application of benzo[*a*]pyrine in acetone to ICR Swiss mice was reduced when the diet was supplemented with 1 ppm of sodium selenite. Schrauzer and co-workers (199) reported that exposure of female virgin C3H/St mice (a highly inbred strain with a high incidence of spontaneous mammary tumors) to 2 ppm Se as selenite in the drinking water for 15 months lowered the incidence of spontaneous mammary tumors to 10% relative to the 82% incidence in controls. In a later lifetime study at higher Se levels (5 and 15 ppm in the drinking water) a significant reduction in mammary tumors was again observed, but the cancer-protecting effect of Se began to be masked by its toxicity, i.e., higher tumor-unrelated mortality. Schrauzer *et al.* (200) had earlier shown that the methylene blue reduction time (MBRT) of human plasma samples is inversely related to total plasma Se content ($r = -0.56, p = 0.002$). A protective effect from dietary Se against *N*-2-fluorenylacetamide (FFA)-induced cancer in Se-depleted rats (94) and an inverse correlation between the Se concentration in the liver and rats fed Se-supplemented rations and the incidence of hepatic and mammary cancer (95) have been demonstrated, although selenium does not affect neoplasia induced by Rauscher leukemia virus (RLV) in mice (49).

The epidemiological evidence began with the examination by Shamberger and Frost (215) of the human cancer death rates in 10 of the cities with populations of 40,000–70,000 from which Allaway *et al.* (4) had taken blood samples. They obtained a high inverse relationship ($r = 0.96$) between the blood Se levels and the human cancer death rates. Later Shamberger *et al.* (217) calculated the ratio of observed to expected cancer rate by organ site for males in 17 paired large cities in high-Se and low-Se areas of the United States. The mortality from cancer of the stomach, esophagus, and rectum was particularly increased in the low-Se areas. The area difference in the cancer death rate was smaller for cancers of the small and large intestine, pharynx, bladder, and kidney and insignificant for cancers of the lung, pancreas, prostate, and lymphatics, and for leukemia. Shamberger's findings are supported by the calculations of Schrauzer (199, 200). The age-adjusted cancer mortality rates for 1960 were found to be inversely correlated with the median Se levels in grains and forage crops in the United States with $r = -0.467$ ($p = 0.01$). The degree of correlation changed little when calculated with cancer mortality data from 1910 ($r = -0.414$) and 1913 ($r = 0.478$). Further, when the data of Kubota *et al.* (124) for the Se distribution in grains and forage crops were compared with the published female breast cancer mortality rates, the incidence was again shown to be higher in the low-Se than in the higher Se areas, as shown in Table 44. Correlations do not necessarily imply causal relationships and regional cancer mortalities correlate with other parameters, but when the epidemiological data are taken in conjunction with the animal data there seems little reason to doubt that selenium can have an inhibitory effect on cancer. A logical corollary to this conclusion is that human cancer incidence and mortality could be lowered by appropriate dietary Se supplementation in low-Se areas. However, it is necessary to point out that neoplasia is not observed among the various lesions attributed to Se deficiency in animals.

TABLE 44
Average Concentration of Selenium in Grain and Forage Crops and Female Breast Cancer Death Rates in the United States[a]

	Breast cancer death rate		
Median Se concentration (ppm)	1959	1961[b]	No. of states
0.02–0.03	1.11	0.16[c]	5
0.05	1.09	0.14	14
0.05–0.10	0.80	0.20	13
0.26	0.75	0.15	16

[a]Excluding Hawaii and Alaska. From Schrauzer and Ishmael (199).
[b]Ratio to U.S. rate per 100,000 population.
[c]Standard deviation from average death rate.

VI. SELENIUM TOXICITY

1. Selenosis in Grazing Animals

The early history of the diseases "alkali disease" and "blind staggers" in grazing livestock in parts of the Great Plains of North America and the demonstration that they are expressions of chronic and acute Se toxicity, respectively, have been thoroughly reviewed (157, 160, 193). Localized seleniferous areas have also been identified in Ireland (50), Israel (189a), Australia (121), the Soviet Union (122), and South Africa (19). In these areas toxic Se intakes by animals arise either from the consumption of Se accumulator plants, the ingestion of more normal forage species with relatively high-Se concentrations due to the presence of above normal levels of available Se in the soils of the affected areas, or from both. Selenium accumulator plants play an important dual role in the incidence of selenosis in grazing stock. They have the ability to absorb Se from soils in which the Se is present in forms relatively unavailable to other plant species, and on their death to return Se to the soil in organic forms that are available to these other species. Hence the alternate name "converter" plants. Accumulator or converter plants thus provide a direct source of Se to animals consuming them, and they convert unavailable forms into available forms, so raising the Se levels in plants which might otherwise contain safe levels of the element. In some areas this may not be important, but in others the presence of converter plants can intensify the severity of selenosis and extend its incidence.

2. Manifestations of Selenosis

All degrees of Se poisoning exist, from a mild, chronic condition to an acute form resulting in death of the animal. Chronic Se poisoning is characterized by dullness and lack of vitality, emaciation and roughness of coat, loss of hair from the mane and tail of horses and the body of pigs, soreness and sloughing of the hoofs, stiffness and lameness due to erosion of the joints of the long bones, atrophy of the heart ("dish-rag" heart), cirrhosis of the liver, and anemia. In acute Se poisoning the animals suffer from blindness, abdominal pain, salivation, grating of the teeth, and some degree of paralysis. Respiration is disturbed and death results from respiration failure. Death also results from starvation and thirst because, in addition to loss of appetite, the lameness and pain from the condition of the hoofs are so severe that the animals are unwilling to move about to secure food and water. This situation is comparable to that of animals in fluorosis areas. With the possible exception of loss of appetite and reduced rate of weight gain, or increased loss of body weight, sheep have not been observed to exhibit signs of selenosis similar to those just described for horses, cattle, and pigs (174, 186).

In the rat and dog, a marked restriction of food intake occurs, together with anemia and severe pathological changes in the liver. This organ becomes necrotic, cyrrhotic, and hemorrhagic to varying degrees. Anemia is a common manifestation of selenosis in all species, but in the rat and dog a microcytic, hypochromic anemia of progressive severity usually develops and animals may die with hemoglobin levels as low as 2 g/100 ml (160, 223).

Growing chicks exhibit a reduction in food intake and growth rate when consuming seleniferous diets, and there is a fall in egg production in hens. Franke and Tully (54) showed that the eggs produced by these hens have low hatchability at Se concentrations in the feed too low to cause manifest signs of poisoning in other animals. The eggs are fertile, but a proportion produce grossly deformed embryos, characterized by missing eyes and beaks and distorted wings and feet.

Consumption of seleniferous diets interferes with the normal development of the embryo in rats (195), pigs (237), sheep (195), and cattle (42). This effect is also apparent from the birth of foals and calves with deformed hoofs in seleniferous areas. According to Olson (174) a reduction in reproductive performance is the main economic effect of Se poisoning of the alkali disease type. He states that the effect on reproduction can be severe without other lesions of selenosis being apparent. The feeding of 10 ppm of selenite Se to young sows lowers the conception rate and increases the number of services per conception, as well as the proportion of piglets small, dead, or weak at birth (237).

3. Factors Affecting Selenium Toxicity

The toxicity of Se to animals varies with the amounts and chemical forms of the Se ingested, the duration and continuity of the Se intake, the nature of the rest of the diet, and to some extent with the species. Limited evidence suggests that monkeys (17) are much more susceptible to Se in the drinking water than rats (77) or hamsters (78). Munsell *et al.* (164) place the minimum dietary levels at which signs of toxicity will ultimately arise at 3–4 ppm Se, but this will clearly vary with the extent to which other dietary components with which Se interacts are present. Levels of 3–4 ppm Se do not adversely affect the growth of chicks or the hatchability of eggs, but at 5 ppm Se hatchability is slightly reduced and at 10 ppm Se is reduced to zero (159). For normal growth of chicks less than 5 ppm Se is necessary (187). In a more recent study 2 ppm of added Se as selenite had no adverse effects on the growth or mortality of chicks or on the production or hatchability of their eggs, but 8 ppm Se significantly reduced hatchability (9).

In rats and dogs given diets containing 5–10 ppm Se signs of chronic poisoning arise, and at 20 ppm there is complete refusal of food and death in a short time (164). Young pigs fed seleniferous diets containing 10–15 ppm Se develop signs of selenosis within 2–3 weeks (195). The minimum toxic levels for grazing stock probably lie close to 4–5 ppm Se. Edible herbage in seleniferous areas commonly contain 5–20 ppm Se. At the lower levels signs of Se toxicity in cattle would take weeks, or even months, to appear. Mature sheep fed regular oral doses of sodium selenite, up to 600 μg Se/kg body weight/day, for periods as long as 15 months revealed no pathological changes in the tissues. After 5–6 months of treatment, a depression in food consumption and body weight increase was observed at intakes of 200 and 300 μg Se/kg/day, with some mortality at 400 μg Se/kg/day (186). The 200-μg/kg/day treatment is equivalent to a dietary intake of approximately 8 ppm Se.

High-protein diets afford some protection against potentially toxic Se intakes. Smith (222) found that 10 ppm Se in a 10% protein diet was highly toxic to rats, whereas with an additional 20% protein as casein scarcely any adverse effects were evident. Linseed oil meal is superior to casein in protecting rats against Se poisoning, and some fraction in the meal, other than protein, is responsible for its superior effect (89). A curious feature of this protective action of linseed oil meal is that it is accompanied by higher Se levels in the liver and kidneys compared with those of rats fed casein-containing seleniferous diets (132). This suggests that the Se is bound in a less toxic form in this organ. Studies by Levander *et al.* (132) show that chelating agents can increase the amount of Se dialyzed from liver homogenates of rats fed the casein diets, but have no such effect when the linseed oil meal seleniferous diet is fed. The S-containing amino acids cannot be implicated with any certainty in the protein-protecting effect (176). Increasing levels of sulfate up to 0.87% of a sulfate-free diet progressively relieve the growth inhibition induced in young rats by 10 ppm Se as selenite or selenate (88). The sulfate supplements are comparatively ineffective against the liver degeneration of selenosis but can alleviate the growth depression by as much as 40%. Inorganic sulfate is less effective against organic forms of Se than against selenate (88). The protective effect of high-protein diets may be related to the production of endogenous sulfate.

The toxicity of Se can be greatly modified by the dietary levels of arsenic, silver, mercury, copper, and cadmium, with each element apparently exerting its protecting action by its own mechanism. The effect of arsenic was first demonstrated by Moxon (156), who showed that 5 ppm As as arsenite in the drinking water prevented all signs of selenosis in rats. Arsenic has since been used successfully to alleviate Se poisoning in pigs, dogs, chicks, and cattle. Sodium arsenite and arsenate are equally effective, arsenic sulfides are ineffective, and such organic forms as arsanilic acid and 3-nitro-4-hydroxyphenylarsonic acid provide partial protection (101, 126, 238). The protection afforded by arsenic is due at least in part to increased biliary (130) and pulmonary (118) Se excretion. The main target organ of Se toxicity, the liver, can apparently rid itself of excess Se in this way. Although 8 or 15 ppm As as arsenite afforded some protection against the toxicity to chicks of 8 ppm Se as selenite, the Se levels in the liver and other tissues were not significantly reduced by the As treatment (9).

The inclusion of mercuric chloride, cupric sulfate, or cadmium sulfate in the diet of chicks has been shown by Hill (102) to partially overcome the growth retardation and mortality induced by feeding 40 ppm Se as selenium dioxide. When the Hg and Se were both fed in inorganic forms the most effective ratio in preventing the Se toxicity was 1:1. At a comparable ratio copper was as effective as mercury in reducing the mortality but did not prevent the growth inhibition as mercury did. Evidence was obtained suggesting that Hg, Cu, and Cd exert their protection by reacting with Se, probably within the intestinal tract, to form relatively innocuous compounds. Jensen (113) similarly found copper to be highly effective in preventing mortality in chicks fed for 2 weeks 20, 40, or 80 ppm Se as selenite, with a much smaller favorable effect on the growth inhibition. The copper sulfate was fed in these experiments at the extremely high level of 1000 ppm Cu. A marked protective effect against both the mortality and growth retardation induced by the same levels of dietary Se from silver fed as silver nitrate was also demonstrated, again at the abnormally high level of 1000 ppm Ag. Evidence was obtained that silver modifies the Se toxicity by interfering with Se absorption and causing the accumulation of a nondeleterious Se compound in the tissues, whereas copper achieves its effect only by the latter process.

A further factor affecting Se toxicity is adaptation by the animal. The adaptive responses to Se seen in bacteria (128, 220) may have a parallel in animals since Jaffe and Mondragon (108) showed that young rats born from mothers fed a Se-containing diet lost Se from their livers, whereas rats bred on the stock diet accumulated this element under the same conditions. These results were confirmed and extended under low-Se (0.5 ppm) or moderate-Se (4.5 ppm) diets to the mothers and high-Se (10 ppm), moderate-Se, or low-Se diets to the young, with results pointing to an adaptation to chronic Se intake (109). There appears to be no information on whether a similar adaptation to chronic Se intake occurs in farm animals.

4. The Mechanism of Selenium Toxicity

The precise ways in which Se at toxic intakes interferes with tissue structure and function are not completely understood. Selenium has long been known to affect certain unicellular organisms and enzyme systems (133) and to inhibit alcoholic fermentation by yeast and some of the enzymes concerned with cellular respiration (188). The inhibition of oxygen consumption by tissues appears to be mediated through a poisoning of succinic dehydrogenase. The liver succinic dehydrogenase levels of rats fed seleniferous diets are reduced below normal and can be maintained at normal levels by appropriate dietary intakes of arsenic (119). It is unlikely that these effects are sufficient to account for the various manifestations of selenosis or their prevention by arsenic. A significant increase in the serum levels of glutamic oxalacetic and glutamic pyruvic transaminases in rats fed high (10 ppm) or moderate-Se (4.5 ppm) diets has been demonstrated (109), but it is doubtful if these enzyme changes have any primary significance. The *in vitro* experiments of Wright (247) with tissues to which sodium selenite or selenate was added indicate a general inhibition of dehydrogenating enzymes and urease, with no impairment of the cytochrome–indophenoloxidase system or of catalase or liver arginase. These findings point to the removal of sulfhydryl groups essential to oxidative processes as possible biochemical sites of the injurious effects of selenium.

5. Prevention and Control of Selenosis

Three possibilities exist for the prevention or treatment of Se poisoning. These are (a) treatment of the soil so that Se uptake by plants is reduced and maintained at nontoxic levels, (b) treatment of the animal so that Se absorption is reduced or excretion increased thus preventing toxic accumulations in the tissues, and (c) modifying the diet of the animal by the inclusion of substances which antagonize or inhibit the toxic effects of selenium within the body tissues and fluids.

The addition of sulfur or gypsum to soils in a toxic area in North America has been unsuccessful in reducing the absorption of Se by cereals (53), probably due to the fact that these soils are mostly already high in gypsum and carry a high proportion of their Se in organic combinations that are relatively little affected by changes in the inorganic S:Se ratio. On the other hand, the addition of sulfur to soils to which selenate has been added can inhibit Se absorption by plants. Selenium uptake by lucerne from a seleniferous soil has been strikingly reduced by additions of calcium sulfate and barium chloride (189a). The latter salt reduced the Se levels in plants by 90–100% when applied in quantities that did not affect plant growth or result in significant concentrations of barium in the tissues. The practical possibilities of these interesting findings appear to be limited.

Urinary loss of Se from the body can be enhanced by the administration of bromobenzene to rats and dogs fed a seleniferous diet, and to steers on seleniferous range (161), but this form of treatment has obvious practical limitations. On the other hand, dietary modifications including high protein and sulfate intakes and the feeding of arsenic at appropriate levels, or mercury or copper as discussed in the previous section, have a possible potential in alleviating Se toxicity, where dietary control of the animals can be achieved. Much more information is required on the quantitative interactions of these elements and their own toxic hazards when used for this purpose before such treatments can be considered practical possibilities. The position is even more difficult under range conditions. Early studies by Moxon *et al.* (162) indicated that 25 ppm As as sodium arsenite added to the salt of cattle on seleniferous range gave some protection, but observations by ranchers and additional studies by Dinkel and co-workers (41) showed that this method of control is ineffective, probably because the As intake is neither high enough nor regular enough. Various management possibilities exist, following mapping of the land into pastures of high- and low-Se contents (174). Furthermore, the more highly seleniferous areas can be used for grain production and the grain sold into normal market channels. In this way it would become so diluted that it should not contribute to a public health problem, and could help in raising Se intakes by animals in Se-deficient areas.

6. Selenium and Dental Caries

The results of epidemiological studies with children and experiments with laboratory animals have led Hadjimarkos (81) to propose that the consumption of small amounts of Se during the development period of the teeth increases the incidence of caries. The epidemiological evidence comes from regional studies of caries incidence in children in the United States. These studies were prompted by early observations that a high prevalance of caries is a characteristic of human populations in the seleniferous areas of that country (225). In the Oregon study, children born and reared in a county where Se deficiency in livestock is common had a lower incidence of caries [9.0 DMF (decayed, missing, and filled) teeth/child] than children from three other counties of known higher Se status (13.4–14.4 DMF teeth/child). Analyses of 24-hr urine specimens from a selection of children in the former county revealed significantly lower mean Se levels than those from comparable children in the other three counties. The Se level in the urine was considered a reliable criterion of total Se ingestion, but later work suggests that the Se levels in nail clippings may be a more convenient and reliable index for assessing Se status in man (84). Ludwig and Bibby (137) similarly reported that children born and reared in high-Se towns in the United States had a significantly higher prevalence of caries than subjects in low-Se towns. Hadjimarkos (81) mentions also a report of an epidemiological study of three population groups living in the Chernovitsi region of the Soviet Union where the prevalence of caries was found to be directly related to the levels of Se found in teeth collected locally, and also to the levels present in the soils.

The effect of Se on caries was studied in young monkeys (*Macaca irus*) for a period of 5 years (17). The monkeys receiving Se at 1 ppm as sodium selenate in the drinking water developed twice as many carious lesions as the controls in the teeth undergoing development at the time Se was ingested, and the carious lesions developed about three times faster in the Se-fed than in the control monkeys. These results are similar to those obtained earlier in rats fed Se in their drinking water during tooth development (29, 79). It seems that the incorporation of dietary Se into the protein fraction of the enamel during the development of teeth (218) may inhibit mineralization of this tissue resulting in an increased susceptibility to caries. Inhibition of the growth and calcification of the teeth and maxillary bones by Se injections as selenite has also been demonstrated in young rats when the Se doses exceed what the authors describe as a "valeur optimale"; (118a).

REFERENCES

1. Allaway, W.H., Cary, E.E., and Ehligh, C.F., *in* "Selenium in biomedicine" (O.H. Muth, ed.), p. 273. Avi. Publ., Westport, Connecticut, 1967.
2. Allaway, W.H., Cary, E.E., Kubota, J., and Ehligh, C.F., *Proc. Cornell. Nutr. Conf., 1964,* p. 9 (1964).
3. Allaway, W.H., and Hodgson, J.F., *J. Anim. Sci.* **23,** 271 (1964).
4. Allaway, W.H., Kubota, J., Losee, F., and Roth, M., *Arch. Environ. Health* **16,** 342 (1968).
5. Allaway, W.H., Moore, D.P., Oldfield, J.E., and Muth, O.H., *J. Nutr.* **88,** 411 (1966).
6. Anderson, P., *Acta Pathol. Microbiol. Scand., Suppl.* **134,** (1960).
7. Andrews, E.D., Hartley, W.J., and Grant, A.B., *N. Z. Vet. J.* **16,** 3 (1968).
8. Arnold, R.L., Olson, O.E., and Carlson, C.W., *Poultry Sci.* **51,** 341 (1972).
9. Arnold, R.L., Olson, O.E., and Carlson, C.W., *Poultry Sci.* **52,** 847 (1973).
10. Aterman, K., *Br. J. Nutr.* **17,** 105 (1963).
11. Barak, A.J., Allison, T.B., Tuma, D.J., and Sorrell, M.F., *Br. J. Nutr.* **34,** 119 (1975).
12. Bienert, H., and Lee, W., *Biochem. Biophys. Res. Commun.* **5,** 40 (1961).
13. Bisjberg, B., Jochumsen, P., and Rasbech, N.O., *Nord. Veterinaermed.* **22,** 532 (1970).
14. Blaxter, K.L., *Br. J. Nutr.* **17,** 105 (1963).
15. Blincoe, C., and Dye, W.B., *J. Anim. Sci.* **17,** 224 (1958).
16. Bowen, H.J.M., and Cawse, P.A., *Analyst* **88,** 721 (1963).
17. Bowen, W.H., *J. Ir. Dent. Assoc.* **18,** 83 (1972).
18. Broderius, M.A., Whanger, P.D., and Weswig, P.H., *J. Nutr.* **103,** 336 (1973).
19. Brown, J.M.M., and De Wet, P.J., *Onderstepoort J. Vet. Res.* **34,** 161 (1967).
20. Brune, D., Samsall, K., and Westov, P.O., *Clin. Chim. Acta* **13,** 285 (1966).
21. Buchanan-Smith, J.G., Nelson, E.C., Osburn, B.I., Wells, M.E., and Tillman, A.D., *J. Anim. Sci.* **29,** 808 (1969).
22. Buescher, R.G., Bell, M.C., and Berry, R.K., *J. Anim. Sci.* **19,** 1251 (1960).
23. Bull, R.C., and Oldfield, J.E., *J. Nutr.* **91,** 237 (1967).
24. Bunyan, J., Diplock, A.T., Cawthorne, M.A., and Green, J., *Br. J. Nutr.* **22,** 165 (1965).
25. Burk, R.F., Pearson, W.N., Wood, R.F., and Viteri, F., *Am. J. Clin. Nutr.* **20,** 723 (1967).
26. Burk, R.F., Whitney, R., Frank, H., and Pearson, W.N., *J. Nutr.* **95,** 420 (1968).
27. Burton, V., Keeler, R.F., Swingle, K.F., and Young, S., *Am. J. Vet. Res.* **23,** 962 (1962).
28. Butler, G.W., and Peterson, P.J., *N. Z. J. Agric. Res.* **4,** 484 (1961).

29. Buttner, W.H., *J. Dent. Res.* **42,** 453 (1963).
30. Cantor, A.H., and Scott, M.L., *Poultry Sci.* **53,** 1870 (1974).
31. Cantor, A.H., Langevin, M.L., Noguchi, T., and Scott, M.L., *J. Nutr.* **105,** 106 (1975).
32. Cantor, A.H., Scott, M.L., and Noguchi, T., *J. Nutr.* **105,** 96 (1975).
33. Caygill, C.P.J., Lucy, J.A., and Diplock, A.T., *Biochem. J.* **125,** 407 (1971).
34. Chow, C.K., and Tappel, A.L., *J. Nutr.* **104,** 444 (1974).
35. Clayton, C.C., and Baumann, C.A., *Cancer Res.* **9,** 575 (1949).
36. Combs, G.F., Jr., and Scott, M.L., *J. Nutr.* **104,** 1292 (1974).
37. Cousins, F.B., and Cairney, I.M., *Aust. J. Agric. Res.* **12,** 927 (1961).
38. Cummins, L.M., and Martin, J.L., *Biochemistry* **6,** 3162 (1967).
39. Dam, H., Nielsen, G.K., Prange, I., and Sondergaard, E., *Nature (London)* **182,** 802 (1958).
40. Dickson, R.C., and Tomlinson, R.N., *Clin. Chim. Acta* **16,** 311 (1967).
41. Dinkel, C.A., Minyard, J.A., and Olson, O.E., *S. D., Agric. Exp. Stn., Circ.* **135,** (1937).
42. Dinkel, C.A., Minyard, J.A., and Ray, D.E., *J. Anim. Sci.* **22,** 1043 (1963).
43. Diplock, A.T., *Proc. Nutr. Soc.* **33,** 315 (1974).
44. Diplock, A.T., and Lucy, J.A., *FEBS Lett.* **29,** 205 (1973).
45. Dornenbal, H., *Can. J. Anim. Sci.* **55,** 325 (1975).
46. Drake, C., Grant, A.B., and Hartley, W.J., *N. Z. Vet. J.* **8,** 4 and 7 (1960).
47. Eggert, R.G., Patterson, E., Akers, W.T., and Stokstad, E.L.R., *J. Anim. Sci.* **16,** 1037 (1957).
48. Ehligh, C.F., Hogue, D.E., Allaway, W.H., and Hamm, D.J., *J. Nutr.* **92,** 121 (1967).
49. Exon, J.H., Koller, L.D., and Elliott, S.C., *Clin. Toxicol.* **9,** 273 (1976).
50. Fleming, G.A., and Walsh, T., *Proc. R. Ir. Acad., Sect. B* **58,** 151 (1957).
51. Flohe, L., *Klin. Wochenschr.* **49,** 699 (1971).
52. Flohe, L., Gunzler, W.A., and Schock, H.H., *FEBS Lett.* **32,** 132 (1973).
53. Franke, K.W., and Painter, E.P., *Cereal Chem.* **15,** 1 (1938).
54. Franke, K.W., and Tully, W.C., *Poultry Sci.* **14,** 273 (1935); **15,** 316 (1936).
55. Fuss, C.N., and Godwin, K.O., *Aust. J. Biol. Sci.* **28,** 239 (1975).
56. Ganther, H.E., and Baumann, C.A., *J. Nutr.* **77,** 208 and 408 (1962).
57. Ganther, H.E., Hafeman, D.G., Lawrence, R.A., Serfass, R.E., and Hoekstra, W.G., *in* "Trace Elements in Human Disease" (A.S. Prasad, ed.), Vol. 2, pp. 165–234. Academic Press, New York, 1975.
58. Ganther, H.E., Gondie, C., Sunde, M.L., Kopecky, M.J., Wagner, P., Hoh, S., and Hoekstra, W.G., *Science* **175,** 1122 (1975).
59. Ganther, H.E., Levander, O.A., and Baumann, C.A., *J. Nutr.* **88,** 55 (1966).
60. Ganther, H.E., Wagner, P.A., Sunde, M.L., and Hoekstra, W.G., *Trace Subst. Environ. Health–6, Proc. Univ. Mo. Annu. Conf., 6th, 1972,* p. 247 (1973).
61. Gardiner, M.R., *Aust. Vet. J.* **38,** 387 (1962).
62. Gardiner, M.R., and Gorman, P.C., *Aust. J. Exp. Agr. Anim. Husb.* **3,** 284 (1966).
63. Gardner, R.W., and Hogue, D.E., *J. Nutr.* **93,** 418 (1967).
64. Godwin, K.O., *Q. J. Exp. Physiol. Cogn. Med. Sci.* **50,** 282 (1965).
65. Godwin, K.O., *Nature (London)* **217,** 1275 (1965).
66. Godwin, K.O., and Frazer, F.J., *Q. J. Exp. Physiol. Cogn. Med. Sci.* **51,** 94 (1966).
67. Godwin, K.O., Fuss, C.N., and Kuchel, R.E., *Aust. J. Biol. Sci.* **28,** 251 (1975).
68. Godwin, K.O., Kuchel, R.E., and Fuss, C.N., *Aust. J. Biol. Sci.* **27,** 633 (1974).
69. Grant, A.B., and Wilson, G.F., *N. Z. J. Agric. Res.* **11,** 733 (1968).
70. Grant, C.A., and Thafvelin, B., *Nord. Veterinaermed.* **10,** 657 (1958).

71. Green, J., and Bunyan, J., *Nutr. Abstr. Rev.* **39**, 321 (1969).
72. Green, R.C., and O'Brien, P.J., *Biochim. Biophys. Acta* **197**, 31 (1970).
73. Gries, C.L., and Scott, M.L., *J. Nutr.* **102**, 1287 (1972).
74. Griffiths, N.M., and Thompson, C.D., *N. Z. Med. J.* **80**, 199 (1974).
75. Hadjimarkos, D.M., *J. Pediatr.* **63**, 273 (1963).
76. Hadjimarkos, D.M., *Lancet* **1**, 605 (1965).
77. Hadjimarkos, D.M., *Experientia* **22**, 117 (1966).
78. Hadjimarkos, D.M., *Nutr. Rep. Int.* **1**, 175 (1970).
79. Hadjimarkos, D.M., *Adv. Oral Biol.* **3**, 252 (1968).
80. Hadjimarkos, D.M., *Caries Res.* **3**, 14 (1969).
81. Hadjimarkos, D.M., *Trace Subst. Environ. Health–7, Proc. Univ. Mo. Annu. Conf., 7th, 1973,* p. 25 (1973).
82. Hadjimarkos. D.M.. and Bonhorst, C.W., *J. Pediatr.* **59**, 261 (1961).
83. Hadjimarkos, D.M., and Shearer, T.R., *Am. J. Clin. Nutr.* **24**, 1210 (1971).
84. Hadjimarkos, D.M., and Shearer, T.R., *J. Dent. Res.* **52**, 389 (1973).
85. Hadjimarkos, D.M., and Shearer, T.R., *Am. J. Clin. Nutr.* **26**, 583 (1973).
86. Hafeman, D.G., Sunde, R.A., and Hoekstra, W.G., *J. Nutr.* **104**, 580 (1974).
87. Hahn, H.K., Williams, R.V., Burch, R., Sullivan, J.F., and Novak, E.A., *J. Lab. Clin. Med.* **80**, 718 (1972).
88. Halverson, A.W., and Monty, K.J., *J. Nutr.* **70**, 100 (1960).
89. Halverson, A.W., Hendrick, C.M., and Olson, O.E., *J. Nutr.* **56**, 51 (1955).
90. Hamilton, E.I., Minski, M.J., and Cleary, J.J., *Sci. Total Environ.* **1**, 341 (1972/1973).
91. Handreck, K.A., and Godwin, K.O., *Aust. J. Agric. Res.* **21**, 71 (1970).
92. Hansson, E., and Jacobson, S.O., *Biochim. Biophys. Acta* **115**, 285 (1966).
93. Harr, J.R., Bone, F.J. Tinsley, I.J., Weswig, P.H., and Yanomoto, R.S., *in* "Selenium in Biomedicine" (O.H. Muth, ed.), p. 153. Avi. Publ., Westport, Connecticut, 1967.
94. Harr, J.R., Exon, J.H., Whanger, P.D., and Weswig, P.H., *Clin. Toxicol.* **5**, 187 (1972).
95. Harr, J.R., Exon, J.H., Weswig, P.H., and Whanger, P.D., *Clin. Toxicol.* **6**, 487 (1973).
96. Harris, P.L., Ludwig, M.I., and Schwa.z, K., *Proc. Soc. Exp. Biol. Med.* **97**, 686 (1958).
97. Hartley, W.J., *in* "Selenium in Biomedicine" (O.H. Muth, ed.), p. 79. Avi. Publ., Westport, Connecticut, 1967.
98. Hartley, W.J., *Proc. N. Z. Soc. Anim. Prod.* **23**, 20 (1963).
99. Hartley, W.J., and Grant, A.B., *Fed. Proc., Fed. Am. Soc. Exp. Biol.* **20**, 679 (1961).
100. Hartley, W.J., and Hansen, B., private communication (1975).
101. Hendrick, C., Klug, H., and Olson, O.E., *J. Nutr.* **51**, 131 (1953).
102. Hill, C.H., *J. Nutr.* **104**, 593 (1974).
103. Hidiroglou, M., Carson, R.B., and Brossard, G.A., *Can. J. Anim. Sci.* **45**, 197 (1965).
104. Hidiroglou, M., Jenkins, R.J., Carson, R.B., and Mackay, R.R., *Can. J. Anim. Sci.* **48**, 335 (1968).
105. Hopkins, L.L., Pope, A.L., and Baumann, C.A., *J. Nutr.* **88**, 61 (1966).
106. Hove, E.L., Fry, G.S., and Schwarz, K., *Proc. Soc. Exp. Biol. Med.* **98**, 1 and 27 (1958).
107. Jacobson, S.O., and Oksanen, H.E., *Acta Vet. Scand.* **7**, 66 (1966).
108. Jaffe, W.G., and Mondragon, M.C., *J. Nutr.* **97**, 431 (1969).
109. Jaffe, W.G., and Mondragon, M.C., *Br. J. Nutr.* **33**, 387 (1975).

110. Jenkins, R.J., *Can. J. Biochem.* **46,** 1417 (1968).
111. Jenkins, R.J., and Hidiroglou, M., *Can. J. Anim. Sci.* **51,** 389 (1971).
112. Jensen, L.S., *Proc. Soc. Exp. Biol. Med.* **128,** 970 (1968).
113. Jensen, L.S., *J. Nutr.* **105,** 769 (1975).
114. Jensen, L.S., *Proc. Soc. Exp. Biol. Med.* **149,** 113 (1975).
115. Jensen, L.S., Walter, E.D., and Dunlap, J.S., *Proc. Soc. Exp. Biol. Med.* **112,** 899 (1963).
116. Jones, G.B., and Godwin, K.O., *Aust. J. Agric. Res.* **14,** 716 (1963).
117. Kaarde, J., *Wien, Tierarztl. Monatsschr.* **52,** 391 (1965); *Vet. Bull. (London)* **36,** 234 (abstr.) (1966).
118. Kamstra, L.D., and Bonhorst, C.W., *Proc. S. D. Acad. Sci.* **32,** 72 (1953).
118a. Kagueler, J.C., and Petrović, A., *Arch. Sci. Physiol.* **24,** 157 (1970).
119. Klug, H.L., Moxon, A.L., Peterson, D.F., and Potter, V.R., *Arch. Biochem.* **28,** 253 (1950).
120. Koeman, J.H., Peeters, W.H.M., Koudstahl-Hol, C.H.M., Tjive, P.S., and Goeij, J.J.M. de, *Nature (London)* **245,** 385 (1973).
121. Knott, S.G., and McCay, C.W.R., *Aust. Vet. J.* **35,** 161 (1959).
122. Kovalsky, V., *Nutr. Abstr. Rev.* **25,** 544 (1955).
123. Ku, P.K., Ely, W.T., Groce, A.W., and Ullrey, D.E., *J. Anim. Sci.* **34,** 208 (1972).
124. Kubota, J., Allaway, W.H., Carter, D.L., Cary, E.E., and Lazar, V.A., *J. Agric. Food Chem.* **15,** 448 (1967).
125. Kuchel, R.E., and Buckley, R.A., *Aust. J. Agric. Res.* **20,** 1099 (1969).
126. Kuttler, K.L., and Marble, D.W., *Am. J. Vet. Res.* **22,** 422 (1961).
127. Latshaw, J.D., *J. Nutr.* **105,** 32 (1975).
128. Letunova, S.V., "Trace Element Metabolism in Animals (C.F. Mills, ed.), Vol. I, p. 432. Livingstone, Edinburgh, 1970.
129. Levander, O.A., and Argrett, L.C., *Toxicol. Appl. Pharmacol.* **14,** 308 (1969).
130. Levander, O.A., and Baumann, C.A., *Toxicol. Appl. Pharmacol.* **9,** 98 & 106 (1966).
131. Levander, O.A., Morris, V.C., and Higgs, D.J., *Biochem. Biophys. Res. Commun.* **58,** 1047 (1974).
132. Levander, O.A., Young, M.L., and Meeks, S.A., *Toxicol. Appl. Pharmacol.* **16,** 79 (1970).
133. Levine, R.E., *J. Bacteriol.* **10,** 217 (1925).
134. Lindberg, P., *Acta Vet. Scand., Suppl.* **23** (1968).
135. Lopez, P.L., Preston, R.L., and Pfander, W.H., *J. Nutr.* **94,** 219 (1968); **97,** 123 (1968).
136. Losee, F., Cutress, T.W., and Brown, R., *Trace Subst. Environ. Health–7, Proc. Univ. Mo. Annu. Conf., 7th, 1973,* p. 19 (1973).
137. Ludwig, T.G., and Bibby, B.G., *Caries Res.* **3,** 32 (1969).
138. Mace, D.L., Tucker, J.A., Bills, C.B., and Ferreira, C.J., *Calif., Dep. Agric., Bull.* No. 1, 63 (1963).
139. Mahan, D.C., Moxon, A.L., and Cline, J.H., *J. Anim. Sci.* **40,** 624 (1975).
140. Majaj, A.S., and Hopkins, L.L., *Lancet* **2,** 592 (1966).
141. Mathias, M.M., and Hogue, D.E., *J. Nutr.* **93,** 418 (1967).
142. McConnell, K.P., *J. Biol. Chem.* **141,** 427 (1941); **145,** 55 (1942); **173,** 633 (1948).
143. McConnell, K.P., *Tex. Rep. Biol. Med.* **17,** 120 (1959).
144. McConnell, K.P., *J. Agric. Food Chem.* **11,** 385 (1963).
145. McConnell, K.P., and Roth, D.M., *Biochim. Biophys. Acta* **62,** 503 (1962); *J. Nutr.* **84,** 340 (1964).
146. McConnell, K.P., and Wabnitz, C.H., *J. Biol. Chem.* **226,** 765 (1957).

147. McConnell, K.P., and Levy, R.S., *Nature (London)* **195**, 774 (1962).
148. McConnell, K.P., Broghamer, W.L., Jr., Blotcky, A.J., and Hurt, O.J., *J. Nutr.* **105**, 1026 (1975).
149. McCoy, K.E.M., and Weswig, P.H., *J. Nutr.* **98**, 383 (1967).
150. McLean, J.W., Thompson, G.G., and Claxton, J.H., *N. Z. Vet. J.* **7**, 47 (1959); McLean, J.W., Thompson, G.G., and Lawson, B.M., *ibid.* **11**, 59 (1963).
151. Millar, K.R., Craig, J., and Dawe, L., *N. Z. J. Agric. Res.* **16**, 301 (1973).
152. Millar, K.R., and Sheppard, A.D., *N. Z. J. Sci.* **15**, 3 (1972).
153. Millar, K.R., Gardiner, M.A., and Sheppard, A.D., *N. Z. J. Agric. Res.* **16**, 115 (1973).
154. Miltmore, J.E., van Ryswyck, A.L., Pringle, W.L., Chapman, F.M., and Kalnin, C.M., *Can. J. Anim. Sci.* **55**, 101 (1975).
155. Morris, V.C., and Levander, O.A., *J. Nutr.* **100**, 1383 (1970).
156. Moxon, A.L., *Science* **88**, 81 (1938).
157. Moxon, A.L., *S. D. Agric. Exp. Stn., Bull.* **311** (1937).
158. Moxon, A.L., Dubois, K.P., and Potter, R.L., *J. Pharmacol. Exp. Ther.* **72**, 184 (1971).
159. Moxon, A.L., and Poley, W.E., *Poultry Sci.* **17**, 77 (1938).
160. Moxon, A.L., and Rhian, M., *Physiol. Rev.* **23**, 305 (1964).
161. Moxon, A.L., Schaefer, A.E., Lardy, H.A., Dubois, K.P., and Olson, O.E., *J. Biol. Chem.* **132**, 785 (1940).
162. Moxon, A.L., Rhian, M., Anderson, H.D., and Olson, O.E., *J. Anim. Sci.* **3**, 299 (1944).
163. Mudd, A.J., and Mackie, I.L., *Vet. Rec.* **93**, 197 (1973).
164. Munsell, H.E., Devaney, G.M., and Kennedy, M.H., *U.S., Dep. Agric., Tech. Bull.* **534** (1936).
165. Muth, O.H., and Allaway, W.H., *J. Am. Vet. Assoc.* **142**, 1379 (1963).
166. Muth, O.H., Weswig, P.H., Whanger, P.D., and Oldfield, J.E., *Am. J. Vet. Res.* **32**, 1603 (1971).
167. Nelson, A.A., Fitzburgh, O.G., and Calvery, H.O., *Cancer Res.* **3**, 230 (1943).
168. Nesheim, M.C., and Scott, M.L., *J. Nutr.* **65**, 601 (1958).
169. Nesheim, M.C., and Scott, M.L., *Fed. Proc., Fed. Am. Soc. Exp. Biol.* **20**, 674 (1961).
170. Noguchi, T., Cantor, A.H., and Scott, M.L., *J. Nutr.* **103**, 1502 (1973).
171. Noguchi, T., Langevin, M.L., Combs, G.F., Jr., and Scott, M.L., *J. Nutr.* **103**, 44 (1973).
172. Oksanen, H.E., *in* "Selenium in Biomedicine" (O.H. Muth, ed.), p. 215. Avi. Publ., Westport, Connecticut, 1967.
173. Oldfield, J.E., Schubert, J.R., and Muth, O.H., *J. Agric. Food Chem.* **11**, 388 (1963).
174. Olson, O.E., *Proc. Ga. Conf. Feed Manuf.* (1969).
175. Olson, O.E., Dinkel, C.A., and Kamstra, L.D., *S. D. Farm Home Res.* **6**, 12 (1954).
176. Olson, O.E., Carlson, C.W., and Leitis, E., *S. D., Agric. Exp. Stn., Tech. Bull.* **20**, (1958).
177. Olson, O.E., Novacek, E.J., Whitehead, E.I., and Palmer, I.S., *Phytochemistry* **8**, 1161 (1970).
178. Omaye, S.T., and Tappel, A.L., *J. Nutr.* **104**, 747 (1974).
179. Ostadius, K., and Aberg, B., *Acta Vet. Scand.* **2**, 60 (1961).
180. Palmer, I.S., Fischer, D.D., Halverson, A.W., and Olson, O.E., *Biochim. Biophys. Acta* **177**, 336 (1969).
181. Parizek, J., Benes, I., Babicky, A., Bênes, J., Proschazkova, V., and Lener, J., *Physiol. Bohemslovac.* **18**, 105 (1969).

182. Patterson, E.L., Milstrey, R., and Stokstad, E.L.R., *Proc. Soc. Exp. Biol. Med.* **95,** 621 (1957).
183. Paulson, G.D., Broderick, G.A., Baumann, C.A., and Pope, A.L., *J. Anim. Sci.* **27,** 195 (1968).
184. Paulson, G.D., Pope, A.L., and Baumann, C.A., *Proc. Soc. Exp. Biol. Med.* **122,** 321 (1966).
185. Peterson, R.P., and Jensen, L.S., *Poultry Sci.* **54,** 795 (1975).
186. Pierce, A.W., and Jones, G.B., *Aust. J. Exp. Agric. Anim. Husb.* **8,** 277 (1968).
187. Poley, W.E., Wilson, W.O., Moxon, A.L., and Taylor, J.B., *Poultry Sci.* **20,** 171 (1941).
188. Potter, V.R., and Elvehjem, C.A., *Biochem. J.* **30,** 189 (1936); *J. Biol. Chem.* **117,** 341 (1937).
189. Quarterman, J., Mills, C.F., and Dalgarno, A.C., *Proc. Nutr. Soc.* **25,** XXIII (1966).
189a. Ravikovitch, S., and Margolin, M., *Emp. J. Exp. Agric.* **27,** 235 (1959).
190. Rhead, W.J., Cary, E.E., Allaway, W.H., Saltzstein, S.L., and Schrauzer, G.N., *Bioinorg. Chem.* **1,** 289 (1972).
191. Riley, J.F., *Experientia* **24,** 1237 (1969).
192. Robinson, M.F., Mckenzie, J.M., Thompson, C.D., and van Rij, A.L., *Br. J. Nutr.* **30,** 195 (1973).
193. Rosenfeld, I., and Beath, O.A., *J. Nutr.* **30,** 443 (1945).
194. Rosenfeld, I., and Beath, O.A., "Selenium." Academic Press, New York, 1964.
195. Rosenfeld, I., and Beath, O.A., *Proc. Soc. Exp. Biol. Med.* **87,** 295 (1954).
196. Rotruck, J.T., Pope, A.L., Ganther, H.E., Swanson, A.B., Hafeman, D.G., and Hoekstra, W.G., *Science* **179,** 588 (1973).
197. Salisbury, R.M., Edmondson, J., Poole, W.H.S., Bobby, F.C., and Birnie, H., *Proc. World's Poultry Congr., 12th, 1962,* p. 379 (1962).
198. Sandholm, M., *Acta Pharmacol. Toxicol.* **33,** 1 and 6 (1973); **36,** 321 (1975).
199. Schrauzer, G.N., and Ishmael, D., *Ann. Clin. Lab. Sci.* **4,** 441 (1974); Schrauzer, G.N., White, D.A., and Schneider, C.J., *Bioinorg. Chem.* **6,** 265 (1976).
200. Schrauzer, G.N., Rhead, W.J., and Evans, G.A., *Bioinorg. Chem.* **2,** 329 (1973).
201. Schroeder, H.A., Frost, D.V., and Balassa, J.J., *J. Chronic Dis.* **23,** 227 (1970).
202. Schroeder, H.A., and Mitchener, M., *J. Nutr.* **101,** 1531 (1971).
203. Schroeder, H.A., and Mitchener, M., *Arch. Environ. Health* **24,** 66 (1972).
204. Schubert, J.R., Muth, O.H., Oldfield, J.E., and Remmert, L.F., *Fed. Proc., Fed. Am. Soc. Exp. Biol.* **20,** 689 (1961).
205. Schwarz, K., *Fed. Proc., Fed. Am. Soc. Exp. Biol.* **20,** 666 (1961).
206. Schwarz, K., *Lancet* **1,** 1335 (1965).
207. Schwarz, K., and Foltz, C.M., *J. Am. Chem. Soc.* **79,** 3293 (1957); *J. Biol. Chem.* **233,** 245 (1958).
207a. Schwarz, K., and Fredga, A., *J. Biol. Chem.* **244,** 2103 (1969).
208. Schwarz, K., and Fredga, A., *Bioinorg. Chem.* **2,** 47 and 171 (1972); **3,** 153 (1974); **4,** 235 (1975).
209. Scott, M.L., *in* "Selenium in Biomedicine" (O.H. Muth, ed.), p. 231. Avi. Publ., Westport, Connecticut, 1967.
210. Scott, M.L., *Proc. Cornell Nutr. Conf.,* p. 123 (1973).
211. Scott, M.L., *J. Nutr.* **103,** 803 (1973).
212. Scott, M.L., Olson, G., Krook, L., and Brown, W.R., *J. Nutr.* **91,** 573 (1967).
213. Scott, M.L., and Thompson, J.N., *Poultry Sci.* **50,** 1742 (1971).
214. Shamberger, R.J., *J. Natl. Cancer Inst.* **44,** 931 (1970).
215. Shamberger, R.J., and Frost, D.V., *Can. Med. Assoc. J.* **100,** 682 (1969).
216. Shamberger, R.J., and Rudolph, G., *Experientia* **22,** 116 (1966).

217. Shamberger, R.J., Tytko, S., and Willis, R.E., *Cleveland Clin. Q.* **39**, 119 (1972).
218. Shearer, T.R., and Hadjimarkos, D.M., *J. Nutr.* **103**, 553 (1973).
219. Shearer, T.R., and Hadjimarkos, D.M., *Arch. Environ. Health* **30**, 230 (1975).
220. Shrift, A., and Kelly, E., *Nature (London)* **195**, 732 (1962).
221. Slen, S.B., Demiruren, A.S., and Smith, A.D., *Can. J. Anim. Sci.* **41**, 263 (1961).
222. Smith, M.I., *Public Health Rep.* **54**, 1441 (1939).
223. Smith, M.I., Stohlman, E.F., and Lillie, R.D., *J. Pharmacol. Exp. Ther.* **60**, 449 (1937).
224. Smith, M.I., Westfall, B.B., and Stohlman, E.F., *Public Health Rep.* **52**, 1171 (1937); **53**, 1199 (1938).
225. Smith, M.I., and Westfall, B.B., *Public Health Rep.* **52**, 1375 (1937).
226. Smith, P.J., Tappel, A.L., and Chow, C.K., *Nature (London)* **247**, 393 (1974).
227. Spallholz, J.R., Martin, J.L., Gerlach, M.L., and Heinzerling, R.H., *Proc. Soc. Exp. Biol. Med.* **143**, 685 (1973); **148**, 37 (1975).
228. Sprinkler, L.H., Harr, J.R., Newberne, P.M., Whanger, P.D., and Weswig, P.H., *Nutr. Rep. Int.* **4**, 335 (1971).
229. Tappel, A.L., *Vitam. Horm. (N.Y.)* **20**, 493 (1962).
230. Taussky, H.H., Washington, A., Zubillaga, E., and Milhorst, A.T., *Nature (London)* **200**, 1211 (1963); **206**, 509 (1965).
231. Thompson, J.N., and Scott, M.L., *J. Nutr.* **97**, 335 (1969).
232. Thompson, J.N., and Scott, M.L., *J. Nutr.* **100**, 797 (1970).
233. Thompson, J.N., Erdody, P., and Smith, D.C., *J. Nutr.* **105**, 274 (1975).
234. Thompson, C.D., and Stewart, R.D.H., *Br. J. Nutr.* **32**, 47 (1974).
235. Trinder, N., Woodhouse, C.D., and Renton, C.P., *Vet. Rec.* **85**, 550 (1969).
236. Tscherkes, L.A., Volgarev, M.N., and Apeteker, S.G., *Acta Unio Int. Cancrum* **19**, 632 (1939).
237. Wahlstrom, R.C., and Olson, O.E., *J. Anim. Sci.* **18**, 141 (1959).
238. Wahlstrom, R.C., Kamstra, L.D., and Olson, O.E., *J. Anim. Sci.* **14**, 105 (1955).
239. Watkinson, J.H., *N. Z. Med. J.* **80**, 199 (1974).
240. Watkinson, J.H., and Davies, E.B., *N. Z. J. Agric. Res.* **10**, 116 and 122 (1967).
241. Weisberger, A.S., and Surland, L.G., *Blood* **11**, 11 (1956).
242. Westfall, B.B., Stohlman, E.F., and Smith, M.I., *J. Pharmacol. Exp. Ther.* **64**, 55 (1938).
243. Whanger, P.D., *Biochem. Med.* **7**, 316 (1973).
244. Whanger, P.D., Pederson, N.D., and Weswig, P.H., *Biochem. Biophys. Res. Commun.* **53**, 1031 (1973).
245. Whanger, P.D., Weswig, P.H., Muth, O.H., and Oldfield, J.E., *J. Nutr.* **99**, 331 (1969).
246. World Health Organization, *W.H.O., Tech. Rep. Ser.* **532**, (1973).
247. Wright, C.I., *J. Pharmacol. Exp. Ther.* **68**, 220 (1940); *Public Health Rep. (U.S.)* **53**, 1825 (1940).
248. Wright, P.L., and Bell, M.C., *J. Nutr.* **84**, 49 (1964).
249. Wright, P.L., and Bell, M.C., *Am. J. Physiol.* **211**, 6 (1966).
250. Wu, S.H., Oldfield, J.E., Muth, O.H., Whanger, P.D., and Weswig, P.H., *Proc. West. Sect., Am. Soc. Anim. Sci.* **20**, 85 (1969).
251. Young, S., and Keeler, R.F., *Am. J. Vet. Res.* **23**, 955 and 966 (1962).
252. Yousef, M.K., Coffman, W.J., and Johnson, H.D., *Nature (London)* **219**, 1173 (1968).
253. Zook, E.G., Greene, F.E., and Morris, E.R., *Cereal Chem.* **47**, 720 (1970).

13

Fluorine

I. FLUORINE IN ANIMAL TISSUES AND FLUIDS

1. Fluorine in Soft Tissues

The F concentrations of the soft tissues of animals are normally low and do not increase with age. Fluorine does not concentrate in any tissues, other than the bones and teeth, although the placenta and the aorta sometimes carry elevated levels of this element, probably as a result of calcification which secondarily holds fluoride (58). The kidneys are usually richer in F than other organs, owing in part to retained urine (Table 45). The thyroid gland has no special ability to concentrate fluorine, even when intakes are high (155).

The data presented in Table 45 reveal only small increases in soft tissue F concentrations in chronic fluorosis. Dairy cows receiving a ration supplemented with 50 ppm F as NaF for $5\frac{1}{2}$ years only increased the fluorine levels in their heart, liver, thyroid, and pancreas 2- to 3-fold above the 2–3 ppm (d.b.) found for these tissues in control cows (149). Even smaller increases were observed in the same tissues of sheep consuming water containing 5 or 10 ppm F for a period of 2 years (51). Similar results were reported for rats receiving 2 mg F/day for 62 days (153), while Schroeder *et al.* (124) were unable to detect any soft tissue F accumulation in mice given 10 ppm F in the drinking water for 2 years.

The tissues of newborn and suckling animals are usually lower in fluorine than those of their mothers. In one study, rats and rabbits were fed varying

TABLE 45
Fluorine Levels in Soft Tissues of Animals[a]

	Rats[b]		Sheep[c]		Cows[d]	
Tissue	Normal	2 mg F/day for 76 days	Normal	10 ppm F in water for 2 yr	Normal	50 ppm F in ration for $5\frac{1}{2}$ yr
Liver	0.21	0.28	3.5	2.4	2.3	3.6
Kidney	0.62	1.50	4.2	20.0	3.5	19.3
Thyroid	–	–	3.0	7.6	2.1	7.3
Heart	2.6	5.4	3.0	2.3	2.3	4.6
Pancreas	–	–	2.8	3.2	2.8	4.2
Muscle	0.53	1.60	–	–	–	–

[a]Expressed as ppm dry basis.
[b]From Venkateswardu and Narayanarao (153).
[c]From Harvey (51).
[d]From Suttie *et al.* (149).

levels of added F, up to 300 ppm of the diet, for several weeks before and during pregnancy. The total fluorine in the fetuses at term was found to be negligible. Some increase occurred at the higher intakes but the amounts were extremely small compared with those in the mothers (80). Rats and puppies subsisting on the milk of fluoride-fed mothers accumulated more F in their bodies than similar animals consuming the milk of normally fed mothers (80, 109), but some of this may have been acquired from the diet or water of the F-fed mothers. Fluorine does not readily pass the placental and mammary barriers. Normal cow's milk contains only 1–2 ppm F (d.b.) (87) and human milk significantly less. Values of 0.02–0.03 ppm F in fresh breast milk have been reported (35, 37).

By contrast, birds consuming high-fluoride diets can readily transfer fluorine to the egg, especially into the yolk. The yolk of eggs from hens on a normal low-F diet was raised from 0.8–0.9 ppm to as high as 3 ppm by supplementing the hen's diet with 2% rock phosphate (114).

2. Fluorine in Blood

For many years analytical inadequacies prevented a proper understanding of the relation of plasma F concentrations to fluoride intakes by the animal. By the use of a suitable micromethod (134) Armstrong and co-workers (5) demonstrated significant increases in rat plasma F values from levels of about 0.1 ppm to as high as 1.0 ppm following large increments in dietary fluoride. Concentrations up to 3.3 ppm F occurred in rats receiving 600 ppm F as NaF (132). Plasma fluorine exists in both ionic and bound forms, with the proportion of each varying with total plasma levels and species. Singer and Armstrong (135) found that in the rat and rabbit ionic fluorine rather than bound fluorine reflects increased total plasma F values, whereas in human and bovine plasma it is the ionic F rather than the bound F concentration that tends to remain low and uniform. An adequate explanation of these species differences in plasma F distribution, or their significance, has not yet appeared.

In a long-term study of dairy cows continuously and periodically exposed to high-fluoride intakes, plasma F concentrations were related to the current level of fluoride ingestion (17). The levels of control animals were consistently below 0.1 ppm F, whereas those of the fluoride-treated animals were significantly higher. A level of 1.0 ppm in the plasma represented a high level, observed only after extended periods of ingestion. A feature of these findings was the rapidity with which plasma F levels changed in response to changes in fluoride intake.

Marked diurnal variation in plasma F levels has been observed in sheep (132), rats (127), dogs, and man (103), following the administration of a single oral dose of fluoride. As Shearer and Suttie (127) have pointed out, plasma samples must be taken very soon after the actual ingestion of the flüoride if they are to reflect the total daily intakes. The loss of fluorine to the urine and skeleton is so rapid that control concentrations are approached within hours of the completion of intake. The magnitude and rapidity of the short-term changes that can take place in plasma F levels, compared with those in the liver and femur, are illustrated in Table 46. No such diurnal variations in plasma fluoride are apparent in man at lower continuous fluoride intakes (133, 138). The teeth, during their period of formation, are exceedingly sensitive to small changes in plasma fluoride concentration. When these concentrations approach 0.5 ppm or more severe dental lesions appear, at levels above 0.2 ppm F less severe damage occurs, while at plasma F levels below 0.2 ppm few adverse effects are apparent

TABLE 46
Tissue F Concentration after Ingestion by Rats of Diets Supplemented with Fluoride[a]

		200 ppm F in diet		450 ppm F in diet	
Tissue	Control	7 A.M.[b]	3 P.M.	7 A.M.	3 P.M.
Plasma (ppm F)	0.09 ± 0.06	0.32 ± 0.09	0.09 ± 0.05	1.31 ± 0.22	0.24 ± 0.16
Liver (ppm wet wt)	0.41 ± 0.05	0.52 ± 0.05	0.48 ± 0.06	0.90 ± 0.05	0.54 ± 0.05
Femur (ppm in ash)	332 ± 19	810 ± 29	809 ± 29	1144 ± 40	1192 ± 43

[a]From Shearer and Suttie (127).
[b]7 A.M. is the end of the normal nocturnal feeding period.

(17). It seems, therefore, that 0.2 ppm F can be regarded as a critical plasma concentration in cattle.

3. Fluorine in Urine

A positive correlation between urinary fluoride levels and the levels of fluoride ingested has been demonstrated in several species. In sheep and cattle not exposed to excess F the urinary F concentration rarely exceeds 10 ppm and is usually closer to 5 ppm. With elevated fluoride intakes the urinary F levels rise quickly to 15–30 ppm and may reach upper limits of 70–80 ppm. Higher values are occasionally observed among animals consuming the same amounts of fluoride, as well as among samples from the same animal taken on different days or at different times on the same day (128, 129, 147). In the experiments of Suttie and co-workers (147), normal cows excreted urine containing less than 5 ppm F; cows that were on the border line of F toxicity, as judged by other criteria, urine with 20–30 ppm F; and those with systemic signs of toxicity, urine containing over 35 ppm F. Essentially similar results were obtained by Shupe *et al.* (128, 129) with heifers fed different amounts and forms of fluoride for prolonged periods.

Comparable data are not, of course available for humans but the level of fluorine in the urine of children 5–14 years of age was shown to rise to 1 ppm 3–5 years after they commenced drinking water fluoridated to 1 ppm F (167). Lower mean values for the urinary fluorine in individuals in Jerusalem consuming water containing 0.5–0.6 ppm F have been reported. These were 0.52 ppm F for adults and 0.27 ppm F for children 3–6 years old. In pregnant women the urinary F levels were found to decrease up to the eighth month of pregnancy, indicating F retention during this period, and then to rise slowly to usual adult values (44).

Urinary fluorine may come from the release of the element from the skeleton, as well as from the food and water supply. Mobilization and excretion of part of the fluoride present in the bones of animals removed from a high-F diet have been observed in rats (98, 122a), man (76), and cattle (116). High urinary F levels can therefore reflect either current ingestion or previous exposure to high intakes.

4. Fluorine in Bones and Teeth

An extremely high proportion of the total body burden of fluorine is present in the skeleton; the higher the intake the higher the amount and the proportion present in this site. In normal adult farm animals not unduly exposed to fluorine, the concentrations in whole dry, fat-free bones usually lie within the range 300–600 ppm F (51, 149, 156). The F concentrations of normal teeth parallel those of the long bones but usually at lower levels. Thus normal enamel is reported to contain 100–270 ppm, normal dentine 240–625 ppm, and normal molar teeth 200–537 ppm F on the dry, fat-free basis (22, 88). It must be emphasized that these levels are highly dependent on the amount, chemical form, duration, and continuity of the F intake and on the age of the animal. Fluorine accumulates with age in human bones, even where intakes from the food and water supply are very low. A rise in the bone fluoride levels with age was observed in a survey in England at all levels of intake and with a plateau at about 55 years of age (62). The height of the plateau was greater with individuals who had consumed throughout life water containing 0.8–1.2 ppm, or 1.9 ppm F, than with those who had consumed water containing virtually no fluorine.

Fluorine levels in bones and teeth many times higher than those cited above as "normal" can occur following prolonged high-F intakes. Swine bones appearing normal have been reported to contain upward of 3000–4000 ppm F (66), and F concentrations in compact bones from cattle below 4500 ppm are considered to be innocuous (149). In dairy cattle fluorine toxicosis is associated with levels in excess of 5500 ppm F in compact and 7000 ppm F in cancellous bone, with a "saturation" point of the order of 15,000–20,000 ppm (115, 149). Concentrations between 4500 and 5500 ppm F indicate a marginal zone (149). The toxic thresholds for fluorine in the bones of sheep have been placed lower, namely 2000–3000 ppm in bulk cortical and 4000–6000 ppm in bulk cancellous bone (62). McClure and co-workers (89) have estimated that a continuous intake of 8 ppm F in the diet for 35 years is necessary before the critical levels of 5000–6000 ppm F are attained in human bones. Hodge and Smith (58) emphasize the remarkable capacity of the skeleton to sequester fluoride without serious pathological change. They distinguish three stages in this process, namely, bone fluoridation or chemical fluorosis, bone mottling, and abnormal bone. Bone fluoridation alone is characteristic of bones containing less than 2500 ppm F and is not accompanied by detectable gross or microscopic abnormalities or effect on cell activity. Bone mottling occurs in bones containing 2500–5000 ppm F, although such bones are normal to both gross and roentgenographic examination. Abnormal bones with gross abnormalities, an enlarged, heavy, chalky white appearance, and irregular surface are characteristic of bones with fluoride levels above 5000–6000 ppm F.

II. FLUORINE METABOLISM

1. Absorption

Soluble fluorides are rapidly and almost completely absorbed from the gastrointestinal tract, even at high intakes. Rats administered small oral doses of $Na^{18}F$ absorbed 75% in 1 (155) and 80–90% within 8 hr (33). When a dilute solution of NaF was given orally to rats and the animals killed immediately and 5, 15, 30, 60, and 90 min later, 12, 22, 36, 50, 72, and 86%, respectively, of the dose was absorbed (31). In humans given small oral doses of soluble fluoride, blood maxima were observed within 1 hr and 20–30% appeared in the urine in the succeeding 3–4 hr (166).

The speed and extent of F absorption varies with the physical and chemical form of the compound ingested. In small amounts the insoluble forms of fluoride are almost as well absorbed as the more soluble. Thus the fluorine of CaF_2 was 96% absorbed and that of cryolite 93% absorbed by the rat when so administered (72). When added to the diet in solid form the absorption was only 60–70% for calcium fluoride and 60–77% for cryolite (78). Even poorer absorption, ranging from 37 to 54%, has been reported for the fluoride in bone. The fluorine in fish protein concentrate (FPC) has also been reported to be only 42–52% as available to young rats as sodium fluoride (168). By contrast, metabolic balance studies in adult men revealed a very high net F absorption for FPC as well as from NaF administered to supply similar amounts of fluorine (137). The net F absorption averaged 88% during the intake of FPC and 94% during the intake of NaF. Whether the difference in availability of fluoride from FPC disclosed by these two studies is related to age or species differences or to differences in the basal diets is not known, but fluoride absorption is known to be affected by several dietary factors.

From the results of mineral balance studies with sheep it appears that aluminum salts exert a protective effect against high intakes of fluoride by reducing its absorption from the intestinal tract (9). Calcium salts function similarly. Thus 3 mg F was found in the whole carcass of rats at the end of a 2-week period when 1% $CaCl_2$ was administered with NaF in the drinking water, compared with 6 and 11 mg, respectively, when this calcium salt was added at the lower levels of 0.1 and 0.01% (158). When $MgCl_2$ and $AlCl_3$ were administered at the same levels F retention was similarly inhibited but to a smaller extent. High dietary Ca and P, independently and together, increase fecal F excretion in man, but the amounts so excreted were small so that the F balance remained unaltered and plasma fluoride levels did not change (139).

The level of fat in the diet also influences F toxicity, although the extent to which this is due to an effect on F absorption is not clear. Raising the level of dietary fat from 5 to 15%, or to 20%, enhances the growth-retarding effect of high-fluoride intakes in rats (15, 98), and chicks (13). The effect is unrelated to the chain length of the fat (98) and has been attributed, in part, to increased F retention in the heart, kidneys, and skeletal tissues (13, 15). More recently McGown and Suttie (91) have found that fluoride in the presence of fat causes delayed gastric emptying. They contend that this effect probably accounts for the increased toxicity of fluoride to rats fed high-fat diets. However, Ericsson (34) reported no differences in femur accumulation of ^{18}F 4 hr after administration of 4 ppm F with 28% olive oil, compared with controls which received only ^{18}F and water. The complexity of the process of absorption of fluorine is illustrated by this worker's further finding that NaCl significantly depressed skeletal F uptake from both water and flour (34).

2. Fluorine Distribution and Retention

Absorbed fluorine is distributed rapidly throughout the body as the fluoride ion, in a pattern similar to that of chloride. It readily crosses cell membranes, including that of the erythrocyte. The speed with which fluoride leaves the blood is illustrated by several experiments. Thus Perkinson *et al.* (110) calculated that 40% of the blood F, following intravenous injection, left the blood per minute in lambs and 32% in cows, while Bell and co-workers (11) found that only 53% of the ^{18}F tracer dose administered to cattle remained in the blood after 2 min. The disappearance curve of ^{18}F from the blood is triphasic in character. In lambs, the first and most rapid phase has a half-time of only 3–4 min, the second about 1 hr, and the third about 3.3 hr (110). The first phase presumably represents mixing of the fluoride with the body water, the second uptake by the skeleton, and the third excretion in the urine.

Uptake of fluoride by bone depends on its vascularity and growth activity, so that in the fully mature bone the rate of F deposition and retention is greatly reduced (112, 155). Hodge (57) postulates that F deposition in bone occurs by two processes. The first involves a rapid exchange of F ions in the tissue fluids with OH or CO_3 ions on the mineral crystal surface. The second process involves slow bone formation leading to storage, in which the fluorine is incorporated into the hydroxylapatite lattice, i.e., the bone salt itself. Since there is no change in the Ca:P ratio of fluorotic bone, the phosphate group cannot be replaced by fluoride ion. However, the carbonate content of fluorotic bone is decreased and the magnesium content increased (66, 85, 150, 159). This suggests a replacement of the carbonate group by fluoride ion and possibly a precipitation of some fluorine as MgF_2.

The level of F intake below which all is excreted and none retained in the bones must be very low indeed, if any such level exists. Metabolic studies with rats suggest that there is no level at which complete elimination occurs (155), while long-term studies with this species indicate continuous retention in the bones at the lowest levels of dietary F employed (119, 162). Machle and co-workers (79) found the input and output of fluorine in man to be approximately equal, over many weeks, when the daily intake was as low as 0.5 mg, whereas definite retention occurred when the intake ranged from 3 to 36 mg daily. In fluoride balance studies in adult men, Spencer *et al.* (138) found positive balances, i.e., F retention, at dietary intakes of 4.5 mg F/day from the diet and drinking water. These balances were increased further when fluoride

TABLE 47
Fluoride Excretions and Balances in Adult Men Expressed as % Fluoride Intake[a]

Fluoride intake (mg/day)	Fluoride intake (%)		Fluoride balance (%)	Net absorption (%)
	Urine	Stool		
4	54	6	40	94
14	53	7	40	94
44–48	58	11	31	89
12[b]	50	12	38	88

[a]From Spencer *et al.* (138).
[b]Fluoride given as fish protein concentrate (FPC).

supplements were given. The amounts retained corresponded to about 40% of a fluoride intake of 4 or 14 mg/day and were about 30% of the highest intake of 44–48 mg F/day. Data from these experiments, which represent maximal values since no account was taken of possible F losses in the sweat, are presented in Table 47.

3. Excretion

Numerous studies have shown that the main pathway of F excretion is via the kidney and that an increase in F intake is associated with a prompt and marked increase in urinary F excretion. This latter point was discussed in Section I,3. Excretion via the intestinal tract is low and even smaller amounts are excreted in the perspiration, except in excessive sweating (90). Average individuals on ordinary diets excrete 80% or more of their ingested fluorine in the urine (79). Adult men consuming a normal diet plus fluoridated water to give an average intake of 4.5 mg F/day excreted 50% of the intake in the urine and 6% in the feces and retained 1.8 mg F/day (138). On a fluoride intake of 45 mg/day added as NaF the urinary F of these men represented nearly 60% of the intake, the fecal F 11%, and the balance was markedly positive (138) (see Table 47). Lower percentages were found in the urine of men given fluoride in the form of bone meal or solid CaF_2 than when fluoridated water of NaF was administered (90), as a consequence of lower absorption. Sheep and cows similarly excrete 50–90% of their dietary fluoride in the urine, the proportion so excreted decreasing as the level of fluoride in the diet increases (54). The method of administration may also influence the route of excretion. Thus Simon and Suttie (132) found that sheep continuously infused with NaF solution into the rumen excreted more in the urine and less in the feces than comparable animals receiving a similar single daily dose. No significant difference between treatments in F retention was observed.

III. FLUORINE FUNCTIONS AND REQUIREMENTS

1. Fluorine as an Essential Element

Several attempts in the past to demonstrate an essential function for fluorine in rats fed diets reported to contain as little as 0.005 ppm F were unsuccessful (29, 84, 126). McClendon and Gershon-Cohen (86) fed weanling rats a basal diet prepared from hydroponically grown materials claimed to be "fluorine-free." These rats grew poorly compared with rats receiving a diet prepared from field-grown crops plus 20 ppm F in the drinking water. These benefits cannot be attributed with any confidence to the added fluorine since the basal diets in the two groups were different and the hydroponically grown diet could have been deficient in elements other than fluorine. Subsequently Schroeder *et al.* (124) reported an increase in body weight and longevity in mice from supplementation of 10 ppm F in the drinking water.

More convincing evidence that fluorine is essential for growth and reproduction in rats and mice has since been obtained. Schwarz and Milne (125), using highly purified diets in a controlled "isolator" system, observed increases in growth of rats of up to 30% when these diets were supplemented with 1, 2.5, and 7.5 ppm F as KF. The added fluorine was also found to improve the deposition of incisor pigment. Subsequently these workers confirmed these findings when KF was added at 2.5 ppm F to a basal diet supplying less than 0.04 ppm F but found that potassium monofluorophosphate, hexafluorosilicate, and monofluoropyruvate, added at the same F level, were much less effective. Discrepancies in the effects of the different compounds with respect to incisor pigmentation were also found, but these differences did not run parallel with the differences in growth (99). A curious feature of these experiments was that all four compounds tested led to the deposition of similar amounts of fluorine in the femurs of the rats. It seems, therefore, that some F compounds, such as hexafluorosilicate and monofluorophosphate, may not be readily available as sources of metabolically active fluorine even though they can lead to F deposition in bone.

In experiments carried out by Messer and co-workers (93, 94) female mice were fed a diet containing 0.1–0.3 ppm F, plus drinking water containing 0, 50, 100, and 200 ppm F as NaF. Toxic effects from the highest dose levels were observed and the mice receiving no supplement developed a progressive infertility in two successive generations. Growth rate and litter size were not affected by the low-F intake, but the percentage of mice producing litters was lower and the age of delivery of the first litter was greater than in mice receiving 50 ppm F in the water. No explanation was given for the very large amounts of fluorine apparently needed. The same group of workers (95, 96) also reported a severe anemia in pregnant mice fed a low-F diet and in their offspring prior to weaning. On the basis of these findings fluorine must be regarded as an essential element, despite the absence of a specific biochemical lesion or a satisfactory explanation of the cause of either the anemia or the infertility. The situation is complicated further by the findings of Weber and Reid (157) with mice. These workers observed no significant differences, through six generations, in body weight, fertility, or intestinal lipase activity between mice fed a low-F diet (0.2 ppm F) and those fed this diet supplemented with 6 ppm F.

2. Fluorine and Dental Caries Inhibition

Long before the demonstration of the essentiality of fluorine, described in the preceding section, the beneficial effects of fluoride on the incidence of dental caries had been recognized, following the early observation of Eager (30) that children with mottled enamel were relatively free from dental decay. A quantitative relationship between caries incidence and fluoride concentration in the drinking water was first clearly shown by Dean (27) in a survey involving 7257 children, 12–14 years old, living in 21 cities in the United States. The caries experienced in the permanent teeth was found to decrease as the fluoride concentration in the drinking water increased above about 1.3 ppm. Hodge (56) subsequently showed that the decrease in dental caries is a linear function of the fluoride expressed on a logarithmic scale. When considered in relation to the incidence of mottling of the enamel it became evident that the two lines intersect at close to 1 ppm F in the water. This point was therefore considered to be optimal, or the point of "maximum health with maximum safety." It was subsequently shown that the benefit of the fluoride is not confined to children provided that the fluoride exposure has occurred throughout or for a substantial part of the period of tooth formation (41, 92, 122).

The above findings stimulated public health authorities in many countries to deliberately fluoridate communal water supplies in an attempt to duplicate mass population exposure to naturally fluoridated waters. The water supplies of hundreds of communities have been treated with sodium fluoride or fluosilicate to maintain F levels ranging from 0.8 to 1.2 ppm, depending on climatic conditions. In several cases, a neighboring city or borough has been maintained as an unfluoridated control. The results of these controlled studies were then assessed after 10 or more years of fluoridation (6, 104). They demonstrate clearly the value of artificially fluoridating water supplies as a means of securing better dental health for a community. Fluoridation does not abolish dental decay. Reductions in the prevalence of caries in the teeth of children born subsequent to the change in water supply ranging from 40 to 70% have been reported in different studies, without observable ill effects on the health of individuals (28, 36). The very slight increase in the mildest forms of teeth mottling that can occur is insufficient to constitute a significant public health problem (6, 36). Furthermore, where the fluoridation is discontinued for any reason, a definite reversal of trend back to the prefluoridation level of caries experience has been demonstrated (104).

The success of mass water fluoridation in promoting community health stimulated attempts to achieve the same benefits without subjecting whole populations to fluoride, when children are the sole or predominant beneficiaries, and when total fluid intakes among individuals are so astonishingly variable. Several studies have emphasized the very wide range in consumption of fluids, giving total F intakes ranging from 2 to 5 mg/day (81) and 2.5–4.2 mg/day (69) in adults in the United States. The fluid intake of normal school children in England has been reported to vary from as little as 200 to 3000 ml daily, with adults similarly variable over a wider range (26). Where a high proportion of the fluid consumed consists of tea, whether made with fluoridated water or not, particularly high-F intakes are assured, as discussed in Section IV,1. For the above reasons, and because large sections of populations do not have access to communal water supplies, the practical possibilities of providing supplemental fluoride by means of fluoride tablets and by fortification of commonly used foods such as salt (82, 160) and flour (31, 82) have been investigated. Administration of fluoride tablets to children and to women during pregnancy and lactation has been found effective, with a decrease in caries incidence comparable with that found in communities with fluoridated water (7, 40, 161). The difficulty with this procedure is to maintain conscientious tablet utilization on a wide enough scale.

Topical applications of fluoride solutions to the teeth can also bring about a significant reduction in caries incidence in children (60), and some success has been achieved with fluoridated dentifrices (38, 68, 101). For example Moller (101) reported that painting of the teeth with a 2% aqueous solution of NaF or an 8% solution of sodium monofluorphosphate reduced the incidence of dental caries by 40%. Painting or brushing once every 2 weeks with a 0.2% aqueous solution of NaF reduced this incidence by 40–50%. Howink and co-workers (60) obtained a 37% reduction in carious lesions in children given a topical application of a 1%F stannous fluoride solution twice yearly over 9 years, compared with untreated twin controls. Five years after the last application the treated individuals, then 21 years of age, revealed significantly superior dental health.

An understanding of the mechanism of the anticaries action of fluoride, as

Hodge and Smith (58) have said, "has been tantalizingly elusive." Evidence obtained since then points increasingly to the conclusion that F exerts its anticariogenic action by enzyme inhibition of cariogenic bacteria in dental plaque (48, 70, 100, 165). Physical and chemical changes in the enamel may also be important. Fluoride is known to be taken up rapidly by the enamel surfaces and to be concentrated more in enamel imperfections than in intact surfaces nearby (49). Furthermore, following F treatment the solution rate, and presumably therefore the solubility of the tooth surface, are reduced (61, 63).

3. Fluoride and Osteoporosis

Equivocal results have been obtained in the treatment of patients with osteoporosis and other demineralizing diseases, following the administration of substantial amounts of fluoride. Some workers have reported beneficial effects on back pains, bone density, and calcium balance (1, 21, 118). Others have observed adverse reactions in fluoride therapy (60a, 77, 120) or failed to show improved Ca balance (53, 121, 136). In a study with dogs in which osteoporosis was induced by feeding a low-Ca, high-P diet for 41 weeks, Henrikson and co-workers (52) found no effect on the degree of osteoporosis from graded doses of dietary fluoride up to 1 mg/kg body weight/day, despite large increases in the F levels in the bones. Similarly, fluoride supplementation increased Ca retention in Japanese quail but had little effect on the integrity and strength of the bones even at high-F levels (23). On the other hand, a 12-month study of 11 human osteoporotics revealed increased new bone formation without any evidence of fluorosis from a combined treatment of 50 mg NaF and 900 mg Ca/day and 50,000 IU of vitamin D twice a week (64).

Limited epidemiological evidence suggests that the level of fluoride ingestion can be important in the maintenance of a normal skeleton. Leone *et al.* (75) observed substantially less osteoporosis in a high-F area in Texas (8 ppm F in the water) than in a low-F area (0.09 ppm F in the water) in Massachusetts. Subsequently Bernstein *et al.* (12) examined approximately 1000 X-rays of the lower lumbar spine of adults over 45 living in two areas of North Dakota. In one area the water supply contained 0.15–0.3 ppm F and in the other 4–6 ppm F. At all ages there was less osteoporosis and less distorted or collapsed vertebrae in women in the high-F area. There was also less osteoporosis in the men in the high-F area, but no such area effect was apparent in the incidence of collapsed and distorted vertebrae in men. The men at all ages, but not the women, revealed a significantly lower incidence of calcification of the aorta. However, in a double-blind study of 460 aged persons, half of whom were given 25 mg F/day as monofluorophosphate for 8 months, no difference was observed in height, hospital admission, or mortality, while fractures and exacerbation of arthrosis were more frequent in the F-treated group (60a).

IV. SOURCES OF FLUORINE

1. Fluorine in Human Foods and Dietaries

Human populations obtain widely varying quantities of fluorine from the food, the water supply, and the atmosphere depending on location. The air does not constitute a significant source of fluorine to man, except in areas of industrial pollution (24). Surface waters, such as are used for drinking and cooking in most communities, generally contain less than 1 ppm, or even 0.1 ppm or less (24). In some parts of the world the population is dependent on water from deep wells or artesian bores which are naturally high in fluoride. Fluoride concentrations of 4–8 ppm are common in such areas, while in endemic fluorosis areas in southern India (108) and South Africa concentrations as high as 20–40 ppm F have been reported (106).

Individuals not exposed to industrial contamination or fluoridated drinking water normally obtain most of their fluorine from food. Hodge and Smith (58) have estimated that "for individuals whose drinking water contains low levels or essentially no fluoride, and for whom there are no special fluoride exposures, a 'normal' daily dietary intake in the United States probably lies in the range of 0.5–1.5 mg." Kramer *et al.* (69) subsequently obtained average intakes of 0.78–1.03 mg F/day from the food for four nonfluoridated cities and 1.73–3.44 mg/day from 12 cities with water supplies fluoridated to approximately 1.0 ppm. Krepkogorsky (71) reported average total fluoride intakes of 2.5 mg/day in England, 3.3 mg/day in the Ukraine, and up to 2.1 mg/day in other parts of the Soviet Union, where 1 ppm F was present in each water supply.

The averages just quoted conceal large individual differences in F intakes, stemming especially from variations in the amounts of fish, tea, total liquids, and certain processed foods consumed. The remarkable individual variation in total daily fluid intake was mentioned in Section III,2. Seafish and fish products may contain 5–10 ppm F (d.b.) compared with 0.5 ppm or less for most other foods (81, 87, 154). The staple diet of Newfoundlanders, for example, has been calculated to supply 2.74 mg F/day in an area where the drinking water was fluoride free (32). The consumption of tea, which commonly contains 100–200 ppm F, two-thirds of which normally passes into the infusion consumed, can be a further major determinant of total F intakes. In fact Cook (26) has reported F intakes *from tea alone* of 1.26 mg/day by English schoolchildren 5–15 years of age and 2.55 mg/day by adults over 15 years of age. Substantial quantities of fluoride might also be expected to be provided where crude sea salt is used domestically in place of refined salt, as in some Asian countries. Samples of such salt imported from India contained 35–55 ppm F (126a).

The effect of cooking and processing with fluoridated water on the F levels of foods and beverages can be substantial. Auermann and Borris (8) reported that

TABLE 48
Increase in Fluoride Content of Vegetables Cooked or Processed in Water Containing 1 ppm F[a]

Food	Open saucepan[b]	Pressure cooker[b]	Commercial processing[c]
Beans	0.75	0.33	0.69
Carrots	0.64	0.33	0.42
Corn	0.26	0.25	0.36
Tomatoes	0.38	0.13	0.34[d]
Average	0.51	0.26	0.45

[a]Values as ppm fresh basis.
[b]From Martin (83).
[c]From Marier and Rose (81).
[d]Tomato soup.

total calculated intakes by the use of fluoridated drinking water were raised from 0.65 to 2.25 mg F/day in children and from 1.3 to 2.93 mg F/day in adults, varying between groups or professions. In one study of canned vegetables and beverages commercial processing with water containing 1 ppm F was found to increase their F concentrations by 0.34–0.75 ppm (81). Comparable increases in the fluoride levels of vegetables subjected to simulated home cooking with similar water were obtained in an earlier study (83). Data from both these investigations illustrating this effect are presented in Table 48. In foods subjected to processing and reconstitution with fluoridated water there can be a multiplying effect on F levels. This has been demonstrated with some processed infant foods (37, 39). A dry milk formula diluted 1:6 with 1 ppm fluoridated water contained 1.33 ppm F, or 53 times the 0.025 ppm F in mother's milk. If the processed infant food contains powdered beef bone, as some do, fluoride concentrations of 4–12 ppm F can be reached *before* domestic mixing with fluoridated water (39, 47). The relatively large amounts of fluoride supplied by such means have not yet been shown to pose a health or cosmetic hazard, such as teeth mottling (37), but they emphasize the need to watch total F intakes by individuals in fluoridated communities, especially with the growing proportion of processed foods in the dietaries of the Western world.

2. Fluorine in Animal Feeds and Pastures

Most plant species have a limited capacity to absorb fluorine from the soil, even when fluoride-containing fertilizers are applied (50). Uncontaminated pasture plants, forages, and grains are characteristically low in fluorine. In two studies of English pastures the fluorine concentrations ranged from 2 to 12 ppm

(2) and from 2 to 16 ppm (mean 5.3) on the dry basis (3). Most of the Australian grasses examined by Harvey (51) contained only 1–2 ppm F (d.b.), with some samples that had been irrigated with naturally fluoridated water reaching 9–13 ppm. Cereals and other grains and their by-products usually contain 1–3 ppm F (67, 87).

Feeds of animal origin, if free from bone, are also low in fluorine since the dry matter of the soft tissues and fluids of the body rarely contains more than 2–4 ppm F. Fish meals provide an exception as they normally contain 5–10 ppm F (d.b.). Most of the mineral phosphates used as dietary Ca and P supplements are substantial sources of fluoride to livestock. North African and North American rock phosphates usually contain 3–4% and those from Pacific and Indian Ocean island deposits half this concentration or less. During the manufacture of superphosphate and dicalcium phosphate 25–50% of the original fluoride is lost, so that these materials are much poorer fluoride sources. Special defluorinating processes can be used to reduce the fluoride levels in mineral phosphates still further.

In endemic fluorosis areas the main fluoride source is either the water or the dust-contaminated herbage, or both. The chronic fluorosis of sheep, cattle, and horses, known locally as "darmous," that occurs in parts of North Africa is caused by contamination of the herbage and water supplies with dusts blown from rock phosphate deposits and mines (151). In other endemic fluorosis areas the main source of fluoride is the naturally fluoridated drinking water derived from deep wells or bores, which commonly contain 3–5 ppm F and not infrequently 10–15 ppm F or more (24, 51).

V. FLUORINE TOXICITY

1. Fluorosis in Animals

a. Clinical and Pathological Signs. The ingestion by animals of amounts of F that may eventually prove toxic at first produce no observable ill effects. During this latent period the animal is protected by two physiological mechanisms that act concurrently. These are (*i*) a prompt rise in urinary F excretion to a maximum that cannot easily be surpassed and, (*ii*) deposition of retained fluoride in the skeleton—a process that initially proceeds rapidly and then more slowly as the F levels in the bones rise, until a stage of saturation is reached. Beyond this point, which approximates 30–40 times the F level in normal bone, "flooding" of the soft tissues with F occurs, plasma F levels rise, and metabolic breakdown occurs. There is then a voluntary refusal of food so that typical starvation phenomena are imposed on the signs of fluorosis (115). This sequence of events refers to chronic fluorosis conditions. At higher F intakes the latent period is reduced and there can be an almost immediate effect on growth and appetite (94).

Clinical signs of fluorosis may appear well before bone saturation is reached, if the F intake is sufficient to produce a significant rise in plasma F levels. Tooth formation and appetite are extremely sensitive to a rise in plasma F concentrations (17, 132). Anemia has also been reported to be characteristic of fluorosis at high intakes (20, 123), but in later studies of chronic fluorosis in cattle no significant changes in blood hemoglobin levels were observed (55, 59).

Animals exposed to excess F prior to the eruption of the permanent teeth develop dental defects which are the most sensitive indicators of elevated intakes and plasma F levels. The teeth become modified in shape, size, color, orientation, and structure. The incisors become pitted and the molars abraded. There may also be exposure of the pulp cavities due to fracture or wear. Once the permanent teeth are fully formed and erupted, their architecture is no longer susceptible to high-fluoride intakes (43, 152). However, the dentine of mature teeth, exposed after maturity, increases in fluoride content with time and dose, as do the bones (149).

Osseous lesions are similarly characteristic of fluorosis and can occur in animals exposed at any age. Exostoses develop, particularly of the jaw and long bones, usually accompanied by a thickening and change in shape of the bones. Such bones appear chalky, rough, and porous compared with normal bones. Mineralization of the tendons at the point of attachment to the long bones may also occur so that the joints become thickened and ankylosed. The animal then becomes stiff and lame, making movement difficult and painful.

From the detailed histopathological studies carried out with the sheep and rabbit by Weatherall and Weidmann (156) it appears that in skeletal fluorosis (a) bone resorption is a marked but not invariable feature of the condition, (b) the newly formed bone of exostoses is histologically similar to normal bone found in the fetus, (c) the exostoses of fluorotic bone are less well calcified than nonfluorotic bone and contain wide areas of osteoid, and (d) no area of hypermineralization occurs in any part of fluorotic bone. Overproduction of osteoid has been consistently observed in skeletal fluorosis (25, 109, 156), and there is evidence of a disturbance of mucopolysaccharide production in the bones and teeth of fluorosed pigs (10). These findings reveal F-induced defects in the mechanism of osteogenesis and chondrogenesis in the functioning bone cells.

Inappetence is not a serious feature of fluorosis during the latent period and neither the digestibility nor the utilization of the energy and protein of the diet is significantly depressed at this stage (129). Appetite, growth, and reproduction are impaired during the more advanced stages of the disease, or earlier if F intakes are high enough to maintain elevated plasma fluoride levels. The reduced milk production of cows receiving toxic levels of fluoride results from, and is in proportion to, the reduction in food consumption that occurs (148). Where dental lesions and bone abnormalities are present the willingness of the animal to gather and masticate fodder may be greatly reduced, so accentuating the inappetence. Reproductive performance, at least in sheep, is extremely resistant to toxic levels of fluoride as long as appetite is not seriously depressed (117). The poor lamb and calf crops of flocks and herds in fluorosis areas result from mortality of the newborn, due to the impoverished condition of the mother rather than to a failure of the reproductive process itself (51).

b. Enzyme Changes in Chronic Fluorosis. The well-known sensitivity of enzymic processes to the fluoride ion has stimulated a number of investigations designed to identify any changes in enzyme activity in fluorosis. Following their discovery that high dietary fat enhances F toxicity in rats (98), Sievert and Phillips (131) observed a marked decline in the fatty acid oxidase activity of the kidney, but not of the liver. Fecal lipid levels were then shown to be greatly increased in the fluorotic rat without any increase in metabolic fat or impairment of efficiency in utilizing free fatty acids (146). It was concluded that there is a partial inhibition of lipase activity in the intestine of the fluorotic rat, which could partly explain the high level of fecal fat. Subsequently it was found that a larger portion of a ^{14}C-labeled palmitate dose remains in the serum and liver lipids of fluoride-fed rats than occurs in normal animals. This suggests that fatty acid utilization is blocked (163). It has also been shown that fluoride in the presence of fat causes delayed gastric emptying (91). An impaired ability to metabolize fat thus appears as one of the biochemical lesions of fluorosis.

Disturbances in carbohydrate metabolism also occur in fluorosis. In fluorotic rats the level of liver glucose-6-phosphate dehydrogenase is decreased and glycogen turnover depressed (18, 163). It seems that these are indirect effects brought about by the reduced food consumption and "continual nibbling" pattern of food intake. Rats consuming as much as 450 ppm F in the diet catabolized glucose at a normal rate, but they were unable to metabolize glycogen normally, due to some effect at the liver enzyme level (164). Since the level of the adaptive hepatic enzymes regulating glycogen turnover can be varied by regulating food intake (140), it seems likely that the level of those enzymes is reduced in fluoride-fed animals as a consequence of the changed pattern of food intake, and that this accounts for the impaired glycogen turnover. Fluoride inhibition of HeLa cell growth is associated with a decreased efficiency of glucose utilization, although conversion of glucose to lactic acid and CO_2 production from glucose are unaffected (19). It appears that these alterations in carbohydrate metabolism may also be related to the slower growth rate brought about by the fluoride. However, fluoride produced a specific metabolic alteration in rapidly growing HeLa cells by decreasing the cellular ATP more rapidly and to a lower level than other metabolic inhibitors. Fluoride inhibition of growth and glycolysis in cultured strain L mouse fibroblasts has also been demonstrated (see Suttie *et al.*, 145). These workers also observed a fluoride-induced depression of enolase activity and a shift in the ratio of DPN and DPNH levels in these cells consistent with the enolase inhibition. They suggest that this shift to a new constant may be of key importance in the growth inhibitory effects of fluoride. An understanding of how the fluoride ion so drastically influences cellular metabolism will no doubt emerge as these and similar studies progress.

Since phosphatase is associated with normal bone formation, and because abnormal bone growth, characterized by excessive periosteal new bone formation, is a feature of fluorosis, the levels of alkaline phosphatase in the blood and tissues in relation to fluoride ingestion have been extensively studied. Some years ago Phillips (111) observed a marked stimulation of alkaline phosphatase in the blood of fluorotic cows. Subsequently, Motzok and Branion (102) reported similar increases in alkaline phosphatase levels in the serum and bones of fluorotic chicks. Olson *et al.* (107) obtained little correlation between the level of fluoride in the ration and the phosphatase activity in the blood of cattle, mainly because of high individual variability. Alkaline phosphatase activity in bone is more securely related to fluoride intakes than is this activity in blood. A close correlation between F ingested and the F content, osseous abnormalities, and alkaline phosphatase activity in the bones of cows and heifers has been demonstrated (97). In an experiment lasting over 7 years, dairy heifers 6 months old at the commencement were fed on rations containing 12, 27, 49, and 93 ppm F as NaF. Osseous abnormalities, excessive accumulation of fluoride in bone, and significant increases in alkaline phosphatase activity in bone occurred at the two higher intakes. No such changes were apparent at fluoride intakes of 12 or 27 ppm.

c. Fluorine Tolerance. Tolerance of fluorine ingestion is influenced by the age and species of the animal, the chemical form, duration, and continuity of the F ingested, and the nature and amount of the diet being consumed. At massive intakes, resulting in immediate F flooding of the soft tissues, these interacting factors are of minor importance. At lower intakes, typical of chronic fluorosis, tolerance levels are highly dependent on the extent to which one or more of the above factors are operating.

Poultry are more F-tolerant than other livestock. Maximum safe dietary levels of 300–400 ppm F as rock phosphate have been reported for growing chicks and 500–700 ppm F for laying hens (45, 46). Similar tolerance levels to those given for chicks have been obtained with growing female turkeys, but 200 ppm F as NaF resulted in decreased weight gains in young male turkeys (4). With sheep, cattle, and pigs 80–100 ppm F of the dry ration is on the border line of toxicity when fed as rock phosphate for long periods (113). When sodium fluoride is similarly administered to dairy cows 40 ppm F is near the margin of tolerance, and signs of fluorosis appear within 3–5 years at 50 ppm F from this source (148). When the cows were first exposed to 50 ppm F as NaF at a mature 4–6 years of age no adverse effects were apparent through three lactations, other than mild exostosis of the long bones (143). From the results of another experiment, beginning with young calves and lasting for 7 years, it was concluded that the tolerance for soluble fluoride is not more than 30 ppm F of the total dry diet (130). Suggested safe ration levels for different farm species taken from the publication of Phillips *et al.* (112) are given in Table 49. Practical permissible limits for cattle dependent on forages contaminated with inorganic fluorides from industrial sources are proposed by Suttie (142) as follows: a yearly average of 40 ppm F (d.b.) in the herbage based on monthly sampling; 60 ppm F for not more than 2 consecutive months; or 80 ppm F for not more than 1 month, even though the yearly average does not exceed 40 ppm F.

The minimum toxic levels of water-borne fluoride for livestock under field conditions vary with the amounts of water consumed, its continuity, and the age of the animals exposed. Artesian bore water containing 5 ppm F induced severe dental abnormalities and other signs of fluorosis in sheep under the hot climatic conditions of Queensland (51). By contrast, Pierce (117) observed no ill effects on health, food consumption, wool production, or dental development in mature sheep initially $2\frac{1}{2}$–$3\frac{1}{2}$ years of age given drinking water containing 20 ppm F as NaF. When fluoridated water was given for $3\frac{1}{2}$ years to younger sheep 10–11 months of age at the start of the experiment, mottling of the teeth was apparent at 5 ppm F or more. These experiments show clearly that F tolerance is vitally affected by the age of exposure of the animals and by the continuity of the intake. In the experiments of Pierce actual consumption of the fluoridated water would be low during the cool, wet months of winter and spring, so that

TABLE 49
Safe Levels of Fluorine in the Total Ration of Livestock[a]

Species	NaF or other soluble fluoride (ppm F)	Rock phosphates or phosphatic limestones (ppm F)
Dairy cow	30–50	60–100
Beef cow	40–50	65–100
Sheep	70–100	100–200
Swine	70–100	100–200
Chickens	150–300	300–400
Turkeys	300–400	–

[a]From Phillips *et al.* (112).

the total annual F intake would be lower than from waters of comparable level under subtropical conditions, as in Queensland.

The importance of continuity of intake to F tolerance is further apparent from experiments comparing continuous and intermittent fluoride dosage in the rat (74), sheep (51), and cow (17). During periods of low intake the exchangeable skeletal stores of fluoride are depleted and excreted in the urine, thus providing the opportunity for further skeletal F immobilization during subsequent periods of high intake. When compared on the basis of similar total yearly intake, skeletal storage of fluoride is similar for continuous and intermittent exposure (17). For example, cattle receiving elevated levels of dietary fluoride (1.5 mg F/kg body weight/day or 40–50 ppm F on the dry ration) for 6 months of the year stored only half as much fluoride, as measured by vertebral biopsy, as animals on this intake for the whole year, but those receiving fluoride in periods of high and then low exposure stored the same amount as those on a continuous intake of the same yearly average (17). Periodic exposure nevertheless resulted in a more dynamic metabolism of fluoride in skeletal tissue, with rapid increases in F content during periods of high ingestion and rapid losses during periods of low-F intake. With short-term ingestion of high levels, systemic reactions such as weight loss and unthriftiness due to decreased appetite may arise, as in the experiments of Carlson (17) during periods of intake at 3.0 mg F/kg body weight/day or 90 ppm F in the dry ration.

Aspects of F tolerance in relation to the chemical form of the element and the nature of the basal diet were considered in Section II,1. The significance of the chemical form was also apparent when considering the greater tolerance of animals to fluoride from rock phosphate than from sodium fluoride. In cattle the F in sodium fluoride is approximately twice as toxic as that in rock phosphate (143, 148). In a direct comparison with dairy cattle of the toxicity of NaF, CaF_2, and the fluoride residue on industrially contaminated hay, in which the F in each source was fed at a level of 65 ppm of the ration, little difference in the toxicity of the NaF and the contaminated hay was observed, but the CaF_2 was only about half as available as the other two sources, as judged by F retention in the skeleton and by blood and urine F levels (129). At a dietary level of 14 ppm F, or similar low levels, there is little difference in tolerance between soluble and insoluble forms of fluorine, judging by their effects on the teeth of rats (73). Apparently under these conditions the volume of digestive fluids is sufficient to dissolve even the more insoluble compounds. Particle size can also have a bearing on toxicity. The finer the particle size the more nearly the toxicity of the F in cryolite approaches that of NaF (73), and both NaF and cryolite promote greater fluoride retention in the rat when provided in the drinking water than the same amounts in the dry diet (74, 158).

The influence of several cations, notably calcium and aluminum, and the fat content of the diet, on tolerance to fluoride has already been mentioned in Section II,1. More recently Suttie and Faltin (144) showed that general under-nutrition also tends to increase the toxic effects of fluoride in cattle. Two groups of heifers 14 weeks old were fed for 58 months on either a full ration or one supplying only 60% of the recommended total digestible nutrients (TDN). Although the fluoride intake on a body weight basis of the heifers on the limited ration was somewhat less during the experiment, this group had a slightly greater skeletal F retention and a greater degree of dental fluorosis after 58 months than those on the full ration. The F level in the total ration was equivalent in each case to 40 ppm F fed as NaF.

d. Diagnosis of Fluorosis. The criteria that are important in recognizing the developing syndrome of fluorine toxicosis have been placed by Phillips *et al.* (112) in the following order of reliability: (*i*) Chemical analyses which indicate an increased amount of fluorine in the diet, urine, bones, and teeth; (*ii*) tooth effects such as chalkiness or mottling, erosion of enamel, enamel hypoplasia, and excessive wear; (*iii*) systemic evidence as reflected by anorexia, inanition, cachexia, exostoses, and bone changes. Plasma fluoride levels and bone alkaline phosphatase activity can now be added to the first of the criteria, as discussed in Section I,2. The F levels in the diet, urine, and bones that relate to freedom from different degrees in fluorosis in the animal were considered in Sections I,3, I,4, and V,1,c. The dental defects mentioned previously as the second criterion apply only to animals exposed to toxic F intakes during the period of tooth formation.

According to Suttie (142), "analysis of bone fluoride content and an estimation of the age of the animal, would perhaps give the best indication of potential damage to an animal. This measurement would be independent of day to day variation in intake and would stress the accumulative nature of the disease process." Fluoride determination on tail bones obtained by a simple and safe biopsy technique provides a useful means of measuring bone fluoride accumulations in cattle, for either diagnostic or experimental purposes (14). Suttie (141) compared the fluoride concentration of the coccygeal vertebrae and the large metacarpal of 114 cattle which had been subjected to a wide range of fluoride intakes. An extremely high correlation was found between the values of the two bones, the fluoride concentration of the dry, fat-free metacarpus being approximately 50% of the vertebrae ash. Since the fluoride content of the large metacarpal bone is well related to other more subjective signs of fluorosis, fluoride determinations of vertebrae biopsy specimens clearly represent a valuable diagnostic aid in bovine fluorosis.

2. Mottled Enamel in Man

Mottled teeth are characterized by chalky white patches distributed irregularly over the surface of the tooth, with a secondary infiltration of yellow to brown staining. The enamel is structurally weak and in severe cases there is a loss of enamel accompanied by "pitting," which gives the tooth surface a corroded appearance. Changes in the size or shape of the affected teeth are rare in mottled enamel areas. The conditions may be mild, moderate, or severe, depending on the level of intake of fluoride and individual susceptibility. Mottled enamel is almost entirely confined to the permanent teeth and develops only during their period of formation. The fully formed enamel of adult teeth is unaffected by fluorine and the deciduous teeth are only affected at high-fluoride intakes where other signs of fluorosis are evident.

In its mild forms mottled enamel has little public health significance. In the more severe forms, involving enamel hypoplasia and pitting, the unsightly appearance of the mouth may be accentuated by excessive wear on the teeth, and mastication can be affected. At the F levels common in most mottled enamel areas, the dental lesions are the only F-induced defects apparent. The growth and health of the individual are not affected, and disturbances seriously affecting the skeletal bones or joints do not develop. In areas with highly fluoridated waters (over 8 ppm F) not only is there a high incidence of severe mottled enamel, but disturbances in ossification and systemic signs of fluorosis can develop, similar to those in individuals suffering from severe industrial skeletal fluorosis. Numerous surveys in several countries have established a relationship between the levels of fluoride in the drinking water and the incidence and severity of mottled enamel in children. Dean (27) studied 5824 children 12–14 years of age, none of whom had been away from their water supply for periods longer than 1 month, in 22 cities in 10 states of the United States. As the fluoride content of the drinking water increased, the "community index," i.e., the average index of mottling based on the two teeth most severely affected in each child, increased along an S-shaped curve. At about 2–3 ppm F in the water the community index corresponded to "very mild" mottling, at about 4 ppm to "mild" mottling, at 5–6 ppm to "moderate" mottling, and even at concentrations of 14 ppm the index lay between "moderate" and "severe" mottling. In a later study Hodge (56) found that the severity of mottling increased in a linear fashion when the community index of dental fluorosis was plotted against the logarithm of the parts per million of fluoride in the drinking water from about 1 ppm to nearly 10 ppm. Below 1 ppm a second straight, nearly horizontal line appeared, indicating practically no difference in the community index whether the water contains 0.1 or 1.0 ppm F.

Classifying drinking waters of differing fluoride concentrations in terms of their mottled enamel potential presents problems similar to those that prevail with respect to caries inhibition. These problems relate to differences in the amounts of fluid consumed and in the amounts and types of foods consumed by individuals or groups, both of which obviously influence total daily fluoride intakes. The former is affected by climate and occupation and the latter most particularly by the degree of dependence of the individual on tea and marine fish in his daily diet. A comparison of the dental situation in two towns in the

United States with water supplies of similar fluoride content, one with a mean annual temperature of 50°F and the other 70°F, revealed a strikingly higher incidence of mottling in children in the town with the hotter climate (42). In a town in Western Australia with a hot, dry climate and a water supply containing 1.5 ppm F a significant proportion of the children examined exhibited mild to moderate dental fluorosis (65). The authors point out that in this environment with a high water consumption and substantial amounts of fish in the diet, a safe and effective fluoride level in the water supply would probably be 0.6 ppm.

The primary defect of the enamel in dental fluorosis is permanent. It is possible to temporarily bleach the disfiguring stain but there is no known cure once defective enamel is formed. Prevention is possible by changing the water supply to low-fluoride sources or by removal of enough fluoride to maintain the concentration below 2 ppm. The great affinity of fluoride for the tertiary phosphates can be exploited as a means of removing fluoride, and practical control of excessive levels has been applied successfully to several community water supplies (see Nichols, 105).

3. Genu Valgum in Man

The dental fluorosis, described in the preceding section, in most parts of the world is not accompanied by other disabilities associated with fluorine. However, in areas of severe fluorosis, as occur in parts of India and South Africa, the dental mottling is accompanied by skeletal manifestations, including osteosclerosis and calcification of ligaments and tendinous insertions, and leading to crippling deformities such as kyphosis, stiffness of the spine, and bony exostoses. These crippling skeletal changes have appeared mainly in adults after 30–40 years residence in the endemic area. More recently the widespread occurrence of genu valgum (knock knee) in individuals 10–25 years of age, but not in the older population, has been reported in endemic fluorosis areas in southern India (71a). The etiology of this rare syndrome of fluoride toxicity and the reasons for its recent appearance are obscure. Genu valgum is seen only among the poor whose staple diet is sorghum, or in some cases rice, and whose diet is low in calcium and other protective foods and high in molybdenum. The relatively high-Mo content of sorghum grain was pointed out in Chapter 4, but its significance, if any, in relation to genu valgum is unknown.

REFERENCES

1. Aeschliman, M.I., Grant, J.A., and Crigler, J.F., *Metab., Clin. Exp.* **15,** 905 (1966).
2. Allcroft, R., *Inst. Biol. (England) Symp.* **8,** 95 (1959).
3. Allcroft, R., Burns, K.N., and Herbert, C.N., "Fluorosis in Cattle," Anim. Dis. Surv. Rep. No. 2. HM Stationery Office, London, 1965.

4. Anderson, J.O., Hurst, J.S., Strong, D.C., Nielsen, H., Greenwood, D.A., Robinson, W., Shupe, J.L., Binns, W., Bagley, R.A., and Draper, C.I., *Poultry Sci.* **34**, 1147 (1955).
5. Armstrong, W.D., Singer, L., and Vogel, J.J., *Fed. Proc., Fed. Am. Soc. Exp. Biol.* **25**, 696 (1966).
6. Arnold, F.A., Jr., *Am. J. Public Health* **47**, 539 (1957).
7. Arnold, F.A., Jr., McClure, F.J., and White, C.L., *Dent. Prog.* **1**, 8 (1960).
8. Auermann, E., and Borris, W., *Caries Res.* **5**, 11 (1971).
9. Becker, D.E., Griffith, J.M., Hobbs, C.S., and McIntyre, W.H., *J. Anim. Sci.* **9**, 647 (1950).
10. Bélanger, L.F., Visek, W.J., Lotz, W.E., and Comar, C.L., *Am. J. Pathol.* **34**, 25 (1958).
11. Bell, M.C., Merriman, G.M., and Greenwood, D.A., *J. Nutr.* **73**, 379 (1961).
12. Bernstein, D.S., Sadowsky, N., Hegsted, D.M., Guri, C.D., and Stare, F.J., *J. Am. Med. Assoc.* **198**, 499 (1966).
13. Bixler, D., and Muhler, J.C., *J. Nutr.* **70**, 26 (1960).
14. Burns, K.N., and Allcroft, R., *Res. Vet. Sci.* **3**, 215 (1962).
15. Buttner, W., and Muhler, J.C., *J. Nutr.* **63**, 263 (1957); **65**, 259 (1958).
16. Carlson, C.H., Armstrong, W.D., and Singer, L., *Am. J. Physiol.* **199**, 187 and 199 (1960).
17. Carlson, J.R., Doctoral Thesis, University of Wisconsin, Madison (1966).
18. Carlson, J.R., and Suttie, J.W., *Am. J. Physiol.* **210**, 79 (1966).
19. Carlson, J.R., and Suttie, J.W., *Exp. Cell Res.* **45**, 415 and 423 (1967).
20. Cass, J.S., *J. Occup. Med.* **3**, 471 and 527 (1961).
21. Cass, R.M., Croft, J.D., Perkins, P., Nye, W., Waterhouse, C., and Terry, R., *Arch. Intern. Med.* **118**, 111 (1966).
22. Chang, C.Y., Phillips, P.H., and Hart, E.B., *J. Dairy Sci.* **17**, 695 (1934).
23. Chann, M.M., Rucker, R.B., Zeman, F., and Riggins, R.S., *J. Nutr.* **103**, 1431 (1973).
24. Cholak, J., *J. Occup. Med.* **1**, 501 (1959).
25. Comar, C.L., Visek, W.J., Lotz, W.E., and Rust, J.H., *Am. J. Anat.* **93**, 361 (1953).
26. Cook, H.A., *Vitalst. Zivilisationskr.* **14**, 244 (1969).
27. Dean, H.T., *in* "Fluorine and Dental Health" (F.R. Moulton, ed.), pp. 6–11 and 23–31. Am. Assoc. Adv. Sci., Washington, D.C., 1942.
28. Dirks, O.B., *Caries Res.* **8**, Suppl. 1, 2 (1974).
29. Doberenz, A.R., Kurnich, A.A., Kurtz, E.B., Kemmerer, A.R., and Reid, B.L., *Proc. Soc. Exp. Biol. Med.* **117**, 689 (1964).
30. Eager, J.M., *Public Health Rep.* **16**, 2576 (1901).
31. Ege, R., *Tandlaegebladet* **65**, 445 (1961), quoted by Ericsson (34).
32. Elliott, C.G., and Smith, M.D., *J. Dent. Res.* **39**, 93 (1960).
33. Ericsson, Y., *Acta Odontol. Scand* **16**, 51 and 127 (1958).
34. Ericsson, Y., *J. Nutr.* **96**, 60 (1968).
35. Ericsson, Y., *Caries Res.* **3**, 159 (1969).
36. Ericsson, Y., *Caries Res.* **8**, Suppl. 1, 16 (1974).
37. Ericsson, Y., and Ribelius, V., *Caries Res.* **5**, 78 (1971).
38. Fanning, E.A., Gotjamanos, T., Vowles, N.J., Cellier, K.M., and Simmons, D.W., *Med. J. Aust.* **1**, 383 (1967).
39. Farkas, C.S., and Farkas, E.J., *Sci. Total Environ.* **2**, 399 (1974).
40. Feltman, R., and Kosel, G., *J. Dent. Med.* **16**, 190 (1961).
41. Forrest, J.R., Parfitt, G.J., and Bransby, E.R., *Mon. Bull. Minist. Health Public Health Lab. Serv. (G.B.)* **10**, 104 (1951).

42. Galagan, D.J., and Lamson, C.G., *Public Health Rep.* **68,** 497 (1953); Galagan, D.J., and Vermillion, J.R., *ibid.* **72,** 491 (1957).
43. Garlick, N.L., *Am. J. Vet. Res.* **16,** 38 (1955).
44. Gedalia, J., Brzezinski, A., and Bercovici, B., *J. Dent. Res.* **38,** 548 (1959).
45. Gerry, R.W., Carrick, C.W., Roberts, R.E., and Hauge, S.M., *Poultry Sci.* **26,** 323 (1947); 28, 19 (1949).
46. Halpin, J.G., and Lamb, A.R., *Poultry Sci.* **11,** 5 (1932).
47. Ham, M.P., and Smith, M.D., *J. Nutr.* **53,** 215 and 225 (1954).
48. Hamilton, I.R., *Can. J. Microbiol.* **15,** 1013 (1969).
49. Hardwick, J.L., Fremlin, J.H., and Mathieson, J., *Br. Dent. J.* **104,** 47 (1958).
50. Hart, E.B., Phillips, P.H., and Bohstedt, G., *Am. J. Public Health* **24,** 936 (1934).
51. Harvey, J.M., *Queensl. J. Agric. Sci.* **9,** 47 (1952); **10,** 127 (1953).
52. Henrikson, P., Lutwak, L., Krook, L., Skogerboe, R., Kallfelz, F., Bélanger, L.F., Marier, J.R., Sheffy, B.E., Romanus, B., Hirsch, C., *J. Nutr.* **100,** 631 (1970).
53. Higgins, B.A., Nassim, J.R., Alexander, R., and Hibb, A., *Br. Med. J.* **1,** 1159 (1965).
54. Hobbs, C.S., Moorman, R.P., Griffith, J.M., Merriman, G.M., Hansard, S.L., and Chamberlain, C.C., *Tenn., Agric., Exp. Stn., Bull.* **235** (1954).
55. Hobbs, C.S., and Merriman, G.M., *Tenn., Agric. Exp. Stn., Bull.* **351** (1962).
56. Hodge, H.C., *J. Am. Dent. Assoc.* **40,** 436 (1950).
57. Hodge, H.C., *Trans. Conf. Metab. Interrelations 4th* (1953).
58. Hodge, H.C., and Smith, F.A., "Fluorine Chemistry" (J.H. Simon, ed.), Vol. IV. Academic Press, New York, 1965.
59. Hoogstraten, B., Leone, N.C., Shupe, J.L., Greenwood, D.A., and Lieberman, J., *J. Am. Med. Assoc.* **192,** 26 (1965).
60. Howink, B., Dirks, O.B., and Kwant, G.W., *Caries Res.* **8,** 27 (1974).
60a. Inkovaara, J., Heikinheimo, R., Jarvinen, K., Kasurinen, U., Hanhijaru, H., and Iisalo, E., *Br. Med. J.* **3,** 73 (1975).
61. Isaac, S., Brudevold, F., Smith, F.A., and Gardner, D.E., *J. Dent. Res.* **37,** 254 (1958).
62. Jackson, D., and Weidmann, S.M., *J. Pathol. Bacteriol.* **76,** 451 (1958).
63. Jenkins, G.N., *Arch. Oral Biol.* **1,** 33 (1959).
64. Jowsey, J., Riggs, B.L., Kelly, P.J., and Hoffman, D.L., *Am. J. Med.* **53,** 43 (1972).
65. Kailis, D.G., and Silva, D.G., *Aust. Dent. J.* **12,** 304 (1967).
66. Kick, C.H., Bethke, R.M., Edginton, B.H., Wilder, O.H.M., Record, R., Wilder, W., Hill, T.J., and Chase, S.M., *Ohio, Agric. Exp. Stn., Bull.* **558,** (1935).
67. Kirchgessner, M., Weser, U., Friesecke, H., and Oelschlager, W., *Z. Tierphysiol. Tierernaehr. Futtermittelkd.* **18,** 251 (1963).
68. Knutson, J.W., and Armstrong, W.D., *Public Health Rep.* **58,** 1701 (1943); **60,** 1085 (1945); **61,** 1683 (1946); **62,** 425 (1947).
69. Kramer, L., Osis, D., Wiatrowski, E., and Spencer, H., *Am. J. Clin. Nutr.* **27,** 590 (1974).
70. Krasse, B., *Arch. Oral Biol.* **11,** 429 (1966).
71. Krepkogorsky, L.N., *Gig. Sanit.* **28,** 30 (1963), quoted by Marier and Rose (81).
71a. Krishnamachari, K.A.V.R., and Krishnaswamy, K., *Lancet* **2,** 877 (1973); *Indian J. Med. Res.* **62,** 1415 (1974).
72. Largent, E.J., and Heyroth, F.F., *J. Ind. Hyg. Toxicol.* **31,** 134 (1949).
73. Lawrenz, M., and Mitchell, H.H., *J. Nutr.* **22,** 451 and 621 (1941).
74. Lawrenz, M., Mitchell, H.H., and Ruth, W.A., *J. Nutr.* **19,** 531 (1940); **20,** 383 (1940).
75. Leone, N.C., Stevenson, C.A., Besse, B., Hawes, L.E., and Dawber, T.R., *AMA Arch. Ind. Health* **21,** 326 (1960).

76. Likins, R.C., McClure, F.J., and Steere, A.C., *Public Health Rep.* **71**, 217 (1956).
77. Lukert, B.P., Bolinger, R.E., and Meek, J.C., *J. Clin. Endocrinol. Metab.* **27**, 828 (1967).
78. Machle, W., and Largent, E.J., *J. Ind. Hyg. Toxicol.* **25**, 112 (1943).
79. Machle, W., Scott, E.W., and Largent, E.J., *J. Ind. Hyg. Toxicol.* **24**, 199 (1942).
80. Maplesden, D.C., Motzok, I., Oliver, W.T., and Branion, H.D., *J. Nutr.* **71**, 70 (1960).
81. Marier, J.R., and Rose, D., *J. Food Sci.* **31**, 941 (1966).
82. Marthaler, M., *Schweiz. Bull. Eidg. Gesundheitsamtes, Part B,* No. 2 (1962).
83. Martin, D.J., *J. Dent. Res.* **30**, 676 (1951).
84. Maurer, R.L., and Day, H.G., *J. Nutr.* **62**, 561 (1957).
85. McCann, H.G., and Bullock, F.A., *J. Dent. Res.* **36**, 391 (1957).
86. McClendon, F.J., and Gershon-Cohen, J., *J. Agric. Food Chem.* **1**, 464 (1953).
87. McClure, F.J., *Public Health Rep.* **64**, 1061 (1949).
88. McClure, F.J., and Likins, R.C., *J. Dent. Res.* **30**, 172 (1951).
89. McClure, F.J., McCann, H.G., and Leone, N.C., *Public Health Rep.* **73**, 741 (1958).
90. McClure, F.J., Mitchell, H.H., Hamilton, T.S., and Kinser, C.A., *J. Ind. Hyg. Toxicol.* **27**, 159 (1945).
91. McGown, E.L., and Suttie, J.W., *J. Nutr.* **104**, 909 (1974).
92. McKay, F.S., *Am. J. Public Health* **38**, 828 (1948).
93. Messer, H.H., Armstrong, W.D., and Singer, L., *Science* **177**, 893 (1972).
94. Messer, H.H., Armstrong, W.D., and Singer, L., *J. Nutr.* **103**, 1319 (1973).
95. Messer, H.H., Wong, K., Wegner, M.E., Singer, L., and Armstrong, W.D., *Nature (London), New Biol.* **240**, 218 (1972).
96. Messer, H.H., Armstrong, W.D., and Singer, L., *in* "Trace Element Metabolism in Animals, 2" (W.G. Hoekstra *et al.*, eds.), p. 425. Univ. Park Press, Baltimore, Maryland, 1974.
97. Miller, G.W., and Shupe, J.L., *Am. J. Vet. Res.* **23**, 24 (1962).
98. Miller, R.F., and Phillips, P.H., *J. Nutr.* **51**, 273 (1953); **56**, 447 (1955); **59**, 425 (1956).
99. Milne, D.B., and Schwarz, K., *in* "Trace Element Metabolism in Animals, 2" (W.G. Hoekstra *et al.*, eds.), p. 710. Univ. Park Press, Baltimore, Maryland, 1974.
100. Molan, P.C., and Hartles, R.L., *Arch. Oral Biol.* **11**, 1163 (1966).
101. Moller, I.J., *Ugeskr. Laeg.* **131**, 2136 (1969), quoted by R. Fond, *Int. J. Environ. Stud.* **5**, 87 (1973).
102. Motzok, I., and Branion, H.D., *Poultry Sci.* **37**, 1469 (1958).
103. Muhler, J.C., Stookey, G.R., Spear, L.B., and Bixler, D., *J. Oral. Ther. Pharmacol.* **2**, 241 (1966).
104. Naylor, M.N., *Health Educ. J.* **28**, 136 (1969).
105. Nichols, M.S., *in* "Fluoridation as a Public Health Measure" (J.H. Shaw, ed.). Am. Assoc. Adv. Sci., Washington, D.C., 1954.
106. Ockerse, T., *in* "Dental Caries and Fluorine." Am. Assoc. Adv. Sci., Washington, D.C., 1946.
107. Olson, L.E., Nielsen, H.M., Shupe, J.L., and Greenwood, D.A., *J. Pharmacol. Exp. Ther.* **122**, 7871 (1958).
108. Pandit, C.G., Raghavacheri, T.N., Rao, D.S., and Krishnamurti, V., *Indian J. Med. Res.* **28**, 533 (1940).
109. Pendborg, J.J., and Plum, C.M., *Acta Pharmacol. Toxicol.* **2**, 294 (1946).
110. Perkinson, J.D., Whitney, I.B., Monroe, R.A., Lotz, W.E., and Comar, C.L., *Am. J. Physiol.* **182**, 383 (1955).
111. Phillips, P.H., *Science* **76**, 239 (1932).
112. Phillips, P.H., Greenwood, D.A., Hobbs, C.S., Huffman, C.F., and Spencer, G.R.,

N.A.S.–N. R. C., Publ. **824** (1960).
113. Phillips, P.H., Hart, E.B., and Bohstedt, G., *Wis. Agric. Exp. Stn., Bull.* **123,** (1934).
114. Phillips, P.H., Halpin, J.G., and Hart, E.B., *J. Nutr.* **10,** 93 (1935).
115. Phillips, P.H., and Suttie, J.W., *Arch. Ind. Health* **21,** 343 (1960).
116. Phillips, P.H., Suttie, J.W., and Zebrowski, E.J., *J. Dairy Sci.* **46,** 513 (1963).
117. Pierce, A.W., *Aust. J. Agric. Res.* **3,** 326 (1952); **5,** 545 (1954); **10,** 186 (1959).
118. Purves, M.S., *Lancet* **2,** 1188 (1962).
119. Ramseyer, W.F., Smith, C., and McCay, C.M., *J. Gerontol.* **12,** 14 (1957).
120. Rich, C.A., Ensinck, J., and Avanovich, P., *J. Clin. Invest.* **43,** 545 (1966).
121. Rose, G.A., *Proc. R. Soc. Med.* **58,** 436 (1965).
122. Russell, A.L., and Elvove, E., *Public Health Rep.* **66,** 1389 (1951).
122a. Savchuk, W.B., and Armstrong, W.D., *J. Biol. Chem.* **193,** 575 (1951).
123. Schmidt, H.J., and Rand, W.E., *Am. J. Vet. Res.* **13,** 38 (1952).
124. Schroeder, H.A., Mitchener, M., Balassa, J.J., Kanisawa, M., and Nason, A.P., *J. Nutr.* **95,** 95 (1968).
125. Schwarz, K., and Milne, D.B., *Bioinorg. Chem.* **1,** 331 (1972).
126. Sharpless, G.R., and McCollum, E.V., *J. Nutr.* **6,** 163 (1933).
126a. Shaw, J.H., and Griffiths, D., *Arch. Oral Biol.* **5,** 301 (1961).
127. Shearer, T.R., and Suttie, J.W., *Am. J. Physiol.* **212,** 1165 (1967).
128. Shupe, J.L., Harris, L.E., Greenwood, D.A., Butcher, J.E., and Nielsen, H.M., *Am. J. Vet. Res.* **24,** 300 (1963).
129. Shupe, J.L., Miner, M.L., Harris, L.E., and Greenwood, D.A., *Am. J. Vet. Res.* **23,** 777 (1962).
130. Shupe, J.L., Miner, M.L., Greenwood, D.A., Harris, L.E., and Stoddart, G.E., *Am. J. Vet. Res.* **24,** 964 (1963).
131. Sievert, A.H., and Phillips, P.H., *J. Nutr.* **68,** 109 (1959).
132. Simon, G., and Suttie, J.W., *J. Nutr.* **94,** 511 (1968); **96,** 152 (1968).
133. Singer, L., and Armstrong, W.D., *J. Appl. Physiol.* **15,** 508 (1960).
134. Singer, L., and Armstrong, W.D., *Anal. Biochem.* **10,** 495 (1965).
135. Singer, L., and Armstrong, W.D., *in* "Trace Element Metabolism in Animals, 2" (W.G. Hoekstra *et al.*, eds.), p. 698. Univ. Park Press, Baltimore, Maryland, 1974.
136. Spencer, H., Lewin, I., Fowler, J., and Samachson, J., *Am. J. Clin. Nutr.* **22,** 381 (1969).
137. Spencer, H., Osis, D., Wiatrowski, E., and Samachson, J., *J. Nutr.* **100,** 1415 (1970).
138. Spencer, H., Osis, D., and Wiatrowski, E., *Trace Subst. Environ. Health–7, Proc. Univ. Mo. Annu. Conf., 7th, 1973,* p. 289 (1974).
139. Spencer, H., *in* "Trace Element Metabolism in Animals, 2" (W.G. Hoekstra *et al.*, eds.), p. 696 Univ. Park Press, Baltimore, Maryland, 1974.
140. Steiner, D.F., Rauda, V., and Williams, R.H., *J. Biol. Chem.* **236,** 299 (1961).
141. Suttie, J.W., *Am. J. Vet. Res.* **28,** 709 (1967).
142. Suttie, J.W., *J. Air Pollut. Control Assoc.* **14,** 461 (1964); **19,** 239 (1969).
143. Suttie, J.W., and Phillips, P.H., *J. Dairy Sci.* **42,** 1063 (1959).
144. Suttie, J.W., and Faltin, E.C., *Am. J. Vet. Res.* **34,** 479 (1973).
145. Suttie, J.W., Drescher, M.P., Quissell, D.O., and Young, K.L., *in* "Trace Element Metabolism in Animals, 2" (G.W. Hoekstra, *et al.,* eds.), p. 327. Univ. Park Press, Baltimore, Maryland, 1974.
146. Suttie, J.W., and Phillips, P.H., *J. Nutr.* **72,** 429 (1960).
147. Suttie, J.W., Gesteland, R., and Phillips, P.H., *J. Dairy Sci.* **44,** 2250 (1961).
148. Suttie, J.W., Miller, R.F., and Phillips, P.H., *J. Nutr.* **63,** 211 (1957); *J. Dairy Sci.* **40,** 1485 (1957).
149. Suttie, J.W., Phillips, P.H., and Miller, R.F., *J. Nutr.* **65,** 293 (1958).

150. Taylor, T.G., and Kirkley, J., *Calcif. Tissue Res.* **1**, 33 (1967).
151. Velu, H., *C.R. Seances Soc. Biol. Ses Fis.* **108**, 750 (1931); **127**, 854 (1938).
152. Venkataramanan, K., and Krishnaswamy, N., *Indian J. Med. Res.* **37**, 277 (1949).
153. Venkateswardu, P., and Narayanarao, D., *Indian J. Med. Res.* **45**, 387 (1957).
154. Waldblott, G.L., *Am. J. Clin. Nutr.* **12**, 455 (1963).
155. Wallace-Durbin, P., *J. Dent. Res.* **33**, 789 (1954).
156. Weatherall, J.A., and Weidmann, S.M., *J. Pathol. Bacteriol.* **78**, 233 (1959).
157. Weber, C.W., and Reid, B.L., *in* "Trace Element Metabolism in Animals, 2" (G. W. Hoekstra *et al.*, eds.), p. 707. Univ. Park Press, Baltimore, Maryland, 1974.
158. Weddle, D.A., and Muhler, J.C., *J. Nutr.* **54**, 437 (1954).
159. Weidmann, S.M., Weatherall, J.A., and Whitehead, R.G., *J. Pathol. Bacteriol.* **78**, 435 (1959).
160. Wespi, H.J., *Schweiz. Bull. Eidg. Gesundheitsamtes, Part B,* No. 2 (1962).
161. Wrodek, G., *Zahnaerztl. Mitt.* **47**, 258 (1959), quoted by Hodge and Smith (58).
162. Wuthier, R.E., and Phillips, P.H., *J. Nutr.* **67**, 581 (1959).
163. Zebrowski, E.J., Suttie, J.W., and Phillips, P.H., *Fed. Proc., Fed. Am. Soc. Exp. Biol.* **23**, 184 (1964).
164. Zebrowski, E.J., and Suttie, J.W., *J. Nutr.* **88**, 267 (1966).
165. Zinner, D.D., Jablon, J.M., Aran, A.P., and Saslaw, M.S., *Proc. Soc. Exp. Biol. Med.* **118**, 766 (1965).
166. Zipkin, I., and Likins, R.C., *Am. J. Physiol.* **191**, 549 (1957).
167. Zipkin, I., Likins, R.C., McClure, F.J., and Steere, A.C., *Public Health Rep.* **71**, 767 (1956).
168. Zipkin, I., Zucas, S.M., and Stillings, B.R., *J. Nutr.* **100**, 293 (1969).

14

Mercury

Mercury occurs widely in the biosphere and has long been known as a toxic element presenting occupational hazards associated with both ingestion and inhalation. No vital function for the element in living organisms has yet been found. The toxic properties of mercury have evoked increasing concern in recent years due to the extent of its use in industry and agriculture, and the recognition that alkyl derivatives of the element are more toxic than most other chemical forms and can enter the food chain through the activity of microorganisms with the ability to methylate the mercury present in industrial wastes.

I. MERCURY IN ANIMAL TISSUES AND FLUIDS

Mercury was detected in all the tissues of human accident victims, with no known abnormal exposure to Hg other than dental repair, examined by a neutron activation technique (20, 31). The mean concentrations mostly fell between 0.5 and 2.5 ppm Hg on the dry basis, or about 0.1 to 0.5 ppm fresh weight. The highest Hg levels were present in the skin, nails, and hair, which are exposed to atmospheric and other contamination. Among the internal organs the kidneys generally carry the highest Hg concentrations. For example, Joselow *et al.* (25) reported a mean of 2.7 ppm Hg wet weight (range 0–26) for kidney, compared with means ranging from 0.05 to 0.30 ppm for 11 other tissues including the liver and lungs. Kosta and co-workers (29) also obtained appreciably higher mean Hg levels in the human kidney than in the liver, brain, thyroid, and pituitary gland of postmortem samples from individuals not exposed to abnormal amounts of mercury. In comparable postmortem samples from mercury mine workers and from the population in which the mine was situated, kidney and brain Hg levels were substantially higher and thyroid and pituitary Hg levels remarkably higher than in the nonexposed group. In the miners the Hg accumulation and increases in the thyroid and pituitary glands were of the order of 1000-fold and in the town population 10-fold compared with the controls (Table 50). In the organs displaying these Hg accumulations Se was also accumulated, giving an approximately 1:1 molar ratio between Hg and Se. The significance of this finding in relation to Hg toxicity is considered in Section IV.

Relatively few data are available on the Hg levels in the tissues of other species, although it is clear that the element is readily retained in the tissues, especially when supplied in methylated forms. For example, Schroeder (47) compared the Hg levels in the organs of mice fed methylmercury at 1 ppm and mercuric chloride at 5 ppm in the diet from 286 to 653 days with those of controls receiving no such supplementary Hg. In the organs of the controls the Hg concentrations varied from 0.013 to 0.034 ppm. In the mercuric chloride fed mice the same organs ranged from 0.16 to 0.93 ppm Hg, with the kidneys having the highest level. In the methylmercury fed mice the organs had 3.8–6.2 ppm Hg. Gardiner *et al.* (13) fed two breeds of chickens from hatching to 8 weeks on a basal diet containing less than 0.08 ppm Hg, supplemented with methylmercury dicyandiamide to give graded increments from 0.33 to 21.16 ppm Hg. On the basal diet the heart, liver, kidneys, and breast muscle were all below 0.03 μg Hg/g. The levels increased progressively with the level of Hg intake, with higher concentrations in the liver and kidney than in the heart or breast muscle tissue at all levels of Hg supplementation. At the highest level of supplementation mean concentrations of 34–39 μg Hg/g were reported for the liver and kidney. Unfortunately the thyroid and pituitary gland levels were not investigated in this study.

The levels of mercury reported for human blood are low and variable. Kellershohn *et al.* (26) found the whole blood of a small group to average 5 ng Hg/g, with extremes ranging from 2 to 9 ng/g. Goldwater (17) reported that 74% of a normal population had blood with less than 5 ng Hg/ml and 95% had less than 50 ng/ml. In a recent study of 679 residents of Saskatchewan blood Hg levels were found to range from 1 to 42 ng/ml, with a mean of 6.7 and a median of 5 ng/ml (7). Significant differences due to age, sex, or occupation were not observed, but recent fish consumption was found to be important. The blood of native residents claiming local freshwater fish consumption within 2 weeks of sampling averaged 10.6 ng Hg/ml while those without recent fish consumption averaged 6.9 ng/ml. The corresponding levels for nonnative residents were 7.6

TABLE 50
Mean Mercury Content of Human Organs in ppm Fresh Weight[a]

Group	Thyroid	Pituitary	Kidney	Liver	Brain
Mercury mine workers	35.2 ± 28.5 (8)[b]	27.1 ± 14.9 (7)	8.4 ± 4.9 (8)	0.26 ± 0.25 (8)	0.70 ± 0.64 (6)
Idrija population	0.70 ± 0.45 (10)	0.46 ± 0.54 (11)	0.66 ± 1.13 (11)	0.107 ± 0.059 (11)	0.038 ± 0.045 (9)
Nonexposed controls	0.03 ± 0.037 (16)	0.04 ± 0.026 (6)	0.14 ± 0.16 (7)	0.03 ± 0.017 (8)	0.0042 ± 0.0026 (5)

[a]From Kosta *et al.* (29).
[b]All figures in parentheses refer to the number of subjects analyzed.

and 5.8 ng/ml. In an extension of this study paired samples of maternal and cord blood from fish-eating and non-fish-eating women were examined. In the former group cord blood Hg levels were higher (mean 26.7 ng/ml) than maternal blood Hg levels (mean 15.0 ng/ml). No such differences were apparent in the non-fish eating women. The respective means were 7.5 and 6.8 ng Hg/ml. In a comparison of 100 blood donors from Michigan and 137 "unaculturated" Indians from southern Venezuela the mean blood Hg levels were 0.5 and 14.3 ng/ml, respectively (19). This difference in favor of the Michigan group contrasts with the findings for blood Cd and Pb levels, where the opposite was found. However, it cannot be taken to indicate a relatively low level of Hg contamination in the Michigan group because the fish consumption of the Indians was either unknown or not stated.

The red cells normally contain about twice the Hg concentration of the plasma (26). After methylmercury administration most of the blood Hg is present in these cells (3). Several *in vitro* studies have shown that alkylmercury penetrates red cells and accumulates in them more readily than does inorganic Hg (15, 18).

The Hg content of human urine is highly variable. Howie and Smith (20) reported a range of 0.1–13.3 μg/100 ml with a mean of 2.3. In industrially exposed individuals urinary Hg concentrations can increase beyond the levels just cited by several orders of magnitude (24). The critical urinary concentration above which Hg poisoning can be suspected has been suggested as 1–2 μg/ml (37).

Studies of hair, nails, and teeth reveal the ease with which these tissues acquire Hg from the environment. Nixon *et al.* (39) reported mean levels of 2.3 and 2.8 ppm Hg for the inner and outer enamel of human teeth and only 0.1

ppm for teeth which had not erupted into the oral cavity. The enamel of teeth in contact with amalgam fillings had Hg contents ranging from 153 to 1600 ppm. The opportunities for contamination are even more evident from comparisons of the Hg content of head hair and pubic or axillary hair, and of fingernails and toenails. For example Rodger and Smith (46) give the following mean Hg concentrations for subjects with no known abnormal exposure to mercury: head hair, 5.5; pubic hair, 1.6; fingernails, 7.3; and toenails, 2.4 ppm. Markedly higher levels occur in these tissues of individuals with dental, accidental, or industrial exposure. The mean Hg concentrations of the head hair and fingernails of a group of 20 dental assistants were 32.3 and 68.8 ppm, respectively, compared with 8.8 and 5.1 ppm for these tissues in 26 control subjects (20). Higher levels, up to 98.6 in head hair and 1068 ppm in the fingernail of one individual, were observed in a group of workers with Hg contamination of their laboratory. Lower head hair Hg levels, ranging from 0.2 to 6.0 ppm, were obtained for Canadian controls, compared with 5–10 ppm in individuals with occupational exposure to mercury (23). The Hg level of hair thus gives some indication of the Hg status of individuals and of their environment, although the "normal" range is extremely wide (2, 46).

The forms in which Hg occurs in the tissues are considered in the next section.

II. MERCURY METABOLISM

The metabolic behavior of mercury varies greatly with the chemical form in which it is presented to the animal, the extent to which other elements with which it interacts are present in the diet, and apparently also with genetic differences. With respect to the last factor, Miller and co-workers (35, 36) observed remarkable strain differences in the ability of chickens to retain and excrete mercury. The preferential Hg retention in the kidneys and liver of one strain was not affected by the dosage or mode of administration. It was suggested that the strain with the smaller retention has a lower renal Hg threshold or less Hg-binding sites.

Marked differences in metabolic behavior exist between inorganic and aryl Hg compounds and the alkyl Hg derivatives, particularly those of short carbon chain such as dimethylmercury, CH_3HgCH_3. Inorganic Hg compounds are relatively poorly absorbed (8, 53). Taguchi (53) reported 73, 45, and 6% absorption of Hg in rats after dosing with methylmercury chloride, phenylmercuric acetate, and mercuric acetate, respectively. Following absorption the inorganic, aryl, and methoxyalkyl Hg compounds behave similarly, due to the rapid degradation of the two latter forms to inorganic Hg (52). Simple alkyl forms of Hg are not only better absorbed, they are much better retained, are more firmly bound in the tissues, and induce higher brain Hg contents than aryl Hg compounds (54). Friberg (11) compared the retention, distribution, and excretion of Hg in rats given subcutaneous injections of methylmercury dicyandiamide or equivalent amounts of mercuric chloride. Almost 100 times as much mercury was found in the blood, 10 times higher Hg concentrations were present in the brain and spleen, and twice as much in the liver of the animals receiving the methyl compound. By contrast, the rats receiving the mercuric chloride excreted about 20 times as much Hg in the urine and about twice as much in the feces as the rats given methylmercury.

Metabolic differences between simple alkyl Hg compounds and other mercurials extend further to their pattern of excretion. With the latter, Hg is excreted via the bile and the feces in inorganic or protein-bound form. With methylmercury much of the biliary Hg is present as a methylmercury–cysteine complex, most of which is subsequently reabsorbed from the intestinal tract and redistributed to the tissues (41), thus contributing to the extremely high retention of methylmercury.

The metabolic differences between methylmercury and inorganic Hg are shown also by studies of placental transfer. The placenta presents an effective barrier against the transfer of inorganic Hg in rats (14), and transfer of methylmercury to the fetus greatly exceeds that of either mercuric chloride or phenylmercuric acetate in mice (51). According to Taguchi (53) over 20% of the methylmercury administered to the pregnant rat can be transferred to the fetus, with a tendency for this Hg to accumulate in the brain. Most of this transfer occurs in the last 4–5 days of gestation and produces a fetal to maternal brain Hg ratio of 2.5–3.5:1 (64). The higher Hg content of fetal blood than of maternal blood in women exposed to methylmercury from fish (7) was detailed in Section I. The ready placental transfer of methylmercury to the fetus, and particularly to its brain, no doubt explains the occurrence of "congenital" Minamata disease in infants, as discussed in Section IV, although ingestion of methylmercury from the mother's milk may be a contributing factor, since mammary transfer of methylmercury has been demonstrated in rats (65).

Much remains to be learned of the chemical forms and intracellular distribution of mercury. Methyl forms do not occur in animal cells and tissues in significant amounts, unless ingested or injected as such. In other words the animal body has an extremely limited capacity to convert inorganic and various organic forms of Hg into the more toxic methyl forms (59). This ability to transform mercury is principally confined to microorganisms, and it is their activity that can introduce dangerous methylated Hg compounds into the food chain.

A nonhistone protein component into which Hg is rapidly incorporated has been isolated from rat kidney nuclei (5), while the main Hg-binding protein in rat kidney cytosol appears to be a stable metallothionein-like protein (61), together with other more labile Hg-binding fractions (21). The Hg-binding capacity of the thioneins is extremely high (45), and Zn and Cd have been completely displaced from metallothionein from rat liver *in vitro*, yielding a purified protein containing 8 g-atoms of Hg^{2+} (49). It is probable that the thioneinlike Hg protein plays a role in the detoxification of mercury.

The metabolic antagonism between mercury and selenium manifested in the protection against Se toxicity displayed by mercury, as discussed in the previous chapter, is paralleled by a comparable protection against Hg toxicity by selenium. A mutual metabolic antagonism between the two elements thus exists. The first evidence for this came from the studies of Parizek and co-workers (42), who showed that selenite protected against the renal necrosis and mortality in rats caused by injected mercuric chloride and counteracted placental Hg transfer. A protective action of dietary Se against chronic Hg toxicity has also been demonstrated in rats and Japanese quail. Rats given a diet to which sodium selenite was added at a level of 0.5 ppm Se grew better and survived longer than rats not so treated with Se, when methylmercury was added to the drinking water of both groups (12). Similarly Japanese quail given 20 ppm Hg as methylmercury in diets containing 17% by weight of tuna fish (2–3 ppm Se) survived longer than quail given this level of Hg in a corn–soya (low-Se) diet. Potter and Matrone (44) fed rats diets containing methylmercury or mercuric chloride with and without selenite supplements. The Se supplements enhanced growth rate with both forms of Hg and protected against the mortality and neurotoxicity caused by methylmercury.

The biochemical mechanisms involved in the prevention or amelioration of Hg toxicity by Se are little understood. It does not appear to be due to increased Hg excretion. In fact there is some evidence of an increase in Hg retention and definite evidence of a changed distribution of retained Hg, with increased levels in the liver and spleen and decreased levels in kidneys (44). Presumably the Se also induces a change in the distribution of Hg on a subcellular level, although this has not been reported. Dietary selenite does not increase the amount of Hg in thionein when this Hg is fed as methylmercury, and methylmercury combines poorly with thionein *in vitro* (6). This suggests that Se does not ameliorate or prevent methylmercury toxicity by increasing the rate of methylmercury breakdown.

The biochemical effects of toxic Hg intakes have been critically reviewed up to 1972 by Vallee and Ulmer (56). Since then Southard and co-workers (50) have shown that toxic levels of mercury suppress mitochondrial oxidative phosphorylation in the kidney of the rat and have directly correlated this inhibition with the death of the animal.

III. SOURCES OF MERCURY

Mercury and methylmercury are naturally occurring substances to which all living organisms have been exposed in varying degrees depending on natural biological, chemical, and physical processes. Modern technological developments involving the use of Hg compounds are responsible for the discharge of large and variable amounts of the element into the environment. The main industrial source is the chloralkali industry, which has been estimated to lose to the environment 0.45 lb Hg per ton of chlorine produced (38). Other major industrial uses of mercury are in the manufacture of electrical apparatus, paint, dental preparations, and pharmaceuticals, and in paper and pulp making as slimicides and algicides. The mercury used in agriculture for seed treatment can be a particularly hazardous source since mostly methylmercury compounds were used. Mercury uses and their possible hazards as sources of Hg contamination in Canada (9) and in the United States (10) have been comprehensively reviewed.

It is apparent that mercury can enter the biosphere from a variety of man-made sources, as just described, and also from the burning of fossil fuels. Methylated Hg compounds enter the food chain mainly through the activity of microorganisms that have the ability to methylate the Hg present in industrial wastes (22). The mercury in fish from polluted waters occurs almost entirely as methylmercury (58). Methylated Hg compounds may also enter the food chain through their use as a disinfectant of grain. Birds fed for 4 weeks on wheat treated with methylmercury dicyandiamide were found to have 8 ppm Hg in their liver and eggs and 4 ppm Hg in the muscle tissue (48). These are high concentrations compared with the negligible Hg levels found in control birds. The mercury concentration in 100 normal shelled chicken eggs from different parts of Canada has recently been reported to range from 0.001 to 0.023 ppm, with a calculated mean of 0.006 ppm (27). The decrease in the methylmercury content of foods of animal origin in Sweden after methoxyethyl mercury was substituted for methylmercury as a seed disinfectant has been documented by Westöö (60). For example, the average Hg content of pig liver fell from 0.060 to 0.025 ppm and that of pork chop and whole egg from 0.03 to 0.01 ppm Hg within 2 years. Prior to this change considerable amounts of mercury were found in the tissues of seed-eating birds and their predators and in small rodents, associated with lethal and sublethal Hg poisoning in Swedish wildlife (4).

Some years ago a typical adult human diet was estimated to provide about 20 μg Hg/day (16). Subsequently Westöö (57) analyzed homogenates of 12 fish-free Swedish diets and found their total Hg content to range from 4 to 20 (mean 11) μg Hg/day. In a more extensive study of Swedish diets the very wide range of 1.0–30.6 μg Hg/day was found (40). The importance of the fish content is evident from the fact that 90 of the diets containing no fish averaged only 3.5 μg Hg/day, whereas the 55 diets that included fish of any kind averaged 9.0 μg Hg/day. The importance of fish is further emphasized by the fact that most of the mercury in fish is present as highly toxic methylmercury compounds. In foods of plant origin little or none of the mercury is normally present in this form, while in meat and dairy products the low levels of Hg can include a small proportion of methylmercury, presumably from residues in feeds containing fish meal or treated cereal grains. In an investigation of Canadian foods, concentrations ranging from 0.005 to 0.075 ppm Hg were reported for a variety of food items not abnormally exposed to Hg at any time, compared with levels of 1 ppm Hg or more in specimens from Hg-contaminated areas (23).

Although man-made Hg pollution of freshwater rivers and lakes can greatly increase the Hg levels in freshwater fish, it should be appreciated that the Hg levels are high in wide-ranging ocean fish where no such pollution could have occurred. Thus Miller *et al.* (34) found the Hg levels of museum specimens of seven tuna caught 62–93 years ago to range from 0.26 to 0.64 ppm wet weight and those of five specimens caught recently from 0.13 to 0.48 ppm wet weight. Much higher Hg levels have been observed in marine mammals and birds (28). The range of Hg concentration in the livers of 22 marine mammals was extremely wide (0.37–326 ppm Hg wet weight). In four marine birds the Hg levels ranged more narrowly from 1.8 to 2.4 ppm in the liver and from 0.35 to 0.58 ppm in the brain. In the marine mammals the Hg was not present largely as methylmercury as it is in fish and was highly correlated with the Se levels. In fact in marine mammals a 1:1 Hg:Se molecular increment ratio and an almost perfect linear correlation between Hg and Se were found. It was suggested that marine mammals are able to detoxify methylmercury by a specific chemical mechanism in which Se in involved. The importance of the high Se levels common in fish and other marine products to the toxicity of the mercury they contain and to the question of safe or maximum tolerable dietary Hg intakes is considered in Section IV.

The contribution of Hg inhaled from the air is negligible compared with intake from the food, except where there is an occupational exposure. Mercury volatilizes so readily that it can constitute a health hazard to laboratory workers and others handling the metal for long periods, unless proper precautions are taken. The water supply is also a relatively insignificant source of Hg, except when contaminated from industrial or geological sources. The tentative upper limit for Hg in drinking water has been placed at 1 μg/liter (62), a figure related to levels found in natural waters for drinking purposes and to the high ingestion rate of 2.5 liters of water/day. At these upper limits 2.5 μg Hg/person/day would be provided, mostly in inorganic form.

Little is known of normal Hg intakes by grazing or even by stall-fed farm animals. Lunde (32) reported a mean Hg concentration of 0.18 ppm (range

0.03–0.40) for 12 commercial fish meals from different sources. This is well above the Hg levels in ordinary cereal grains or protein supplements of plant origin.

IV. MERCURY TOXICITY

Mercury poisoning has been prominent at times among goldsmiths and mirror makers, and the term "mad as a hatter" derives from the symptoms shown by workers in the treatment of furs with mercuric nitrate. The manifestations of such subacute Hg poisoning are primarily neurological, with tremors, vertigo, irritability, moodiness, and depression, associated with salivation, stomatitis, and diarrhea. In poisoning from the ingestion of inorganic Hg salts, the liver and kidneys are the tissues most affected and there may also be proteinuria, necrosis of the intestinal tract, and diarrhea. When the Hg is ingested as the more toxic alkyl derivative the symptoms include progressive incoordination, loss of vision and hearing, and mental deterioration, arising from a toxic neuroencephalopathy in which the nerve cells of the cerebral and cerebellar cortex are selectively involved. These clinical and pathological changes, with particular impairment of scotopic vision, have been demonstrated in the squirrel monkey, a species reported to be a valuable model for the study of methylmercury poisoning in humans (3a).

The changes just described were evident in victims of a tragic occurrence of methylmercury poisoning in Japan (Minamata disease) following the dumping into the Minamata Bay of Hg-containing factory wastes and the consumption of fish caught in the bay (30, 55). The outbreak was characterized by a high incidence of "congenital" cases in infants. The mental retardation, cerebral palsy, and mortality that were evident resulted from the ease with which methylmercury passes the placental barrier and concentrates preferentially in fetal tissues, particularly the brain. The sensitivity of the pregnant women and her offspring to methylmercury is apparent from these observations.

Outbreaks of methylmercury poisoning have also occurred from the accidental consumption of bread made from seed grain treated with mercurial fungicides (1) and meat from animals fed on such grain.

The minimum safe levels of dietary Hg, or the maximum intakes compatible with the long-term health of humans, depend on pregnancy and age as just mentioned, on the form or forms in which the mercury is ingested, and on dietary Se intakes. The joint FAO/WHO expert committee on food additives (63) established a provisional tolerable weekly intake of 0.3 mg Hg per person, of which no more than 0.2 mg Hg should be present as the methylmercury ion CH_3Hg^+. These amounts are equivalent to 0.5 μg and 0.33 μg, respectively, per kg body weight, or assuming an average dry matter intake of 400 g/day, a total Hg content of the dry diet close to 0.08 and 0.06 ppm, respectively. It was stated further that "where the total Hg in the diet is found to exceed 0.3 mg/week the level of methylmercury compounds should also be investigated. If the excessive intake is attributable entirely to inorganic Hg, the above provisional limit for total Hg no longer applies and will need to be reappraised in the light of all the prevailing circumstances."

It is now apparent that one of these circumstances is the level of dietary Se intake, since Se compounds counteract the toxicity of both inorganic and organic Hg compounds in animals. It seems likely that the protection afforded by Se also applies to man. In the studies of Kosta and co-workers (29) of mercury mine workers a coaccumulation of Hg and Se in the organs and tissues was observed, with an approximately 1:1 molar ratio. In these circumstances the abnormally high-Hg levels found in the tissues were apparently without deleterious effects on the individuals, several of whom had been 10–16 years in retirement following long periods of Hg exposure in the mines. The Se intakes from the diet were not reported but were said not to be abnormally high, suggesting that the coaccumulation with Hg is a natural or autoprotective effect. Where the main source of dietary Hg is fish or marine mammals the diet is naturally enriched in Se relative to Hg, and would be expected to provide some protection. It seems reasonable to suggest, further, that in areas naturally low in Se individuals would be at greater risk from Hg poisoning than those in areas of high-Se status. A logical corollary from this is that maximum tolerable Hg intakes, or minimum safe Hg intakes, would be lower in low-Se status than in high-Se status areas. It is obvious that the quantitative aspects of Hg–Se interactions warrant further study in both man and animals.

In addition to dietary Hg determinations, measurements of the Hg levels in hair, urine, blood, and saliva have been given consideration as diagnostic procedures in the prediction of incipient Hg poisoning. An expert group that made an evaluation of the risks from methylmercury in fish contends that the best available index of the degree of exposure is the level of methylmercury or Hg in the red blood cells, and that the level in whole blood or hair is also valuable (33). Clinically manifest poisoning of adults sensitive to methylmercury, it is claimed, "may occur at a level in whole blood down to 0.2 μg Hg/g., which level seems to be reached on exposure to about 0.3 mg Hg as methylmercury/day, or about 4 μg Hg/kg bodyweight." If, as the group believed, a factor of 10 gives a sufficient margin of safety, the acceptable level in whole blood would be 0.02 μg/g, corresponding to about 0.04 μg/g in the red cells and about 6 μg/g in the hair.

In long-term studies with mice fed 5 ppm Hg as mercuric chloride and 5 ppm Hg followed by 1 ppm Hg as methylmercuric acetate in the drinking water, marked differences in relative toxicity were apparent (47). The mercuric chloride at this level induced no observable toxic effects. The methylmercuric acetate at 1 ppm increased body weights of both males and females. At 5 ppm it decreased body weight and was toxic but was not tumorogenic. Perry and Erlanger (43) observed a small increase in systolic pressure in rats given 5 and 10 ppm Hg as mercuric chloride in the drinking water for 12 months, while 2.5 and 25 ppm had no such significant effect.

The high toxicity of methylated Hg compounds compared with inorganic forms of the element is further apparent from studies with chickens and Japanese quail. Scott *et al.* (47a) found that mercury as $HgSO_4$ or $HgCl_2$ fed at dietary levels up to 200 mg Hg/kg had little effect on egg production, hatchability, egg shell strength, morbidity, and mortality, whereas methylmercury chloride at levels which provided 10 or 20 mg Hg/kg diet severely affected all these parameters.

REFERENCES

1. Bakir, F., Damluji, S.F., Amin-Zaki, L., Mutadha, M., Khalidi, A., Al-Rawi, N.Y., Tikriti, S., Dhahir, H.I., Clarkson, T.W., Smith, J.C., and Doherty, R.A., *Science* **181**, 230 (1973).
2. Bate, L.C., and Dyer, F.F., *Nucleonics* **23**, 74 (1965).
3. Berglund, F., and Berlin, M., *in* "Chemical Fallout" (M.W. Miller and G.C. Berg, eds.), p. 258. Thomas, Springfield, Illinois, 1969.

3a. Berlin, M., Grant, C.A., Hellstrom, J., and Schutz, A., *Arch. Environ. Health* **30,** 340 (1975).

4. Borg, K., Wanntorp, H., Erne, K., and Hanko, E., *J. Appl. Ecol.* **3**, Suppl., 171 (1966).
5. Chanda, S.K., and Cherian, M.G., *Biochem. Biophys. Res. Commun.* **50,** 1013 (1973).
6. Chen, R.W., Ganther, H.E., and Hoekstra, W.G., *Biochem. Biophys. Res. Commun.* **51,** 383 (1973).
7. Dennis, C.A.R., and Fehr, F., *Sci. Total Environ.* **3,** 267 and 275 (1975).
8. Ellis, R.W., and Fang, S.C., *Toxicol. Appl. Pharmacol.* **11,** 104 (1967).
9. Fimreite, N., *Environ. Pollut.* **6,** 119 (1970).
10. Fishbein, L., *Sci. Total Environ.* **2,** 341 (1974).
11. Friberg, L., *Arch. Ind. Health* **20,** 42 (1959).
12. Ganther, H.E., Gondie, C., Sunde, M.L., Kopecky, M.J., Wagner, P., Hoh, S., and Hoekstra, W.G., *Science* **175,** 1122 (1972).
13. Gardiner, E.E., Hironaka, R., and Slen, S.B., *Can. J. Anim. Sci.* **51,** 657 (1971).
14. Garrett, N.E., Garrett, R.J.B., and Archdeacon, J.W., *Toxicol. Appl. Pharmacol.* **22,** 649 (1972).
15. Garrett, R.J.B., and Garrett, N.E., *Life Sci.* **15,** 733 (1974).
16. Gibbs, O.S., Pond, H., and Hansmann, G.A., *J. Pharmacol. Exp. Ther.* **72,** 16 (1941).
17. Goldwater, L.J., *R. Inst. Public Health Hyg. J.* **27,** 279 (1964).
18. Guirgis, H.A., Stewart, W.K., and Taylor, I.W., *in* "Environmental Mercury Contamination" (R. Hartung and B.D. Dinman, eds.), p. 239. Science Publ. Inc., Ann Arbor, Michigan, 1972.
19. Hecker, L.H., Allen, H.E., Dinman, B.D., and Neel, J.V., *Arch. Environ. Health* **29,** 181 (1974).
20. Howie, R.A., and Smith, H., *J. Forensic Sci. Soc.* **7,** 90 (1967).

21. Jakubowski, M., Piotrowski, J., and Trojanowska, B., *Toxicol. Appl. Pharmacol.* **16**, 743 (1970).
22. Jensen, S., and Jernelov, A., *Nature (London)* **223**, 753 (1969).
23. Jervis, R.E., Debrun, D., Le Page, W., and Tiefenbach, B., report Dep. Chem. Eng. & Appl. Chem. University of Toronto, Canada, 1970.
24. Joselow, M.M., and Goldwater, L.J., *Arch. Environ. Health* **15**, 155 (1967).
25. Joselow, M.M., Goldwater, L.J., and Weinberg, S.B., *Arch. Environ. Health* **15**, 64 (1967).
26. Kellershohn, C., Comar, D., and Lopoec, C., *J. Lab. Clin. Med.* **66**, 168 (1965).
27. Kirkpatrick, D.C., and Coffin, D.E., *J. Sci. Food Agric.* **26**, 99 (1975).
28. Koeman, J.H., van de Ven, W.S.M., Goeij, J.J.M. de, Tjioe, P.S., and van Haaften, J.L., *Sci. Total Environ.* **3**, 279 (1975).
29. Kosta, L., Byrne, A.R., and Zelenko, V., *Nature (London)* **254**, 238 (1975).
30. Kurland, L.T., Faro, S.N., and Siedler, H.S., *World Neurol.* **1**, 320 (1960).
31. Lenihan, J.M., and Smith, H., *in* "Nuclear Activation Techniques in the Life Sciences." IAEA, Vienna, 1967.
32. Lunde, G., *J. Sci. Food Agric.* **19**, 432 (1968).
33. Methylmercury in Fish. A Toxicological-Epidemiological Evaluation of Risks: *Nord. Hyg. Tidskr., Suppl.* **4**, (1971).
34. Miller, G.E., Grant, P.M., Kishore, R., Steinkruger, F.J., Rowland, F.S., and Guinn, V.P., *Science* **175**, 1121 (1972).
35. Miller, V.L., Bearse, G.E., and Hammermeister, K.E., *Poultry Sci.* **38**, 1037 (1959).
36. Miller, V.L., Larkin, D.V., Bearse, G.E., and Hamilton, C.M., *Poultry Sci.* **46**, 142 (1967).
37. Monier-Williams, G.W., "Trace Elements in Food." Chapman & Hall, London, 1949.
38. Murozumi, M., *Electrochem. Technol.* **5**, 236 (1967).
39. Nixon, G.S., Smith, H., and Livingstone, H.D., *in* "Nuclear Activation Techniques in the Life Sciences." IAEA, Vienna, 1967.
40. Norden, A., Dencker, I., and Schutz, A., *Naeringsforskning* **14**, 40 (1970).
41. Norseth, T., and Clarkson, T.W., *Biochem. Pharmacol.* **19**, 2775 (1970).
42. Parizek, J., Bênes, I., Ostadalova, I., Bâbicky, A., Bênes, J., and Pitha, J., *in* "Mineral Metabolism in Pediatrics" (D. Barltrop and W.J. Burland, eds.). Blackwell, Oxford, 1969.
43. Perry, H.M., Jr., Erlanger, M.W., *J. Lab. Clin. Med.* **83**, 541 (1974).
44. Potter, S., and Matrone, G., *J. Nutr.* **104**, 638 (1974).
45. Pulido, P., Kägi, J.H.R., and Vallee, B.L., *Biochemistry* **5**, 1768 (1966).
46. Rodger, W.J., and Smith, H., *J. Forensic Sci. Soc.* **7**, 86 (1967).
47. Schroeder, H.A., and Mitchener, M., *J. Nutr.* **105**, 452 (1975).
47a. Scott, M.L., Zimmerman, J.R., Marinsky, S., Mullenhof, P.A., Rumsey, G.L., and Rice, R.W., *Poultry Sci.* **54**, 350 (1975).
48. Smart, N.A., and Lloyd, M.K., *J. Sci. Food Agric.* **14**, 734 (1963).
49. Sokolowski, G., Pilz, W., and Weser, U., *FEBS Lett.* **48**, 222 (1974).
50. Southard, J., Nitisewojo, P., and Green, D.E., *Fed. Proc., Fed. Am. Soc. Exp. Biol.* **33**, 2147 (1974).
51. Suzuki, T., Matsumoto, N., Miyama, T., and Kalsunuma, H., *Ind. Health* **5**, 149 (1967).
52. Swensson, A., and Ulfvarson, U., *Acta Pharmacol. Toxicol.* **26**, 259 (1968).
53. Taguchi, Y., *Nippon Eiseigaku Zasshi* **25**, 563 (1971).
54. Takeda, Y., Konugi, T., Hoshino, I.O., and Ukita, T., *Toxicol. Appl. Pharmacol.* **13**, 156 (1968).

55. Takeuchi, T., *in* "Environmental Mercury Contamination" (R. Hartung and B.D. Dinman, eds.), p. 247. Science Publ. Inc., Ann Arbor, Michigan, 1972.
56. Vallee, B.L., and Ulmer, D.D., *Annu. Rev. Biochem.* **41,** 91 (1972).
57. Westöö, G., *Var foeda,* No. 4 (1965).
58. Westöö, G., *Acta Chem. Scand.* **20,** 2131 (1966).
59. Westöö, G., *Acta Chem. Scand.* **22,** 2277 (1968).
60. Westöö, G., *in* "Chemical Fallout" (M.W. Miller and G.C. Berg, eds.), p. 75. Thomas, Springfield, Illinois, 1969.
61. Wisniewska, J.M., Trojanowska, B., Piotrowski, J., and Jakubowski, M., *Toxicol. Appl. Pharmacol.* **16,** 754 (1970).
62. World Health Organization, "International Standards for Drinking Water." World Health Organ., Geneva, 1971.
63. World Health Organization, *W.H.O., Tech. Rep. Ser.* **505** (1972).
64. Yang, M.G., Krawford, K.S., Garcia, J.D., Wang, J.H.C., and Lei, K.Y., *Proc. Soc. Exp. Biol. Med.* **141,** 1004 (1972).
65. Yang, M.G., Wang, J.H.C., Garcia, J.D., Post, E., and Lei, K.Y., *Proc. Soc. Exp. Biol. Med.* **142,** 722 (1973).

15

Vanadium

I. VANADIUM IN ANIMAL TISSUES AND FLUIDS

Data on the vanadium content of animal tissues are meager and discordant, presumably as a consequence of analytical difficulties. It is clear, nevertheless, that vanadium is widely distributed in very low concentrations, and on present evidence, is not concentrated in any particular organ or tissue in the higher animals. Extremely high V concentrations, ranging from 3 to 1900 ppm, have long been known to occur in the blood of ascidian worms (22, 53), mainly as the V protein compound hemovanadin (9). Hemovanadin cannot act as an oxygen carrier and it is not known if the V performs any vital function in this species (8). The blood cells of *Ascidia nigra* contain the remarkably high concentration of 1.45% V (10), and the V in *Ascidia aspersa* is in a dynamic equilibrium of V(III) and V(IV) in the blood cells (40), which suggests a role for this element in an oxidation–reduction reaction in these cells. Vanadium also occurs in large concentrations in some holothurians (4), and up to 150 ppm (d.b.) has been reported in the mollusk *Pleurobranchus plumula* (52).

Bertrand (4) reported a variety of vertebrate tissues to range from 0.02 (the limit of his method) to 0.3 ppm, with a mean of 0.1 ppm V. Tipton and Cook (51) found average concentrations of the order of 0.02–0.03 ppm V (d.b.) for adult liver, spleen, pancreas, and prostate gland. The only organ carrying consistently higher levels was the lung, which averaged close to 0.6 ppm V. Significantly higher V levels in adult human lung than in other tissues were also obtained by Schroeder and co-workers (43). More recently Hamilton *et al.* (18)

reported the following values for adult human tissues: brain, 0.03 ± 0.008; muscle, 0.01 ± 0.003; liver, 0.04 ± 0.01; testis, 0.20 ± 0.08; lung, 0.10 ± 0.02; and lymph nodes, 0.40 ± 0.2 μg V/g wet weight. Vanadium is present in human dental enamel in low concentrations, i.e., below 0.1 μg/g (28).

In a study of blood from male donors in 19 U.S. cities, over 90% of the samples were found to contain less than 1 μg V/100 ml of whole blood and 2 μg/100 ml was the highest level observed (2). Several other groups of workers have reported substantially higher blood levels (6, 37, 43). For example Nozdryukina *et al.* (37), in a study of trace elements in the blood of ishemic heart disease patients, found no immediate drop after infarction from a "normal" of 4.6 μg V/100 ml, but by 70 days and over the level had fallen to 1.96 ± 0.5 μg/100 ml. Söremark (46), using neutron activation, has obtained lower V concentrations for most biological materials than those just cited. Calf livers from two sources were reported to average 2.4 and 10 ppb wet weight and fresh cow's milk from five locations to range from the astonishingly low level of 0.07 to 0.11 ppb V.

II. VANADIUM METABOLISM

Little is known of V metabolism in animals at physiological levels. Very small amounts (0–8 μg V/day) are normally excreted in the urine of man (38, 43). These amounts are increased greatly when V salts are administered orally at toxic or subtoxic levels (13, 43). Absorption appears to be poor but more evidence is needed on this point. In one study, only 0.1–1.0% of the V in 100 mg of the soluble diammonium oxytartarovanadate was found to be absorbed from the human gut (11). In another study, 60% of absorbed V was excreted in the urine in the first 24 hr, the remainder being retained in the liver and bone (49). The V present in the bone was mobilized and excreted much more slowly than that in the liver. Söremark and Ullberg (47) found the highest retention of injected ^{48}V in the bones and teeth of mice and Hathcock *et al.* (19) the greatest retention in the bones and kidneys of chicks.

In timed distribution studies of intravenously injected ^{48}V in rats no significant difference was seen in the rate or amount of uptake of the three oxidation states of vanadium (26). Liver, kidney, spleen, and testis accumulated ^{48}V up to 4 hr and retained most of this radioactivity up to 96 hr, at which time other major organs retained less. At this time also 46% of the ^{48}V had been excreted in the urine and 9% in the feces. Evidence was obtained that the marked liver retention of ^{48}V was due to its movement into the mitochondrial and nuclear fractions of the cells. It was further apparent from these studies that most of the ^{48}V was present in the noncellular portion of the blood, probably bound to transferrin.

The metabolic interaction between vanadium and chromium, or more properly between vanadate and chromate anions, is considered in Section V.

III. VANADIUM FUNCTIONS AND REQUIREMENTS

Vanadium deficiency, manifested in impaired growth and reproduction and disturbed lipid metabolism, has been demonstrated in chicks (24, 36) and rats (44, 48).

1. Growth

The original observation indicative of V deficiency was a significantly reduced wing and tail feather growth in chicks consuming a diet containing less than 10 ppb V (24). Reduced body growth was subsequently demonstrated in chicks on diets containing 30–35 ppb V, with a significant growth response from 3 ppm of supplementary V (35). In rats Strasia (48) obtained a growth stimulation from 0.5 ppm V added to a diet containing less than 100 ppb V. Unfortunately Schwarz and Milne (44) did not report the V concentration of their basal diet, but they obtained significant growth responses from 250 and 500 ppb of V as sodium orthovanadate. It therefore seems that on purified diets the V requirement for growth probably lies between 50 and 500 ppb. Whether the requirements are higher on natural diets, as Hopkins and Mohr (25) have tentatively suggested, is unknown.

2. Reproduction

Rats consuming diets of less than 10 ppb V over several generations exhibit a marked impairment of reproductive performance and increased pup mortality (25). Fertility was reduced slightly in third generation females and markedly in fourth generation females, with no such effects on control V-supplemented animals. Pregnancies per mating period and pup survival at 21 days were both reduced in the V-deficient animals in two different strains of rats, as shown in Table 51. These effects were not accompanied by significant differences in organ cholesterol, fat, total protein, dry weight, and phosphorus of the adult first, second, and third generations. The mechanism of action of V on the reproductive processes of the female, and the point or points in the reproductive cycle at which V exerts its inhibiting effect, remain to be determined.

3. Changes in Erythrocyte and Iron Levels

In the experiments of Strasia (48) with rats fed diets containing less than 100 ppb V significant increases in packed cell volume of blood and blood and bone

TABLE 51
Effect of Vanadium Deficiency on Rat Reproduction and Pup Morality[a]

Diet	No. litters from matings	Total no. of pups	% mortality at 21 days
	2 matings		
V-deficient[b]	4	22	32
V-supplemented[b,c]	9	65	1.5
	3 matings		
V-deficient[d]	3	21	38
V-supplemented[c,d]	6	30	7

[a]From Hopkins and Mohr (25).
[b]Five Sprague–Dawley rats per group.
[c]1 ppm as ammonium vanadate.
[d]Six BHE rats per group (first generation).

iron were observed, compared with controls receiving 0.5, 2.5, and 5.0 ppm of supplemental vanadium. Increased hematocrits in chicks consuming a 30–35 ppb V diet have similarly been reported (36). The significance of these findings remains obscure.

4. Disturbances in Lipid Metabolism

An inhibition of cholesterol synthesis by vanadium has been observed *in vivo* in human and animal tissues when the vanadium is used at pharmacological levels (see Curran and Burch, 12). This inhibition is accompanied by decreased plasma phospholipid and cholesterol levels and by reduced aortic cholesterol concentrations.* In older individuals and in patients with hypercholesteremia or ischemic heart disease no such effect from vanadium is apparent (12, 13, 43, 45), while in older rats the inhibition can be demonstrated *in vitro* but not *in vivo.* The site of the inhibition by vanadium is the microsomal enzyme system referred to as squalene synthetase.

Reports of altered blood lipid levels in V-deficient chicks are difficult to evaluate. Thus Hopkins and Mohr (24) initially reported lowered serum cholesterol levels in V-deficient chicks at 4 weeks. This observation was repeated in a further experiment, but after 7 weeks on the deficient diet the serum cholesterol levels were slightly but significantly higher than the V-supplemented controls (25). Nielsen and Ollerich (36) also observed increased serum cholesterol levels

*Author's note: Vanadium fed as ammonium vanadate to young chicks at 100 ppm V *increased* liver and plasma total lipid and cholesterol levels, and plasma cholesterol turnover rate [Hafez, Y., and Kratzer, F.H., *J. Nutr.* **106,** 249 (1976)].

in their V-deficient chicks, but in their case the increase was apparent after 4 weeks on the deficient diet. These results are confusing but point strongly to an influence of vanadium on cholesterol metabolism at deficiency levels, as well as pharmacological levels. Furthermore, evidence has been obtained that plasma triglyceride levels are greatly increased in V-deficient chicks (25). The mean plasma triglyceride level of nine V-deficient chicks at 4 weeks of age was reported as 48.7 ± 2.4 mg/100 ml, compared with 25.4 ± 3.0 mg/100 ml for nine chicks of the same age which had received supplementary V at 1 ppm as ammonium metavanadate. These interesting findings highlight the need for further research on vanadium in relation to lipid metabolism.

5. Vanadium and Dental Caries

Radiovanadium injected subcutaneously into mice is concentrated in the areas of rapid mineralization of bones and teeth dentine (47) and is incorporated into the tooth structure of rats and retained in the molars up to 90 days after injection (50). The addition of vanadium (and strontium) to specially purified diets has also been reported to promote mineralization of the bones and teeth and to reduce the numbers of carious teeth in rats and guinea pigs (27). Geyer (15) obtained a high degree of protection against caries in hamsters fed a cariogenic diet, when V was administered as V_2O_5 either orally or parenterally. Kruger (27) similarly reported that vanadium, administered intraperitoneally to rats during the period of tooth development, is effective in reducing the incidence of caries.

It is difficult to reconcile the above findings with those of other experiments in which the administration of vanadium in the drinking water at varying levels has either been unsuccessful in decreasing caries incidence (7, 33) or has actually increased caries incidence (5, 21). Thus Bowen (5) gave water containing 2 ppm V to monkeys for a period of 5 years and found that this increased the incidence of caries compared with controls with no added V in their water. No explanation of the highly divergent results obtained by different investigators has appeared. The whole question of a possible relation between vanadium and tooth development and decay warrants further critical investigation. The evidence to date has been evaluated by Hadjimarkos (16).

IV. SOURCES OF VANADIUM

Reliable data on the vanadium content of human foods and animal feeds are even more limited than those on animal tissues and fluids. Söremark (46), using activation analysis, found a range of human foods to vary greatly in V concentration, from less than 0.1 ppb for pea, beet, carrot, and pear (ash weight) to 140 ppb for dill and 790 ppb for radish (wet weight). Liver, fish, and meat contained from 2 to 10 ppb (fresh basis). More than half of 34 samples of wheat grain from 12 different locations in North America were found to contain less than 6.5 ppb V and the highest level reported was 20.0 ppb (54). Higher V concentrations (28–55 ppb) were found for the seeds of wheat, barley, oats, and peas from plants grown in nutrient solution, and these were substantially increased when vanadate was added to the nutrient solution. To what extent the vanadium in wheat is lost in milling to white flour as other trace elements are lost is unknown. Hamilton and Minsky (17) have shown that most of the V in unrefined sugar is lost in the refining process. Barbados brown sugar was reported to contain 0.4 μg V/g, compared with 0.002 μg/g in white sugar.

Estimates of the normal or average daily V intakes by man cannot yet be made with any confidence. Schroeder *et al.* (43) reported that an institutional diet supplied about 1.2 mg V/day and that a good well-balanced diet should provide 1–4 mg/day. It was contended further that the major determinant of V intake is the type of fat in the diet, with diets high in unsaturated fatty acids from vegetable sources being much richer in V than diets containing saturated fats from animal sources. However, Welch and Cary (54) later examined 10 commercial vegetable oils and found no particular affinity of V for such oils. The V concentrations reported ranged from 14 to 139 ppb. It seems probable, therefore, that V intakes from ordinary Western-type diets are well below the 1–4 mg/day estimated by Schroeder. Whether they are likely to supply insufficient vanadium for human needs can only be answered as research with this element proceeds.

Information on the V content of animal feeds is limited. Mitchell (30, 31) reported V concentrations ranging from less than 30 to 160 ppb for red clover and from less than 30 to 110 ppb for ryegrass. More than half the pasture samples contained 30–70 ppb V on the dry basis. A level of 60 ppb was obtained for oats (grain) and 120 ppb for oat straw. Berg (3a) reports 50 ppb V for corn, 80 ppb for soybean meal, and 2700 ppb for herring fish meal.

V. VANADIUM TOXICITY

Vanadium is a relatively toxic element. Some years ago Franke and Moxon (14) found the relative toxicity of five different elements, fed at 25 ppm of the diet, to lie in the increasing order As, Mo, Te, V, and Se. Dietary concentrations of 25 ppm V were toxic to rats and at intakes of 50 ppm V the animals exhibited diarrhea and mortality. The toxicity of ingested vanadium (as vanadate) is similar in chicks. Thus 30 ppm V as calcium vanadate added to practical chick rations depressed their rate of gain and 200 ppm resulted in high mortality (41). However, Nelson *et al.* (34) reported that chicks tolerated V intakes of

TABLE 52
Effects of Dietary Vanadium and Chromium on Growth and Mortality of Chicks[a]

Chromium as $CrCl_3$ (ppm Cr)	Vanadium as NH_4VO_3 (ppm):	Body weight at 3 wk of age (g)[b]		% mortality at 3 wk of age	
		0	20	0	20
0		238.1	98.5	6.7	86.6
500		256.6	125.2	6.7	66.7
1000		232.6	158.9	10.0	40.0
2000		182.7	193.4	6.7	13.3
Duncan's multiple range test:[c]		98.5 125.2 158.9 182.7 193.4 232.6 238.1 256.6		6.7 6.7 6.7 10.0 13.3 40.0 66.7 86.6	

[a]From Wright (55) as reported by Hill (23).
[b]Averages of from 5 to 15 chicks.
[c]Values on different lines are significantly different; $p = 0.05$.

20–35 ppm and that further amounts induced growth depression. A total intake of 13 ppm V as ammonium metavanadate was then shown to result in growth depression in this species (3). Hathcock *et al.* (19) demonstrated growth depression and mortality in chicks from 25 ppm V fed either as ammonium metavanadate or vanadyl sulfate. The related elements, scandium, titanium, and niobium were not toxic even when fed at 200 ppm. Vanadium toxicity can be

TABLE 53
Effect of Vanadate and Chromate on the Uptake of $^{51}CrO_4$ and $^{48}VO_4$ by Respiring Mitochondria[a]

Labeled ion uptake[b]	m*M*	cpm[c]
$^{51}CrO_4$ vs VO_4		
VO_4	0	15,559
	1	9,803
$^{48}VO_4$ vs CrO_4		
CrO_4	0	7,979
	1	3,880

[a]From Wright (55) as reported by Hill (23).
[b]1 m*M* CrO_4 and VO_4.
[c]$p < 0.05$.

completely prevented by EDTA, apparently by inhibiting its absorption from the gastrointestinal tract. Hill (23) has reported that high intakes of chromium can prevent the growth depression and mortality of chicks associated with feeding 20 ppm V as vanadate (Table 52). Vanadium toxicity is also greatly affected by the diet composition. Thus Berg (3a) found 20 ppb V as sodium metavanadate to be much more toxic, as judged by growth depression of chicks, on sucrose–soybean meal, sucrose–herring fish meal, and corn–soybean diets than on a corn–herring fish meal diet. It was further found the toxicity of the sucrose-containing diets could be increasingly reduced by replacing the sucrose with graded levels of corn, and that the inclusion of 5% of cottonseed meal or dehydrated grass of 0.5% ascorbic acid in the sucrose–fish meal diet markedly reduced the V toxicity. The reasons for these differences are not yet known.

The hair of rats fed 100 ppm V exhibits a reduction in cystine content, suggesting that this element affects the reaction of S-containing compounds (32). Vanadium also reduces coenzyme A (29) and coenzyme Q (1) levels in rats and stimulates monoamine oxidase activity (39). Of further significance is the finding that vanadium, at dietary levels of 25 ppm or less, uncouples oxidative phosphorylation both *in vivo* and *in vitro* (20). This effect of V can partially be prevented by chromate both *in vivo* and *in vitro* (55). Vanadate inhibits the uptake of chromate by respiring mitochondria and chromate inhibits the uptake of vanadate (Table 53). An antagonism between vanadate and chromate is thus apparent from experiments on chick growth and mortality, and on uncoupling of oxidative phosphorylation. A mutual antagonism between the two anions is evident from the studies with respiring mitochondria.

On the evidence from two studies with small numbers of human subjects it seems that vanadium is not a particularly toxic metal to man. Thus Dimond *et al.* (13) gave ammonium vanadyl tartrate orally to six subjects for 6–10 weeks in amounts ranging from 4.5 to 18 mg V/day with no toxic effects other than some cramps and diarrhea at the larger dose levels, and Schroeder *et al.* (43) fed patients 4.5 mg V/day as the oxytartarovanadate for 16 months, with no signs of intolerance becoming apparent but with increased V excretion in the urine. However, 4.5 mg V/day represents a dietary V level of only 11 ppm, assuming that the subjects were consuming about 400 g of dry matter daily. Toxic effects at such intakes would not be expected on the basis of animal experiments with vanadium, as considered previously in this section.

REFERENCES

1. Aiyar, A.S., and Sreenivason, A., *Proc. Soc. Exp. Biol. Med.* **107**, 914 (1961).
2. Allaway, W.H., Kubota, J., Losee, F., and Roth, M., *Arch. Environ. Health* **16**, 342 (1968).

3. Berg, L.R., *Poultry Sci.* **42**, 766 (1963).
3a. Berg, L.R., *Poultry Sci.* **45**, 1346 (1966); Berg, L.R., and Lawrence, W.W., *ibid.* **50**, 1399 (1971).
4. Bertrand, D., *Bull. Soc. Chim. Biol.* **25**, 36 (1943); *Bull. Am. Mus. Nat. Hist.* **94**, 403 (1956).
5. Bowen, W.H., *J. Ir. Dent. Assoc.* **18**, 83 (1972).
6. Butt, E.M., Nusbaum, R.E., Gilmour, T.C., DiDio, S.L., and Mariano, S., *Arch. Environ. Health* **8**, 52 (1964).
7. Buttner, W., *J. Dent. Res.* **42**, 453 (1963).
8. Califano, L., and Boeri, E., *J. Exp. Biol.* **27**, 253 (1950).
9. Califano, L., and Caselli, P., *Pubbl. Stn. Zool. Napoli* **21**, 261 (1948).
10. Ciereszko, L.S., Ciereszko, E.M., Harris, E.R., and Lane, C.A., *Comp. Biochem. Physiol.* **8**, 137 (1963).
11. Curran, G.L., Azarnoff, D.L., and Bolinger, R.E., *J. Clin. Invest.* **38**, 1251 (1959).
12. Curran, G.L., and Burch, R.E., *Trace Subst. Environ. Health–1, Proc. Univ. Mo. Annu. Conf., 1st, 1967,* p. 96 (1968).
13. Diamond, E.G., Caravaca, J., and Benchimol, A., *Am. J. Clin. Nutr.* **12**, 49 (1963).
14. Franke, K.W., and Moxon, A.L., *J. Pharmacol. Exp. Ther.* **61**, 89 (1937).
15. Geyer, G.F., *J. Dent. Res.* **32**, 590 (1953).
16. Hadjimarkos, D.M., *Adv. Oral Biol.* **3**, 253 (1968); *Trace Subst. Environ. Health–8, Proc. Univ. Mo. Annu. Conf., 8th, 1973,* p. 25 (1973).
17. Hamilton, E.I., and Minski, M.J., *Sci. Total Environ.* **1**, 375 (1972/1973).
18. Hamilton, E.I., Minski, M.J., and Cleary, J.J., *Sci. Total Environ.* **1**, 341 (1972/1973).
19. Hathcock, J.N., Hill, C.H., and Matrone, G., *J. Nutr.* **82**, 106 (1964).
20. Hathcock, J.N., Hill, C.H., and Tove, S.B., *Can. J. Biochem.* **44**, 983 (1966).
21. Hein, J.W., and Wisotsky, J., *J. Dent. Res.* **34**, 756 (1955).
22. Henze, M., *Hoppe-Seyler's Z. Physiol. Chem.* **72**, 494 (1911); **83**, 340 (1913).
23. Hill, C.H., *in* "Trace Elements and Human Disease" (A.S. Prasad, ed.), Vol. 2, pp. 281–300. Academic Press, New York, 1975.
24. Hopkins, L.L., Jr., and Mohr, H.E., *in* "Newer Trace Elements in Nutrition" (W. Mertz and W.E. Cornatzer, eds.), p. 195. Dekker, New York, 1973.
25. Hopkins, L.L., Jr., and Mohr, H.E., *Fed. Proc., Fed. Am. Soc. Exp. Biol.* **33**, 1773 (1974).
26. Hopkins, L.L., Jr., and Tilton, B.E., *Am. J. Physiol.* **211**, 169 (1966).
27. Kruger, B.J., *J. Aust. Dent. Assoc.* **3**, 298 (1958).
28. Losee, F., Cutress, T.W., and Brown, R., *Trace Subst. Environ. Health–7, Proc. Univ. Mo. Annu. Conf., 7th, 1973,* p. 192 (1973).
29. Mascitelli-Coriandoli, E., and Citterio, C., *Nature (London)* **183**, 1527 (1959).
30. Mitchell, R.L., *Research (London)* **10**, 357 (1957).
31. Mitchell, R.L., *in* "Trace Analysis" (J.H. Yoe and H.J. Koch, eds.), p. 398. Wiley, New York, 1957.
32. Mountain, J.T., Delker, L.L., and Stokinger, H.E., *Arch. Ind. Hyg. Occup. Med.* **8**, 406 (1953).
33. Muhler, J.C., *J. Dent. Res.* **36**, 787 (1957).
34. Nelson, T.S., Gillis, M.B., and Peeler, H.T., *Poultry Sci.* **41**, 519 (1962).
35. Nielsen, F.H., personal communication (1975).
36. Nielsen, F.H., and Ollerich, D.A., *Fed. Proc., Fed. Am. Soc. Exp. Biol.* **32**, 329 (1973).
37. Nozdryukina, L.R., Grinkevich, N.I., and Gribovskaya, I.F., *Trace Subst. Environ. Health–7, Proc. Univ. Mo. Annu. Conf., 7th, 1973,* p. 353 (1973).
38. Perry, H.M., Jr., and Perry, E.F., *J. Clin. Invest.* **38**, 1452 (1959).

39. Perry, H.M., Jr., Tietelbaum, S., and Schwartz, P.L., *Fed. Proc., Fed. Am. Soc. Exp. Biol.* **14,** 113 (1955).
40. Rezayera, L.T., *Zurmal Obscej, Biol. XXV* **5,** 347 (1964), cited by Söremark (46).
41. Romoser, G.L., Dudley, W.A., Machlin, L.J., and Loveless, L., *Poultry Sci.* **40,** 1171 (1961).
42. Rygh, O., *Bull. Soc. Chim. Biol.* **31,** 1052 and 1408 (1949); **33,** 133 (1953); *Research (London)* **2,** 340 (1949).
43. Schroeder, H.A., Balassa, J.J., and Tipton, I.H., *J. Chronic Dis.* **16,** 1047 (1963).
44. Schwarz, K., and Milne, D.B., *Science* **174,** 426 (1971).
45. Somerville, J., and Davies, B., *Am. Heart J.* **64,** 54 (1962).
46. Söremark, R., *J. Nutr.* **92,** 183 (1967).
47. Söremark, R., and Ullberg, S., *in* "Use of Radioisotopes in Animal Biology and the Medical Sciences" (M. Fried, ed.), Vol. 2. Academic Press, New York, 1962.
48. Strasia, C.A., Thesis, University Microfilms, Ann Arbor, Michigan, 1971.
49. Talvitie, N.A., and Wagner, W.D., *Arch. Ind. Hyg.* **9,** 414 (1954).
50. Thomassen, P.R., and Leicester, H.M., *J. Dent. Res.* **39,** 473 (1960).
51. Tipton, I.H., and Cook, M.J., *Health Phys.* **9,** 103 (1963).
52. Webb, D.A., *Proc. R. Soc. Dublin* **21,** 505 (1937).
53. Webb, D.A., *Publ. Stn. Zool. Napoli* **28,** 273 (1956).
54. Welch, R.M., and Cary, E.E., *Agric. Food Chem.* **23,** 479 (1975).
55. Wright, W.R., Doctoral Thesis, North Carolina State University, Chapel Hill, North Carolina (1968).

16

Silicon

I. SILICON IN ANIMAL TISSUES AND FLUIDS

Reliable data on the distribution of silicon in the animal body are now becoming available with the development of improved methods of analysis (36, 42) and with care to minimize contamination from glass and dust (27). Some years ago the normal range for human fetal tissues was given as 18–180 ppm Si (dry basis) compared with 23–460 ppm for adult human tissues (36). The results of two more recent studies for a range of tissues in adult man (23), rats, and rhesus monkeys (39) are presented in Table 54. The very high levels in the human lymph nodes were shown to be associated with the presence of clusters and grains of quartz (23). Comparable levels of silicon in rat tissues to those given in Table 54 have been reported by McGavack *et al.* (42) and Carlisle (15). Normal whole blood levels in man (42), monkeys (39), and bovines (8) lie close to 1 μg Si/ml, while 5 μg Si/ml may be considered normal for ovine blood (28). This silicon occurs mostly in solution as monosilicic acid and varies little in concentration except after massive oral administration of soluble silicates.

The highest Si concentrations generally occur in the skin and its appendages and in the aorta, trachea, and tendon (15). The concentrations in these tissues decline with age, whereas most other tissues display no such age change. For example, Carlisle (15) compared the tissue Si levels in rabbits at 12 weeks and at 18–24 months of age. The levels in the heart, liver, and muscle remained between 5 and 15 μg Si/g (d.b.) at the two ages, while those of the aorta, thymus, and skin declined from 80 to 15, from 56 to 2, and from 46 to 9 μg/g,

TABLE 54
Silicon Concentrations in Animal Tissues[a]

Tissue	Adult man[b]	Adult rat[c]	Rhesus monkey[c]
Brain	23 ± 4.4	0.8 ± 0.9	1.4 ± 0.7
Kidney	40 ± 11	0.5 ± 0.7	1.6 ± 1.5
Liver	33.6 ± 13.8	1.6 ± 1.5	1.2 ± 1.2
Lung	57.4 ± 10.7	1.6 ± 1.4	194 ± 183.2
Muscle	41 ± 0.9	0.9 ± 0.7	1.2 ± 0.5
Testis	3.1 ± 1.6	1.1 ± 1.1	2.0 ± 1.2
Lymph nodes	489 ± 215	4.1 ± 5.5	21.9 ± 10.6

[a] μg Si/g wet weight.
[b] From Hamilton *et al.* (23).
[c] From LeVier (39).

respectively. Similarly, the mean level in fetal pig hair was reported as 95 μg Si/g, compared with 10 μg/g in mature pig hair. A comparable decrease in the Si content of rat skin with age has also been reported (38). The Si content of the normal human aorta decreases considerably with age, and the level in the arterial wall has been shown to decrease with the development of atherosclerosis (40). High levels of silicon (mean 243 ± 18, range 100–450 μg/g dry weight) have been reported in human dental enamel (41) and in the head of the femur, containing the epiphysis, of monkeys (456.3 ± 71.0 μg Si/g dry weight) (39).

The high-Si content of epithelial and connective tissues arises from the occurrence of this element as an integral part of the mucopolysaccarides which constitute the essential structural components. Thus Schwarz (50) detected no less than 330–554 ppm bound Si in purified hyaluronic acid from umbilical cord, chondroitin 4-sulfate, dermatan sulfate, and heparan sulfate. The corresponding figures for chondroitin 6-sulfate, heparin, and keratan sulfate-2 from cartilage were 57–191 ppm Si, while hyaluronic acids from vitreous humor and keratan sulfate-1 from cornea were Si-free. Even larger amounts of bound silicon were found in pectin (2580 ppm) and alginic acid (451 ppm). Using emission spectroscopy to avoid possible losses by ashing, Schwarz and Chen (51) found the collagens of mouse skin, calf skin, and articular cartilage to contain 4201, 4369, and 3859 μg Si/g, respectively. Several collagens were then prepared under conditions which would eliminate any Si derived from water, chemicals, or glass. Under these conditions the level of Si obtained in three times reprecipitated salt soluble collagen from rat skin was 897, acid soluble rat skin 1997, and rat tail tendon 1108 μg Si/g. These levels indicate the presence of at least 3–6 atoms of Si per each protein α chain in the collagen molecule.

Strong alkali and acid hydrolysis frees the Si–polysaccharide bond, giving free, dialyzable silicate, but enzymatic hydrolysis of hyaluronic acid or pectin does not liberate silicic acid. It leads to products of low molecular weight still containing Si in bound form. Carlisle (15), who had independently shown silicon to be a component of mucopolysaccharides (14), found that disaccharides enzymatically derived from chondroitin sulfate A contained considerably more Si than disaccharides obtained from chondroitin sulfate C. The finding that Si is a constituent of disaccharide units led her to suggest that it may be added at the state of formation of the polysaccharide chain from smaller units. Schwarz (50) concluded that "Si is present as silanolate, i.e., an ether (or esterlike) derivative of silicic acid, and that R_1–O–Si–O–R_2 or R_1–O–Si–O–Si–O–R_2 bridges play a role in the structural organization of glycosaminoglycans and polyuronides."

Few data are available on the silicon content of milk in which acceptable modern methods of analysis have been employed and care taken to avoid contamination from glass. It seems that with cow's milk there is considerable individual variation, a marked decrease from the levels in colostrum to those in milk (37), and little influence of dietary Si intakes. Thus Archibald and Fenner (3) reported that the milk of six cows alternately fed a control ration and one containing added sodium silicate at the rate of 1 g/day (230 mg Si) averaged 1.4 μg/ml, irrespective of treatment.

II. SILICON METABOLISM

Silicon enters the alimentary tract from the food as monosilicic acid, as solid silica, and in the organic bound forms with pectin, mucopolysaccharides, and other such compounds, as discussed in the previous section. Little is known of the extent or mechanism of Si absorption from these sources. In guinea pigs absorption apparently occurs mainly as monosilicic acid (49). Some of this comes from the solid silica of the plant materials consumed, which is partly dissolved by the fluids of the gastrointestinal tract. In sheep the extent of solution and absorption of solid silica as monosilicic acid varies with the silica content of the diet. Jones and Handreck (28) found that the amounts excreted in the urine increased with the increasing silica content of the diet from 0.10 to 2.84%, but reached no more than a maximum of 205 mg SiO_2/day. This amount represented less than 4% of the total intake.

Increased urinary Si output with increasing intake, up to fairly well-defined limits, has been demonstrated in man (26), rats (32), guinea pigs (49), and cows (4). In sheep Nottle (45) found urinary excretion of silica to increase with rising dietary silica intakes up to an intake of 8 g SiO_2/day. Thereafter urinary excretion leveled off at 200–250 mg/day. The upper limits of urinary Si excretion do not seem to be set by the ability of the kidney to excrete more, because much greater urinary excretion can occur after peritoneal injections (49). Those limits are determined by the rate and extent of Si absorption from the gastrointestinal tract into the blood. In the ruminant this is influenced by the solubility of silica in the rumen fluid (28). Once it has entered the bloodstream, Si must pass rapidly into the urine and tissues because, even at widely divergent Si intakes, the Si level in the blood remains practically constant (28, 35). In the experiments of Jones and Handreck (28), cited earlier in this section, the sum of the amounts appearing in the feces and urine was within 1% of the amounts ingested, indicating that body retention was small. Urinary excretion was also low. In three separate experiments with sheep the proportion excreted in the urine, but not the amounts, decreased progressively from 3.3 to 0.55% as the intake increased from 0.4 to 14 g Si/day (19, 28, 45).

Changes in the absorption and resulting levels of Si in the blood and intestinal tissues of rats in relation to age, sex, and the activity of various endocrine glands have also been reported (16). These might well account for some of the changes in Si in the tissues with age, mentioned in Section I.

Microscopic solid particles of silica from plants have been demonstrated in the lymph nodes and urinary calculi of sheep. These particles are absorbed as such, to a small extent, from the alimentary tract (5). The gut wall of man is also permeable to particles the size of diatoms. Volkenheimer (58) showed that diatomaceous earth particles are absorbed through the intact intestinal mucosa, pass through the lymphatic and circulatory systems, and reach other tissues in arterial blood via the alveolar region of the lung. Examination of human organs has revealed silicons diatoms in lungs, liver, and kidney, as a consequence of their presence in atmospheric dust and their movement from the respiratory tract (21). The capacity of these particles to travel in the blood and to penetrate body membranes, including the placenta, is illustrated further by their presence in the organs of stillborn and premature infants (21).

III. SILICON DEFICIENCY AND FUNCTIONS

Silicon is essential for growth and skeletal development in rats and chicks, and a mechanism and site of action have been identified.

1. Growth

By means of specially purified diets and a plastic isolator environment Schwarz and Milne (52) were able to demonstrate significant increases in the growth rate of rats from the addition of 50 mg Si/100 g of diet (5 ppm) as sodium metasilicate ($Na_2SiO_3 \cdot 9H_2O$) in aqueous solution. The increases in weight of the weanling rats over that of the controls were 33.8% on one basal

TABLE 55
Growth Effects of Dietary Silicon in Rats and Chicks Maintained in a Trace Element Controlled Environment on Low-Silicon Diets

	No. of animals	Average daily weight gain (g)	% increase	p
Rat studies[a]				
Basal diet A				
Control	15	1.51 ± 0.11		
50 mg% Si	11	2.02 ± 0.08	33.8	< 0.005
Basal diet B				
Control	12	1.19 ± 0.06		
50 mg% Si	11	1.49 ± 0.06	25.2	< 0.005
Chick studies[b]				
Study no. 1				
Control	36	2.37 ± 0.11		
Si-supplemented	36	3.10 ± 0.10	30	< 0.01
Study no. 2				
Control	30	3.25 ± 0.09		
Si-supplemented	30	4.20 ± 0.09	30	< 0.02
Study no. 3				
Control	48	2.57± 0.09		
Si-supplemented	48	3.85 ± 0.11	49.8	< 0.01

[a]From Schwarz and Milne (52).
[b]From Carlisle (13).

diet and 25.5% on another, each over a 26-day period (Table 55). Lower levels of supplementary silicon gave statistically insignificant responses. The unsupplemented animals also exhibited an impaired incisor pigmentation which was significantly improved but not prevented by Si supplements. Similar responses in the growth of chicks from Si supplementation of purified diets were independently demonstrated by Carlisle (13). The results of three studies, in which 30, 30, and 49.8% increases in average daily weight gain over a 23-day period were obtained, are given in Table 55. The deficient chicks were smaller but in proportion with all organs appearing relatively atrophied, and with the legs and comb particularly pale. The deficient chicks had no wattles and the combs were severely attenuated. Skeletal development was significantly retarded as discussed in the next section.

2. Calcification and Bone Development

The first indications of a physiological role for silicon came from the *in vitro* electron microprobe studies of Carlisle (11) showing silicon to be localized in active growth areas in the bones of young mice and rats. The amount present in specific very small regions within the growth areas appeared to be uniquely related to the "maturity" of the bone mineral. In the earliest stages of calcification in these regions both the Si and Ca contents of the osteoid tissue were found to be very low, but as mineralization progressed the Si and Ca contents rose congruently. In a more advanced stage the amount of Si fell markedly so that as Ca approached the proportion present in bone apatite the Si was present only at the detection limit. In other words, the more "mature" the bone mineral the smaller the amount of Si. Further studies of the Ca:P ratio in Si-rich sites gave values below 1.0 compared with a Ca:P ratio of approximately 1.67 in mature bone apatite. These findings suggested strongly that Si is involved with P in an organic phase during the series of events leading to calcification.

In subsequent *in vivo* studies of bones from weanling rats on a low-Si (< 5 ppm) diet or on supplements of 10, 25, or 250 ppm Si, the silicon was found to hasten the rate of bone mineralization (12). The tibia of the rats on the 250 ppm Si reached a higher degree of mineralization in a shorter time than the tibia from the low- and medium-Si diets. Abnormalities involving articular cartilage and connective tissue were also observed in Si-deficient chicks. Long bone joints were smaller and the bones contained 34–35% less water than those of Si-supplemented chicks. The supplemented chicks also revealed significantly higher total percentage hexosamine contents in their articular cartilage and higher Si and hexosamine concentrations in their combs. These findings point clearly to an involvement of silicon in mucopolysaccharide synthesis in cartilage and connective tissue. The occurrence of this element as an integral component of the mucopolysaccharides of cartilage and collagens (50, 51) was mentioned in Section I. They indicate that Si probably functions as a biological cross-linking agent contributing to the strength, structure, and resilience of connective tissue. The unique properties of Si atoms in respect to bonding and macromolecular structures are well known (44). The bone abnormalities of Si-deficient rats (52) and chicks (13) arise from an impairment of mucopolysaccharide synthesis in the formation of articular cartilage. Silicon must participate in other processes in which mucopolysaccharides are involved apart from bone formation, including the growth and maintenance of the arterial wall and the skin and its appendages. Apart from the absence of wattles and the severely attenuated combs of Si-deficient chicks, marked changes in the integument have not been reported in Si deficiency, either in rats (52) or chicks (15).

IV. SILICON REQUIREMENTS AND SOURCES

The minimum dietary Si requirements compatible with satisfactory growth and health are largely unknown, although it is clear from the limited evidence obtained from experiments with rats and chicks that the amounts are large relative to those of other trace elements. The Si levels in the basal diets used in those experiments were not reported but are obviously extremely low. On this assumption the Si requirements for growth and satisfactory skeletal development in rats approximate 50 μg/g of dry diet (52), where the silicon is provided as water-soluble sodium metasilicate. Whether the silicon compounds occurring in natural materials would be more available or less available and the Si requirement therefore lower or higher than the 50 ppm tentatively given must await the results of further research. Evidence that the dietary Si requirement is relatively high is further apparent from the experiments of Carlisle (12) on the rate of bone mineralization in chicks. Silicon supplementation of a low-Si diet at the rate of 250 ppm Si increased the ash content of the tibia significantly more at 2 weeks than did either 25 or 10 ppm, although the differences had largely disappeared at 5 weeks. Unfortunately a supplementation level of 50 ppm Si was not included in this experiment, so that one can only speculate whether such a level would have been as effective as the 250 ppm.

The demonstration of the essentiality of silicon for the higher animals is so recent that reliable data on the Si content of human foods and dietaries are meager indeed. Also silicon is so plentiful and ubiquitous in the environment that analyses rightly conducted to minimize contamination could give misleading information on actual amounts ingested. By the same reasoning it is difficult to imagine a Si deficiency ever arising under natural conditions in man or domestic animals.

Foods of plant origin are normally much richer in silicon than those of animal origin, although it is obvious from Section I that muscle and organ meats contain substantial Si concentrations. In plants the amounts or proportions of silicon present as monosilicic acid and solid silica vary with the species, stage of growth, and soil conditions under which the plant has grown. Whole grasses and cereals may contain 30–40% of their total ash, or 3–4% of the whole dry plants, as SiO_2, with levels up to 6% silica in some range grasses (10). In leguminous plants total Si concentrations are appreciably lower, with a high proportion of the relatively low amounts present as monosilicic acid. Solid silica is only sparsely deposited in these species (7, 10). Cereal grains high in fiber such as oats are much richer in silicon than low fiber grains such as wheat or maize (46). This would suggest that the Si content of patent white flour is significantly lower than that of the whole wheat from which it is made, although this does not appear to have been specifically studied. Substantial losses of silicon occur in the refining of sugar. Thus Hamilton and Minski (22) reported the following values: Barbados brown sugar, 735; Demerara sugar, 60; refined sugar, 2; and granulated sugar, 4 μg Si/g dry weight. These workers also obtained a mean total Si intake from human adult diets consumed in Great Britain of 1.2 ± 0.1 g/day. This level places silicon in the major element class in terms of magnitude of dietary intake. The authors were careful to point out that their data for silicon are subject to possible contamination because of contact between the samples and glass surfaces prior to receipt by the laboratory, although they state that "previous studies suggest that the degree of contamination is slight."

In mature gramineous plants most of the silicon present is in the form of solid mineral particles, known as opal phytoliths ($SiO_2 \cdot H_2O$) (7). The marked species difference among plants in the total amount of Si absorbed and subsequently secreted as phytoliths is illustrated by one study in which ryegrass (*Lolium perenne*) was found to contain 23 times as much insoluble ash as lucerne (*Medicago sativa*). This difference was reflected in the Si and opal phytolith contents of the two species (7). In another study, prairie grass hay (mainly *Festuca scabrella*) averaged 2.92% total Si (d.b.) compared with only 0.18% Si in alfalfa hay (4).

Since the opal phytoliths of pasture plants are harder than the dental tissues of sheep, and the amounts ingested by grazing animals are so large and continuous, it has been suggested that these minerals may be a major cause of wear in sheep's teeth (6). At a level of 4% SiO_2, and at a dry matter intake of 1 kg/day, a sheep would ingest 40 g SiO_2/day, or 14 kg over a period of a year. This represents a very large amount of abrasive material. Moreover, it excludes the large quantities of silica that can occur as quartz particles on the surface of plants from soil contamination. This source of adventitious silica has been implicated as a significant source of wear of teeth in grazing sheep (24).

V. SILICON TOXICITY

1. Silicosis in Man

Detailed consideration of the disease silicosis, which occurs in certain classes of miners due to the continued inhalation of silical particles into the lungs, lies outside the scope of this text. Particles of silica and asbestos (fibrous silicates of complex composition) have long been known to stimulate a severe fibrogenic reaction in the lungs and elsewhere in the body. This reaction arises initially from phagocytosis of silica particles by alveolar macrophages. Collagen synthesis by neighboring fibroblasts is stimulated by the death of these macrophages. The particular toxicity of silica to macrophages derives from the fact that the particles are taken up into lysosomes and readily damage lysosomal membranes through H-bonding reactions (2, 43). Heppleston and Styles (25) have provided evidence that the macrophage–silica interaction results in the release of a factor of unknown nature that stimulates collagen formation, a finding of great interest in light of the involvement of Si in collagen synthesis disclosed by recent research.

Malignant tumors of the pleura and peritoneum constitute a further manifestation of the toxicity of silica and asbestos to man and experimental animals (59). Allison (1) has suggested that lysosomes may be involved in the malignant transformations brought about by silica and that enzymes released from lysosomes damage chromosomes, with a chromosome mutation leading to malignancy.

2. Silica Urolithiasis

Under some conditions part of the silicon of the urine is deposited in the kidney, bladder, or urethra to form calculi or uroliths. Small calculi may be excreted without harm while large calculi can block the passage of urine and cause death of the animal. Urinary calculi can be composed of various predominent minerals, particularly Ca, Mg, P, and Si. Silica urolithiasis is a serious problem in grazing wethers in Western Australia (47), and in grazing steers in the western regions of Canada (17, 60) and the northwestern parts of the United States (48, 55).

The silica of ovine and bovine calculi has been identified as amorphous opal (5, 20), most of which is derived from the absorbed monosilicic acid. A small proportion of the opal occurs as phytoliths from plants, with occasional fragments of sponge spicules and diatoms embedded in the calculi (5). In addition to hydrated silica, siliceous uroliths contain small amounts of accessory elements and organic material. The exact nature of the organic material, or matrix, in siliceous calculi has not been determined, but the chemical studies of Keeler (30, 34) indicate the presence of a glycoprotein containing a neutral carbohydrate moiety.

The factors responsible for the formation of siliceous calculi are poorly defined. Attempts to produce them in sheep and cattle by adding silicates to the diet (9, 60) or by restricting water consumption (56) have not been successful, even when the urinary excretion level achieved was 2- to 3-fold greater than the 70–80-μg Si/ml level at which silicon normally precipitates in bovine urine (33). The concentration in the urine of sheep (45) and cattle (4) usually exceeds that of a saturated solution of amorphous silica and may reach 467 ppm Si in sheep. It is clear that high dietary intakes and high silica outputs in the urine, associated with supersaturation of the urine, are insufficient to explain the polymerization of the monosilicic acid, deposition of silica, and formation of calculi.

The glycoprotein component of the organic matrix has been assigned a critical role in the formation of urinary calculi in man, through acting as a primary matrix which becomes secondarily mineralized. A similar theory has been adopted to explain the formation of siliceous calculi in cattle (31) and phosphatic calculi in sheep, cattle, and dogs (18). Jones and Handreck (29) have disputed this theory on the grounds that the main mechanism involved is "precipitation of the inorganic components which, in turn, depends upon both supersaturation and nucleation." They tend to favor the theory that foreign particles of silica and other foreign particles, such as have been found in calculi from sheep, act as nuclei for the deposition of silica. Solid (amorphous) silica is known to accelerate the polymerization and deposition of silica from supersaturated solution, but whether solid particles of silica are consistently implicated in the formation of siliceous calculi is unknown. In view of the marked dependence of urinary silica concentration on the rate of urine excretion in cattle, it has been suggested that any method which increases urine output could also be used to prevent urolith formation (4).

3. Silica and Forage Digestibility

The digestibility of forage dry matter *in vivo* has been shown to be significantly depressed by increasing levels of silica (SiO_2 in ash from plant tissues) (53, 57). These observations were investigated further by Smith and co-workers (54), who found that aqueous sodium silicate added to rumen cultures significantly depressed the organic matter digestibility *in vitro* of siliceous forages, already known to exhibit depressed organic matter digestibility due to silica accumulated in the plant tissues. The depression generally amounted to about one percentage unit of organic matter digestibility for each increase of 100 mg/liter in "soluble silica" concentration but was greater when a highly siliceous grass was used as the substrate. The effect of silicate was modified by adding glucose, urea, and/or a mixture of minerals (Mg, Mn, Zn, Co, and Cu), suggesting that availability of minerals to sustain cellulolytic microbial activity may be a major factor influencing the effect of soluble silica on forage digestion by ruminants.

REFERENCES

1. Allison, A.C., *Proc. R. Soc. London, Ser. B* **171**, 19 (1968).
2. Allison, A.C., Harington, J.S., and Birbeck, M., *J. Exp. Med.* **124**, 141 (1966).
3. Archibald, J.G., and Fenner, H., *J. Dairy Sci.* **40**, 703 (1957).
4. Bailey, C.H., *Am. J. Vet. Res.* **28**, 1743 (1967).
5. Baker, G., Jones, L.H.P., and Milne, A.A., *Aust. J. Agric. Res.* **12**, 473 (1961).
6. Baker, G., Jones, L.H.P., and Wardrop, I.D., *Nature (London)* **184**, 1583 (1959).
7. Baker, G., Jones, L.H.P., and Wardrop, I.D., *Aust. J. Agric. Res.* **12**, 426 (1961).
8. Baumann, H., *Hoppe-Seyler's Z. Physiol. Chem.* **319**, 38 (1960); **320**, 11 (1960).
9. Beeson, W.M., Pence, J.W., and Holan, G.C., *Am J. Vet. Res.* **4**, 120 (1943).
10. Bezeau, L.M., Johnston, A., and Smoliak, S., *Can. J. Plant Sci.* **46**, 625 (1966).
11. Carlisle, E.M., *Fed. Proc., Fed. Am. Soc. Exp. Biol.* **28**, 374 (1969); *Science* **167**, 279 (1970).
12. Carlisle, E.M., *Fed. Proc., Fed. Am. Soc. Exp. Biol.* **29**, 565 (1970).
13. Carlisle, E.M., *Fed. Proc., Fed. Am. Soc. Exp. Biol.* **31**, 700 (1972).

14. Carlisle, E.M., *Fed. Proc., Fed. Am. Soc. Exp. Biol.* **32,** 930 (1973).
15. Carlisle, E.M., *Fed. Proc., Fed. Am. Soc. Exp. Biol.* **33,** 1758 (1974).
16. Charnot, Y., and Péres, G., *Ann. Endocrinol.* **32,** 397 (1971).
17. Connell, R., Whiting, F., and Forman, S.A., *Can. J. Comp. Med. Vet. Sci.* **23,** 41 (1959).
18. Cornelius, C.E., and Bishop, J.A., *J. Urol.* **85,** 842 (1961).
19. Emerich, R.J., Embay, L.B., and Olson, O.E., *J. Anim. Sci.* **18,** 1025. (1959).
20. Forman, S.A., Whiting, F., and Connell, R., *Can. J. Comp. Med. Vet. Sci.* **23,** 157 (1959).
21. Geissler, U., and Gerloff, J., *Nova Hedwigia* **10,** 565 (1965).
22. Hamilton, E.I., and Minski, M.J., *Sci. Total Environ.* **1,** 375 (1972/1973).
23. Hamilton, E.I., Minski, M.J., and Cleary, J.J., *Sci. Total Environ.* **1,** 341 (1972/1973).
24. Healey, W.B., and Ludwig, T., *N. Z. J. Agric. Res.* **8,** 737 (1965).
25. Heppleston, A.W., and Styles, J.A., *Nature (London)* **214,** 521 (1967).
26. Holt, P.F., *Br. J. Ind. Med.* **7,** 12 (1950).
27. Jankowiak, M.E., and LeVier, R.R., *Anal. Biochem.* **44,** 462 (1971).
28. Jones, L.H.P., and Handreck, K.A., *J. Agric. Sci.* **65,** 129 (1969).
29. Jones, L.H.P., and Handreck, K.A., *Adv. Agron.* **19,** 107 (1967).
30. Keeler, R.F., *Am. J. Vet. Res.* **21,** 428 (1960).
31. Keeler, R.F., *Ann. N.Y. Acad. Sci.* **104,** 592 (1963).
32. Keeler, R.F., and Lovelace, S.A., *J. Exp. Med.* **109,** 601 (1959).
33. Keeler, R.F., and Lovelace, S.A., *Am. J. Vet. Res.* **22,** 617 (1961).
34. Keeler, R.F., and Swingle, K.F., *Am. J. Vet. Res.* **20,** 249 (1959).
35. King, E.J., and Belt, T.H., *Physiol. Rev.* **18,** 329 (1938).
36. King, E.J., Stacy, B.D., Holt, P.F., Yates, D.M., and Pickles, D., *Analyst* **80,** 441 (1955).
37. Kirchgessner, M., *Z. Tierphysiol., Tierernahr. Futtermittelkd.* **14,** 270 and 278 (1959).
38. Leslie, J.G., Kung-Ying, T.K., and McGavack, T.H., *Proc. Soc. Exp. Biol. Med.* **110,** 218 (1962).
39. LeVier, R.R., *Bioinorg. Chem.* **4,** 109 (1975).
40. Loeper, J., Loeper, J., and Lemaire, A., *Presse Med.* **74,** 865 (1966).
41. Losee, F., Cutress, T.W., and Brown, R., *Trace Subst. Environ. Health–7, Proc. Univ. Mo. Annu. Conf., 7th, 1973,* p. 19 (1973).
42. McGavack, T.H., Leslie, J.G., and Tang Kao, K., *Proc. Soc. Exp. Biol. Med.* **110,** 215 (1962).
43. Nash, T., Allison, A.C., and Harington, J.S., *Nature (London)* **210,** 259 (1966).
44. Needham, A.E., "The Uniqueness of Biological Materials." Pergamon, Oxford, 1965.
45. Nottle, M.C., *Aust. J. Agric. Res.* **17,** 175 (1966).
46. Nottle, M.C., private communication (1962).
47. Nottle, M.C., and Armstrong, J.M., *Aust. J. Agric. Res.* **17,** 165 (1966).
48. Parker, K.G., *J. Range Manage.* **10,** 105 (1957).
49. Sauer, F., Laughland, D.H., and Davidson, W.M., *Can. J. Biochem. Physiol.* **37,** 183 and 1173 (1959).
50. Schwarz, K., *Proc. Natl. Acad. Sci. U.S.A.* **70,** 1608 (1973).
51. Schwarz, K., and Chen, S.C., *Fed. Proc., Fed. Am. Soc. Exp. Biol.* **33,** Abstr. No. 2795, p. 704 (1974).
52. Schwarz, K., and Milne, D.B., *Nature (London)* **239,** 333 (1972).
53. Smith, G.S., Nelson, A.B., and Boggino, E.J.A., *J. Anim. Sci.* **33,** 466 (1971).
54. Smith, G.S., and Urquhart, N.S., *J. Anim. Sci.* **41,** 882 (1975); Smith, G.S., and Nelson, A.B., *ibid.* **41,** 891 (1975).

55. Swingle, K.F., *Am. J. Vet. Res.* **14**, 493 (1953).
56. Swingle, K.F., and Marsh, H., *Am. J. Vet. Res.* **14**, 16 (1953).
57. Van Soest, P.J., and Jones, L.H.P., *J. Dairy Sci.* **51**, 1544 (1965).
58. Volkenheimer, G.Z., *Gastroenterol.* **2**, 57 (1964).
59. Wagner, C., *Perugia Quadrenn. Int. Conf. Cancer* **3**, p. 589 (1966).
60. Whiting, F., Connell, R., and Forman, S.A., *Can J. Comp. Med. Vet. Sci.* **22**, 332 (1958).

17

Lead

Biological interest in lead has centered principally on its properties as a highly toxic cumulative poison in man and animals. In recent years the problem of long-term exposure to increased amounts of lead in highly urbanized and motorized environments has engaged particular attention. The possibility that lead in low concentrations performs some vital functions cannot be excluded, especially in the light of the suggestive evidence obtained by Schwarz (77) that lead is required for growth in rats. However, more clear-cut data are required before lead can be included among the essential trace elements.

I. LEAD IN ANIMAL TISSUES AND FLUIDS

The total body burden of lead in "normal" adult man ranges from 90 to 400 mg (45, 57, 76). Schroeder and Tipton (76) found the mean total body Pb content of 150 U.S. accident victims to be 121 mg, of which about 90% was present in the skeleton. The affinity of bone for lead and the much higher Pb concentrations in bone than in soft tissues are apparent from numerous studies (40, 45, 82). Human soft tissues were reported some years ago to range from 0.13 to 0.50 ppm Pb in the brain and from 1.3 to 1.7 ppm Pb (wet weight) in the liver (45, 82). The levels in four fetuses of 7–8 months gestation ranged from 0.17 ppm in the brain to 0.68 ppm Pb in the liver. Similar relatively high-Pb concentrations in the tissues of stillborn infants were later demonstrated (76). Lead was shown to accumulate in the tissues with age, up to 50–60 years in U.S.

residents, particularly in the bones, aorta, kidney, liver, lung, and spleen. No comparable increase with age, except in the aorta, was observed in tissues from Africa and the Middle East, and their median values were generally lower than those in the United States. The mean concentrations in the tissues of English residents were recently reported as follows: brain, 0.3 ± 0.1; kidney, 1.4 ± 0.2; liver, 2.3 ± 0.6; lung, 0.4 ± 0.05; lymph nodes, 0.4 ± 0.1; ovary, 0.09 ± 0.03; testis, 0.10 ± 0.03; and muscle, 0.02 ± 0.006 μg Pb/g wet weight (34).

The pattern of distribution of lead, with the highest levels in the bones and the lowest in the muscles, is similar in normal laboratory and farm animals to that in man, and concentrations are of the same order as those just cited for human tissues (35, 40, 41). For example, Hammond *et al.* (35) found the whole fresh livers of 14 healthy calves and five cows to range from 0.2 to 1.9 μg Pb/g (mean 0.5). These values compare well with the range of 0.3–1.5 μg/g reported by Allcroft (1) for the livers of healthy heifers and calves. Hsu and co-workers (40) obtained the following levels in the kidneys, liver, humerus, and femur of weanling pigs fed a low-Pb, low-Ca diet and a low-Pb, high-Ca diet, respectively: kidney, 15 and 8; liver, 10 and 7; humerus, 75 and 50; and femur, 83 and 50 ppm Pb dry basis.

Lead concentrations increase at high-Pb intakes in all tissues, except the muscles, and especially in the bones, liver, kidney, and hair. The hair of normal children has been reported to range from 2 to 95 (mean 24) μg Pb/g, compared with a range of 42–975 (mean 282) μg Pb/g for the hair of patients with chronic plumbism (48). In a separate study the hair of 78 normal males averages 17.8 ± 2.17 and that of 47 normal females 19.0 ± 2.95 μg Pb/g (75). Lead does not accumulate significantly in the tissues at moderately high dietary Pb intakes because of corresponding increases in excretion, but at higher intakes, substantial tissue deposition occurs. For example, Dinius *et al.* (19) observed no significant differences in the Pb concentrations of the liver, kidneys, and cerebral cortex of calves fed diets containing 0.9, 11.1, and 10.5 ppm Pb for 100 days, whereas comparable calves fed similar diets containing 102 ppm Pb as added lead chromate revealed markedly increased Pb concentrations in the liver and kidneys, but not in the cerebral cortex or longissimus muscle. In the experiments of Hsu and co-workers (40) a dietary Pb intake of 1000 ppm as lead acetate produced still higher Pb concentrations in the liver and kidneys and especially in the bones of the pigs on the low-Ca basal diet.

The subcellular distribution of lead in the tissues, the association of the lead within intranuclear inclusion bodies, and the nature and significance of other associations of lead within the cells and tissues of the body are considered in Section II.

The levels of lead in the blood of inhabitants of 16 countries ranged from 0.15 to 0.40 μg/ml, with an overall mean of 0.17 μg/ml (27). A mean of 0.25 ± 0.03 μg Pb/ml was recently reported for the blood of 103 individuals in England (34).

The upper limit of "normal" is often given as 0.50 μg Pb/ml, and according to Kehoe (44) clinical Pb poisoning will not occur below 0.8 μg Pb/ml. However, cases of Pb poisoning, particularly in children, have been reported in which the blood levels have been below 0.8 μg/ml, and even as low as 0.4 μg/ml (see Hicks, 38). Blood levels of more than 0.4 μg/ml are now considered to represent an undesirable level of exposure (61), and 0.25 μg Pb/ml has been suggested as the "danger" blood level in children (16).

Average whole blood lead levels have been given as 0.26 μg/g for rats and 0.25 μg/g for guinea pigs, when both are consuming normal low-Pb diets (41). Lower mean blood levels of 0.14 ± 0.01 and 0.13 ± 0.01 μg Pb/ml have been reported for normal sheep and cattle, respectively (1). Weanling pigs fed a low-Pb, low-Ca diet for 13 weeks carried blood Pb concentrations of 0.4 μg/ml, while the blood of those on a low-Pb, high-Ca diet contained only 0.2 μg Pb/ml. Comparable pigs given these diets plus 1000 ppm Pb as lead acetate for the same period developed blood Pb levels of 1.0 and 0.6 μg/ml, respectively (40).

Most of the lead in blood is present in the erythrocytes, both at high and at low total blood levels. In fact Rosen and Trinidad (68) found plasma Pb levels to be constant between 1 and 7 μg/100 ml in normal and Pb-intoxicated children over a wide range of red cell Pb concentrations. It seems that plasma lead has a ceiling value, possibly related to the binding capacity of a low molecular weight protein or polypeptide of the serum (47), and that the red cells represent a large repository for Pb, maintaining plasma Pb concentrations within narrow limits (68).

The urinary Pb concentration in the inhabitants of the 16-country survey mentioned previously centered about a mean of 35 μg/liter, with 95% of the samples containing less than 65 μg/liter (27). These values compare well with the mean of 27 μg Pb/liter obtained much earlier by Kehoe and co-workers (45). A significant positive correlation exists between urinary lead levels and urinary ALA (aminolevulinic acid) levels in children, and determination of urinary ALA has been suggested as a screening procedure for the detection of early lead exposure in asymptomatic children (18). However, erythrocyte δ-aminolevulinic acid dehydrase (ALA-D) is more sensitive than ALA in urine as an indicator of blood Pb levels, particularly at levels below 40 μg/ml (37). These aspects are considered further in relation to the diagnosis of lead poisoning.

The level of lead in normal cow's milk is reported to be 0.02–0.04 mg/kg and that of ewe's milk in early lactation 0.11–0.15 mg/kg (9, 45). Lead readily passes the mammary barrier so that dosing of the animals with lead salts produces a marked increase in the level in the milk (9, 89). The lead content of market milk in U.S. cities ranged from 0.02 to 0.08 mg/kg, with no significant differences between cities and with a national weighted average close to 0.05 mg Pb/kg or about 50 μg/liter (60). A later survey of bulk milk in that country revealed a mean level of 40 μg Pb/liter for 270 samples, with no samples greater than 200 μg/liter (55). Human milk does not appear to have been so extensively studied, but 22 samples examined by Murthy and Rhea (59) averaged only 0.012 ± 0.004 ppm, or close to 12 μg Pb/liter.

II. LEAD METABOLISM

Gruden and Stantic (30) showed that ^{203}Pb is absorbed equally well into and through the wall of the duodenum, jejunum, and ileum of rats and that there is no significant difference in this absorption between 6- and 26-week-old females. In younger suckling rats 13–20 days old Pb absorption is high (83–89%), with an abrupt drop to adult levels (15–16%) at about the time of weaning (21). Kostial *et al.* (50) also found intestinal Pb absorption to be high in young suckling rats, compared with a reported 1% in adult females 4 months old. The tendency of the young of a species to be more prone to Pb poisoning than the adult (6) is therefore due, in part, to higher Pb absorption from the diet.

Alimentary absorption of lead approximates 5–10% in man (43, 52). A lower apparent absorption of 1.3 ± 0.8% was found by Blaxter (9) in sheep and rabbits. This occurred over a wide range of Pb intakes, with little reduction of true absorption with increasing amounts ingested. The retention of inhaled lead is much higher, of the order of 30–50%, and is even higher when the particle size is very small (43).

The absorption and retention of ingested Pb is greatly affected by the dietary levels of Ca, P, Fe, Cu, and Zn. Subnormal intakes of Ca and P increase Pb retention in body tissues (78), and such retention decreases as dietary Ca increases from below to above requirements (64a). A reduction in dietary Ca intake from 0.7 to 0.1% markedly increased Pb retention in rats given 200 μg Pb/ml in their drinking water (80). Similar effects have been demonstrated in rats given much smaller amounts of lead (53), and the loss of Pb already incorporated into the carcass is greater when the dietary Ca is near to requirement than when it is greater or less than requirement (64a). A marked reduction in Pb retention in the tissues and some protection against the toxic effects of 1000 ppm Pb has been reported in pigs when the dietary Ca was increased from 0.7 to 1.1% (40). Lowering the dietary P intakes similarly increased Pb retention, and the effects of Ca and P deficiency have been shown to be additive (64). However, phosphate supplements to a diet normal in P have little effect on Pb metabolism in young lambs or in protecting against the growth inhibition induced by moderately toxic Pb intakes (58a).

Lead toxicity symptoms are exacerbated in Fe-deficient rats given 200 μg Pb/ml in their drinking water (81), and tissue Pb concentrations increased up to 20-fold compared with controls. The increase in the apparent absorption of Pb in Fe deficiency is far greater than the increase in the absorption of Co or Mn induced by the same deficiency (22). The effect of Cu deficiency on Pb absorption and toxicity does not appear to have been similarly investigated, but very high dietary levels of lead (0.5% Pb as lead acetate) reduce plasma Cu and ceruloplasmin levels in rats and decreased dietary Cu levels have been associated with increased erythrocyte Pb concentrations (46). These results imply a mutual metabolic antagonism, but further studies with lower Pb intakes and a wider range of Cu intakes are needed to clearly establish such a relationship, especially as lead-exposed workers have been shown to have *increased* erythrocyte Cu concentrations (69).

Lead and zinc fed together at 1000 and 4000 ppm, respectively, compared with this level of Pb alone, enhance the toxic effects of Pb as shown by reduced weight gains, severe clinical signs and pathological changes in growing pigs, and increase Pb levels in blood, soft tissues, and bone (40). On the other hand, Willoughby *et al.* (90) found a decreased susceptibility to Pb toxicity in young horses fed toxic amounts of zinc, despite increases in hepatic and renal Pb concentrations. The possibility that the increased Zn intake increased renal and hepatic metallothionein contents and caused detoxication through Pb-binding in this form has been raised by Bremner (10). Alternatively, or in addition, the Zn could be inhibiting Ca and P assimilation which, in turn, would enhance the absorption of Pb (40).*

The lead which is absorbed enters the blood and reaches the bones and soft tissues of the body, from which it is gradually excreted via the bile into the small intestine and thence eliminated in the feces. Fecal lead thus consists largely of unabsorbed lead, together with a small proportion that has been absorbed and excreted by this route (2). Predominant excretion of lead in the feces is apparent in all species studied. The mean daily Pb excretion by adult man in the feces and urine was reported some years ago as 0.32 and 0.03 mg, respectively, in one study (45) and 0.22 and 0.05 mg, respectively, in two other studies (57, 82). Up to certain intakes, Pb excretion keeps pace with ingestion, so that retention in the tissues is slight or nonexistent. In the sheep, no lead is retained if less than 3 mg is ingested daily, which is equivalent to about 3 ppm of the dry diet (9). In the calf no Pb retention was observed over a period of 100 days at a dietary intake of 0.9 ppm Pb, very slight retention occurred at 10 or 11 ppm Pb, and appreciable retention was apparent (3 mg over 100 days or 0.03 mg Pb/day) at 102 ppm Pb (19) (See Table 56).

The increases in Pb concentrations that occur with age in human tissues under U.S. environmental conditions indicate that excretion is not quite keeping pace

*Author's note: Protection against Pb toxicity by excess dietary Zn, mediated by an inhibition of intestinal Pb absorption, has recently been demonstrated in rats [Cerklewski, F.L., and Forbes, R.M., *J. Nutr.* **106**, 689 (1976)].

TABLE 56
Lead Intake and Excretion in Calves over a 100-Day Period[a]

	Control diet (0.9 ppm Pb)	Newsprint diet (11.1 ppm Pb)	Added[b] lead diet (10.5 ppm Pb)	Added[b] lead diet (102 ppm Pb)
Lead intake (g)	0.42	4.22	5.40	49.27
Fecal excretion (g)	0.41	4.02	5.07	45.20
Fecal excretion (%)	97.10	95.36	93.78	91.73
Urinary excretion (g)	0.01	0.08	0.07	1.17
Urinary excretion (%)	1.80	1.79	1.31	2.38

[a]From Dinius *et al.* (19).
[b]Lead added as lead chromate.

with total intakes, resulting in a small retention over time (76). There is no evidence that tissue Pb accumulations of this magnitude are either harmful or harmless to man. In the rat, similar tissue Pb concentrations, brought about by the consumption of water containing 5 ppm Pb as lead acetate, were nontoxic as judged by the length of the life-span, although some loss of hair and body weight was evident (72).

The subcellular distribution of lead and its turnover in the kidneys and other tissues are of interest with respect to both its toxicity and possible detoxication mechanisms. Barltrop *et al.* (7) found renal Pb to accumulate in subcellular fractions 48 hr after injection in the order supernatant > microsomal > mitochondrial = nuclear. Mitochondrial accumulation of lead in kidneys is potentially damaging to normal renal function, because ADP-stimulated respiration in mitochondria is completely inhibited on incubation with lead (28). However, these renal cells are still viable at cellular Pb concentrations that are lethal to mitochondria *in vitro,* suggesting that a protective mechanism operates within the cell to limit mitochondrial Pb uptake. This is probably achieved through the formation of renal intranuclear inclusion bodies, since over 50% of the additional lead found in the nuclear fraction of renal tubular cells as their Pb concentrations increase is present in this form (29). Lead-induced inclusion bodies in the renal tubular cells of rats are insoluble in physiological media and contain about 40–50 μg Pb/mg protein, of which only about 10% is tightly bound. They also contain Ca, Fe, Zn, Cu, and Cd (58). Moore and Goyer (58) have suggested that the inclusion bodies function as an intracellular depot of nondiffusible lead. These workers further found small amounts of a soluble Pb-containing protein which may have a role in the formation of the nuclear inclusion bodies. Lead forms a relatively stable metallothionein complex, but the amounts of the metal that occur in this form are small. Thionein does not therefore appear to be involved in the mechanism of Pb detoxication as it is in the detoxication of cadmium (84).

The lead present in the soft tissues is readily mobilized by chelating agents, such as EDTA and penicillamine (15), whereas the lead stored in bone is less easily released by such means. Increased levels of circulating corticosteroids (65) and of parathormone (5) enhance the mobilization of lead from bone. Release of bone Pb into the bloodstream also occurs in some types of physiological stress, including pregnancy, trauma, and infection (5, 13). The mobilization of stored lead from bone during pregnancy can be important in view of the readiness with which lead can cross the placental barrier and affect the fetus (6). Recent studies with rodents indicate that lead crosses the placental membranes rapidly and in significant amounts even at relatively low maternal blood levels (14).

III. SOURCES OF LEAD

1. Lead in Air and Water

The "natural" level of lead in air, if there were no contribution from man-made pollution, has been estimated by Patterson (62) to be 0.0005 $\mu g/m^3$. Actual atmospheric Pb levels ranging from 0.4 to 7.6 $\mu g/m^3$ in different cities, at sites with varying motor traffic densities, are cited by Hicks (38). Most of this lead comes from the exhaust fumes of cars burning petrol containing lead alkyl additives, so that intakes from this source will clearly be much greater in motorized urban communities than in rural areas. In such urban communities the amount of lead absorbed via the lungs can be as much or greater than the amount retained from the diet, although most people ingest very much more from the diet than they inhale from the air (26, 61). Atmospheric lead contributes to total intakes by dustfall as well as by inhalation. High concentrations of lead have been found on roadside soil and grass with the Pb content declining with increasing distance from the road (12). Some of this lead fallout from the atmosphere can be incorporated into vegetable crops (31) and rainwater.

Surface waters used for domestic purposes vary greatly in lead content. A survey of such waters in the United States gives a range from 0 to 55 μg/liter (20). The upper limit of safety given by the World Health Organization is 100 μg Pb/liter. Many domestic water supplies can exceed this limit where the water is soft and comes from lead-lined tanks and water pipes. For example, Goldberg (25) studied the following three groups of households in Glasgow: those with a lead-lined storage tank and lead piping, those with no lead tank but lead piping in excess of 60 feet, and those with less than 60 feet of lead piping. The mean lead contents of the cold tap water for the three groups were, in the order given, approximately 1000, 220, and 100 μg/liter. The blood lead of the inhabitants showed a significant positive correlation and the erythrocyte δ-aminolevulinic dehydrase activity a significant negative correlation with water lead content.

2. Lead in Human Foods and Dietaries

Published values for the lead content of individual food items are so variable that it is difficult to provide a meaningful classification into high, medium, and low groups. Thus Warren and Delavault (87) obtained the following values for English and Canadian grown vegetables from a range of locations: lettuce, 0.3–56 (mean 12); cabbage, 0.2–2.3 (mean 1); potato, 0.2–7.6 (mean 1.6); carrot, 0.2–11 (mean 4); and bean, 1–12 (mean 4) ppm Pb dry basis. Garcia *et al.* (24) observed a smaller range of 0.20–0.34 (mean 0.27) ppm Pb d.b. for 11 samples of whole kernel dent corn. The levels of lead in wheat, flour, and bread (North American grown) have been reported as follows: common hard wheat, 0.50 ± 0.22; common soft wheat, 1.00 ± 0.61; flour, baker's patent, 0.92 ± 0.43; flour, soft patent, 1.02 ± 0.59; and bread, white, 0.41 ± 0.29 ppm d.b. (92). On the basis of these figures lead is not concentrated in the germ and bran and lost in the milling of flour, as with other minerals. Average lead levels in organ meats and milk were given in Section I.

Mitchell and Aldous (55) have drawn attention to the potential health hazard of the large amounts of lead that can occur in canned baby foods. Canned evaporated milk averaged 202 μg Pb/liter, compared with 40 μg/liter in a survey of bulk milk. Of 256 baby foods, mostly fruit juices, 62% of those in metal cans contained 100 μg Pb/liter or more, 37% contained 200 μg/liter or more, and 12% contained 400 μg/liter or more. Of products in glass or aluminum containers only 1% had Pb levels in excess of 200 μg/liter. The lead content of 168 samples of canned fruit retailed in England ranged widely from 0.02 to 8.16 ppm, with a mean of 0.94 ppm (81a). Sixteen of the samples were found to be above the statutory limit of 2 ppm. A further hazardous source of lead in young children is high-Pb content paint peeling from walls and in household dust, particularly under poor housing conditions.

Estimates of average total daily lead intakes from food and beverages by human adults have been made in several countries. Some years ago Monier-Williams (57) estimated that "normal, healthy individuals" in England would obtain about 0.32 mg Pb/day from the food and water supply. This is identical with the mean of 320 μg ± 150 μg Pb/day obtained more recently for total adult U.K. diets (33). Schroeder and Tipton (76) estimate an average total intake from food and beverages in the United States of 280 μg Pb/day, and Horiuchi (39) gives a range of 239–318 μg/day for Japanese adults. A much lower estimate of daily intake per capita of the whole Dutch population (105 μg Pb) has recently been made (66). This low intake is consistent with the unusually low levels of lead reported by the authors for Dutch foods.

3. Lead in Animal Feeds and Pastures

The lead content of pastures from areas remote from industrial or auto exhaust contamination ranged from 0.3 to 1.5 ppm in the dry matter during the period of actual growth (56). When these plants were mature or senescent much higher Pb concentrations were observed. On soils normally low in lead, pasture plants rarely contain more than 2–3 ppm of the element, unless contaminated by dust (3). As Jones and Clement (42) have shown, only a small proportion of total soil lead is available to plant roots and only a small proportion of the lead that is absorbed is transported from the roots to the shoots and leaves. The Pb levels in the roots and shoots of perennial ryegrass grown in a soil low in extractable Pb were 10.0 and 5.1 ppm, respectively, and the corresponding levels for plants grown on a high-Pb soil were 37.8 and 7.4 ppm. Much higher Pb levels have been found in pastures from lead-rich areas, up to several hundred parts per million due to surface contamination by dust from lead mines (3). The cereal grains and other seeds used in pig and poultry feeding rarely contain more than 1 ppm Pb.

IV. LEAD TOXICITY

1. Toxic Effects of Lead

The symptoms and pathology of acute lead poisoning lie outside the scope of this text and have been well documented elsewhere (5, 13, 32). Chronic lead poisoning is characterized particularly by neurological defects, renal tubular dysfunction, and anemia. Damage to the central nervous system—causing lead encephalopathy and neuropathy—is a marked and common feature, especially in children with their low Pb tolerance. In children, chronic Pb poisoning involves physical brain damage, with permanent sequelae, including behavioral problems, intellectual impairment, and hyperactivity (8, 17, 63). Hyperactivity and behavior disturbances have been induced by lead in infant mice (79) and monkeys (4). The mechanism by which lead affects the nervous system is largely obscure, although it is known to block both impulse transmission and acetylcholine release (49). Retardation of brain growth (51) and a reduction in the DNA content of the cerebellum (54) have also been demonstrated in lead-intoxicated suckling rats.

The anemia that is a common feature of chronic Pb poisoning arises from effects on heme synthesis and red blood cells (86), and perhaps also from an effect of lead on iron and copper metabolism. Heme synthesis is affected primarily by the inhibition of δ-aminolevulinic acid (δ-ALA) dehydratase and ferrochelatase, the enzyme controlling the incorporation of Fe into the heme molecule. Lead also affects the fragility of red cells which have a shortened life-span. These effects of lead are the main cause of the anemia, but there are also reports of synergistic effects of Pb and Fe deficiency on impairment of hematopoiesis (88).

Chronic Pb toxicity is manifested further by the occurrence of renal tubular dysfunction with aminoaciduria and glycosuria. Goyer (28) has suggested that an impairment of energy metabolism may be responsible for the reduced transport function of the kidney and therefore the aminoaciduria and glycosuria in Pb toxicity, and that the respiratory abnormalities in renal mitochondria may result from their reduced cytochrome content (67).

The life-span of rats exposed to low levels of lead is shorter than controls maintained on "lead-free" diets (73), and the breeding potential of mice is reduced (74). Exposure to relatively low levels of lead also reduces the resistance of animals to infectious diseases. For example, a markedly reduced resistance to infection with *Salmonella typhimurium* has been observed in mice receiving Pb intakes insufficient to produce any overt signs of toxicity (36), while Pb-treated rats (71) and chicks (83) are more susceptible to bacterial infection than normal. Lead also aggravates viral diseases in mice, probably in part through reduced interferon synthesis (23).

The function of certain of the endocrine glands has been shown by Sandstead (70) to be adversely affected in a small series of patients lead-intoxicated through drinking "moonshine" whiskey. For example, six of 18 patients displayed a decreased excretion of pituitary gonadotropic hormones, nine of 16 had low ($< 10\%$) 24-hr thyroid uptakes of iodine, and three were hyporesponsive to TSH. The adrenal response to exogenous ACTH was also low in several of the patients. The author contends that lead is the most likely cause of the above abnormalities, although the influence of other dietary and environmental factors, including alcohol, cannot entirely be dismissed.

The biochemical effects of lead have been critically reviewed by Vallee and Ulmer (85) and will therefore not be repeated here. However, it is pertinent to mention that certain ATPases are sensitive to low-Pb concentrations and that lead strongly inhibits lipoamide dehydrogenase, an enzyme crucial to cellular oxidation (84). Lead also inhibits the enzyme aminolevulinic acid dehydrase, which governs the condensation of two molecules of ALA to form porphobilinogin and requires the presence of reduced glutathione (GSH) for its activation. Since lead is believed to affect enzyme systems by its ability to combine with sulfhydryl groups (86), it seems possible that the depression in ALA-D may be due to the metal combining with GSH, thus preventing or reducing the activity of this enzyme. Support for this concept has recently been obtained by Howard (39a).

2. Tolerance to Lead and Criteria of Safety

Tolerance to lead varies with the age of the animal, the forms and sources of the lead, and the composition of the diet being consumed. Species comparisons in which these variables were adequately controlled are insufficient to permit a secure definition of differences among species in tolerance to lead, but the young animal is less tolerant than the adult in all species studied. As indicated in Section II, this relates at least partly to the higher absorption of dietary lead in the young. Lower tolerance to inhaled lead than to ingested lead, particularly when the former is of low particle size, similarly relates to the substantially higher retention of inhaled than ingested lead.

The exacerbation of Pb toxicity symptoms induced by dietary deficiencies of Ca and Fe in the rat, and the further interactions of Pb with P, Zn, and Cu, also discussed in Section II, indicate clearly that tolerance to particular, potentially toxic, Pb intakes is greatly influenced by the dietary levels of these elements relative to that of lead. In this connection Six and Goyer (80, 81) have pointed out that Ca and Fe intakes are most likely to be inadequate in children and pregnant women, the population groups most susceptible to Pb poisoning. Extrapolation from the effects of severe deficiencies of Ca and Fe on Pb absorption and toxicity in rats to the effects of marginal deficiencies of these elements in man is perhaps open to question, but the provision of adequate intakes of Ca and Fe appears to be highly desirable to minimize the potential hazards of chronic exposure of children and women to lead.

Because tolerance to lead is influenced by so many factors, a series of maximum, long-term, "safe" tolerances exists, depending on the extent to which one or more of these factors continues to operate. Definition of the threshold or thresholds at which adverse affects arise is therefore difficult and data and criteria on which such thresholds could be based are meager. It can be stated that about 100 μg Pb must be assimilated daily by adult man for blood Pb concentrations to exceed 40 μg/100 g, the level at which most individuals reveal increased urinary ALA excretion. On this basis, a blood lead of 40 μg/100 g can tentatively be taken to indicate the threshold at which the body burden of lead exceeds the homeostatic mechanisms and adverse effects could become evident. The World Health Organization (91) gives a provisional tolerable weekly intake of lead by man as 3 mg per person or 0.05 mg/kg body weight. It is pointed out that these intake levels do not apply to infants and children. Efforts to decrease exposure to such readily assimilable forms of lead as atmospheric lead appear to be thoroughly justified by the available evidence, limited though this is (11, 38).

In sheep 3 mg Pb/day or about 3 ppm of the dry diet is the level below which excretion keeps pace with ingestion (9). A dietary level of 3 ppm Pb can therefore be regarded as a threshold level in this species. However, the extent to which 3 ppm Pb can be exceeded without *harmful* body retention and accumulation in the tissues is unknown. Allcroft (2) considers that levels of 10 ppm Pb in the liver and 40 ppm in the kidney cortex of sheep and cattle are "of definite diagnostic significance where there is collateral evidence of lead poisoning" and are "highly suggestive" where there is no such evidence.

REFERENCES

1. Allcroft, R., *J. Comp. Pathol. Ther.* **60,** 190 (1950).
2. Allcroft, R., *Vet. Rec.* **63,** 583 (1951).
3. Allcroft, R., and Blaxter, K.L., *J. Comp. Pathol. Ther.* **60,** 209 (1950).
4. Allen, J.R., McWey, P.J., and Suomi, S.J., *Environ. Health Perspect. Exp.* Issue No. 7, p. 239 (1974).
5. Aug, J., Fairhall, L., Minot, A., and Reznikoff, P., *in* "Medicine Monographs," Vol. 7. Williams & Wilkins, Baltimore, Maryland, 1926.
6. Barltrop, D., *Postgrad. Med. J.* **45,** 129 (1969).
7. Barltrop, D., Barrett, A.J., and Dingle, J.T., *J. Lab. Clin. Med.* **77,** 705 (1971).
8. Berg, J., and Zapella, M., *J. Ment. Defic. Res.* **8,** 44 (1964).
9. Blaxter, K.L., *J. Comp. Pathol. Ther.* **60,** 140 (1950).
10. Bremner, I., *Q. Rev. Biophys.* **7,** 75 (1974).
11. Bryce-Smith, D., *Chem. Br.* **8,** 240 (1973).
12. Cannon, H.L., and Bowles, J.M., *Science* **137,** 765 (1962).
13. Cantarow, A., and Trumper, M., "Lead Poisoning." Williams & Wilkins, Baltimore, Maryland, 1944.
14. Carpenter, S.J., *Environ. Health Perspect. Exp.* Issue No. 7, p. 129 (1974).
15. Chisholm, J., *J. Pediatr.* **73,** 1 (1968).
16. Chow, T.J., *Chem. Br.* **9,** 258 (1973).
17. David, O.J., *Environ. Health Perspect. Exp.* Issue No. 7, p. 17 (1974).
18. Davis, J.R., and Andelman, S.L., *Arch. Environ. Health* **15,** 53 (1967).
19. Dinius, D.A., Brinsfield, T.H., and Williams, E.E., *J. Anim. Sci.* **37,** 169 (1973).
20. Durum, W.H., and Haffty, J., *U.S., Geol. Surv., Circ.* **445** (1961).
21. Forbes, G.B., and Reina, J.C., *J. Nutr.* **102,** 647 (1972).
22. Forth, W., and Rummel, W., *in* "Intestinal Absorption of Metal Ions, Trace Elements and Radionuclides" (S.C. Skoryna and D. Waldon-Edward, eds.), p. 173. Pergamon, Oxford, 1971.
23. Gainer, J.H., *Environ. Health Perspect. Exp.* Issue No. 7, 113 (1974).
24. Garcia, W.J., Blessin, C.W., and Inglett, G.E., *Cereal Chem.* **51,** 788 (1974).
25. Goldberg, A., *Environ. Health Perspect. Exp.* Issue No. 7, p. 103 (1974).
26. Goldsmith, J.R., *J. Air Pollut. Control Assoc.* **19,** 714 (1969).
27. Goldwater, L.J., and Hoover, A.W., *Arch. Environ. Health* **15,** 60 (1967).
28. Goyer, R.A., *Am. J. Pathol.* **64,** 167 (1971).
29. Goyer, R.A., May, P., Cates, M.M., and Krigman, M.R., *Lab. Invest.* **22,** 245 (1971).
30. Gruden, N., and Stantic, M., *Sci. Total Environ.* **3,** 288 (1975).
31. Haar, G. Ter, *Environ. Sci. Technol.* **4,** 226 (1970).
32. Hamilton, A., and Hardy, H.O., *in* "Industrial Toxicology." Harper (Hoeber), New York, 1949.
33. Hamilton, E.I., and Minski, M.J., *Sci. Total Environ.* **1,** 375 (1972/1973).
34. Hamilton, E.I., Minski, M.J., and Cleary, J.J., *Sci. Total Environ.* **1,** 341 (1972/1973).
35. Hammond, P.B., Wright, H.N., and Roepke, M.H., *Minn., Agric. Exp. Stn., Bull.* **221** (1956).
36. Hemphill, F.E., Kaeberle, M.L., and Buck, W.B., *Science* **172,** 1031 (1971).
37. Hernberg, S., Nikkanen, J., Mellin, G., and Lilluis, H., *Arch. Environ. Health* **21,** 140 (1970).
38. Hicks, R.M., *Chem.-Biol. Interact.* **5,** 361 (1972).
39. Horiuchi, K., *Osaka City Med. J.* **11,** 265 (1965).

39a. Howard, J.K., *Clin. Sci. Molec. Med.* **47**, 515 (1974).
40. Hsu, F.S., Krook, L., Pond, W.G., and Duncan, J.R., *J. Nutr.* **105**, 112 (1975).
41. Ishikawa, I., *Osaka City Med. J.* **5**, 99, 109, and 117 (1959).
42. Jones, L.H.P., and Clement, C.R., *in* "Lead in the Environment." Institute of Petroleum, London, 1972.
43. Kehoe, R.A., *R. Inst. Public Health Hyg. J.* **24**, 81, 101, 129, and 177 (1961).
44. Kehoe, R.A., *U.S., Public Health Serv. Publ.* **1440**, 54 (1966).
45. Kehoe, R.A., Cholak, J., and Storey, R.V., *J. Nutr.* **19**, 579 (1940).
46. Klauder, D.S., Murthy, L., and Petering, H.G., *Trace Subst. Environ. Health–6, Proc. Univ. Mo. Annu. Conf., 6th, 1972,* p. 131 (1972).
47. Kochen, J., and Greener, Y., *Pediatr. Res.* **7**, 937 (1974).
48. Kopita, L., and Byers, R.K., and Shwachman, H., *N. Engl. J. Med.* **276**, 949 (1967).
49. Kostial, K., and Vouk, V.B., *Br. J. Pharmacol. Chemother.* **12**, 219 (1957).
50. Kostial, K., Simonovic, I., and Pisonic, M., *Nature (London)* **233**, 564 (1971).
51. Krigman, M.R., and Hogan, E.L., *Environ. Health Perspect. Exp.* Issue No. 7, p. 187 (1974).
52. Langham, W., and Anderson, E.C., *U.S. A.E.C., Health Safety Lab.* **42**, 282 (1958).
53. Mahaffney, K.R., Goyer, R.A., and Haseman, J.K., *J. Lab. Clin. Med.* **82**, 92 (1973).
54. Michaelson, I.A., and Sauerhoff, M.W., *Environ. Health Perspect. Exp.* Issue No. 7, p. 201 (1974).
55. Mitchell, D.G., and Aldous, K.M., *Environ. Health Perspect. Exp.* Issue No. 7, p. 59 (1974).
56. Mitchell, R.L., and Reith, J.W.S., *J. Sci. Food Agric.* **17**, 437 (1967).
57. Monier-Williams, G.W., "Trace Elements in Food." Chapman & Hall, London, 1949.
58. Moore, J.F., and Goyer, R.A., *Environ. Health Perspect. Exp.* Issue No. 7, p. 121 (1974).
58a. Morrison, J.N., Quarterman, J., and Humphries, W.R., *Proc. Nutr. Soc.* **33**, 88A (1974).
59. Murthy, G.K., and Rhea, U.S., *J. Dairy Sci.* **54**, 1001 (1971).
60. Murthy, G.K., Rhea, U.S., and Peeler, J.T., *J. Dairy Sci.* **50**, 651 (1967).
61. National Academy of Sciences (U.S.), Report. Comm. Biol. Effects Atmos. Pollutants, 1971.
62. Patterson, C.C., *Arch. Environ. Health* **11**, 344 (1965).
63. Perlstein, M.A., and Attala, R., *Clin. Pediatr.* **5**, 292 (1966).
64. Quarterman, J., Morrison, J.N., and Carey, L.F., *Trace Subst. Environ. Health–7, Proc. Univ. Mo. Annu. Conf., 7th, 1974,* p. 347 (1974).
64a. Quarterman, J., Morrison, J.N., and Humphries, W.R., *Proc. Nutr. Soc.* **34**, 89A (1975).
65. Reifenstein, C., and Albright, F., *J. Clin. Invest.* **26**, 24 (1947).
66. Reith, J.F., Engelsma, J., and Ditmarsch, M., *Z. Lebensm.-Unters.-Forsch.* **156**, 271 (1974).
67. Rhyme, B.C., and Goyer, R.A., *Exp. Mol. Pathol.* **14**, 386 (1971).
68. Rosen, J.F., and Trinidad, E.E., *Environ. Health Perspect. Exp.* Issue No. 7, p. 139 (1974).
69. Rubino, B.E., Pagliandi, E., Prato, V., and Giangrandi, E., *Br. J. Haematol.* **4**, 103 (1971).
70. Sandstead, H.H., *Trace Subst. Environ. Health–6, Proc. Univ. Mo. Annu. Conf., 6th, 1973,* p. 223 (1973).
71. Selye, H., Tuchweber, B., and Bertok, L., *J. Bacteriol.* **91**, 884 (1966).
72. Schroeder, H.A., and Balassa, J.J., *J. Nutr.* **92**, 245 (1967).

73. Schroeder, H.A., Balassa, J.J., and Vinton, W.H., *J. Nutr.* **86,** 51 (1965).
74. Schroeder, H.A., and Mitchener, M., *Arch. Environ. Health* **23,** 102 (1971).
75. Schroeder, H.A., and Nason, A.P., *J. Invest. Dermatol.* **53,** 71 (1969).
76. Schroeder, H.A., and Tipton, I.H., *Arch. Environ. Health* **17,** 965 (1968).
77. Schwarz, K., *in* "Trace Element Metabolism in Animals-2" (W.G. Hoekstra *et al.,* eds.), p. 355. Univ. Park Press, Baltimore, Maryland, 1974.
78. Shields, J.B., and Mitchell, H.H., *J. Nutr.* **21,** 541 (1941).
79. Silbergeld, E.K., and Goldberg, A.M., *Environ. Health Perspect. Exp.* Issue No. 7, p. 227 (1974).
80. Six, K.M., and Goyer, R.A., *J. Lab. Clin. Med.* **76,** 933 (1970).
81. Six, K.M., and Goyer, R.A., *J. Lab. Clin. Med.* **79,** 128 (1972).
81a. Thomas, B., Edmunds, J.W., and Curry, S.J., *J. Sci. Food Agric.* **26,** 1 (1975).
82. Tompsett, S.L., and Anderson, A.N., *Biochem. J.* **29,** 1851 (1935).
83. Truscott, R.B., *Can. J. Comp. Med.* **34,** 134 (1970).
84. Ulmer, D.D., and Vallee, B.L., *Trace Subst. Environ. Health–2, Proc. Univ. Mo. Annu. Conf., 2nd, 1968,* p. 7 (1969).
85. Vallee, B.L., and Ulmer, D.D., *Annu. Rev. Biochem.* **41,** 91 (1972).
86. Waldron, H.A., *Br. J. Ind. Med.* **23,** 83 (1966).
87. Warren, H.V., and Delavault, R.E., *Mem. Geol. Soc. Am.* **123,** (1971).
88. Waxman, H.S., and Rabinowitz, M., *Biochim. Biophys. Acta* **129,** 369 (1966).
89. White, W.B., Clifford, P.A., and Calvery, H.O., *J. Am. Vet. Med. Assoc.* **102,** 292 (1943).
90. Willoughby, R.A., MacDonald, E., Sherry, B.J., and Brown, G., *Can. J. Comp. Med.* **36,** 348 (1972).
91. World Health Organization, *W.H.O., Tech. Rep. Ser.* **505,** (1972).
92. Zook, E.G., Greene, F.E., and Morris, E.R., *Cereal Chem.* **47,** 720 (1970).

18

Arsenic

I. ARSENIC IN ANIMAL TISSUES AND FLUIDS

Arsenic is widely distributed throughout the tissues and fluids of the body in low and highly variable concentrations. In a study of healthy adult human tissues Smith (33), using neutron activation, reported the mean As concentrations of most tissues to lie between 0.04 and 0.09 ppm on the dry basis. The variability was extremely high and the skin (mean 0.12), nails (0.36), and hair (0.65 ppm) were substantially and relatively consistently higher in arsenic than other tissues, with no evidence of marked accumulation in any internal organ or tissue. This last point is also evident from a later study of human tissues by Hamilton *et al.* (16). These workers obtained much higher mean concentrations than Smith for all tissues examined, other than the muscles. No systematic studies of the As levels in the tissues of normal domestic or laboratory animals have appeared.

The arsenic content of human hair has excited considerable interest because of its value in the diagnosis of As poisoning. Normal hair always contains arsenic in small amounts which are greatly increased by excessive intakes of the element in certain forms. The levels remain high for some days after cessation of intake and then return rapidly to normal (8, 32). In a study of over 1000 hair samples taken at random from living subjects Smith (32) obtained concentrations ranging from 0.03 to 74 ppm As, with a mean of 0.81 and a median of 0.51. This worker contends that hair samples with an As concentration greater than 3 ppm should always be suspect, those with 2–3 ppm require further examination, and those with less than 2 ppm should not be dismissed where As poisoning is suspected. Significantly higher As levels were obtained in this investigation for male than for female hair. The median value for males was 0.62 and for females 0.37 ppm As. Higher hair As levels in men than in women are also apparent from the more limited study of Schroeder and Balassa (26).

The levels of As reported for human blood vary widely among individuals and among investigations using different analytical methods. Vallee and co-workers (35) quote values ranging from 0.01 to 0.64 μg As/g whole fresh blood, and Iwataki and Horiuchi (18) reported levels ranging from 0.0 to 0.37 μg/g whole blood. In a later study of blood samples from 163 healthy human adults in England Hamilton *et al.* (16) obtained a mean of 0.2 ± 0.02 μg As/g wet weight. This is of the order of 10 times the level recently obtained by Wagner and Weswig (36) for a small group of individuals not exposed to any abnormal source of arsenic. These workers found no evidence of As accumulation in the blood of individuals exposed to cacodylic acid (dimethylarsenic acid) used as a forest herbicide over a 2-month period and concluded that blood As level is not a good indicator of As exposure.

The distribution of blood arsenic between the cellular elements and the plasma requires further study. In the rat 80% of the blood As is concentrated in the red cells (17). The affinity of the erythrocytes for As in this species is further apparent from studies with radioactive As (20).

Normal cow's milk contains 0.03–0.06 ppm As (2, 15, 17), with values up to 1.5 ppm in the milk of cows grazing As-contaminated areas in New Zealand (14). High levels have also been found in the milk of women receiving As therapy for syphilis (10).

II. ARSENIC METABOLISM

Arsenic absorption and retention, and the routes of its excretion, are influenced by the level and the chemical form in which it is ingested. Arsenic in the forms in which it ordinarily occurs in foods, including the organically bound As of shrimp, is well absorbed and rapidly eliminated, mainly in the urine (6). Less than 10% of the usual soluble forms of As appear in the feces. Arsenic trioxide is also well absorbed but more of it is retained in the tissues of man and rats (6).

Urinary As excretion rises with increasing As intakes so that total urinary As excretion provides a useful index of exposure (4, 36). Occupational studies have indicated that workers chronically exposed to inorganic As salts will excrete an average of 70 μg As/liter without symptons of As intoxication, and that levels as high as 5 mg/liter may be reached (4).

The arsenic of such organic compounds as arsanilic acid, used as growth stimulants for pigs and poultry, is similarly well absorbed and disappears rapidly from the tissues, mostly into the feces (24). When these forms of As are fed in the recommended amounts the element does not accumulate in the tissues to excessive concentrations (12).

III. ARSENIC AS AN ESSENTIAL ELEMENT

In the early studies with arsenic no improvement was obtained in the growth, hemoglobin levels, or red cell numbers of rats consuming a purified diet supplying 2 μg As daily, when this diet was supplemented with As (17). Slight indications of a favorable effect on hemoglobin production were observed by these workers and by Skinner and McHargue (31). Subsequently a diet containing even smaller amounts of arsenic was fed to mice and rats over long periods without any observable impairment of growth and development (26). Recently Nielsen *et al.* (23) have produced strong evidence that arsenic has an essential physiological role in the rat. The offspring of dams placed on a purified diet containing approximately 30 ppb As, and maintained in a plastic isolator environment, displayed a rough coat and a significantly slower rate of growth than controls receiving a supplement of 4.5 ppm As as sodium arsenate (4 ppm) and arsenite (0.5 ppm). The males were more affected than the females. At 12–15 weeks the deficient males exhibited significantly decreased hematocrits and greatly enlarged and blackened spleens containing 50% more iron on a per gram basis. The spleens of deficient females were also enlarged and both sexes revealed an increase in erythrocyte osmotic fragility.

The beneficial effects of various organic arsenicals on the growth, health, and feed efficiency of pigs and poultry are well established (11, 12). The four compounds found to be of particular value are arsanilic acid, 4-nitrophenylarsonic acid, 3-nitro-4-hydroryphenylarsonic acid, and arsenobenzene (phenylarsenoxide). No clear relation between structure and growth-promoting effect is discernible, although the phenylarsenoxides are more potent than the arsonic acids as coccidiostats, and only the arsonic acids are recognized as growth stimulants for pigs and poultry. The action of arsonic acids closely resembles that of antibiotics and is to some extent complementary to the action of these substances.

Arsenic as a Se antagonist is discussed in Chapter 12.

IV. SOURCES OF ARSENIC

Arsenic is widely distributed in the biosphere. It occurs in the air in areas where coal is burned, particularly near smelters and refineries, in seawater to the extent of 2–5 ppb, and in public water supplies in concentrations which may exceed that of seawater (26). Arsenic occurs in normal soils at levels ranging usually from 1 to 40 ppm, although much higher levels can result from the continued use of arsenical sprays for insect control (1) or on disused mine tips (24a). The amounts absorbed from these soils by the aerial parts of plants have been reported to be small (1, 13), but recent evidence indicates that certain plants growing on As-enriched soils can accumulate extremely high-As levels (24a). Concentrations as high as 2080 and 3470 ppm As (dry basis) were obtained for the foliage of samples of *Agrostis tenuis.* Surface contamination of herbage, fruits, and vegetables with spray residues can also raise their As concentrations well above normal levels.

Most human foods contain less than 0.5 ppm As and rarely exceed 1 ppm on the fresh basis (26). This applies to fruits, vegetables, cereals, meats, and dairy products. Hamilton and Minski (15), using neutron activation, reported the following mean values for English food types: cereals, 0.18 ± 0.05; fats, 0.05 ± 0.01; root vegetables, 0.08 ± 0.01; fruits and preserves, 0.07 ± 0.01; milk, 0.05 ± 0.01; meat (uncooked pork, beef, and lamb), 0.10 ± 0.05; and fish, 2.0 ± 0.8 μg As/g fresh weight. Foods of marine origin are much richer in arsenic than other foods. Fish contains 2–8 ppm, oysters 3–10 ppm, and mussels as high as 120 ppm As (5, 7, 26). Up to 174 ppm has been found in prawns from the coastal waters of Britain (5) and 42 ppm in shrimp from the southeastern coastal waters of the United States (6). The As content of fish meals used in animal feeding ranged from 2.6 to 19.1 ppm, with a mean of 6 ppm (21). Fish and crustacea from freshwater usually contain lower As concentrations than those just cited.

The total amounts of arsenic ingested daily are obviously greatly influenced by the amounts and proportions of seafoods included in the diet. An institutional diet containing no such foods was reported to supply 0.4 mg As/day and an average U.S. diet to supply 0.9 mg/day (26). These calculated daily intakes are substantially higher than the 0.07–0.17 mg As/day reported for Japanese individuals (22), and the 0.14 mg/day obtained by Duggan and Lipscombe (9) in their U.S. "market basket" survey.

V. ARSENIC TOXICITY

The symptoms of acute arsenic poisoning in man by the oral route—nausea, vomiting, diarrhea, burning of the mouth and throat, and severe abdominal pains—have frequently been described. Chronic exposure to smaller toxic doses results in weakness, prostation, and muscular aching with few gastrointestinal symptoms. Skin and mucosal changes usually develop, together with a peripheral neuropathy and linear pigmentations in the fingernails (3). Headache, drowsiness, confusion, and convulsions are seen in both acute and chronic As intoxication. The biochemical basis for these disturbances is probably an inhibition by arsenite of a wide range of enzyme systems. Enzymes containing active thiol groups are effectively inhibited through combination of arsenic with these groups (19).

The maximum long-term As intake compatible with health and well-being in man cannot be given with any precision because variation in individual susceptibility is high and because the chemical form of the arsenic greatly affects its toxicity. Thus orchardists have been found to ingest as much as 6.8 mg As/day without signs of intoxication, due presumably to the prior oxidation of the arsenic from the trivalent to the much less toxic pentavalent form. By contrast, 30 mg As_2O_3 has been found to be fatal (19).

Body weight gain in turkey poults is significantly decreased by either 0.04% (400 ppm) of arsanilic acid in the diet, or about four times the dose required for growth stimulation, or by 0.05–0.06% sodium arsanilate (34). At twice these levels mortality of the young turkeys to age 28 days was high. Rats have been fed diets supplemented with 50 ppm As as As_2O_3 without any toxic effects, whereas at 200 ppm a significant growth depression was observed (30). Sodium arsenite fed in the drinking water at a level of 5 μg As/ml to mice and rats from weaning to natural death had no effect on growth, health, and longevity, despite some As accumulation in the tissues, especially in the aorta and red blood cells. There was no evidence of tumorogenicity or carcinogenicity (27, 28). Arsenic in the drinking water in the form of arsenite, at a level of 10 ppm As for a period of 15 months, has recently been shown to reduce mammary tumor incidence in female virgin C_3H/St mice to 27%, compared with 82% of spontaneous tumors in untreated controls. However, a significant enhancement of the growth rate of spontaneous or transplanted mammary tumors was apparent (25).

High levels of arsenic can induce goiter in rats (30), and reports of an arsenic–thyroid antagonism in man have appeared (29). A high incidence of goiter, with deaf-mutism, occurs in the Styrian Alps, the home of arsenic eaters, and in the Cordoba Province of Argentina, where chronic As poisoning is endemic. Arsenic levels in the drinking water of this latter region as high as 1.4 mg/liter have been reported (29). This concentration is one to two orders of magnitude higher than that of most natural waters consumed by man.

There is little agreement about what constitute dangerous levels of As in body fluids. As indicated in Sections I and II, both total daily urinary As excretion and the levels of As in hair and nails provide useful indices of exposure. Precise diagnostic criteria of potentially harmful As intakes or "safe," long-term dietary As intakes are unavailable.

REFERENCES

1. Anastasia, F.B., and Kender, W.J., *J. Environ. Qual.* **2**, 335 (1973).
2. Archibald, J.G., *Dairy Sci. Abstr.* **20**, 712 (1958).

3. Bennett, I.L., and Heyman, A., *in* "Principles of Internal Medicine" (T.R. Harrison *et al.*, eds.), 5th ed., p. 1405. McGraw-Hill, New York, 1966.
4. Browning, E., "Toxicology of Industrial Metals." Butterworth, London, 1961.
5. Chapman, A.C., *Analyst* **51**, 548 (1926).
6. Coulson, E.J., Remington, R.E., and Lynch, K.M., *J. Nutr.* **10**, 255 (1935).
7. Cox, H.E., *Analyst* **50**, 3 (1925); **51**, 132 (1926).
8. Dewar, W.A., and Lenihan, J.M., *Scott. Med. J.* **1**, 236 (1956).
9. Duggan, R.E., and Lipscomb, G.Q., *Pestic. Monit. J.* **2**, 153 (1969).
10. Fordyce, J.A., Rosen, I., and Myers, C.N., *Am. J. Syph.* **8**, 65 (1924).
11. Frost, D.V., *Fed. Proc., Fed. Am. Soc. Exp. Biol.* **26**, 194 (1967).
12. Frost, D.V., Overby, L.R., and Spruth, H.C., *J. Agric. Food Chem.* **3**, 235 (1955).
13. Greaves, J.E., *Soil Sci.* **38**, 355 (1934).
14. Grimmett, R.E.R., *N. Z. J. Agric.* **58**, 383 (1939).
15. Hamilton, E.I., and Minski, M.J., *Sci. Total Environ.* **1**, 375 (1972/1973).
16. Hamilton, E.I., Minski, M.J., and Cleary, J.J., *Sci. Total Environ.* **1**, 341 (1972/1973).
17. Hove, E., Elvehjem, C.A., and Hart, E.B., *Am. J. Physiol.* **124**, 205 (1938).
18. Iwataki, N., and Horiuchi, K., *Osaka City Med. J.* **5**, 209 (1959).
19. Johnstone, R.M., *Metab. Inhibitors* **2**, (1963).
20. Lowry, O.H., Hunter, F.T., Kip, A.F., and Irvine, J.W., *J. Pharmacol. Exp. Ther.* **76**, 221 (1942).
21. Lunde, G., *J. Sci. Food Agric.* **19**, 432 (1968).
22. Nakao, M., *Osaka Shiritzu Daikaku Igaku Zasshi* **9**, 541 (1960), cited by Schroeder and Balassa (26).
23. Nielsen, F.H., Givand, S.H., and Myron, D.R., *Fed. Proc., Fed. Am. Soc. Exp. Biol.* **34**, 923 (abstr.) (1975).
24. Overby, L.R., and Frost, D.V., *Toxicol. Appl. Pharmacol.* **4**, 38 (1962).
24a. Porter, E.K., and Peterson, P.J., *Sci. Total Environ.* **4**, 365 (1975).
25. Schrauzer, G.N., and Ishmael, D., *Ann. Clin. Lab. Sci.* **4**, 441 (1974).
26. Schroeder, H.A., and Balassa, J.J., *J. Chronic Dis.* **19**, 85 (1966).
27. Schroeder, H.A., and Balassa, J.J., *J. Nutr.* **92**, 245 (1967).
28. Schroeder, H.A., Kanisawa, M., Frost, D.V., and Mitchener, M., *J. Nutr.* **96**, 37 (1968).
29. Scott, M., *Trans Int. Goitre Conf., 3rd,* p. 34 (1938).
30. Sharpless, G.R., and Metzger, M., *J. Nutr.* **21**, 341 (1941).
31. Skinner, J.T., and McHargue, J.S., *Am. J. Physiol.* **143**, 85 (1945); **145**, 500 (1946).
32. Smith, H., *J. Forensic Sci. Soc.* **4**, 192 (1964).
33. Smith, H., *J. Forensic Sci. Soc.* **7**, 97 (1967).
34. Sullivan, T.W., and Al-Timimi, A.A., *Poultry Sci.* **50**, 1635 (1971).
35. Vallee, B.L., Ulmer, D.D., and Wacker, W.E.C., *AMA Arch. Ind. Health* **21**, 132 (1960).
36. Wagner, S.L., and Weswig, P., *Arch. Environ. Health* **28**, 77 (1974).

19

Other Elements

I. ALUMINUM

There is no conclusive evidence that aluminum performs any essential function in plants, animals, or microorganisms. An attempt to produce a purified Al-deficient diet for rats was unsuccessful and led to the conclusion that "if aluminum is required by the rat the requirement can be met by as little as 1 μg daily" (77). Aluminum may be involved in the succinic dehydrogenase–cytochrome c system and promotes the reaction between cytochrome c and succinic dehydrogenase *in vitro* (76, 163), but it is not known to be involved in this reaction or in any other enzyme system in the living body.

Ingested Al is very poorly absorbed and is excreted mostly in the feces. Ondreicka *et al.* (121) showed that increasing the Al in food 2-fold greatly increased fecal excretion without significantly affecting Al retention or tissue levels. When high doses of aluminum sulfate (200 mg Al/kg) were given to rats retention was greatly increased and increased Al levels appeared in the urine and the body tissues, notably the liver, testes, and bones. It appears that the animal organism can cope with a moderate increase in Al intake by increased fecal elimination, but at extreme increases in intake absorption and retention are raised. Absorption and retention of Al from the food is affected by the fluoride intake. The administration of 1 mg F to rats was followed by increased Al elimination in feces and urine and decreased retention in the tissues (121). It was further shown that simultaneously dosing Al (50 mg/kg) and 1 mg F^- perorally per rat for 24 days resulted in decreased Al levels in all tissues except the muscles, compared with animals receiving the Al alone. Ondreicka and co-workers (121) suggest that a mutual reaction between Al and F occurs, resulting in the formation of a readily soluble complex $(AlF_6)^{3-}$ which is more soluble than CaF_2, with the result that Al is not retained in the organism. The reverse aspect of the metabolic antagonism between the two elements is mentioned further in Chapter 13, in connection with the ameliorative action of Al salts against F toxicity in animals.

The Al concentrations of normal human tissues obtained spectrographically nearly 40 years ago by Kehoe *et al.* (85) agree well with the more recent figures of Tipton and Cook (167), and with those reported for the dog (184). A high proportion of their values lie between 0.2 and 0.6 ppm on the fresh basis, except for the lungs in which 20–60 ppm or higher were observed. Similar levels for normal human tissues, with accumulation in the lungs and lymph nodes, presumably from inhaled atmospheric dust, have recently been obtained by Hamilton and co-workers (65). Their mean levels were as follows: testis, 0.4 ± 0.2; ovary, 0.4 ± 0.1; muscle, 0.5 ± 0.2; brain, 0.4 ± 0.1; kidney, 0.4 ± 0.1; liver, 2.6 ± 1.3; lung, 18.2 ± 9.7; and lymph nodes, 32.5 ± 18.0 μg Al/g wet weight. Rib bone from hard water areas averaged 73.4 ± 16 and from soft water areas 60 ± 10 μg/g ash. The bone levels are substantially lower than those reported by Koch *et al.* (91) and Stitch (163). Whether this difference arises from contamination or analytical errors, or is an expression of variation of environmental origin, is unknown.

The Al concentration in human hair of individuals from three locations ranged from 1.2 to 9.2 ppm with a mean close to 4 ppm and no significant difference due to location (9). These values are of a similar order to the range of 3–11 ppm, with a mean of 7 ppm Al, obtained for sheep's wool in New Zealand (68). The Al content of milk is subject to considerable individual and month to month variation. Kirchgessner (89) obtained a mean of 0.7 ± 0.9 μg/ml for the milk of 18 cows under normal dietary conditions and Archibald (4) a mean of 0.5 μg Al/ml, with a range from 0.15 to 0.97 μg/ml. Supplementing the cows' ration with alum at the rate of 114 mg Al/day raised the milk of these cows to a mean of 0.8 μg Al/ml, with a range of 0.4–1.4 μg Al/ml.

A mean level of 0.13 μg Al/ml for normal whole human blood was observed some years ago (85). Twenty years later a mean of 0.17 μg/ml (range 0.05–0.50) was obtained for 536 samples of human blood serum (156). A higher mean of 0.4 ± 0.6 μg/g for 94 samples of whole human blood was recently reported (65).

Levels of 10–50 ppm Al (dry basis) have been reported for grasses and clovers (141, 157, 158). Data on the Al content of human foods and dietaries are highly variable, due almost certainly to the abundance of this element in the lithosphere and the ease of contamination that this makes possible. A range of vegetables varied from 0.5 to 5.0 ppm Al (75), and of fruits more from sample to sample than from one producing area to another (188). In this latter study the edible fresh portions of citrus and stone fruits were very low in Al content, usually less than 0.1–0.2 ppm Al, and those of berries and stone fruits were mostly much higher (2–4 ppm Al on the fresh basis). A comparison of peeled and unpeeled apples and pears revealed significantly higher Al concentrations in the peel than in the flesh. Meat and meat products and milk and dairy products are invariably poor sources of Al.

Total dietary Al intakes by man will clearly vary with the amounts and proportions of foods of plant origin relative to those of animal origin. Kehoe *et al.* (85) found the mean Al intake from the food and beverages of a normal adult North American diet to be 36.4 mg/day over 28 days, and Hamilton and Minski (64) obtained a mean of 23.3 ± 10.6 mg Al/day in total English diets. The latter figure compares well with the mean daily intakes over 30 days of 22 mg Al/day obtained for one U.S. male adult and 18 mg/day for one U.S. female adult (168). In a study of total diet composites calculated to supply the high intake of 4200 cal/day (189) the average intake was shown to be 24.6 mg Al/day, with a 14-fold variation from as low as 3.8 to as high as 51.6 mg/day in composites from different geographical locations.

The amounts of Al in human foods and dietaries are increased by contamination from aluminum vessels used in cooking. Ondreicka *et al.* (121) examined the Al levels in a range of foods cooked and stewed in aluminum and in stainless steel utensils. In every case the Al concentrations were higher when Al utensils were used, but in most cases they still fell within the same limits as those of the raw materials. The use of aluminum sulfate baking powders can also increase Al intakes but the increases from all these sources are usually small, poorly absorbed, and do not constitute a health hazard (13).

The relatively low toxicity of aluminum is apparent from the lifetime studies conducted with rats by Schroeder and Mitchener (149). These workers found that 5 ppm Al (as aluminum sulfate) in the drinking water was innocuous as measured by median life-span, longevity, incidence of tumors, serum cholesterol, glucose, and uric acid.

At much larger intakes, several orders of magnitude higher than those normally encountered, Al produces gastrointestinal irritation and can produce rickets by interfering with phosphate absorption (39). A negative phosphorus balance, associated with decreased P absorption and increased P output in the feces, occurs in rats receiving a high-Al intake, apparently as the result of the formation of insoluble, nonabsorbable, phosphate–aluminum complexes in the gastrointestinal tract (121). Chronic and acute peroral intoxication by $AlCl_3$ also caused a decrease in incorporation of ^{32}P into the phospholipid fraction, as well as into ribonucleic and deoxyribonucleic acids in the liver, spleen, and kidneys of rats (121). It appears that the Al interferes with or disturbs the activity of the phosphorylating mechanisms in a manner not yet fully understood. A study of adenosine mono-, di-, and tri-phosphates in rats' serum has shown that chronic and acute intoxication with $AlCl_3$ causes a decrease in ATP and an increase in ADP and AMP (120). The shift of equilibrium to the right in the system

$$ATP \rightleftharpoons ADP + \text{inorganic P} \rightleftharpoons AMP + \text{inorganic P}$$

is possibly connected with the decrease in inorganic P in the blood of animals intoxicated with Al compounds (83).

II. ANTIMONY

Antimony has no known function in living organisms and is not among the more toxic elements. Data on the distribution and metabolism of this element are therefore meager. Smith (164) examined a range of human tissues by neutron activation and found antimony to be present in all tissues. Most of the median values fell between 0.05 and 0.15 ppm Sb (dry basis), with the highest levels in the lungs (0.28) and hair (0.34 ppm). In a more detailed study of human lungs and some other tissues similar Sb levels were obtained, with the highest concentrations in the apex and the lowest in the base of the lungs (108). A feature of this investigation was the high-Sb levels found in the lymph glands (0.34 and 0.43 μg Sb/g wet weight). Hamilton *et al.* (65) also reported the presence of antimony in all human tissues examined with mean concentrations generally lower than those of Smith (164), but similar to those obtained for mouse organs and tissues by Molokhia and Smith (109). The lungs and lymph nodes revealed similar high values. The mean level in 75 samples of blood was 0.005 ± 0.0009 μg Sb/g wet weight, or close to 0.02 μg/g on the dry basis (65). Rib samples from 22 hard water areas in England averaged 1.3 ± 0.2 μg Sb/g ash, and from 22 soft water areas 1.7 ± 0.3 μg/g ash (65). In two separate studies of human dental enamel similar Sb levels were obtained. Nixon *et al.* (118) reported a range of 0.005–0.67 ppm, with a mean of 0.034 for Scottish subjects and 0.070 for Egyptian subjects who had received antimony treatment for bilharzia. Losee *et al.* (97) obtained a range of 0.02–0.34 μg Sb/g for dental enamel from 28 U.S. subjects, with the higher mean concentration of 0.13 ± 0.01 μg/g.

Little has been published on the Sb levels in individual foods or animal feedstuffs, although it has long been known that foods stored in enamel vessels and cans may contain appreciable Sb concentrations (112). Two recent studies of total diets have given widely divergent results. Murthy *et al.* (117) analyzed the total 7-day diets of children in institutions in 28 localities in the United States, including between-meal snacks. The total intakes varied from 0.247 to 1.275 mg Sb/day and the total dietary concentrations from 0.209 to 0.693 mg Sb/kg. Hamilton and Minski (64) analyzed English total adult diets and obtained the much smaller mean intake of 34 ± 27 μg Sb/day. How far this difference reflects variations related to location, differences in particular food items, or analytical errors is unknown. However, significant seasonal and geographical variations were observed in the former investigation. Furthermore Hamilton and Minski (64) reported 0.08 μg Sb/g in brown sugar compared with < 0.002 μg/g in refined white sugar. This suggests that the proportion of refined foods in the diet would influence dietary intakes of Sb, as occurs with many other trace elements.

Interest in the metabolism of antimony developed following the discovery by Christopherson (28) that the trivalent Sb compound tartar emetic is effective in the treatment of schistosomiasis. Tartar emetic and other antimonials when regularly injected significantly raise the Sb levels of most tissues and blood and particularly that of the red cells (8, 28, 94). Several workers have demonstrated the affinity of trivalent antimony for red blood cells (110, 124) both *in vivo* and *in vitro.* It has been suggested that the therapeutic value of the Sb compounds lies in part in this fact since the adult worms digest the erythrocytes. By contrast the red cells are almost impermeable to pentavalent Sb compounds in man (123). Trivalent Sb can readily be demonstrated in schistosomes, especially female schistosomes, following injection of infected mice with tartar emetic (88, 111).

Lifetime studies with mice fed 5 ppm Sb as antimony potassium tartrate in the drinking water revealed no demonstrable toxic effects on males and a slight decrease in life-span and longevity and some suppression of growth of older animals in females (150). No evidence of carcinogenesis or tumorogenesis was obtained. On this basis antimony must be considered to have a low inherent toxicity.

III. BARIUM

There is no conclusive evidence that barium performs any essential function in living organisms. The experiments of Rygh (142) with rats and guinea pigs carried out over 25 years ago do not appear to have been either confirmed or invalidated. This worker supplemented specially purified diets with a "complete" mineral mixture and obtained satisfactory growth and development of both animal species. The omission of either Ba or Sr from the mineral supplement was reported to result in depressed growth.

According to Schroeder *et al.* (151) the "standard reference man" contains 22 mg Ba, of which 93% is present in the bones. The remainder is widely distributed throughout the soft tissues of the body in very low concentrations which do not increase with age, except in the lungs presumably from atmospheric dust. Tipton and Cook (167) report the following mean concentrations for normal adult human tissues: adrenals, 0.02; brain, 0.04; heart, 0.05; kidney, 0.10; liver, 0.03; lung, 0.10; spleen, 0.08; and muscle, 0.05 ppm Ba (d. b.). Hamilton *et al.* (65) obtained generally similar low values for human tissues, except for the lymph nodes which were reported to contain 0.8 ± 0.3 μg Ba/g wet weight. The mean level in 103 samples of whole human blood was 0.1 ± 0.06 μg Ba/g. The relatively high-Ba content of bone is evident from two studies. Sowden and Stitch (160) obtained a mean of 7 ppm of ash for 35 samples of bone from normal men and women. Hamilton *et al.* (65) reported a mean concentration of 18.0 ± 2.8 μg/g ash for 22 rib samples from hard water areas and 19.3 ± 20 μg/g ash for 22 rib samples from soft water areas in England.

Barium is poorly absorbed from ordinary diets with little retention in the tissues or excretion in the urine. Thus in rats fed barium, 24-hr urinary Ba excretion was 7% of the amount ingested, while three human subjects excreted 1.8, 1.9, and 5.8% of their dietary Ba intake in the urine (23). Barium obviously crosses the placental barrier since the element has been found in the tissues of all newborn infants examined (151).

The mean Ba intake from English diets has been estimated as 603 ± 225 μg/day (64) and from a U.S. hospital diet 750 μg/day (151). Higher intakes by five adult human subjects (0.65–1.77 mg/day) were reported by Tipton and co-workers (168, 169), and lower amounts (< 0.30–< 0.59 mg Ba/day) were present in the hospital diets studied by Gormican (61). The 20 school lunches from 300 schools in 19 U.S. states analyzed by Murphy *et al.* (114) ranged widely from 0.09 to 0.43 mg Ba, with a mean of 0.17 mg. On the basis of the lunches supplying one-third of the daily food intake the total Ba intakes would range from 0.27 to 1.29 mg/day and would average 0.51 mg (Ba/day).

Comprehensive studies of the Ba levels in human foods do not appear to have been carried out, so that the influence of location or of varying amounts and proportions of various food items on total human intakes cannot be estimated. However, Brazil nuts are exceptionally rich in barium. The levels vary with locality but concentrations of 3000–4000 ppm Ba are common and are not accompanied by unusual concentrations of strontium (136, 155). Certain other plant species can accumulate large amounts of barium from Ba-rich soils. For example, *Juglans regia* and *Fraxinus pennsylvanica* were reported to contain 2600 and 1700 ppm Ba, respectively (138). Wide variation in Ba levels appears to be a characteristic of plants both within and among species. Bowen and Dymond (21) reported a range of 0.5–40 ppm Ba (mean 10 ppm) for different plant species growing on different soils, while 10–90 ppm Ba on the dry basis was reported in an earlier study for Kentucky hay and 1–20 ppm for cereal grains (74). In a comparison of red clover and ryegrass grown together on different soils in Scotland, Mitchell (106) found the clover to contain 12–134 ppm (mean 42) and the grass 8–35 (mean 18) ppm Ba.

Barium has a low order of toxicity by the oral route. For example, Hutcheson *et al.* (81) fed barium sulfate to mice at various levels up to 8 ppm dietary Ba for three generations with no significant effects on growth, mortality, morbidity, or reproductive or lactational performance. Schroeder and Mitchener (149) fed mice on 5 ppm Ba (as barium acetate) in the drinking water in a life-term study with no observable effects on longevity, mortality, and body weights or on the incidence of tumors.

IV. BORON

Boron has been known to be essential for the higher plants for over half a century (178), but no evidence has been obtained indicating that it is required by animals. Rajaratnam *et al.* (132) suggest that this is due to the fact that boron is required only or mainly for flavonoid synthesis, compounds that are found only in the higher plants. Several unsuccessful attempts have been made to induce B deficiency in rats by the use of purified diets containing 0.15–0.16 ppm, but the rats grew and reproduced as well as those receiving additional B (78, 122, 166). If this element is required by the rat it must therefore be at a dietary level below 0.15 ppm. Earlier indications that B is beneficial to K-deficient rats were not supported by subsequent investigation (47).

Boron is distributed throughout the tissues and organs of the animal at concentrations mostly between 0.5 and 1.5 ppm dry basis in the soft tissues and several times these levels in the bones (1, 49, 65, 78). In a recent study of human tissues (65) the following mean B concentrations were reported, in μg/g wet weight: blood, 0.4; liver, 0.2; kidney, 0.6; muscle, 0.1; brain, 0.06; testis, 0.09; lung, 0.6; and lymph nodes, 0.6. The mean B level in 22 samples of human rib from hard water areas in England was 10.2 ± 5 μg/g ash and in 22 samples from soft water areas 6.2 ± 2 μg/g ash. Human dental enamel has been found to vary widely in B content from 0.5 to 69 μg/g dry weight (97). The median level in this study, which included 56 individual samples, was 9.1 and the mean 18.2 ± 2.65 μg B/g. The ingestion of large amounts of boric acid results in a marked increase in the B levels of the tissues, particularly the brain (129).

Cow's milk normally contains 0.5–1.0 ppm B, with little variation due to breed or stage of lactation (78, 125), but with ready response to changes in dietary B intake. Owen (125) raised the B level in the milk from 0.7 to over 3 ppm by adding 20 g of borax daily to the cow's normal ration.

The boron in food and B added as sodium borate or boric acid is rapidly and almost completely absorbed and excreted, largely in the urine (87, 125, 168). Where high intakes occur, either accidentally or from the treatment of large burns with boric acid, similar high absorption and excretion take place but sufficient B can be temporarily retained in the tissues to produce serious toxic effects (129). However, B has a low order of toxicity when administered orally.

Schroeder and Mitchener (149) found that 5 ppm B as sodium metaborate fed in the drinking water of mice for life has no effect on their life-span, longevity, or the incidence of tumors.

Some years ago Kent and McCance (87) reported daily B intakes by human adults of 10–20 mg, the higher amounts being associated with the consumption of large quantities of fruits and vegetables. These levels contrast greatly with the mean B intakes of 0.42 and 0.35 mg/day found by Tipton *et al.* (168) for two adults consuming their normal diets. Zook and Lehmann (189), in their study of the minerals in total diets made up from various sources to supply 4200 cal/day, obtained an overall average B content of 3.1 mg, with individual composites varying relatively little from 2.1 to 4.3 mg/day. These levels conform well with the mean of 2.8 ± 1.5 mg/day more recently reported for English total diets (64). Boron occurs in foods of plant origin in much higher concentrations than in animal tissues and products under normal conditions. Great variation in the B content of tropical and subtropical fruits has been observed, both within and among different producing areas (188). Of the fruits examined, avocados were the highest (7–10 ppm B on the fresh edible basis), followed by stone fruits (1.4–3.5 ppm), and pome and citrus fruits and berries (0.3–2.4 ppm). The cereal grains mostly contain 1–5 ppm B (12). As indicated earlier, muscle meats and milk are normally much lower in B content.

Boron intakes by grazing animals vary with the soil type and plant species consumed because the B content of plants is greatly influenced by the species and B status of the soil. Levels of 4–7 ppm B have been reported for the dry matter of European pasture grasses (90), and there is evidence that legumes are richer in this element than grasses (12). The pastures of the solonetz and solonchak soils of the Kulindisk steppe in the Soviet Union are so high in B that gastrointestinal and pulmonary disorders occur in lambs (130). The water supplies in this region are also unusually high in B (0.2–2.2 mg/liter), which probably contributes to the B toxicity.

V. BROMINE

Bromine is one of the most abundant and ubiquitous of the recognized trace elements in the biosphere. It has not been conclusively shown to perform any essential function in plants, microorganisms, or animals. Bromide can completely replace chloride in the growth of several halophytic algal species (104) and can substitute for part of the chloride requirement of chicks (93). A small significant growth response to trace additions of Br has been reported in chicks fed a semisynthetic diet (79) and in mice fed a smiliar diet containing iodinated casein to produce a hyperthyroid-induced growth retardation (18). The Br content of the basal diet was not given, and these indications of a growth requirement for

Br do not appear to have been investigated further. Winnek and Smith (182) were earlier unable to demonstrate any effect on the growth, health, or reproductive performance of rats fed a diet containing less than 0.5 ppm Br over a period of 11 weeks, or any improvement from a supplement of 20 ppm Br as KBr. The opportunities for contamination of the animal's dietary and physical environment with Br are so great that further studies with this element employing the plastic isolator environment are desirable.

All animal tissues, other than the thyroid where the position is reversed, contain 50–100 times more Br than I. Species differences in tissue Br concentrations are small and the element does not accumulate to any marked degree in any particular organ or tissue (31, 65, 69). Claims that Br is concentrated in the thyroid and pituitary glands have not been substantiated (10, 35, 41). In a study of adult human tissues the following mean levels were reported in μg Br/g wet weight: brain, 1.7 ± 0.05; liver, 4.0 ± 0.3; lung, 7.5 ± 0.7; lymph nodes, 0.9 ± 0.3; testis, 5.1 ± 1.7; and ovary 3.3 ± 1.1 (65). The mean concentration in 163 samples of whole human blood was 4.6 ± 0.06 μg Br/g. This conforms well with the mean level of 3.7 μg Br/ml reported much earlier using a delicate micromethod (35) and the 2.9 μg/ml obtained later by activation analysis (19). Brune *et al.* (24) reported a mean value of 3.9 μg Br/g of whole blood in eight healthy subjects and 3.7 μg/g for eight uremic patients.

The Br levels in all tissues and fluids can be substantially raised by increasing dietary Br intakes (100, 182). The Br level in milk is extremely sensitive to differences in Br intakes by the cow. Lynn *et al.* (100) increased the Br concentration of milk from a pretreatment level of 10 ppm to as high as 60 ppm by feeding 12.5 g NaBr to cows for 5 days. Smaller increases in milk Br levels and a high correlation between milk and blood Br levels were observed when the cows were fed a methylbromide-fumigated grain ration. Much lower milk Br levels were mostly reported in several other investigations (27, 37, 51). How far these lower values are a reflection of lower Br intakes by the cows or result from analytical difficulties is uncertain.

Little information is available on Br metabolism, although the element appears to be well absorbed and is mostly excreted in the urine, whether ingested or injected (31, 69). Bromide and chloride readily exchange to some degree in the body tissues, so that administration of bromide results in some displacement of body chloride and vice versa (103). This occurs as a consequence of feeding large amounts (182) or injecting physiological quantities (69). Rabbit thyroid glands rendered hyperplastic by lack of iodine are richer in Br than the blood (10). This suggests that the thyroid distinguishes imperfectly between Br and I and seizes some Br in the absence of sufficient I. The Br accumulated in the thyroid is quickly lost when I is supplied and cannot be used for hormone synthesis (134). There is some evidence that injected bromide reduces ^{131}I uptake by the rat thyroid and that goiter can occur in rats fed bromide during their first year of life (29).

The chemical form, or forms, in which Br exists in the tissues is largely unknown. Many years ago Zondek and Bier (187) reported the possible presence of a sleep hormone containing Br in dog pituitary gland. Later Gruner *et al.* (63) observed a significant incorporation of radioactive Br into the pituitary gland and epiphysis when the isotope was given to rats. A new Br compound has now been isolated from human cerebrospinal fluid, with properties corresponding well to 1-methylheptyl-γ-bromoacetoacetate (synonym of 2-octylbromoacetoacetate) (186). The physiological significance of this organic Br compound remains to be determined. It has been shown to "provoke paradoxical sleep" when administered intravenously to cats (170).

Human dietary intakes of bromine are large and variable. Duggan and Lipscomb (43) obtained an average intake over 2 years from U.S. diets of 24 mg Br/day, while Hamilton and Minski (64) reported a mean intake from English total diets of 8.4 ± 0.9 mg Br/day. Actual intakes will be much higher where organic Br compounds are used as fumigants for soils and stored grains (44, 71). Bromine concentrations as high as 53–220 ppm in grain (oats and corn) fumigated with methyl bromide and marked increases in the Br content of milk from cows fed such grain, as indicated earlier, have been reported (100). Even in untreated grain samples Br levels ranging widely from 0.5–25 ppm have been observed (71). In earlier studies whole cereal grains were reported to contain 1–11 ppm, flour 5–8 ppm, and white bread 1–6 ppm Br (38, 51, 182).

VI. GERMANIUM

The biology of germinium has excited little interest, despite its relative abundance in the lithosphere, its chemical properties, and its position in the Periodic Table within the range of the biologically active trace elements. There is no evidence that Ge is essential in mammalian nutrition, and germanates have a low order of toxicity in mice and rats (140, 146, 148). Schroeder and Balassa (146) used the photometric method of Luke and Campbell (99), following a special low temperature ashing technique, to examine the Ge concentrations of a range of biological materials. Of 125 samples of foods and beverages analyzed, almost all revealed detectable germanium. Only four of these samples contained more than 2 μg/g and 15 others more than 1 μg/g. Germanium was not detected in refined white flour, although it was present in whole wheat and was concentrated in bran. The mean levels in groups of vegetables and leguminous seeds were 0.15–0.45 μg/g on the fresh basis. Similar concentrations were observed in meat and in dairy products.

Few data on the Ge content of normal human organs have appeared. The element was not detected in 56 samples of human dental enamel examined by spark source mass spectrometry (97). The liver, kidneys, heart, lungs, and spleen of laboratory mice and rats fed normal diets contained Ge in concentrations ranging from 0.10 to 2.79 μg/g wet weight. When mice were given 5 ppm Ge in the drinking water for their lifetime, higher concentrations were found in these organs, especially in the spleen, but rat tissues accumulated very little Ge under these conditions (146, 148). The Ge in the drinking water at this level gave slight evidence of toxicity in both species, as indicated by reduced life-span and increased incidence of fatty degeneration of the liver. The element was neither tumorogenic nor carcinogenic.

Little is known of the metabolism of the Ge ingested from ordinary diets. Rosenfeld (139) found that oral doses of sodium germanate were rapidly and almost completely absorbed from the gastrointestinal tract within a few hours, and were excreted largely in the urine during 4–7 days. The data obtained by Schroeder and Balassa (146) for the Ge content of the urine of four individuals (mean 1.26 μg/ml) suggest that dietary Ge is also well absorbed and excreted largely via the kidneys in man. These workers calculated that adults ingest approximately 1.5 mg Ge in the daily diet, of which 1.4 mg appear in the urine and 0.1 mg in the feces. A lower mean level of 367 ± 159 μg Ge/day was obtained for total English adult diets (64).

VII. LITHIUM

Lithium has no known vital function in living organisms. Biological interest in this element developed rapidly following the original demonstration in 1949 by Cade (25) of the value of lithium carbonate in treating manic depressive psychosis. The effectiveness of this treatment has since been widely confirmed (see Gershon, 57). A further stimulus to studies of the biology of lithium came from epidemiological evidence suggesting, but not proving, that lithium ions may exert a protective influence on atherosclerotic heart disease in man (36, 176).

Lithium, a light metal with an atomic number of 3 and an atomic weight of 6.94, is widely distributed throughout the geosphere and biosphere. In two separate studies the Li content of soils ranged from 8 to 40 ppm (164) and from 15–32 ppm (172). Much lower concentrations are common in plants used as foods and forage. Bertrand (14) reported an average of 0.85 ppm Li (dry weight) in monocotyledons and 1.3 ppm in dicotyledons. Borovik-Romanova (17) analyzed 138 plants from eight soil types by emission spectrography. Average levels between 0.15 and 0.3 ppm Li were generally found, although plants rooted in saline soils can contain 2–10 times these levels. The values reported by Robinson *et al.* (137) were mostly lower. The Li content of timothy grass varied from 0.07

to 0.28 and of clover from 0.023 to 0.23 ppm, while single samples of apples and beets gave values of 0.023, cabbage 0.093, and onions 0.23 ppm dry weight.

Lithium occurs in all the tissues and fluids of the body in low concentrations and with some evidence of slight concentration in the brain, thyroid, and ovaries (183). No such concentration is evident from the later studies of human tissues by Hamilton *et al.* (65). These workers reported the following mean concentrations in μg/g wet weight: lymph nodes, 0.2 ± 0.07; lung, 0.06 ± 0.01; liver, 0.007 ± 0.003; muscle, 0.005 ± 0.002; brain, 0.004 ± 0.001; testis, 0.003 ± 0.001; and ovary, 0.002 ± 0.004. The mean Li concentrations in 100 samples of whole human blood was 0.006 ± 0.002 μg/g. Further data on human tissues obtained by modern acceptable analytical techniques are obviously necessary to establish the normal distribution of Li throughout the body and the possible influence of geographical differences in Li intakes from the food and water supplies. For example Bowen (20) estimated the average total adult dietary to supply 2 mg Li/day, whereas Hamilton and Minski (64) later reported the much lower mean level of 107 ± 53 μg Li/day for the total English diet. The latter figure conforms more to the estimate of 20 μg for the daily intestinal Li absorption by man (183).

Little is known of the metabolic behavior of Li from foods and dietaries, i.e., at physiological levels. Medicinal Li salts in drinking water are believed to be mostly absorbed and largely excreted in the urine (95, 144). Since all the tissues carry some Li the metal is obviously absorbed from foods and beverages to some extent and retained in small quantities (171). Daily urinary Li excretion would thus normally provide a reasonable index of daily intestinal absorption.

Consideration of the pharmacological action of lithium in the relatively massive doses required in the treatment of manic depressive psychosis in man (3 mg or more ionic Li/kg bodyweight/day) lies outside the scope of this text since such intakes are several thousand times the probable normal dietary levels of the element. The mechanism of this action is still obscure. Many changes in brain biochemistry and electrolyte distribution have been reported (143), but it is not clear whether these are causal or consequential. Because of possible toxic side effects, including polyuria, ataxia, hypothyroidism, and weight gain, the serum Li levels must be controlled within narrow limits. For prophylaxis against depression the serum concentration should be kept between 0.5 and 1.0 mEq Li/liter and for treatment of manic episodes the goal is a serum concentration of 1.0–1.5 mEq/liter until the end of the attack (131). Toxic symptoms have been reported from concentrations as low as 1.6 mEq Li/liter (2), and severe toxicity requiring heroic therapeutic measures have been observed in patients with Li concentrations between 2.4 and 4.2 mEq/liter (30). In a proportion of patients, particularly women, the lithium treatment depresses the thyroid function, either by blocking thyrotropic stimulating release hormone, or thyroid stimulating hormone itself, or most probably by blocking thyroxine release (26). In rats lithium has been shown to lower the uptake and turnover rate of ^{125}I by the thyroid (80).

The epidemiological evidence suggesting that Li may exert a protective effect on the incidence of atherosclerotic heart disease (36, 176) raises the possibility that it may act through an effect on the serum lipids (176). Fleischman *et al.* (45) have obtained data with rats indicating that exogenous Li stimulates triglyceride metabolism. Liver triglyceride levels were found to decrease concomitant with an increase in serum free fatty acid as the dietary Li increased from 0.008, 0.02, and 0.08%. No changes in serum cholesterol or triglyceride levels were observed as an outcome of the treatment. In seeking to relate these findings to the negative correlation between lithium ion and atherosclerotic heart disease it should be appreciated that the lowest Li supplementary level employed (0.008% or 80 ppm) is much higher than any conceivable Li intake from normal foods and beverages, except in areas with very high levels of Li in the water supplies. Waters with such high Li levels ($>$ 100 ppb) occur in several of the western states of the United States, but these are not the areas where the incidence of heart disease is lowest (3, 55).

VIII. RUBIDIUM

Biological interest in Rb has been stimulated by its close physicochemical relationship to potassium and its presence in living tissues in higher concentrations, relative to those of potassium, than in the terrestrial environment. Nearly a century ago Ringer (135) observed that Rb affected the contractions of the isolated frog's heart in a manner similar to potassium. A general relationship between these two elements, and between cesium and potassium, exist for a variety of physiological processes. These include such diverse actions as their ability to neutralize the toxic action of lithium on fish larvae and to affect the motility of spermatozoa, the fermentative capacity of yeast, and the utilization of Krebs cycle intermediates by isolated mitochondria. Their extracellular ionic concentrations also influence the resting potential in nerve and muscle preparations and the configuration of the electrocardiogram (133).

The metabolic interchangeability evident in these processes suggests that Rb might act to some extent as a nutritional substitute for potassium. Rubidium, and to a lesser extent cesium, can replace K as a nutrient for the growth of yeast (92) and of sea urchin eggs (96). This nutritional replaceability can be extended to bacteria (102), but the higher animals are more discriminating. Additions of Rb or Cs to K-deficient diets prevent the lesions in the kidneys and muscles characteristic of K depletion in rats (46) and, for a short period, permit almost normal growth until death inevitably supervenes (70). Using purified diets with varying supplements of Rb, Na, and K, Glendenning *et al.* (59) obtained no evidence that Rb is essential for rats, although there were indications that it may substitute partially for potassium and is more toxic on low- than on high-K diets. Purified diets containing up to 200 ppm Rb were nontoxic, whereas levels of 1000 ppm or more resulted in decreased growth, reproductive performance, and survival time.

Rubidium resembles potassium in its pattern of distribution and excretion in the animal body (15, 59, 70, 133, 163). All plant and animal cells are freely permeable to Rb and Cs ions, at rates comparable with those of potassium (133). All the soft tissues of the body carry Rb concentrations that are high compared with many trace elements, with a total body content approximating 360 mg in adult man (185). It does not accumulate in any particular organ or tissue and is normally relatively low in the bones (59, 157, 162) and enamel of the teeth (97). The plasma also contains lower concentrations than the red cells (15, 52, 59). The soft tissues of man and other vertebrates contain 20–60 ppm Rb on the dry basis (15, 157). Higher average values for these tissues have been observed in rats (59) and adult man, with lower levels common in infants and children (163). In a more recent study of human tissues the following mean Rb concentrations, expressed as μg/g wet weight, were obtained: brain 4.0 ± 1.1; muscle, 5.0 ± 0.5; ovary, 5.0 ± 0.9; kidney cortex, 5.2 ± 0.5; kidney medulla, 5.0 ± 0.3; lung, 3.5 ± 0.4; lymph nodes, 5.5 ± 1.1; liver, 7.0 ± 1.0; and testis, 19.6 ±6.2. The mean level in 165 samples of whole blood was 2.7 ± 0.04 μg Rb/g wet weight (65). Concentrations ranging from 100 to 200 ppm of dry tissues were found in the muscles, liver, lungs, kidneys, heart, and brain of rats fed a standard diet, and much higher concentrations, up to 8000 and 12,000 ppm, when the animals were fed toxic Rb levels (59). The Rb retained in tissues as a result of such high dietary intakes, or from injection, is slowly lost from the body, mainly in the urine.

Human and animal foods were examined by Glendenning *et al.* (58) using a spectrographic method sensitive to 1 ppm Rb. Soybeans contained 160–225 ppb Rb, and bromegrass and sorghum silage 130 ppm on the dry basis. The human foods of plant origin averaged about 35 ppm, with the cereal grains below this average and white flour the lowest in Rb ($<$ 1 ppm) of those analyzed. Meat foods were relatively rich in Rb, with beef muscle containing no less than 140 ppm on dry tissue. Market milk collected from cities in the United States averaged between 0.57 and 3.39 ppm Rb on the fresh basis, with significant differences among the various cities (116). On the basis of the available data it seems that human diets high in refined cereals would supply much less Rb than diets rich in animal foods and fruits and vegetables. English total diets have been reported to supply an average of 4.35 ± 1.54 mg Rb/day (64) and Japanese diets from four different locations to average 1.9–2.6 mg Rb/day (185).

IX. SILVER

There is no evidence that silver is essential for any living organisms, nor is it ranked among the more toxic trace elements. It occurs naturally in very low concentrations in soils, plants, and animal tissues and can gain access to foods from silver-plated vessels, silver–lead solders, and silver foil used in decorating cakes and confectionery. In 1942 Kent (86), using a spectrographic method, reported 0.3 ppm Ag in wheat flour and 0.9 ppm in bran. These levels now appear suspiciously high in light of the 60–80 (85) and 27 ± 17 μg Ag/day (64) reported for total human dietary intakes. Cow's milk from different U.S. cities varied from 0.027 to 0.054 mg Ag/liter, with no significant variation among cities and sampling periods (115). The national weighted average in this study was 0.047 ± 0.007 mg Ag/liter. Earlier spectrographic analyses of cow's milk taken directly into glass containers revealed similar Ag concentrations, ranging from 0.015 to 0.037 ppm (54).

In the early spectrochemical study of Kehoe *et al.* (85) of human tissues and fluids Ag was not detected in the blood and urine or in the kidney, heart, spleen, muscle, and stomach. The mean values for brain, liver, and lung were 0.03, 0.05, and 0.04 ppm, respectively. Much lower levels for human tissues have recently been reported, using spark source mass spectrometry (65). The mean values were brain, 0.008; kidney, 0.002; liver, 0.006; lung, 0.002; muscle, 0.002; testis, 0.002; and ovary, 0.002 μg Ag/g wet weight. The mean concentration found for 93 samples of whole human blood was 0.008 ± 0.0002 μg Ag/g. In another study the silver content of 56 samples of human dental enamel ranged from 0.01 to 0.77 μg/g dry weight, with a median of 0.06 and a mean of 0.14 μg Ag/g (97).

Silver is known to interact metabolically with copper and selenium. Of the known Cu antagonists tested by Whanger and Weswig (181) in the rat dietary Ag was the strongest, with Cd, Mo, Zn, and sulfate following in descending order. Hill *et al.* (72) first showed that Ag accentuates the Cu deficiency induced by a low-Cu basal diet. Later Jensen and co-workers (82) found that the depressed growth rate, reduced packed cell volume, and cardiac enlargement induced in turkey poults by adding 900 ppm Ag as acetate or nitrate to practical diets could be prevented by 50 ppm Cu. In similar studies with chicks (128) the cardiac enlargement induced by the high-Ag diet was prevented by 50 ppm Cu but the growth retardation was only partially corrected, probably because of inadequate levels of Se and vitamin E relative to the large amounts of Ag. The manner in which Ag interferes with Cu metabolism is not clear. In the chick experiment just described the high-Ag intake markedly reduced the Cu levels in the tissues which would suggest reduced Cu absorption, but total Cu excretion was apparently not affected by the treatment. Furthermore, as mentioned in Chapter 3, Van Campen (173) observed very little effect of Ag on ^{64}Cu uptake in the rat.

However, a greater proportion of the absorbed isotope was retained in the liver and less by the blood of the Ag-treated rats.

Interactions of silver with selenium are described in Chapter 13.

X. STRONTIUM

The metabolic behavior of strontium, and particularly its interaction with calcium, have attracted considerable attention since it became obvious that ^{90}Sr is an abundant and potentially hazardous by-product of nuclear fission. There is no conclusive evidence that Sr is essential for living organisms, although over 25 years ago Rygh (142) reported that the omission of this element from the mineral supplement fed to rats and guinea pigs consuming a purified diet resulted in growth depression, an impairment of the calcification of bones and teeth, and a higher incidence of carious teeth. This report has been neither confirmed nor invalidated. However, Colvin and co-workers (32) showed that at least 25% of the Ca requirement of growing chicks could be spared by strontium (see Table 57).

Many years ago Gerlach and Muller (56) reported that the Sr concentration of a wide variety of animal tissues ranged from 0.01 to 0.10 ppm, with no evidence of accumulation in any particular species, soft organ, or tissue. More recently the following two sets of mean values for adult human organs have been reported: adrenals, 0.06; muscle, 0.07; liver, 0.10; brain, 0.12; heart, 0.15; kidney, 0.35; and lung, 0.50 ppm Sr *dry* weight (167): brain, 0.08 ± 0.01; kidney, 0.10 ± 0.02; liver, 0.10 ± 0.03; muscle, 0.05 ± 0.02; testis, 0.09 ± 0.002; ovary, 0.14 ± 0.06; lung, 0.20 ± 0.02; and lymph nodes, 0.30 ± 0.08 μg Sr/g *wet* weight (65). In this latter study 103 samples of whole human blood were found to average 0.02 ± 0.002 μg Sr/g.

The Sr content of bone has attracted particular interest because of the affinity of this tissue for Sr and therefore its relevance to the problem of ^{90}Sr retention from radioactive fallout. The total Sr content of the standard reference man has been reported to be 323 mg, of which 99% is present in the bones (151). The bone ash of human fetuses and adults in an early study averaged 160 and 240 ppm Sr, respectively (73). These levels are similar to those recently reported for human rib bone, i.e., 155.9 ± 14.6 and 138.7 ± 9.0 μg Sr/g ash (65), but are appreciably higher than the mean of 100 ppm Sr in the ash of human bones examined by others (84, 160, 167). There is some evidence that the Sr levels in bones, as well as in the lungs and aorta, increase with age in man and that they vary in different geographical regions (64, 151, 160, 175). For example, significantly higher rib ash Sr concentrations were found in Far East adults (190 ppm) and children (320 ppm) than in U.S. adults (110 ppm Sr median value) and children (96 ppm median value) (151). The extent to which these differences are a reflection of the lower Ca intakes of the Oriental people or of possible higher Sr intakes is unknown. The different Sr levels in the bones and teeth of human fetuses from different parts of Israel have been attributed to differences in the Sr levels in the water supplies (53, 175).

Strontium occurs in the enamel and dentin of teeth in concentrations that parallel the levels in the bones of the same individuals and with similar geographical variations (161, 175). This strontium is deposited primarily before eruption, during tooth calcification, is mostly permanently retained, and is not affected by F in the drinking water (162). Fifty-six samples of dental enamel from U.S. residents ranged from 21 to 280 μg Sr/g dry weight, with a median level of 96 and a mean of 121 ± 11 μg/g (97).

The bones of chicks exhibit a remarkable capacity to retain Sr, mainly at the expense of Ca, when high dietary Sr levels are fed (32, 180). Such incorporation of Sr is accompanied by a growth depression and a reduction in bone ash content (Table 57) which are more severe at lower Ca intakes. An interaction with Ca is further apparent from the finding that the bone ash Sr concentrations in the chicks were lower at higher Ca intakes. For example at 3000 ppm dietary Sr the bone ash contained 5.7 and 4.1 ppm Sr at 0.75% dietary Ca and at 1.0% Ca, respectively. At 6000 ppm dietary Sr the comparable bone ash Sr levels were 9.4 and 7.7 ppm. When similar levels of Sr up to 12,000 ppm were fed to laying hens the Sr content of eggshell was found to increase stepwise from 0.1 to 6.4% (179). An inverse relationship between Ca and Sr in eggshell formation is apparent from this study because the Ca decreased as the Sr increased, with the total Ca + Sr remaining approximately constant at 31% for all treatments.

The absorption of Sr from the small intestine is increased under fasting conditions (6) and is decreased in the presence of food (161), a Ca carrier (119), and with advancing age. In rats ^{85}Sr absorption drops abruptly at about the time

TABLE 57
Effects of Strontium Supplementation on Body Weight and Bone Composition of Chicks[a]

		Tibiotarsus		
Dietary treatment	Body wt 15 days (g)	Ash (%)	Ca (%)	Sr (%)
1% Ca, 0.0% Sr	247 ± 23	42.2 ± 1.4	14.5 ± 0.9	0.01 ± 0.003
0.75% Ca, 0.25% Sr	249 ± 17	39.8 ± 1.2	13.0 ± 0.1	2.45 ± 0.06
0.50% Ca, 0.50% Sr	220 ± 28	35.3 ± 2.9	11.1 ± 2.2	3.67 ± 0.63
0.25% Ca, 0.75% Sr	175 ± 26	25.8 ± 2.5	7.5 ± 1.2	5.35 ± 0.47
0.00% Ca, 1.0% Sr	183 ± 21	18.5 ± 2.5	4.6 ± 1.0	–

[a]From Colvin *et al.* (32).

of weaning to the low level characteristic of the adult (48). Intestinal Sr absorption by adults of various mammalian species ranges from 5 to 25%, with age changes in the same species varying from more than 90% in very young to less than 10% in old individuals (33, 174). Raising dietary Ca intakes from low to normal reduces Sr retention (126), and supplementation with Ca plus P is more effective than Ca alone (66). Increasing the dietary levels of the alkaline earths similarly depresses radiostrontium retention, with Sr being the least effective (113). Various other dietary factors that can influence Sr absorption and retention are reviewed by Wolf (174). The biological discrimination that occurs against Sr compared with Ca has been stressed by Comar and co-workers (34). These workers have shown that the Sr:Ca ratios in the bones and bodies of animals are lower than the Sr:Ca ratios of their diets, mainly as a consequence of lower Sr than Ca absorption and higher Sr than Ca urinary excretion. This discrimination against Sr extends to pregnancy and lactation, as a consequence of smaller placental and mammary transfers of Sr than Ca.

Absorbed Sr is carried in the blood to the tissues, where it is preferentially deposited in the bones and teeth by two distinct processes. MacDonald *et al.* (101) have described these as (a) a rapid incorporation phase attributed to the blood Sr deposited by ionic exchange, surface absorption, and/or preosseous protein binding, and (b) a slow incorporation of Sr into the lattice structure of the bone crystals during their formation. Both processes are believed to depend mainly on the Sr concentration of the blood, including the cord blood in view of the early appearance of Sr in the bones and teeth of human fetuses (175).

Strontium is poorly absorbed and retained from ordinary human diets, which generally supply from 1 to 3 mg Sr/day. Harrison *et al.* (67) found one adult to ingest 1.99 mg Sr/day over an 8-day period and to excrete 1.58 mg in the feces and 0.39 mg in the urine. Tipton and co-workers (168) obtained the following mean values for two adults over a 30 day period: ingestion, 1.37 and 1.2; fecal excretion, 0.81 and 0.97; and urinary excretion, 0.24 and 0.42 mg Sr/day. In another study of children of different ages the Sr intakes ranged from 0.67 to 3.57 mg/day, most of which appeared in the feces (11). The daily Sr intakes from the food of adults in seven regions of India ranged from 3.1 to 4.7 mg/day (159), a level much higher than the mean of 858 ± 144 μg Sr/day obtained recently for total English diets (64). In two studies of U.S. hospital diets average Sr intakes of 1.2–2.9 (61) and 2.1–2.4 mg/day (151) were observed.

The Sr levels in a small number of samples of a wide range of foods comprising U.S. diets have been reported (61, 151), but there appear to be no systematic analyses which would permit a classification of food groups in accordance with their normal stable Sr status. In general, foods of plant origin are appreciably richer sources of Sr than animal products, except where the latter include bone. Strontium tends to be concentrated in the bran rather than the endosperm of grains and in the peel of root vegetables (42). In some areas the drinking water contributes a substantial proportion of total Sr intakes. Wolf *et al.* (175) contend that the differences found in Sr concentrations of the hard tissues of fetuses in three locations in Israel are due to differences in the Sr content of the drinking water supplied during pregnancy. They reported levels of 1.4–1.6 mg Sr/liter in one area, compared with 1.0–1.1 mg/liter in the location with the lower fetal bone Sr concentrations. At a daily water consumption of 1500–2500 ml these levels would supply no less than 1.5–4.0 mg Sr/day, which is as much or more than the amount probably ingested in the food. The drinking waters studied by Schroeder *et al.* (151) were highly variable in Sr content, but most of the values were much lower than those just quoted, with a high proportion lying between 0.02 and 0.06 mg Sr/liter.

Little information is available on the normal Sr intakes of farm animals. Mitchell (106) found 15 samples of red clover, growing on different soils, to vary from 53 to 115 (mean 74) ppm Sr, and 15 comparable samples of ryegrass to range from 5 to 18 (mean 10) ppm Sr on the dry basis. On this evidence Sr intakes would be much higher from leguminous than from gramineous forages. These intakes would also be greatly influenced by soil types on which the plants are grown. For example Bowen and Dymond (21) found plants growing on a variety of normal soils in England ranged widely from 1 to 169 (mean 36) ppm Sr dry weight. Certain species growing on Sr-rich soils contained much higher concentrations, up to 26,000 ppm Sr, or 2.6%. These are regarded as true accumulator plants, although other species will take up large amounts of Sr from culture solutions enriched with this element (177).

A wide margin of safety exists between dietary levels of stable Sr likely to be ingested from ordinary foods and water supplies and those that induce toxic effects. For example, neither 3000, 6000, nor 12,000 ppm Sr as carbonate fed to laying hens for 4 weeks on a diet containing 3.6% Ca affected body weights, feed conversion, egg weight, or egg production, despite marked increases in eggshell Sr contents (179). The importance of dietary Ca to the toxicity of Sr is evident from studies with growing chicks and pigs. Feeding 6000 ppm Sr as carbonate to chicks reduced their growth rate slightly at 1.0% Ca and severely at 0.72% Ca (180). Piglets were fed rations containing 0.16, 0.55, or 0.89% Ca and either 0, 0.47, or 0.67% Sr (6700 ppm) in factorial experiments (7). The pigs fed Sr and 0.16% Ca were the most severely affected by incoordination and weakness followed by posterior paralysis, while mild toxic effects were only occasionally observed in the pigs fed Sr and 0.55 or 0.89% Ca. The bone deformities were also more marked in the pigs fed Sr and 0.16% Ca. These findings conform well with those of Forbes and Mitchell (50), who found little effect on the daily weight gain or feed efficiency when up to 0.1% Sr was fed to rats in the presence of adequate Ca.

XI. TIN

Biological interest in tin initially focused on its toxic potential to man through the contact of foods with tin-coated cans and tinfoil. Tin has now been shown to be an essential nutrient for the growth of rats. Schwarz *et al.* (153) found the growth rate to be enhanced by nearly 60% if 1–2 ppm Sn was added as stannic sulfate to a highly purified diet fed in a plastic isolator environment (Table 58). Several tin compounds including organic tin derivatives were reported to be effective, indicating that the organism is capable of utilizing tin covalently bound to carbon. Specific biochemical lesions associated with Sn deficiency have not been reported and the biological chemistry of the element remains to be determined. It is also highly desirable that the findings of Schwarz be confirmed and extended in other laboratories and with other species.

In early spectrographic studies tin was demonstrated in most but not all human and animal tissues examined, in concentrations generally lying between 0.5 and 4.0 ppm of dry tissue (22). These studies included human and cow's milk (40), human teeth (98a), and spinal fluid (154). Kehoe and co-workers (85) found tin in 80% of human tissues and fluids with a fairly even distribution among 10 of the 11 tissues studied. No tin was detected in the brain and the mean values for other tissues ranged from 0.1 ppm in muscle and 0.2 ppm in kidney, heart, spleen, and intestines to 0.8 ppm Sn in bone, all on the fresh basis. In a later study carried out by Schroeder *et al.* (147) a wide but variable distribution of tin in human tissues was demonstrated, with significant differences related to geographical location and, with the lungs, to age. Little or no tin was observed in the tissues of stillborn infants but, as Schwarz *et al.* (153) have pointed out, tin is readily lost in drying and ashing due to its volatility, particularly when present as organic derivatives. They point out further that various organic tin compounds are lipid soluble and a substantial proportion of the tin in tissues is lipid extractable. Many reported values for tin in biological

TABLE 58
Growth Effects of Tin in Trace Element Controlled Environment[a]

Compounds	Dose level (μg Sn/g)	No. of animals	Av daily weight gain (g)	% increase	p
Control	–	5	1.10 ± 0.05[b]	–	–
Stannic sulfate	0.5	8	1.37 ± 0.10	24	0.002
Stannic sulfate	1.0	8	1.68 ± 0.10	53	0.001
Stannic sulfate	2.0	8	1.75 ± 0.10	59	0.001

[a]From Schwarz *et al.* (153).
[b]Mean plus standard error.

materials may therefore be too low. In the investigation of adult human tissues made by Hamilton *et al.* (65), using spark souce mass spectrometry following ashing by a low temperature (~ 100°C) nascent oxygen technique, Sn concentrations were obtained that are mostly comparable with those reported in the earlier investigations just mentioned. Their mean values were muscle, 0.07 ± 0.01; kidney, 0.2 ± 0.04; brain, 0.3 ± 0.04; testis, 0.3 ± 0.1; ovary, 0.32 ± 0.19; liver, 0.4 ± 0.08; lung, 0.8 ± 0.2; and lymph nodes, 1.5 ± 0.6 μg Sn/g wet weight. The mean Sn level reported for 102 samples of whole human blood was very low (0.009 ± 0.002 μg/g). The mean Sn concentration of 22 samples of rib bones from hard water areas was 4.1 ± 0.6 μg/g ash and for 22 such samples from soft water areas 3.7 ± 0.6 μg/g ash. Losee *et al.* (97), also using spark source mass spectrometry, found the Sn levels in human dental enamel to vary widely from 0.03 to 7.10 μg/g dry weight, with a median value of 0.23 and a mean of 0.53 ± 0.4.

Data on the tin content of foods obtained by reliable methods of treatment and analysis, in which losses of the element are avoided or minimized, are exceedingly meager. Mean daily intakes by U.S. adults have been variously reported from 1.5 (168) and 3.5 mg (147) to 17 mg (85), while the average English total diet was reported to supply only 187 ± 42 μg Sn/day (64). Schroeder *et al.* (147) calculated that a diet composed largely of fresh meats, cereals, and vegetables, "which usually contain less than 1 ppm Sn," would supply about 1 mg Sn/day, whereas a diet that included substantial amounts of canned vegetables, fruit juices, and fish could supply as much as 38 mg/day. However, cereal Sn levels well above 1 ppm have been observed (190). The reported mean concentrations were as follows: common hard wheat, 5.6 ± 0.6; common soft wheat, 7.9 ± 0.9; baker's patent flour, 4.1 ± 0.4; and soft patent flour, 3.7 ± 0.7 μg/g dry weight. Two points about this study should be noted. The first is that the samples were dry ashed at 480°C and then analyzed by atomic absorption. The second point is that substantially higher mean Sn concentrations were obtained for the dry matter of white bread (8.9 ± 1.0 and 9.5 ± 1.2 μg/g) than for white flour, indicating that less than one-half of the tin was contributed by the flour.

Large amounts of tin can accumulate in foods in contact with tin plate unless these are lacquered or coated with resin. In a recent comparison of the Sn levels of various types of foodstuffs in either (a) cans partially, or not resin coated, (b) cans entirely resin coated, or (c) nontin packing, the complete resin coating of tin-plated steel reduced the amount of tin in the foods by at least a factor of 50, as long as the coating remained intact (60). In group (a) the Sn levels ranged from 9 to 700 ppm, with median and mean values of 100 and 150 ppm, respectively. In group (c) the Sn levels were some 10 times lower than those of group (b). They ranged from 0.005 to 30 ppm, with a median value of 0.08 ppm. The highest levels related to tomato soup and tomato ketchup (above 1 ppm Sn) and were much greater than the low-Sn levels reported for fresh, unprocessed tomatoes (0.02 ppm). Humans and experimental animals are extremely tolerant of fruit juices and other foods contaminated with tin from containers, due to its poor absorption. No toxic signs were evident in five human volunteers after consuming fruit juices containing 498, 540, or 740 ppm Sn derived from the containers (12a). All five had some gastrointestinal disturbance after drinking a juice containing 1370 ppm Sn but even at this high level no such effect was apparent in a further trial with four of the five volunteers.

Little interest has been shown in the tin content of forage and pasture plants or in the intakes of this element by farm animals. Pasture herbage growing in Scotland was reported to contain 0.3–0.4 ppm Sn on the dry basis (105). Lichens have been found to concentrate this element, so that those growing on silicic rocks contained no less than 72 ± 4.7 ppm and those on ultrabasic rocks 37 ± 8.0 ppm (98).

Tin is poorly absorbed and retained by man and is excreted mainly in the feces. In three separate studies adults were found to excrete only 23 (127), 16 (85), 11, and 8 μg Sn/day in the urine (167). The fecal Sn excretion in these investigations, by contrast, approximated the total amount ingested with the food.

Ingested tin has a low toxicity, no doubt due in part to its poor absorption and retention in the tissues. Convincing evidence of human poisoning in tin from canned or other foods is difficult to find. A few instances of tin to the amount of 200–3000 ppm in food have been reported (112). Mice and rats given 5 ppm Sn as stannous ions in the drinking water from weaning to natural death grew normally, and the life-span of mice of both sexes and of male rats was unaffected (146, 148). Female rats displayed a reduced longevity and an increased incidence of fatty degeneration of the liver, and vacuolar changes in the renal tubules in both sexes. Inorganic tin fed to rats at subacute toxicity levels (250, 500, 1500, and 5300 mg Sn/kg diet) reduced growth, feed efficiency, and hemoglobin levels at low Fe and Cu dietary levels (62). Increasing the Fe and Cu levels prevented the hematological effect but not the growth inhibition.

XII. TITANIUM

Titanium resembles aluminum in being abundant in the lithosphere and soils, and in being poorly absorbed and retained by plants and animals, so that the levels in their tissues are generally much lower than those in the environment to which the organisms are exposed. The Ti levels in herbage samples can be used as an index of soil contamination (5), because the Ti concentration of most soils is some 10,000 times greater than in uncontaminated herbage (165). Bertrand (16) examined a variety of plants for titanium and recorded levels ranging from 0.1 to

5 ppm, with a high proportion of the values lying close to 1 ppm. Similar concentrations were reported by Mitchell (106) in his later study of the mineral composition of red clover and ryegrass grown on different soils. A mean of 1.8 ppm on the dry basis (range 0.7–3.8) was obtained for the former species and 2.0 ppm (range 0.9–4.6) for the latter. Very little is known of the titanium content of human foods. Tipton *et al.* (168) reported the 30-day mean total dietary Ti intakes of two individuals to be 0.37 and 0.41 mg/day. A feature of this investigation was the high urinary excretion, suggesting either considerable absorption from the diet or loss from previously retained tissue titanium. Both individuals were in negative balance, with approximately equal excretion via the feces and urine. No evidence has appeared that absorbed Ti performs any vital function in animals, or that it is a dietary essential for any living organism.

The reported Ti concentrations of human and animal tissues are extremely variable, with high levels commonly appearing in the lungs, probably as a result of inhalation of dust (16, 65, 167). Tipton and Cook (167) found most of the soft tissues of the adult human body to contain from 0.1 to 0.2 ppm, but the lungs averaged over 4 ppm, with some samples containing more than 50 ppm Ti wet weight. Hamilton *et al.* (65) reported the following mean levels in human tissues in μg/g wet weight: muscle, 0.2 ± 0.01; brain, 0.8 ± 0.05; kidney cortex, 1.3 ± 0.2; kidney medulla, 1.2 ± 0.2; liver, 1.3 ± 0.2; and lung, 3.7 ± 0.9. High variability in Ti concentration is also apparent from a study of 29 samples of human dental enamel (97). The levels reported ranged from 0.1 to 4.8 μg Ti/g dry weight, with a median value of 0.12 and a mean of 0.46 ± 0.13.

XIII. ZIRCONIUM

Zirconium is relatively abundant in the lithosphere but has no known biological significance. Reported values for the Zr levels in plant and animal tissues are highly discordant. Schroeder and Balassa (145), using a spectophotometric procedure stated to be sensitive to 0.5–0.8 μg/g ash, or about 0.01–0.02 ppm wet weight of most samples, and highly specific for Zr, found Zr to be widely distributed in soils and in plant and animal tissues. The element was present in all the organs examined from four male accident victims, including the brain, and was reported to be especially high in fat (mean 18.7 μg/g wet weight), liver (6.3 μg/g), and red blood cells (6.2 μg/g). In other tissues and blood serum the levels were reported to lie between 1 and 3 μg/g wet weight. Hamilton *et al.* (65), using spark source mass spectrometry, obtained much lower Zr levels for human tissues than those just cited. The brain, muscle, testis, ovary, liver, kidney, and lung all averaged between 0.01 and 0.06 μg Zr/g wet weight, with only the lymph nodes exhibiting a significantly higher mean concentration (0.3 ± 0.6 μg/g wet weight). A series of 98 samples of whole human blood revealed a mean of

0.02 ± 0.008 μg Zr/g. Examination of 53 samples of human dental enamel disclosed the wide range of $<$0.02–2.6 μg Zr/g dry weight, with a median of 0.2 and a mean of 0.53 ± 0.08 μg/g (97).

Schroeder and Balassa (145) estimated a daily oral Zr intake by man of about 3.5 mg. It was claimed that this quantity could be increased by high intakes of vegetable oils (3–6 μg/g) or tea (11.7 μg/g), the highest in Zr of the items examined. Meat, dairy products, vegetables, cereal grains, and nuts were reported to generally contain 1–3 μg/g on the fresh basis, with appreciably lower levels in most fruits and seafoods. A further selection of human foods and animal feedstuffs needs to be analyzed for Zr by alternative analytical procedures.

The metabolic movements of Zr in the animal body do not appear to have been directly studied. Its presence in the blood and tissues indicates that it is absorbed from ordinary diets, while its virtually absence from the urine (145) suggests that it is excreted by the intestine. More detailed and direct evidence of the sites of absorption, retention, and excretion of zirconium is clearly desirable.

Zirconium compounds have a low order of toxicity for rats and mice, whether injected (152) or orally ingested (150, 152). Schroeder *et al.* (150) studied the lifetime effects on mice of adding 5 ppm Zr as zirconium sulfate to the drinking water. No effects on growth were observed but there was a small reduction in survival time. The element was neither carcinogenic nor tumorogenic, with little evidence of accumulation in the tissues.

REFERENCES

1. Alexander, G.V., Nusbaum, R.E., and MacDonald, N.S., *J. Biol. Chem.* **192,** 489 (1951).
2. Amdisen, A., *Scand. J. Clin. Lab. Invest.* **20,** 104 (1967).
3. Anderson, B.M., *U.S., Geol. Surv., Open-File Rep.* (1972).
4. Archibald, J.G., *J. Dairy Sci.* **38,** 159 (1955).
5. Barlow, R.M., Purves, D., Butler, E.J., and McIntyre, I.J., *J. Comp. Pathol. Ther.* **70,** 396 (1960).
6. Barnes, D.W.H., Bishop, M., Harrison, G.E., and Sutton, A., *Int. J. Radiat. Biol.* **3,** 637 (1961).
7. Bartley, J.C., and Reber, E.F., *J. Nutr.* **75,** 21 (1961).
8. Bartter, F.C., Cowie, D.B., Most, H., Ness, A.T., and Firbush, S., *Am J. Trop. Med.* **27,** 403 (1947).
9. Bate, L.C., and Dyer, F.F., *Nucleonics* **23,** 74 (1965).
10. Baumann, E.J., Sprinson, D.B., and Marine, D., *Endocrinology* **28,** 793 (1941).
11. Bedford, J., Harrison, G.E., Raymond, W.H.A., and Sutton, A., *Br. Med. J.* **1,** 589 (1960).
12. Beeson, K.C., *U.S., Dep. Agric., Misc. Publ.* **369** (1941).

12a. Benoy, C.J., Hooper, P.A., and Schneider, C.A., *Food Cosmet. Toxicol.* **9,** 645 (1971).

13. Bernheim, F., and Bernheim, M.L.C., *J. Biol. Chem.* **127,** 353 (1939); **128,** 79 (1939).
14. Bertrand, D., *C. R. Hebd. Seances Acad. Sci.* **234,** 2102 (1952); **249,** 787 (1959).
15. Bertrand, G., and Bertrand, D., *Ann. Inst. Pasteur, Paris* **72,** 805 (1946); **80,** 339 (1951).
16. Bertrand, G., and Varonea-Spirt, C., *C.R. Hebd. Seances Acad. Sci.* **188,** 119 (1929); **189,** 73 and 122 (1929).
17. Borovik-Romanova, T.F., *in* "Inst. Geokem. Akad. Nauk, U.S.S.R." (N.I. Khitarov, ed.), 1965.
18. Bosshardt, D.K., Huff, J.W., and Barnes, R.H., *Proc. Soc. Exp. Biol. Med.* **92,** 219 (1956).
19. Bowen, H.J.M., *Biochem. J.* **73,** 381 (1951).
20. Bowen, H.J.M., "Trace Elements in Biochemistry." Academic Press, New York, 1966.
21. Bowen, H.J.M., and Dymond, J.A., *Proc. R. Soc. London, Ser. B* **144,** 355 (1955).
22. Boyd, T.C., and De, N.K., *Indian J. Med. Res.* **20,** 789 (1933).
23. Browning, E., "Toxicity of Industrial Metals." Butterworth, London, 1969.
24. Brune, D., Samsall, K., and Wester, P.O., *Clin. Chim. Acta* **13,** 285 (1966).
25. Cade, J.F.J., *Med. J. Aust.* **2,** 349 (1949).
26. Cade, J.F.J., *Med. J. Aust.* **1,** 684 (1975).
27. Casini, A., *Ann. Chim. Appl.* **36,** 219 (1946).
28. Christopherson, J.B., *Lancet* **2,** 325 (1918).
29. Clode, W., Sobral, J.M., and Baptista, A.M., *Adv. Thyroid Res., Trans. Int. Goitre Conf., 4th, 1960,* p. 65 (1961).
30. Coats, D.A., Trautner, E.M., and Gershon, S., *Australas. Ann. Med.* **6,** 11 (1957).
31. Cole, B.T., and Patrick, H., *Arch. Biochem. Biophys.* **74,** 357 (1958).
32. Colvin, L.B., Creger, C.R., Ferguson, T.M., and Crookshank, H.R., *Poultry Sci.* **51,** 576 (1972).
33. Comar, C.L., and Wasserman, R.H., *Miner. Metab.* **2,** Part A, 523 (1964).
34. Comar, C.L., Scott-Russell, R., and Wasserman, R.H., *Science* **126,** 485 (1957).
35. Conway, E.J., and Flood, J.C., *Biochem. J.* **30,** 716 (1936).
36. Correa, P., and Strong, J.P., *Ann. N.Y. Acad. Sci.* **199,** 217 (1972).
37. Curli, G., and Coppini, D., *Lait* **38,** 497 (1938).
38. Damiens, M.A., and von Blaignan, S., *C.R. Hebd. Seances Acad. Sci.* **193,** 1460 (1931); **194,** 2077 (1932).
39. Deobald, H.J., and Elvehjem, C.A., *Am. J. Physiol.* **111,** 118 (1958).
40. Dingle, H., and Sheldon, J.H., *Biochem. J.* **31,** 837 (1937).
41. Dixon, T.F., *Biochem. J.* **28,** 86 (1935).
42. Duckworth, R.B., and Hawthorn, J., *J. Sci. Food Agric.* **11,** 218 (1960).
43. Duggan, R.E., and Lipscomb, G.Q., *Pestic. Monit. J.* **2,** 153 (1969).
44. Duggan, R.E., and Weatherwax, J.R., *Science* **157,** 1006 (1957).
45. Fleischman, A.I., Lenz, P.H., and Bierenbaum, M.L., *J. Nutr.* **104,** 1242 (1974).
46. Follis, R.H., Jr., *Am. J. Physiol.* **138,** 246 (1943).
47. Follis, R.H., Jr., *Am. J. Physiol.* **143,** 385 (1945).
48. Forbes, G.B., and Reina, J.C., *J. Nutr.* **102,** 647 (1972).
49. Forbes, R.M., Cooper, A.R., and Mitchell, H.H., *J. Biol. Chem.* **209,** 857 (1954).
50. Forbes, R.M., and Mitchell, H.H., *AMA Arch. Ind. Health* **16,** 489 (1957).
51. Ford, W.P., Kent-Jones, D.W., Maiden, A.M., and Spalding, R.C., *J. Soc. Chem. Ind., London* **59,** 177 (1940).

52. Freedberg, A.S., Pinto, H.P., and Zipser, A., *Fed. Proc., Fed. Am. Soc. Exp. Biol.* **11**, 49 (1952).
53. Gedalia, I., Yariv, S., Nayot, H., and Eidelman, E., *in* "Trace Element Metabolism in Animals-2" (W.G. Hoekstra *et al.*, eds.), p. 461. Univ. Park Press, Baltimore, Maryland, 1974.
54. Gehrke, C.W., Baker, J.W., and Affsprung, H.E., *J. Dairy Sci.* **37**, 643 (1954).
55. "Geochemistry and the Environment," Vol. I. Natl. Acad. Sci., Washington, D.C., 1974.
56. Gerlach, W., and Muller, R., *Virchow's Arch. Pathol. Anat. Physiol.* **294**, 210 (1934).
57. Gershon, S., *Annu. Rev. Med.* **23**, 439 (1972).
58. Glendenning, B.L., Parrish, D.B., and Schrenk, W.G., *Anal. Chem.* **7**, 1554 (1954).
59. Glendenning, B.L., Schrenk, W.G., and Parrish, D.B., *J. Nutr.* **60**, 563 (1956).
60. Goeij, J.J.M. de, and Kroon, J.J., "IAEA/FAO/WHO Symposium on Nuclear Techniques in Comparative Studies of Food and Environmental Contamination, Otaniemi, Finland, 1973." IAEA, Vienna.
61. Gormican, A., *J. Am. Diet. Assoc.* **58**, 397 (1970).
62. Groot, A.P. de, *Food Cosmet. Toxical.* **11**, 955 (1973).
63. Gruner, J., Sung, S.S., Tubiania, M., and Segarra, J., *C.R. Seances Soc. Biol. Ses Fil.* **145**, 203 (1951).
64. Hamilton, E.I., and Minski, M.J., *Sci. Total Environ.* **1**, 375 (1972/1973).
65. Hamilton, E.I., Minski, M.J., and Cleary, J.J., *Sci. Total Environ.* **1**, 341 (1972/1973).
66. Harrison, G.E., Howells, G.R., Pollard, J., Kostial, K., and Manitasevic, R., *Br. J. Nutr.* **20**, 561 (1966).
67. Harrison, G.E., Raymond, W.H.A., and Tretheway, H.C., *Clin. Sci.* **14**, 681 (1955).
68. Healy, W.B., Bate, L.C., and Ludwig, T.G., *N. Z. J. Agric. Res.* **7**, 603 (1964).
69. Hellerstein, S., Kaiser, C., Darrow, D.D., and Barrow, D.C., *J. Clin. Invest.* **39**, 282 (1960).
70. Heppel, L.A., and Schmidt, C.L.A., *Univ. Calif., Berkeley, Publ. Physiol.* **8**, 189 (1938).
71. Heywood, B.J., *Science* **152**, 1408 (1966).
72. Hill, C.H., Starcher, B., and Matrone, G., *J. Nutr.* **83**, 107 (1964).
73. Hodges, R.M., MacDonald, N.S., Nusbaum, R., Stearns, R., Ezmirlian, F., Spain, P., and Macarthur, S., *J. Biol. Chem.* **185**, 519 (1950).
74. Hodgkiss, W.S., and Errington, B.J., *Trans. Ky. Acad. Sci.* **9**, 17 (1940).
75. Hopkins, H., and Eisen, J., *J. Agric. Food Chem.* **7**, 633 (1959).
76. Horecker, B.L., Stotz, E., and Hogness, T.R., *J. Biol. Chem.* **128**, 251 (1939).
77. Hove, E., Elvehjem, C.A., and Hart, E.B., *Am. J. Physiol.* **123**, 640 (1938).
78. Hove, E., Elvehjem, C.A., and Hart, E.B., *Am. J. Physiol.* **127**, 689 (1939).
79. Huff, J.W., Bosshardt, D.K., and Barnes, R.H., *Proc. Soc. Exp. Biol. Med.* **92**, 216 (1956).
80. Hullin, R.P., and Johnson, A.W., *Life Sci.* **9**, 9 (1970).
81. Hutcheson, D.P., Gray, D.H., Venugopal, B., and Luckey, T.D., *J. Nutr.* **105**, 670 (1975).
82. Jensen, L., Peterson, R.P., and Falen, J., *Poultry Sci.* **53**, 57 (1974).
83. Jones, J.H., *Am. J. Physiol.* **124**, 230 (1938).
84. Jury, R.V., Webb, M.S., and Webb, R.J., *Anal. Chim. Acta* **22**, 145 (1960).
85. Kehoe, R.A., Cholak, J., and Storey, R.V., *J. Nutr.* **19**, 579 (1940).
86. Kent, N.L., *J. Soc. Chem. Ind., London* **61**, 183 (1942).

87. Kent, N.L., McCance, R.A., *Biochem. J.* **35,** 837 and 877 (1941).
88. Khayyal, M.T., *Br. J. Pharm. Chemother.* **22,** 342 (1964).
89. Kirchgessner, M., *Z. Tierphysiol., Tierernahr. Futtermittelkd.* **14,** 270 and 278 (1959).
90. Kirchgessner, M., Merz, G., and Oelschlager, W., *Arch. Tierernahr.* **10,** 414 (1960).
91. Koch, H.J., Smith, E.R., Shimp, N.F., and Connor, J., *Cancer* **9,** 499 (1956).
92. Laznitski, A., and Szorenyi, E., *Biochem. J.* **28,** 1678 (1934).
93. Leach, R.M., Jr., and Nesheim, M.C., *J. Nutr.* **81,** 193 (1963).
94. Lippincott, S.W., Ellerbrook, L.D., Rhees, M., and Mason, P., *J. Clin. Invest.* **26,** 370 (1947).
95. Ljumberg, S., and Paalzow, L., *Acta Psychiat. Scand., Suppl.* **207,** 68 (1969).
96. Loeb, R.F., *J. Gen. Physiol.* **3,** 229 (1920).
97. Losee, F., Cutress, T.W., and Brown, R., *Trace Subst. Environ. Health–7, Proc. Univ. Mo. Annu. Conf., 7th, 1973,* p. 19 (1974).
98. Lounamaa, J., *Ann. Bot. Soc. Zool. Bot. Fenn. "Vanamo"* **29,** 4 (1956).
98a. Lowater, F., and Murray, M.M., *Biochem. J.* **31,** 837 (1937).
99. Luke, C.L., and Campbell, M.D., *Anal. Chem.* **28,** 1273 (1956).
100. Lynn, G.E., Shrader, S.A., Hammer, O.H., and Lassiter, C.A., *J. Agric. Food Chem.* **11,** 87 (1963).
101. MacDonald, N.S., Nusbaum, R.E., Stearns, R., Ezmirlian, F., Spain, P., and McArthur, C., *J. Biol. Chem.* **185,** 519 (1951).
102. MacLeod, R.A., and Snell, E.E., *J. Bacteriol.* **59,** 783 (1950).
103. Mason, M.J., *J. Biol. Chem.* **113,** 61 (1936).
104. McClachlan, J., and Craigie, J.S., *Nature (London)* **214,** 604 (1967).
105. Mitchell, R.L., *Commonw. Bur. Soil Sci. Tech. Commun.* No. 44 (1948).
106. Mitchell, R.L., *Research (London)* **10,** 357 (1957).
107. Mole, R.H., *Br. J. Nutr.* **19,** 13 (1965).
108. Molokhia, M.M., and Smith, H., *Arch. Environ. Health* **15,** 745 (1967).
109. Molokhia, M.M., and Smith, H., *Bull. W. H. O.* **40,** 123 (1969).
110. Molokhia, M.M., and Smith, H., *J. Trop. Med. Hyg.* **72,** 222 (1969).
111. Molokhia, M.M., and Smith, H., *Ann. Trop. Med. Parasitol.* **62,** 158 (1968).
112. Monier-Williams, G.W., "Trace Elements in Foods." Chapman & Hall, London, 1949.
113. Mraz, F.R., *Proc. Soc. Exp. Biol. Med.* **110,** 273 (1962).
114. Murphy, E.W., Page, L., and Watt, B.K., *J. Am. Diet. Assoc.* **59,** 115 (1971).
115. Murthy, G.K., and Rhea, U., *J. Dairy Sci.* **51,** 610 (1968).
116. Murthy, G.K., Rhea, U., and Peeler, J.T., *J. Dairy Sci.* **50,** 651 (1967).
117. Murthy, G.K., Rhea, U., and Peeler, J.T., *Environ. Sci. Technol.* **5,** 436 (1971).
118. Nixon, G.S., Livingstone, H.D., and Smith, H., *Caries Res.* **1,** 327 (1967).
119. Nordin, B.E.C., Smith, D.A., MacGregor, J., and Nisbett, J., Tech. Rep. Ser. No. 32. IAEA, Vienna, 1964.
120. Ondreicka, R., Ginter, E., and Kortus, J., *Br. J. Ind. Med.* **23,** 305 (1966).
121. Ondreicka, R., Kortus, J., and Ginter, E., *in* "Intestinal Absorption of Metal Ions, Trace Elements and Radionuclides" (S.C. Skoryna and D. Waldron-Edward, eds.), p. 293. Pergamon, Oxford, 1971.
122. Orent-Keiles, E., *Proc. Soc. Exp. Biol. Med.* **44,** 199 (1941).
123. Otto, C.F., Jackrowski, L.A., and Wharton, J.D., *Am. J. Trop. Hyg.* **2,** 495 (1953).
124. Otto, C.F., Maren, T.H., and Brown, H.W., *Am. J. Hyg.* **46,** 193 (1947).
125. Owen, E.C., *J. Dairy Res.* **13,** 243 (1944).
126. Palmer, R.F., and Thompson, R.C., *Am. J. Physiol.* **207,** 561 (1974).
127. Perry, H.M., Jr., and Perry, E.F., *J. Clin. Invest.* **38,** 1452 (1959).

128. Peterson, R.P., and Jensen, L.S., *Poultry Sci.* **54,** 771 (1975).
129. Pfeiffer, C.C., Hallman, L.F., and Gersh, I., *J. Am. Med. Assoc.* **128,** 266 (1945).
130. Plotnikov, K.I., *Veterinariya* **37,** 35 (1960); *Nutr. Abstr. Rev.* **30,** 1138 (1960).
131. Pybus, G., and Bowers, G.N., Jr., *Clin. Chem.* **16,** 139 (1970).
132. Rajaratnam, J.A., Lowry, J.B., Avadhani, P.N., and Corley, R.H.V., *Science* **172,** 1142 (1971).
133. Relman, A.S., *Yale J. Biol. Med.* **29,** 248 (1956/1957).
134. Richards, C.E., Brady, R.O., and Riggs, D.S., *J. Clin. Endocrinol.* **9,** 1107 (1949).
135. Ringer, S., *J. Physiol.* **4,** 370 (1882).
136. Robinson, W.O., and Edginton, G., *Soil Sci.* **60,** 15 (1945).
137. Robinson, W.O., Steinkoenig, L.A., and Miller, C.F., *U.S., Dep. Agric., Bull. No. 600,* Washington, D.C., 1971.
138. Robinson, W.O., Whetstone, R.R., and Edginton, G., *U.S., Dep. Agric. Tech. Bull.* **1013** (1950).
139. Rosenfeld, G., *Arch. Biochem.* **48,** 54 (1954).
140. Rosenfeld, G., and Wallace, E.D., *Arch. Ind. Hyg.* **8,** 466 (1953).
141. Rubins, E.J., and Hagstrom, G.R., *J. Agric. Food Chem.* **7,** 722 (1959).
142. Rygh, O., *Bull. Soc. Chim. Biol.* **31,** 1052, 1403, and 1408 (1949); **33,** 133 (1953); *Research (London)* **2,** 340 (1949).
143. Samuel, D., and Gottesfeld, Z., *Discovery* **32,** 122 (1973).
144. Schou, M., *Pharmacol. Rev.* **9,** 17 (1957).
145. Schroeder, H.A., and Balassa, J.J., *J. Chronic Dis.* **19,** 573 (1966).
146. Schroeder, H.A., and Balassa, J.J., *J. Nutr.* **22,** 245 (1967); *J. Chronic Dis.* **20,** 211 (1967).
147. Schroeder, H.A., Balassa, J.J., and Tipton, I.H., *J. Chronic Dis.* **17,** 483 (1964).
148. Schroeder, H.A., Kanisawa, M., Frost, D.V., and Mitchener, M., *J. Nutr.* **96,** 37 (1968).
149. Schroeder, H.A., and Mitchener, M., *J. Nutr.* **105,** 421 and 452 (1975).
150. Schroeder, H.A., Mitchener, M., Balassa, J.J., Kanisawa, M., and Nason, A.P., *J. Nutr.* **95,** 95 (1968).
151. Schroeder, H.A., Tipton, I.H., and Nason, A.P., *J. Chronic Dis.* **25,** 491 (1972).
152. Schubert, J., *Science* **105,** 389 (1947).
153. Schwarz, K., Milne, D.B., and Vinyard, E., *Biochem. Biophys. Res. Commun.* **40,** 22 (1970).
154. Scott, G.H., and McMillen, J.H., *Proc. Soc. Exp. Biol. Med.* **35,** 287 (1936).
155. Seaber, W., *Analyst* **58,** 575 (1930).
156. Seibold, M., *Klin. Wochenschr.* **38,** 117 (1960).
157. Sheldon, J.H., and Ramage, H., *Biochem. J.* **25,** 1068 (1931).
158. Shorland, F.B., *Proc. R. Soc. N.Z.* **64,** 35 (1934).
159. Soman, S.D., Panday, V.K., Joseph, K.T., and Raut, S.J., *Health Phys.* **17,** 35 (1969).
160. Sowden, E.M., and Stitch, S.R., *Biochem. J.* **67,** 104 (1957).
161. Spencer, H., Li, M., Samachson, J., and Laszlo, D., *Metab. Clin. Exp.* **9,** 916 (1960).
162. Steadman, L.T., Brudevold, F., and Smith, F.A., *J. Am. Dent. Assoc.* **57,** 340 (1958).
163. Stitch, S.R., *Biochem. J.* **67,** 97 (1957).
164. Smith, H., *J. Forensic Sci. Soc.* **7,** 97 (1967).
165. Swaine, D.J., *Commonw. Bur. Soils Tech. Commun.* No. 48 (1955).
166. Teresi, J.D., Hove, E., Elvehjem, C.A., and Hart, E.B., *Am J. Physiol.* **127,** 689 (1939).
167. Tipton, I.H., and Cook, M.J., *Health Phys.* **9,** 103 (1963).
168. Tipton, I.H., Stewart, P.L., and Martin, P.G., *Health Phys.* **12,** 1683 (1966).

169. Tipton, I.H., Stewart, P.L., and Dickson, J., *Health Phys.* **16,** 455 (1969).
170. Torii, S., Mitsumori, K., Inubushi, S., and Yanagisawa, I., *Psychopharmacologia* **29,** 65 (1973).
171. Trautner, E.M., Morris, R., Noack, C.H., and Gershon, S., *Med. J. Aust.* **2,** 280 (1955).
172. U.S. Geological Survey, Open-File Rep. Denver, Colorado, 1972.
173. Van Campen, D.R., *J. Nutr.* **88,** 125 (1966).
174. Volf, N., *in* "Intestinal Absorption of Metal Ions, Trace Elements and Radionuclides" (S.C. Skoryna and D. Waldron-Edward, eds.), p. 277. Pergamon, Oxford, 1971.
175. Wolf, N., Gedalia, I., Yariv, S., and Zuckerman, H., *Arch. Oral Biol.* **18,** 233 (1973).
176. Voors, A.W., *Lancet* **2,** 1337 (1969); *Am. J. Epidemiol.* **93,** 259 (1971).
177. Walsh, T., *Proc. R. Ir. Acad., Sect. B* **50,** 287 (1945).
178. Warrington K., *Ann. Bot. (London) [N.S.]* **37,** 629 (1923); **40,** 27 (1926).
179. Weber, C.W., Doberenz, A.R., and Reid, B.L., *Poultry Sci.* **47,** 1731 (abstr.) (1968).
180. Weber, C.W., Doberenz, A.R., Wyckoff, R.W.G., and Reid, B.L., *Poultry Sci.* **47,** 1318 (1968).
181. Whanger, P.D., and Weswig, P.H., *J. Nutr.* **100,** 341 (1970).
182. Winnek, P.S., and Smith, A.H., *J. Biol. Chem.* **119,** 93 (1937); **121,** 345 (1937).
183. Wittrig, J., Anthony, E.J., and Lucarno, H.E., *Dis. Nerv. Syst.* **31,** 408 (1970).
184. Würker, J., *Biochem. Z.* **265,** 169 (1933).
185. Yamagata, N., *J. Radiat. Res.* **3,** 9 and 158 (1962).
186. Yanagisawa, I., and Yoshikawa, H., *Biochim. Biophys. Acta* **329,** 283 (1973).
187. Zondek, H., and Bier, A., *Klin. Wochenschr.* **11,** 760 (1932).
188. Zook, E.G., and Lehmann, J., *J. Am. Diet. Assoc.* **52,** 225 (1968).
189. Zook, E.G., and Lehmann, J., *J. Assoc. Off. Agric. Chem.* **48,** 850 (1965).
190. Zook, E.G., Green, F.E., and Morris, E.R., *Cereal Chem.* **47,** 720 (1970).

20

Soil—Plant—Animal Interrelations

I. INTRODUCTION

All plants and animals, including man, depend ultimately on the soil for their supply of mineral nutrients. With plants this relationship is direct and is simplified by the fact that the plant is stationary. On the other hand, grazing animals may derive their minerals from a variety of soil types and plant species, so long as they are free to range freely over wide areas. In this way any disabilities associated with particular soils would tend to be minimized or even eliminated. Intensification of production imposes restrictions on movements until some animals become dependent on a single soil type, or a narrower range of soil types which may be incapable, without appropriate treatment, of sustaining the health, fertility, and productivity of stock. The fact that animals did not thrive or suffered various disorders when restricted to some areas, and remained healthy in other areas, was recognized in Europe as early as the eighteenth century. These observations focused attention on the soil factors involved because areas considered satisfactory for stock and those classed as unthrifty or unhealthy were often adjacent, which would minimize climatic differences as causal factors. Furthermore, animals transferred from unhealthy to healthy areas usually recovered, which suggested the existence of nutritional rather than infectious disorders.

Investigations carried out during the last half-century have shown that many of the nutritional maladies of the type just mentioned result from the inability of the soils of the affected areas to supply, through the plants that grow on them, the mineral needs of man and his domestic animals in adequate, safe, or nontoxic amounts, and in proper proportions. Nutritional abnormalities involving the trace elements may arise as simple, gross deficiencies or excesses of single elements. More usually they occur as deficiencies or excesses conditioned by the extent to which other mineral elements, nutrients, or organic factors are present in the environment and are capable of modifying the ability of the animal to utilize the deficient or toxic element. The conditioning factors may themselves be a reflection of the soils on which the herbage grows, since the soil conditions affect the chemical composition of plants in a variety of ways, additional to their effect on the primary element itself. The presence of particular plant species, such as Se accumulator or goitrogenic plants, may also influence the incidence or severity of various trace element deficiencies or excesses, because of their effect on the amounts or the availability of particular elements consumed by the animal. Soil–plant–animal interrelations are thus complex and transcend the simple concept that deficiency or toxicity conditions in the animal are merely a reflection of a deficiency or excess of a particular mineral in the soil and therefore in the herbage that this soil supports.

The incidence of the disease *Phalaris* staggers in sheep and cattle provides a further example of the complexity of soil–plant–animal trace element interrelationships, as discussed in Chapter 5. This condition is unknown in most areas where the grass *Phalaris tuberosa* is grown. On Co-deficient or marginally Co-deficient soils, the neurotoxic substance present in this plant species induces demyelination and a “staggers” syndrome in animals consuming the plant, unless they are treated with Co salts or pellets or the soils are fertilized with Co salts or ores. It seems that normal soils produce herbage with Co levels adequate to meet the normal needs of ruminants, plus sufficient Co to enable them to detoxicate the neurotoxic substance in *Phalaris.* Marginally Co-deficient soils are capable of fulfilling the former needs but not the latter. Severely Co-deficient soils result in the growth of plants carrying insufficient cobalt to fulfill either of these needs satisfactorily. The incidence of *Phalaris* staggers thus depends on the interaction between a soil factor and a plant factor, each of which can vary independently in a particular environment.

The incidence of chronic Cu poisoning in sheep and cattle and its relation to the presence or absence of the hepatotoxic alkaloid-containing plant *Heliotropium europaeum*, described in Chapter 3, provide a different example of complex trace element interactions involving the soil, plant, and animal. The field occurrence and severity of deficiency or toxicity conditions in animals, involving the four trace elements iodine, cobalt, copper, and selenium, are therefore influenced by complex interrelations between the soil, plant, and animal. Comparable interrelations involving other elements and other plant species will no doubt emerge as research proceeds.

The food, and not the water or the atmosphere, normally supplies a major proportion of total daily trace element intakes by animals and man. This generalization does not apply in endemic fluorosis regions where the water supplies constitute the principal source of the high-F intakes. Even in these circumstances the toxic quantities of fluoride present in the water seldom bear any relationship tò the F status of the soils and herbage of the affected area because the water usually comes from deep wells or bores drawing from other soil or rock formations. A high inverse correlation between the I content of the drinking water and the incidence of goiter has long been known, but only some 10% of the total intakes of this element by man comes from the water supply, in goitrous and nongoitrous regions alike. In a study of the concentration of 17 trace elements in the public water supplies located in 44 states of the United States, it was calculated that 0.3–10.1% of the total daily intakes came from the drinking water (28). Similar conclusions were reached for individuals living in certain large cities of the United States (61). Such conclusions do not necessarily apply in other areas where the water supplies are not subject to comparable purification and control processes. Areas have been reported in which the drinking water is naturally and unusually high in arsenic (64), boron, (55), strontium (75), and lithium (22) and which therefore contributes substantial quantities of these elements to total human intakes.

Insignificant quantities of trace elements, compared with the amounts in foods, are also normally contributed by the atmosphere, except in areas adjacent to mines and factories, where substantial atmospheric pollution can occur. With the rise of modern industrial technology and with the increasing urbanization and motorization of large sections of the population, these sources of trace

TABLE 59
Lead Contents of Plants and Soils (0–5 cm) as a Function of Distance from the Cunningham Highway[a]

	Lead content (μg/g)			
	Site 1		Site 2	
Distance (m)	Plant	Soil	Plant	Soil
2	41	84	64	207
5	26	34	48	128
25	20	27	27	49
125	21	23	21	31
250	10	23	9	21

[a]From Wylie and Bell (76).

elements, together with contamination of the water supplies, may constitute an increasingly significant source of a number of elements with possible long-term dangers to human health. For example, the deposition of lead from automobile exhausts on soil and plants along highways and in urban areas is well established (15, 57, 66, 76) (see Table 59).

II. SOIL AND WATER RELATIONS IN HUMAN HEALTH

Trace element deficiencies, toxicities, and imbalances are more difficult to relate to the soil in man than in farm stock. Many factors conspire to minimize the effects on man of soil-induced variations in the trace element contents of human foods. The geographical sources of human foods and beverages are ever-widening in most modern communities, so that the overall diet comprises materials grown on a range of soil types. Modern human dietaries also contain a considerable variety of types of foods. Trace element abnormalities which may be present in particular plants, parts of plants, or animal tissues and fluids can therefore be offset by the consumption of other foods not so affected.

Industrial treatment of an increasing proportion of materials destined for human consumption provides the opportunity for both gain and loss of trace elements during storage, transport, preservation, and processing. The impact of such processes on dietary intakes varies markedly with different elements. They all tend to reduce the directness of the relationship between the soils of a given area and the actual intakes of mineral elements by man. The refining of sugar to the white form commonly consumed by man results in striking reductions in the levels of all mineral elements. On the other hand the pasteurization and drying of milk, where these involve contact with metal containers, can substantially increase the levels of iron and copper. The milling of grain into white flour results in large but variable losses of most of the minerals present in the original whole grain. With iron, zinc, and chromium such losses can be nutritionally important. This loss applies to elements that are initially high, as well as to those that are present in low concentrations. For instance, the Se level in white flour made from seleniferous wheat is much lower than in the whole grain. This may be one of the factors contributing to the absence of signs of selenosis in man in seleniferous areas. Finally, the domestic treatment of human foods can affect the position with several elements. Significant losses of F and Se, which may be present in toxic quantities in raw vegetables, can occur in the cooking water. Rising standards of cleanliness and hygiene at the retail and domestic level and decreasing use of iron cooking vessels and containers reduce the opportunities for contamination of foods with iron and other trace elements.

Where the choice of foods is poor because of poverty, ignorance, or prejudice, so that the general quality of the diet is low, and where the dependence on locally grown foods is high, local soil deficiencies or excesses are likely to accentuate any dietary disabilities and adversely affect human health and nutrition. Poor choice of foods, involving diets high in white sugar and refined cereals both low in chromium, rather than the source of these foods, may be the causative factor in the incidence of the Cr-responsive impaired glucose tolerance which exists in a proportion of old people. The conditioned zinc deficiency which occurs in Middle East dwarfs is also unrelated to the soil conditions under which their foods are grown. Total dietary Zn intakes by these individuals compare well with those from other parts of the world where no such disorders arise in man. The zinc deficiency in these cases results from a combination of circumstances, including excessive dependence on a high-phytate, whole wheat or corn bread and beans diet, with its low-Zn availability, plus intestinal parasitism, which induces high-Zn losses. Nevertheless, there is ample evidence that the content of zinc and other trace elements in plants used as foods by man and animals is influenced by the soil type and fertilizer treatment applied, as discussed in Section III,2. Regional differences in plasma Zn levels in human adults have also been demonstrated in the United States (41) which presumably reflect differences in dietary Zn intakes. These are more likely to arise from local differences in the zinc content of the foods and beverages consumed than from variation in dietary habits.

The Se levels in plants are influenced so markedly by the available levels of Se in soils that dramatic evidence is provided by this element of soil effects on the health of animals, and to a lesser extent man. These effects range from deficiency to toxicity conditions. The high inverse correlation found between the incidence of various forms of cancer in man and the Se status of the soils and crops in various areas and the Se level in the blood of residents of those areas was described in Chapter 13. The positive correlation reported between such Se status and the incidence of dental caries in children was also considered in that chapter.

The relationship between the incidence of goiter and soil and plant iodine levels provides direct evidence of a link between the composition of the soil and human disease. Subnormal levels of iodine in the soils from which the food is derived have been correlated with the incidence of goiter in man and animals in many areas. In New Zealand, Hercus and co-workers (33) found that "variation in the average amount of iodine in soils containing more than 10 ppm has little effect on the small incidence of goiter; but as the amount of soil iodine decreases, so the incidence of goiter rises."

Normal, mature soils contain about 10 times the I concentration of the rocks from which they are formed. Since the solubility of most iodine compounds is high and concentration during the process of soil formation would therefore be difficult, an extraneous source of iodine is implied. This is believed to be the iodine borne in on ocean winds. Long periods of time would be necessary for this I source to effect substantial increases in soil iodine. Goldschmidt (26) contends that the low-I status of the soils of goitrous areas is associated with the removal of I-enriched surface soils by recent glaciation, coupled with insufficient time for replenishment with postglacial airborne oceanic iodine during the subsequent soil-forming period. The principal factors determining a regional I deficiency in soils would therefore be recent glaciation, distance from the sea, and low annual precipitation. With respect to the last two factors, it has been calculated that from 22 to 50 mg I/acre will fall annually in the rain on the Atlantic coastal plain, compared with 0.7 mg/acre in the goitrous Great Lakes region of North America (34).

The relationship between soil I levels and the incidence of goiter is subject to several modifying influences. The nature of the plant species or strains present can be important because of their variable uptake of soil iodine. The amounts and types of goitrogens in the plants consumed as food can also be significant. Dietary habits, involving a greater diversity of foods and wider sources of supply, also reduce the dependence of human populations on foods and drinking water derived from local soils. The opportunities for substantially increasing I intakes by natural means are limited, except where they include consumption of foods of marine origin.

Numerous links between human health and the soil of a more tenuous nature have been proposed. These associations rest more securely on correlation than on causation. Thus differential mortality from cancer of the stomach in different parts of England (43, 65) and Holland (71) have been correlated with soil type. In New Zealand, Saunders (60) has similarly related the highest prevalence of total cancer, total digestive tract cancer, and stomach cancer with particular types of soil. In Finland, Marjanen (47) has shown that with increasing Mn content in the cultivated soils there is a significant decrease in the incidence of cancer.

Regional variations in the prevalence of human dental caries have repeatedly been demonstrated (6, 45). In many of these studies the caries differences follow the distribution of F levels in the drinking waters. They cannot therefore necessarily be related to the soils of the areas because the water supplies may be unrelated in origin to the local soils. A similar reasoning applies to the geographical variations in the incidence of mottled enamel. However, broad relationships exist between variations in caries incidence and the nature of soils which occur in particular geographical areas in the United States (45, 53), New Zealand (14, 35, 46) and Papua–New Guinea (6). In a study of 1876 twelve to fourteen-year-old lifelong residents of 19 communities with populations of 3000–15,000 in the eastern United States, situated on four different soil types and all using water containing 0.3 ppm F or less, caries prevalence was highest on the podzol soils of New England. It was of descending prevalence on gray-brown and red-yellow podzols and of lowest prevalence on the subhumic gray soils of the South Atlantic states (45).

Comparable differences in caries incidence on different soil types have been observed in New Zealand (14, 35, 46). The adjacent cities of Napier and Hastings have significantly different caries prevalence rates, despite similar socioeconomic conditions, dietary habits, and F content of the drinking water (46). In an attempt to identify the cause of the difference between the two towns, the composition of the vegetables grown on each of the soils was examined. Those grown on the Napier soil were found to be generally richer in Mo, Al, and Ti and poorer in Cu, Mn, Ba, and Sr than those from the Hastings soil. The higher Mo content of the Napier vegetables was suggested as a likely factor in the lower prevalence of caries among the children of that city (45). Direct evidence in support of this suggestion is not available.

The situation is broadly similar in the primitive village communities of Papua–New Guinea. The mean prevalence of dental caries in 21 villages in the Sepik and Fly River regions of that country was shown by Barmes and co-workers (6) to range from 0 to 29.5% decayed teeth per person. Analyses of soils and vegetables from these villages have correlated the caries prevalence with soil associations, but attempts to identify a direct causal link with a particular element or elements have so far been elusive. A strong inverse association between the caries prevalence and the concentrations of alkali and alkaline earth elements, especially Sr, Ba, K, Mg, Ca, and Li, in the garden soils has been demonstrated, and consistent evidence of inverse associations with concentrations of V, Mo, Mn, Al, Ti, and P in the staple foodstuffs has also been observed. By contrast, direct associations with the levels of Pb, Cu, Cr, Zn, and Se in the staple foods, namely, sago, sweet potato, and Chinese taro, and a possible direct association with Pb and Cu levels in village garden soils, were established. The amounts of fluorine in the village foodstuffs were reported to be sufficient to account for the overall low frequency of caries in the areas under study but could not explain the differences among villages in caries incidence.

The quality of the drinking water as a possible etiological factor in the incidence of cardiovascular disease has received considerable attention since Schroeder (62, 63) demonstrated a significant negative correlation between water hardness and certified death rates from total cardiovascular disease and from the main subgroups: cerebrovascular, coronary heart disease, and hypertensive heart disease. This investigation was stimulated by the earlier observation of Kobayashi (39) that cerebrovascular disease in the Japanese was related to acidity of river water. Very similar results to those of Schroeder were reported from England (50) and Sweden (11). In some population groups the statistical association was not highly significant (52), or no inverse correlation between water hardness and cardiovascular death rates was apparent (44, 51). As Perry *et al.* (54) have pointed out, these latter studies do not effectively refute the relationship because of the small geographical areas and the relatively narrow ranges of water hardness involved.

The closest associations between water hardness and cardiovascular death rates have been found in England, where in the 61 larger country boroughs the correlation coefficients between cardiovascular death rates during middle age and the hardness/Ca content of the drinking water was around −0.6 or −0.7 ($p < 0.0001$). Cardiovascular death rates were found to be some 50% higher in towns with very soft water than in those with very hard water (16). Of great significance is the further finding of Crawford and co-workers (17) that *change* in water hardness is associated with *change* in cardiovascular death rates. In five boroughs the water became harder, in six it became softer, and in 72 it remained unchanged between 1925 and 1955. Between 1950 and 1960 the noncardiovascular death rates declined similarly and there was a general increase in cardiovascular mortality in all boroughs, but in boroughs where the water had become softer the increase in cardiovascular mortality was higher and in those where the water had become harder this increase was lower than the average.

Efforts to incriminate individual substances in hard or soft water with cardiovascular mortality have not yet been successful. The finding of Crawford (see Perry *et al.*, 54) that the median Cd in bones is more than twice as high in autopsy samples from individuals who had lived in soft water areas than in hard water areas may be significant. The relation of Cd to hypertension in rats, discussed in Chapter 9, and the fact that human hypertension predisposes to myocardial infarction give special interest to this association. Furthermore, Perry *et al.* (54) have shown that hard water substantially inhibits the hypertension induced by Cd in rats. Whether this is a real metabolic effect or simply results from a loss of Cd from solution by the hard water has not been determined. Of further interest are the findings of Punsar and co-workers (56a), who studied the quality of the drinking water in two rural areas in Finland, in one of which the death rate from coronary heart disease (CHD) was twice that of the other. In both areas the water was soft. In the high CHD area the water was significantly lower in pH, F, Ca, Mg, Na, and Cr and significantly higher in NO_3, K, Cu, Co, and Ni. Assuming that a single water constituent is responsible for the difference in CHD mortality, the elements most suspect were stated to be chromium and, to a lesser degree, copper. The authors concluded that "CHD was associated with low concentrations of chromium and high concentrations of copper in drinking water."

III. FACTORS AFFECTING TRACE ELEMENT LEVELS IN PLANTS

Plant materials provide the main source of minerals to animals and to most members of the human race. The factors influencing the trace element content of plants are therefore major determinants of dietary intakes of these elements.

The concentration of all minerals in plants depends on four basic, interdependent factors: (*i*) the genus, species, or strain of plant, (*ii*) the type of soil on which the plant has grown, (*iii*) the climatic or seasonal conditions during growth, and (*iv*) the stage of maturity of the plant. The relative impact of these variables depends on the element in question and can be modified by man through the use of fertilizers and soil amendments, weedicides and pesticides, and by irrigation and husbandry practices. In addition, the inherent capacity of particular plant species to absorb and retain trace elements from the soil can be changed by crossbreeding and selection.

1. Genetic Differences

Certain plant species have the ability to accumulate uniquely high concentrations of particular elements. The Se accumulator plants, described in Chapter 12, provide an outstanding example of this genetic effect. Sr, Al, As, and Co accumulator plant species are also known. Several species growing on Sr-rich soils in England contained Sr concentrations as high as 26,000 ppm, compared with 100–200 ppm in other species from similar soils. Black gum (*Nyssa sylvatica*) similarly takes up about 100 times as much cobalt as broom sedge growing on the same soils of the coastal plain in the eastern United States (10).

When growing together on the same soils and at similar stages of maturity, leguminous plants commonly contain higher concentrations of Co, Ni, Fe, Cu, and Zn than grasses or cereals (4, 5, 23, 48, 69), although the differences in favor of legumes are rarely as large or consistent as with Ca and Sr. Grasses and cereals are commonly higher than legumes in Mn and Mo, and especially in silicon (5, 7, 48, 69). Most of the evidence available indicates that this is also true for selenium, although the difference is less obvious when the soil Se status is low (74). The more limited data for I, Cr, Ti, and V do not indicate consistent differences in concentration in favor of either grasses or legumes.

Within both the grasses and legumes substantial species differences have been observed. Thus Beeson *et al.* (9), in an investigation of 17 grass species grown together on a sandy loam soil and sampled at the same time, found the Co concentration to range from 0.05 to 0.14, the Cu from 4.5 to 21.1, and the Mn from 96 to 815 ppm on the dry basis. Intraspecies differences have been highlighted by New Zealand investigations of the iodine content of pasture plants. The total I levels of two strains of white clover were significantly different, and 10-fold differences were observed among strains of ryegrass growing on similar soils (36). Furthermore, differences in the I content of single plants within strains of ryegrass were found to be large. When diallele crosses were made, analysis of the progenies revealed that herbage I content is a strongly inherited character (13).

Genetic differences in the trace element levels in plants and in their capacity to respond to soil applications can have important consequences for the grazing animal. Any agronomic practice which results in changes in the botanical composition of the herbage must clearly be considered in assessing the adequacy or safety of any environment with respect to trace element levels. Plant breeding and selection, resulting in the development of higher yielding or otherwise superior strains, can also lead to significant changes in the concentrations of trace elements present in their tissues. Plant breeders and agronomists therefore need to give consideration to this aspect of their efforts, especially in areas deficient or marginal in a particular element.

The genetic differences in the trace element composition of the vegetative parts of plants are not necessarily paralleled by comparable differences in the seeds of those plants. The seeds of leguminous plants and the oilseeds are almost invariably higher in most trace elements than the seeds of grasses or cereals. With Co this difference is usually substantial. Significant species differences within leguminous and gramineous seeds also exist. This is more striking with manganese than with other trace elements. Thus wheat and oat grains are normally five times higher in Mn than corn or maize grain and some three times higher than barley (72) or sorghum grain (21). Species differences in the Mn concentrations of the seeds of various species of lupins are even greater. Thus Gladstones and Drover (23) found the seeds of *Lupinus albus* to contain Mn concentrations ranging from 817 to 3397 ppm, or 10–15 times those of other lupin species growing on the same sites. The seeds of some samples of *Lathyrus sativus* have similarly been reported to contain high-Mn levels, up to 500 ppm (58). Apart from manganese, the cereal grains do not reveal significant species differences in Cu, Co, Zn, or Se concentrations when grown under comparable conditions. However, recent evidence from India suggests that sorghum grain is relatively rich in molybdenum (18).

2. Soil and Fertilizer Effects

Plants normally react to a lack of an available element in the soil either by limiting their growth, reducing the concentration of the element in their tissues, or usually both simultaneously. Conversely, plants respond to soil applications of the deficient element or of soil amendments that increase the plant availability of this element either by increasing their growth, raising their tissue concentrations of the element, or both. The extent to which one or more of these responses takes place varies among different trace elements and different plant species. Soil composition is nevertheless the basic factor determining the level of trace elements in plants and therefore their capacity to supply adequate, or nontoxic, amounts of these elements to animals and man.

The composition of the soil is influenced primarily by the nature of the rocks from which the soil is derived. Different parent rocks contain the various trace elements in different amounts and proportions and in differing degrees of stability and are subject to varying influences over variable periods of time during the soil-forming processes. In this way trace elements initially present in the parent rock can be lost, concentrated, or changed in chemical form. Highly leached (podzolized) soils have usually lost an appreciable proportion of their original trace elements. Where such leaching has taken place in soils derived from granitic rocks, low in the trace element-bearing minerals, deficiencies affecting plants and animals can be predicted with some confidence.

The amounts of certain plant-extractable trace elements are often much greater on poorly drained than on well-drained soils on the same soil association. The marked influence of this soil factor on the uptake of cobalt and nickel by pasture plants was first demonstrated by Mitchell (48). A wet soil condition similarly favors the uptake of Co and Mo by clover on some soils, but has no such effect on Cu (42). Adams and Honeysett (1) showed that the Co concentrations, and to a lesser extent the Cu concentrations, of subterranean clover and ryegrass are sensitive to periods of water-logging in some soils. In several cases the Co content of the plants on water-logged soils was 10–20 times greater than that of the controls. Climatic effects on soil–water status in different years could therefore provide one explanation of the seasonal differences in the incidence of Co deficiency in grazing sheep and cattle observed in some areas. The high levels of Se in plants in the seleniferous areas of Ireland have also been claimed to result from concentration of this element in the soil, due to poor drainage over long periods of time (20).

The uptake of trace elements by plants is influenced further by the acidity of the soil. Co and Ni, and to a lesser extent Cu and Mn, are poorly absorbed from calcareous soils, whereas Mo uptake is greater from such soils than from those which are acid or neutral in reaction. Teart soils, which carry herbage exceptionally high in Mo, are mostly derived from clays and limestones, calcareous and alkaline in character. These differences provide the opportunity for modifying the levels of individual trace elements in pasture and forage plants by the use of soil amendments such as sulfur, which can lower soil pH, or lime which can raise soil pH. The depressing effects of incremental dressings of calcium carbonate on the levels of Co, Ni, and Mn in red clover and ryegrass and the enhancing effects of such dressings on the Mo levels are illustrated in Table 60.

Although the use of soil amendments or treatments such as drainage or aeration that alter the availability of particular trace elements are of practical value in some circumstances, the application of trace element-containing fertilizers is widely practiced as a means of raising herbage concentrations from deficient to satisfactory levels for livestock. With some soils and elements these

TABLE 60
Effects of Liming on Trace Element Content of Plants Grown on a Granitic Soil[a]

Soil treatment	Element content (ppm dry basis)				
	Co	Ni	Mo	Mn	Soil pH
			Red clover		
Unlimed	0.22	1.98	0.28	58	5.4
115 cwt $CaCO_3$/acre	0.18	1.40	1.48	41	6.1
216 cwt $CaCO_3$/acre	0.12	1.10	1.53	40	6.4
			Ryegrass		
Unlimed	0.35	1.95	0.52	140	5.4
115 cwt $CaCO_3$/acre	0.20	1.16	1.44	120	6.1
216 cwt $CaCO_3$/acre	0.12	0.92	1.53	133	6.4

[a] From Mitchell (48).

applications also increase yields of herbage. The effects may be wholly or largely on yield or plant composition, depending on the element. Thus in I-deficient and Se-deficient soils, treatment with I or Se salts can markedly increase the concentrations of these elements in the herbage to levels well beyond those required by animals, without increasing or decreasing plant growth (3, 27). With cobalt the effect is mainly on the Co level in the plants, although growth responses to Co-containing fertilizers have been reported on certain soils. On Cu-deficient soils applications of copper often raise both herbage yields and herbage Cu concentrations, but the ability of most plant species to respond to such applications with high-Cu concentrations is much less than with most other trace elements. Thus Gladstones and Loneragan (24), in a study of a range of crop and pasture species, found substantial increases in Zn and Mn concentrations following soil treatment with those elements, whereas Cu increases were very small following treatment with comparable amounts of that element. On Mo-deficient soils spectacular increases in the yield of pasture legumes can be achieved by small applications of molybdenum, accompanied usually by small increases in the levels of Mo in the plant tissues. The latter are of no direct significance to the grazing animal because of its low-Mo requirement, but they can be of great importance in areas of high-Cu status because of their ability to reduce Cu retention in animals to safe levels. Conversely the increased herbage Mo concentrations could accentuate or precipitate Cu deficiency in animals in areas of low-Cu status. The excessively high-Mo levels and the very low-Cu and Mn levels in plants grown on the humic peat soils of Europe have been stressed by Szalay (59, 68).

3. The Influence of Season and Stage of Maturity

Plants mature partly in response to internal factors inherent in their genetic constitution, and partly in response to external factors, among which climatic and seasonal effects are of major importance. The effects of the latter can, of course, be modified greatly by irrigation and grazing management practices.

The marked decline in whole plant concentration with advancing maturity which occurs with P and K is not paralleled by comparable declines in the trace elements. Whole plant concentrations of these elements may increase, decrease, or show no consistent change with stage of growth, depending on the element, the plant species, and the soil or seasonal conditions. Most investigators have observed a rise with advancing maturity of the plant in the concentrations of Si, Al, and Cr and a fall in Cu, Zn, Co, Ni, Mo, Fe, and Mn, together with fluctuations in I not clearly related to the stage of growth (4, 5, 8, 25, 37, 38, 48, 73). Bisbjerg and Gissel-Nielsen (12) found the Se concentration of the grain and straw of barley to be lower than that of the green plant, but with mustard this decline varied with the oxidation state of the added selenium, i.e., whether selenate or selenite. Large increases in the lead content of mixed pasture herbage at senescence, i.e., when active growth ceases, have been observed for a range of Scottish soils where no problem of external contamination exists (49). The extremely high-As levels, up to 4000 ppm or more, especially in the mature leaves, that can occur in plants grown on As-rich (mine-waste) soils were described in Chapter 18.

Changes in the trace element concentrations of forages or pastures related to the stage of growth of the plants are more likely to be of significance to the animal in areas of marginal status of particular elements than elsewhere. In these circumstances management can have important consequences. Seasonal differences in the severity of both deficiency and toxicity states in animals exist arising either from seasonal changes in (*i*) the moisture status of the soil, affecting trace element availability to plants, (*ii*) the botanical composition of the herbage, (*iii*) the morphological characteristics of the plants, such as proportions of leaf to stem and seed, (*iv*) the palatability and hence level of consumption by the animal, (*v*) the relative amounts and proportions of other elements or compounds in the plant which affect the utilization by the animal of the element in question, or (*vi*) the chemical forms of the elements in the plants. Little is yet known of the chemical forms and availability to animals of trace elements in plants at different growth stages and different total concentrations. Observations of the Cu status of grazing and housed cattle in the Netherlands indicate that differences exist in the availability of Cu from herbage at different growth stages. Hartmans and Bosman (29) found that feeding grass at an older growth stage, despite its lower Cu content, resulted in higher liver Cu levels in cattle than did young grass.

IV. THE DETECTION AND CORRECTION OF DEFICIENCIES AND TOXICITIES IN ANIMALS AND MAN

1. Detection or Diagnosis

Mild trace element deficiencies and toxicities are difficult to diagnose because their effects on the animal are often indistinguishable from those arising from a primary dietary energy deficit, and because they are seldom accompanied by specific clinical signs. Loss of appetite and subnormal growth are common manifestations of most trace element deficiencies and excesses. The extent to which these occur and take precedence over other expressions of the dietary abnormality varies greatly with different trace elements. For instance, inappetence and growth failure are not conspicuous features of Cu deficiency, as they are in Zn and Co (vitamin B_{12}) deficiencies. Furthermore, reduced food consumption can arise either as a direct physiological response to some metabolic defect, or it can be secondary to a structural abnormality of teeth and joints which limit the animal's ability and willingness to masticate and graze.

The appearance of particular lesions or abnormalities in animals and man has long been used to define the limiting or precipitating factors in diseases of nutritional origin. Clinical and pathological studies have therefore become essential diagnostic tools in the investigation of all trace element deficiencies, imbalances, and toxicities. It is important, nevertheless, to recognize their limitations. Various functional and structural disorders apparent to the clinician and pathologist may be merely the final expression of a defect arising early or late in a chain of metabolic events. Trace element x can be vital at one point in this chain, and trace element y at another. A simple or conditioned deficiency of either element would therefore lead to the same end result in the animal, although the cause would obviously be different. For instance, anemia can be a manifestation of Fe, Cu, Ni, or Co deficiency or of Se, Zn, or Mo toxicity. Similarly, abnormalities in the size, shape, strength, and composition of bones, amounting in some cases to gross deformities, can occur in Cu, Mn, and Zn deficiencies and in F and Mo toxicities. Furthermore, the nature as well as the severity of the signs of deficiency or toxicity can vary greatly with the age and sex of the animal.

The diagnostic limitations of clinical and pathological observations, particularly in mild deficiency and toxicity states, can be minimized by concurrent biochemical studies of appropriate body tissues and fluids. The concentrations of the trace elements in the tissues, or of their functional forms such as thyroxine or vitamin B_{12}, must be maintained within fairly narrow limits if the growth, health, and productivity of the animal are to be sustained. Departures from these normal limits, which are now well defined for most trace elements, therefore constitute useful diagnostic aids, especially as they frequently arise prior to the appearance of clinical or pathological signs of deficiency or toxicity. The organ, tissue, or fluid chosen for analysis varies with the element, but estimations of whole blood or plasma trace element and enzyme concentrations have wide applicability and represent the most valuable diagnostic criteria. The normal range of concentration in the blood of man and his domestic animals and the levels which can be regarded as indicative of marginal, mild, or severe deficiencies or toxicities of particular elements are presented in the individual chapters dealing with those elements. The levels of certain trace elements in hair are also of great value in the detection of deficiencies and toxicities, despite considerable individual variability. Hair analyses have proved useful indicators of Zn and Cr deficiencies and of Se and As toxicities.

Chemical determination of the trace element levels in human and animal diets and their components provide the best indication of levels of intake in relation to minimum needs and toxic potential. Reasonably satisfactory standards of adequacy and safety assessed against such criteria as growth, health, performance, and tissue concentrations have been developed for most of the trace elements. The duration of intake, as well as the magnitude, and the criteria of adequacy employed can be particularly important in any such assessment. In human dietaries the impact of food preparation and cooking on actual and available daily trace element intakes also requires further evaluation. In stall-fed animals the actual feed intakes are usually known and there is little opportunity for feed selection or discrimination. Total dietary trace element determinations are therefore particularly meaningful, so long as these are not confined to a single element and the significance of dietary balance is appreciated. With grazing stock, diagnoses of deficiency or excess based on herbage analyses cannot be made with the same confidence because the samples analyzed may not represent the material actually consumed and because variations in total concentrations may not correspond with variations in the availability of the element to the animal.

Analyses of washed herbage samples can give misleading estimates of trace element intakes by animals, because of contamination with soil and heavy soil ingestion when grazing intensity is high or when pasture availability is low. Under such conditions soil ingestion by sheep and cattle can constitute 10–25% of the total dry matter intake (19, 30, 31, 70), and even 40% with sheep in the winter months in Great Britain (67). This ingestion can be beneficial with elements such as Co that occur in soils in concentrations much higher than those of the plants growing on them, and in some circumstances with Cu, Zn, Se (32), and I (66a). By contrast, Suttle *et al.* (67) showed that the Cu antagonists Mo and Zn are biologically available in soils. When three different soils were ingested at 10% of the dry matter of the diet the response in plasma Cu was inhibited in initially hypocupremic sheep and the availability of Cu probably reduced by more than 50%. It was suggested that soil ingestion may be involved in the etiology of hypocuprosis in cattle and swayback in sheep.

Attempts to correlate soil type and soil trace element content with the incidence of nutritional disabilities in animals have met with varied success. This is hardly surprising in view of the many factors that affect mineral uptake by plants and the levels in mixed herbage, as discussed in Section III. Nevertheless, the correlation with particular soil types can be high, as has been shown for cobalt by Mitchell (48) in Scotland, by Kubota (40) in the United States, and by Andrews in New Zealand (4). Correlations of this kind are required for other elements, not only for mapping the location and extent of known problem areas but for predicting their likely occurrence elsewhere.

None of the criteria mentioned in earlier paragraphs of this section is completely satisfactory when considered in isolation. When these are used together and their combined evidence is assessed, deficiency and toxicity states can be securely recognized and confidently predicted, even when these are mild. The ultimate criterion is, of course, the improvement in health or productivity which occurs in response to changes in the intake or utilization by animals or man of the element or elements in question.

2. Prevention and Control

Various direct and indirect means exist for the prevention and control of trace element deficiencies and toxicities in man and animals. The method of choice varies with different elements and different animal species and their normal feeding practices. For instance, with I deficiency in man direct supplementation through iodination of the domestic salt supplies has proved the most convenient and effective procedure. With iron deficiency in man the direct approach is also accepted, either through iron fortification of a staple food such as flour and bread or through the prescription of iron tablets to infants and pregnant women during their periods of special iron need. Direct supplementation of the diet of animals normally housed and hand fed for extended periods is also the cheapest and most convenient means of preventing trace element deficiencies in farm livestock. The element can be incorporated into the whole diet, or more commonly, can be included in prescribed proportions in the mineral mixtures required for other reasons. This is now common practice for Zn and Mn in pig and poultry rations and for the prevention of Se deficiency in these species. Direct supplementation may also be achieved by oral dosing, injections of slowly absorbed organic forms of the element, or the provision of trace-mineralized salt "licks," available for voluntary consumption. With ruminants, the administration of heavy Co or Se pellets into the reticulorumen constitutes an additional direct means of preventing deficiencies of those elements, which is of particular value for grazing animals not normally subject to frequent handling.

Indirect means of controlling trace element intakes by animals, i.e., by raising or lowering the concentrations in the plant materials as grown and consumed, are successfully practiced in many environments. The agronomic practices which offer alternative opportunities toward achieving such control have been categorized by Allaway (2) as follows: (a) soil selection, (b) trace element fertilization, (c) soil management or practices directed toward increasing or decreasing the availability to plants of the trace elements in the soil, including the use of competitive elements, (d) crop selection, and (e) crop management and utilization. The applicability of one or more of these procedures varies with different elements and in different environments.

Trace element fertilization of the soil is widely practiced as a means of raising herbage concentrations of Co and Cu to satisfactory levels for animals. Under sparse grazing or range conditions trace element fertilization is usually uneconomical and unreliable because of low herbage productivity per unit area, variable uptake of the element, and high application and transport costs. Under these circumstances direct supplementation through the use of trace element-mineralized salt licks, periodic oral dosing or injections, or, with Co and Se, the use of heavy pellets are the preferred methods of control. Mineral licks are generally the easiest and cheapest form of treatment but they suffer from two disadvantages, namely, irregular consumption and loss of the element due to leaching or volatilization.

Trace element toxicities in animals are usually more difficult to control than deficiencies, especially under grazing or range conditions, although various procedures have been successfully adopted. Soil selection, involving the actual elimination of particularly toxic soils from use for grazing, then becomes an important control procedure. In endemic fluorosis areas the only practical form of protection is periodic removal of the animals from dependence on the fluoridated waters. The control of molybdenosis in animals can be achieved by regular oral doses of copper sulfate or periodic injections of this salt. Where the herbage Mo levels are lower but still potentially toxic, either moderate dosage with copper or treatment of the soil with Cu-containing fertilizers to raise the Cu concentrations in the herbage provides a satisfactory means of control. In areas where chronic Cu poisoning in sheep occurs as a consequence of normal to high copper, accompanied by very low-Mo levels in the pastures, the provision of molybdate-containing salt licks to achieve a better Cu:Mo dietary ratio can reduce or eliminate the incidence of the disease. Finally, the "dilution" technique offers some possibilities for control. This involves the importation of feeds known to be low in the toxic element which can be used in conjunction with the local toxic feeds, thus reducing overall intakes to safe levels.

Prevention and control of trace element deficiencies or excesses in human dietaries can often be achieved by an intelligent choice of foods, involving either a restriction on intakes of items known to be high or low in the element in question, or an increase in the proportion of such food items consumed. For example, a reduction in the content of refined carbohydrates will raise dietary intakes of several essential trace elements, notably iron, zinc, and chromium, while a reduced content of fish will lower the intakes of several potentially toxic elements, notably mercury in its dangerous methylated forms. The opportunities provided by food choice for safe and satisfactory trace element intakes by man should increase greatly as continued research offers more information on the distribution of these elements in foods and beverages, and particularly on their biological availability and metabolic interactions with other elements and compounds, organic and inorganic.

REFERENCES

1. Adams, S.N., and Honeysett, J.L., *Aust. J. Agric. Res.* **15**, 357 (1964).
2. Allaway, W.H., *Adv. Agron.* **20**, 235 (1968).
3. Allaway, W.H., Moore, D.P., Oldfield, J.E., and Muth, O.H., *J. Nutr.* **88**, 411 (1966).
4. Andrews, E.D., *N. Z. J. Agric.* **92**, 239 (1956).
5. Anke, M., *Z. Acker- Pflanzenbau* **112**, 113 (1961).
6. Barmes, D.E., Adkins, B.L., and Schamschula, R.G., *Bull. W. H. O.* **43**, 769 (1970).
7. Beck, A.B., *Aust. J. Exp. Agric. Anim. Husb.* **2**, 40 (1960).
8. Beeson, K.C., and McDonald, A.H., *Agron. J.* **43**, 589 (1951).
9. Beeson, K.C., Gray, L.C., and Adams, M.G., *J. Am. Soc. Agron.* **39**, 356 (1947).
10. Beeson, K.C., Lazar, V.A., and Boyce, S.G., *Ecology* **36**, 155 (1955).
11. Biorck, G., Bostrom, H., and Widstrom, A., *Acta Med. Scand.* **178**, 239 (1965).
12. Bisjberg, B., and Gissel-Nielsen, G., *Plant Soil* **31**, 287 (1969).
13. Butler, G.W., and Johns, A.T., *J. Aust. Inst. Agric. Sci.* **27**, 123 (1961).
14. Cadell, P.D., *Aust. Dent. J.* **9**, 32 (1964).
15. Cannon, H.L., and Bowles, J.M., *Science* **137**, 765 (1962).
16. Crawford, M.D., Gardner, M.J., and Morris, J.N., *Lancet* **1**, 827 (1968).
17. Crawford, M.D., Gardner, M.J., and Morris, J.N., *Lancet* **3**, 327 (1971).
18. Deosthale, Y.G., and Gopalan, C., *Br. J. Nutr.* **31**, 351 (1974).
19. Field, A.C., and Purves, D., *Proc. Nutr. Soc.* **23**, XXIV (1964).
20. Fleming, G.A., and Walsh, T., *Proc. R. Ir. Acad., Sect. B* **58**, 151 (1957).
21. Gartner, R.J.W., and Twist, J.O., *Aust. J. Exp. Agric. Anim. Husb.* **8**, 210 (1968).
22. "Geochemistry and the Environment," Vol. I, p. 36. Natl. Acad. Sci., Washington, D.C., 1974.
23. Gladstones, J.S., and Drover, D.P., *Aust. J. Exp. Agric. Anim. Husb.* **2**, 46 (1962).
24. Gladstones, J.S., and Loneragan, J.F., *Proc. Int. Grassl. Congr., 11th, 1969* Sect. IV/18 (1970).
25. Gladstones, J.S., and Loneragan, J.F., *Aust. J. Agric. Res.* **12**, 427 (1967).
26. Goldschmidt, V.M., "Geochemistry." Oxford Univ. Press (Clarendon), London and New York, 1954.

27. Gurevich, G.P., *Fed. Proc., Fed. Am. Soc. Exp. Biol.* **23,** T511 (1964).
28. Hadjinarkos, D.M., *J. Pediatr.* **70,** 967 (1967).
29. Hartmans, J., and Bosman, M.S., *in* "Trace Element Metabolism in Animals–1" (C.F. Mills, ed.), p. 362. Livingstone, Edinburgh, 1970.
30. Healy, W.B., *N. Z. J. Agric. Res.* **11,** 487 (1968).
31. Healy, W.B., *N. Z. J. Agric. Res.* **17,** 59 (1974).
32. Healy, W.B., McCabe, W.J., and Wilson, G.F., *N. Z. J. Agric. Res.* **13,** 503 (1970).
33. Hercus, C.E., Aitken, H.H., Thompson, H.M., and Cox, G.H., *J. Hyg.* **31,** 493 (1931).
34. Hercus, C.E., Benson, W.N., and Carter, G.L., *J. Hyg.* **24,** 321 (1925).
35. Hewat, R.E.T., and Eastcott, D.F., *Rep. Med. Res. Counc. N. Z.* (1955).
36. Johnson, J.M., and Butler, G.W., *Physiol. Plant.* **10,** 100 (1957).
37. Kirchgessner, M., Merz, G., and Oelschläger, W., *Arch. Tierernahr.* **10,** 414 (1960).
38. Kirchgessner, M., Muller, H.L., and Voigtländer, G., *Z. Wirtsch. Futter.* **3,** 179 (1971).
39. Kobayashi, J., *Ber. Ohara Inst. Landwirtsch. Biol., Okayama Univ.* **11,** 12 (1957).
40. Kubota, J., *Soil. Sci. Soc. Am., Proc.* **28,** 246 (1964); *Soil Sci.* **106,** 122 (1968).
41. Kubota, J., Lazar, V.A., and Losee, F., *Arch. Environ. Health* **16,** 788 (1966).
42. Kubota, J., Lemon, E.R., and Allaway, W.H., *Soil. Sci. Soc. Am., Proc.* **27,** 679 (1963).
43. Legon, C.D., *Br. Med. J.* **2,** 700 (1952).
44. Lindeman, R.D., and Assenzo, J.R., *Am. J. Public Health* **54,** 1071 (1964).
45. Ludwig, T.G., and Bibby, B.G., *Caries Res.* **3,** 32 (1969).
46. Ludwig, T.G., Healy, W.B., and Ludwig, T.G., *Nature (London)* **186,** 695 (1960).
47. Marjanen, H., *Ann. Agric. Fenn.* **8,** 326 (1969); Marjanen, H., and Soini, S., *ibid.* **11,** 391 (1972).
48. Mitchell, R.L., *Research (London)* **10,** 357 (1957).
49. Mitchell, R.L., and Reith, J.W.S., *J. Sci. Food Agric.* **17,** 437 (1966).
50. Morris, J.N., Crawford, M.D., and Heady, J.A., *Lancet* **1,** 860 (1961).
51. Morton, W.E., *J. Chronic Dis.* **23,** 537 (1971).
52. Muleahy, R., *Br. Med. J.* **1,** 861 (1966).
53. Nizel, A.E., and Bibby, B.G., *J. Am. Dent. Assoc.* **31,** 1619 (1944).
54. Perry, H.M., Jr., Perry, E.F., and Erlanger, M.W., *Trace Subst. Environ. Health–8, Proc. Univ. Mo. Annu. Conf., 8th,* p. 51 (1975).
55. Plotnikov, K.I., *Veterinariya* **37,** 35 (1960); *Nutr. Abstr. Rev.* **30,** 1138 (1960).
56. Porter, E.K., and Pederson, P.J., *Sci. Total Environ.* **4,** 365 (1975).
56a. Punsar, S., Erametsa, D., Karvonen, M.J., Ryhanen, A., Hilska, P., and Votnamo, H., *J. Chronic Dis.* **28,** 259 (1975).
57. Purves, D., *Plant Soil* **26,** 380 (1967).
58. Sadavisan, T.S., Gulochana, C.B., John, V.T., Subbarain, M.R., and Gopalan, C., *Curr. Sci.* **29,** 86 (1960).
59. Samsoni, Z., Szalay, A., and Szilagyi, M., *Agrochem. Soil Sci.* **20,** 350 (1971).
60. Saunders, J.L., Medical Statistics Branch, Dept. of Public Health, Wellington, New Zealand, 1960 (private communication).
61. Schroeder, H.A., *J. Am. Med. Assoc.* **195,** 125 (1966).
62. Schroeder, H.A., *J. Chronic Dis.* **12,**586 (1960).
63. Schroeder, H.A., *J. Am. Med. Assoc.* **195,** 81 (1966).
64. Scott, M., *Trans. Int. Goitre Conf., 3rd,* p. 34 (1938).
65. Smith, G.W., "Soil and Cancer." Medical Press, London, 1960.
66. Smith, W.H., *Science* **176,** 1237 (1972).
66a. Statham, M., and Bray, A.C., *Aust. J. Agric. Res.* **26,** 751 (1975).
67. Suttle, N.F., Alloway, B.J., and Thornton, I., *J. Agric. Sci.* **84,** 249 (1975).

68. Szalay, A., Samsoni, Z., and Szilagyi, M., *Agrochem. Soil Sci.* **19,** 13 (1970).
69. Thomas, B., Thompson, A., Oyenuga, V.A., and Armstrong, R.H., *Emp. J. Exp. Agric.* **20,** 10 (1952).
70. Thornton, I., *in* "Trace Element Metabolism in Animals–2" (W.G. Hoekstra *et al.,* eds.), p. 451. Univ. Park Press, Baltimore, Maryland, 1974.
71. Tromp, S.W., and Diehl, J.C., *Br. J. Cancer* **9,** 349 (1955).
72. Underwood, E.J., Robinson. T.J., and Curnow, D.H., *J. Dep. Agric., West. Aust.* **24,** 259 (1947).
73. Voigtländer, G., Lang, V., and Kirchgessner, M., *Z. Acker- Pflanzenbau* **135,** 204 (1972).
74. Watkinson, J.H., and Davies, E.B., *N. Z. J. Agric. Res.* **10,** 122 (1967).
75. Wolf, N., Gedalia, I., Yariv, S., and Zuckerman, H., *Arch. Oral Biol.* **18,** 233 (1973).
76. Wylie, P.B., and Bell, L.C., *Search* **4,** 161 (1973).

Author Index

Numbers in parentheses are reference numbers and indicate that an author's work is referred to although his name is not cited in the text.

Numbers in italics show the page on which the complete reference is listed.

C

F

H

K

M

N

O

P

S

W

Y

Z

Subject Index

J

K

L

M

T

U

V